DIFFERENTIAL AND INTEGRAL EQUATIONS

DIFFERENTIAL AND INTEGRAL EQUATIONS

Second Edition

B.P. PARASHAR
Professor of Mathematics
Rani Durgawati Vishwa Vidyalaya
JABALPUR (M.P.)

CBS

CBS Publishers & Distributors Pvt. Ltd.

New Delhi • Bengaluru • Chennai • Kochi • Kolkata • Mumbai
Hyderabad • Nagpur • Patna • Pune • Vijayawada

ISBN: 81-239-0130-5

First Indian Reprint: 1992
Reprint: 2000, 2003, 2004, 2008

This edition has been reprinted in India by arrangement with
Worth Publishers Inc., USA

Published by **Satish Kumar Jain** and produced by **Varun Jain** for
CBS Publishers & Distributors Pvt. Ltd.,
4819/XI Prahlad Street, 24 Ansari Road, Daryaganj, New Delhi - 110002
delhi@cbspd.com, cbspubs@airtelmail.in • www.cbspd.com
Ph.: 23289259, 23266861, 23266867 • Fax: 011-23243014

Corporate Office: 204 FIE, Industrial Area, Patparganj, Delhi - 110 092
Ph: 49344934 • Fax: 011-49344935
E-mail: publishing@cbspd.com • publicity@cbspd.com

Branches:
- ***Bengaluru:*** 2975, 17th Cross, K.R. Road, Bansankari 2nd Stage, Bengaluru - 70 • Ph: +91-80-26771678/79 • Fax: +91-80-26771680 E-mail: cbsbng@gmail.com, bangalore@cbspd.com
- ***Chennai:*** No. 7, Subbaraya Street, Shenoy Nagar, Chennai - 600030 Ph: +91-44-26681266, 26680620 • Fax: +91-44-42032115 E-mail: chennai@cbspd.com
- ***Kochi:*** Ashana House, 39/1904, A.M. Thomas Road, Valanjambalam, Ernakulum, Kochi • Ph: +91-484-4059061-65 Fax: +91-484-4059065 • E-mail: cochin@cbspd.com
- ***Kolkata:*** 6-B, Ground Floor, Rameshwar Shaw Road, Kolkata - 700014 Ph: +91-33-22891126/7/8 • E-mail: kolkata@cbspd.com
- ***Mumbai:*** 83-C, Dr. E. Moses Road, Worli, Mumbai - 400018 Ph: +91-9833017933, 022-24902340/41 • E-mail: mumbai@cbspd.com

Representatives:
- Hyderabad: 0-9885175004
- Patna: 0-9334159340
- Jharkhand: 0-9811541605
- Nagpur: 0-9021734563
- Pune: 0-9623451994
- Uttarakhand: 0-9716462459

Printed at:
J.S. Offset Printers, Delhi (India)

Preface to the Second Revised Edition

The first edition started on a positive note and attracted a few suggestions for improvement, We, therefore, take this opportunity to remove the lacunae detected and amplify the content to meet the requirements in different quarters. Hopefully, the book, now presents a wellrounded account of the major topics of ordinary and partial differential equations and linear integral equations.

In this edition we have thoroughly investigated the theory, construction and application of Green's function in all its ramifications. Its role as the interface between annihilating and summing operators has been brought into sharp focus. It may not be long before the curriculum makers in the academia will give this versatile concept the preeminance it deserves.

Partial differential equations of second order have been described in greater detail than before. Potential theory has received our special attention. Boundary value problems of the wave and heat equations have been considered to some extent.

Some of the powerful techniques like spline collocation, integral transforms and variational methods have their own individual status. We have brought them into the mainstream of differential and integral equations. They now hyphen the numerous concepts of the subject without making the book an omnibus. We hope the subject will have fewer grey areas than before.

The orginal edition served its educational objectives well, However, some fresh adherents to the subject insisted on a more leisurely exposition. This I have done. To give a hint as to how I did it, I would say it is the exact replay of how I went about learning those topics. Some day, some discerning reader would tell me if that learning mode of mine was right or flawed.

The content of the book can now be addressed to the following categories of users :—

—Veterans who are well-conversant with the subject and are teaching it.

—Newcomers who have either opted for or were forced into or have merely drifted into the subject.

—Youngmen about to appear for a profuse variety of competitive examinations.

The author profoundly thanks his past students who by their devotion and interest year after year, provided the necessary feedback and motivated him to learn and develop the content in the from of this book.

JABALPUR

B.P. PARASHAR

Contents

1

Ordinary and Partial Differential Equations—A Re'sume'

§1·1. Ordinary Differential Equations of the First Order

The most general equation of the first order is of the form

$$p^n+P_1p^{n-1}+\dots+P_{n-1}p+P_n=0 \qquad \dots(1)$$

where $p\equiv\frac{dy}{dx}$ and the coefficiedts $P_1,\dots, P_n$ are given functions of x, y. For clarity of description we consider the solutions of this equation according to the following categories :

(A) **First order and first degree :**

The most general form of such an equation is

$$M+N\frac{dy}{dx}=0 \qquad \dots(2)$$

or in the form

$$M\,dx+N\,dy=0 \qquad \dots(3)$$

where M, N are known functions of x, y.

It is difficult to integrate such an equation in its most general form in terms of known functions. But the following canonical forms of such equations lend themselves to precise solutions :

(*i*) Variables separable

(*ii*) Homogeneous equations

(*iii*) The linear equations

(*iv*) Exact equations.

(*i*) **Variables Separable.** In this case $M=f(x)$, $N=g(y)$ and solution is obtained by a quadrature in the form

$$\int f(x)dx+\int g(y)dy=C.$$

A change of variables sometimes reduces an equation to this form.

Example 1. Solve the equation

$$xp+2y=pxy.$$

Solution. Put $x^2y=Y$. Then $x^2p+2xy=\frac{dY}{dx}$

From the given equation

$$\frac{dY}{dx}=p\,x^2y=\frac{dy}{dx}\,Y,$$

i.e.,
$$\frac{dY}{Y}=dy.$$

Integrating $\log Y=C+y$ *i.e.,* $Y=x^2y=Ae^y$.

Exercise 1

Solve the equations :

1. $\cos(x+y)\,dy=dx$
2. $y^3+x^3p+x+yp=0$ (Put $x^2+y^2=u$, $xy=v$)
3. $(x+y)^2\frac{dy}{dx}=a^2$
4. $(x-y^2)dx+2xy\,dy=0$ (Divide by x^2)
5. $y'-\cos^2(x-y)=0$ $\left(y'\equiv\frac{dy}{dx}\right)$
6. $(x^2-yx^2)\frac{dy}{dx}+y^2+y^2x^2=0$
7. $(1+x^2)\frac{dy}{dx}=2x\cot y$
8. $x^2(xdx+ydy)+y(xdy-ydx)=0$
9. $y'=y^2/(1+xy)$ $\left(\text{Put } u=\frac{x}{y},\ v=y\right)$

(*ii*) **Homogenous Equations.** A function $f(x, y)$ is said to be **homogeneous** of degee n in x, y if

$$f(tx, ty)=t^nf(x, y)$$

i.e.,
$$f(x, y)=x^n\phi(y/x).$$

For example, $f(x, y)=2x^2+3xy-y^2$ is a homogeneous polynomial of degree 2 ;

$$g(x, y)=\frac{x^2+xy+y^2}{x^2-y^2}+\sin^{-1}\left(\frac{x}{y}\right)$$

is a homogeneous polynomial of degree zero.

$$Mdx+Ndy=0$$

is called a **homogeneous** equation if M and N are homogeneous functions of the same degree. Such equations are reduced to the variables separable form by changing the dependent variable by substituting

$$y=vx.$$

If M and N are non-homogeneous linear, *i.e.*, the equation is of the form

$$\frac{dy}{dx}=\frac{ax+by+c}{a'x+b'y+c'}$$

where c and c' are not both zero, then we put $x=X+h$, $y=Y+k$ and choose h, k in such a way that

$$ah+bk+c=0$$
$$a'h+b'k+c'=0$$

If $ab'-a'b\neq 0$, the two equations are enough to determine h, k and the given equation is reduced to the homogeneous form

$$\frac{dY}{dX}=\frac{aX+bY}{a'X+b'Y}$$

If $ab'-a'b=0$, the equation reduces to the form

$$\frac{dy}{dx}=\frac{m(a'x+b'y)+c}{a'x+b'y+c'}$$

which is transformed to variables separable form by the substitution

$$a'x+b'y=v.$$

Example 2. Solve : $(x^2-3xy+y^2)y'+xy=0$.

Solution. The substitution $y=vx$, changes the equation to

$$\frac{v^2-3v+1}{v(v-1)(v-2)}\,dv=-\frac{dx}{x}.$$

Resolving into partial fractions and integrating we have

$$\frac{v(v-1)^2}{v-2}=A/x^2$$

i.e.,
$$y(x-y)^2=A(y-2x).$$

Exercise 2

Solve the equations :

1. $y'=\frac{y}{x}+\sin\frac{y}{x}$
2. $(\sqrt{x^2+y^2}-y)dx+xdy=0$
3. $y^2+x^2\,y'=xy\,y'$
4. $y'=\frac{y(2y-x)}{x(2y+x)}$
5. $5(x-3)^2\frac{dy}{dx}=(2x-5y-1)^2$
6. $y'=\frac{2x+2y+3}{2x+2y-1}$
7. $y'=\frac{x+y}{x-y+z}$

(*iii*) Linear Equations $\frac{dy}{dx}+P(x)y=Q(x)$

where $P(x)$ and $Q(x)$ are continuous functions.

Multiplying both sides by $\mu(x)$ to get

$$\mu(x)\frac{dy}{dx}+\mu(x)P(x)y=\mu(x)Q(x).$$

Choose $\mu(x)$ so that $\mu P=\frac{d\mu}{dx}$,

i.e.,
$$\mu(x)=e^{\int P\,dx}$$

The left side of (4) is thus changed to

$$\frac{d}{dx}(\mu\, y)=\frac{d}{dx}\left(e^{\int P dx}\, y\right)$$

The factor $\mu(x)=e^{\int P dx}$, the multiplication of which renders the left side of (4) an exact differential, is called an **integrating factor** (I.F.).

We obtain the general solution in the form

$$y\, e^{\int P dx}=\int e^{\int P dx}\, Q(x) dx+A.$$

Sometimes a change of variable reduces an equation to linear form :

Example 3. Solve the equation :

$$\sin y \frac{dy}{dx}=\cos y(1-x \cos y)$$

Solution. Dividing both sides of the equation by $\cos^2 y$, we get

$$\sec y \tan y \frac{dy}{dx}-\sec y=-x.$$

Put $\sec y=z$ to get

$$\frac{dz}{dx}-z=-x.$$

I.F.$=e^{-x}$. Multiplying both sides of this equation by e^{-x}, we get

$$\frac{d}{dx}(z\, e^{-x})=-xe^{-x}.$$

Integrating $z\, e^{-x}=e^{-x}+xe^{-x}+A$.
Finally, $\sec y=1+x+Ae^{x}$.

Bernoulli's Equation

$$\frac{dy}{dx}+P(x)y=Q(x)y^{n} \quad \text{..(5)}$$

or

$$y^{-n}\frac{dy}{dx}+P(x)y^{1-n}=Q(x)$$

as reduced to a linear equation by the change of variable

$$y^{1-n}=z,$$

i.e.,

$$y^{-n}\frac{dy}{dx}=\frac{1}{1-n}\cdot\frac{dz}{dx}.$$

The equation is reduced to

$$\frac{dz}{dx}+(1-n)P(x)z=Q(x).$$

Riccati's Equation

$$\frac{dy}{dx}=P(x)+Q(x)y+R(x)y^{2} \quad \text{...(6)}$$

in the general form cannot be solved by quadrature. But if a particular solution $y_1(x)$ is known, then assuming $y=y_1+z$ the equation is transformed to Bernoulli's equation. Indeed, we shall have

$$\frac{dz}{dx}-(Q+2Ry_1)z=Rz^2.$$

Example 4. Solve the equation :

$$(x^2y^3+xy)\frac{dy}{dx}=1.$$

Solution. The equation can be written as

$$\frac{dx}{dy}-yx=y^3x^2,$$

i.e.,
$$x^{-2}\frac{dx}{dy}-y\frac{1}{x}=y^3.$$

The substitution $-\frac{1}{x}=u$ changes this equation to

$$\frac{du}{dy}+yu=y^3.$$

$$\text{I.F.}=e^{\int y\,dy}=e^{\frac{1}{2}y^2}.$$

Multiplying both sides of the last equation by $e^{\frac{1}{2}y^2}$ we get

$$\frac{d}{dx}\left(u\,e^{\frac{1}{2}y^2}\right)=y^3\,e^{\frac{1}{2}y^2}$$

Integration leads to $u\,e^{\frac{1}{2}y^2}=A_1+y^2e^{\frac{1}{2}y^2}-2e^{\frac{1}{2}y^2}$.

Finally, $\frac{1}{x}=2-y^2+A\,e^{\frac{1}{2}y^2}$.

Example 5. Solve the Riccati's equation

$$y'=y^2-\frac{2}{x^2}$$

Solution. It is easy to see that $y_1=\frac{1}{x}$ is a particular solution of the equation.

Put $y=\frac{1}{x}+z$ in the equation to get

$$z'-\frac{2}{x}z=z^2,$$

which is Bernoulli's equation.

Dividing by z^2 and putting $-\frac{1}{z}=v$, we get $v'+\frac{2}{x}v=1$.

$$\text{I.F.}=e^{2\int\frac{dx}{x}}=x^2.$$

Multiplying both sides by x^2, we get

$$\frac{d}{dx}(vx^2)=x^2.$$

Integrating $$vx^2=\frac{x^3}{3}+A.$$

Finally,

$$\frac{1}{y-\frac{1}{x}}=\frac{c_1}{x^2}-\frac{x}{3}$$

Exercise 3

Solve completely the equations :

1. $(x+2y^3)\frac{dy}{dx}=y$
2. $x\cos x\frac{dy}{dx}+(x\sin x+\cos x)y=1$
3. $y\log y\,dx+(x-\log y)\,dy=0$
4. $xy'+y\log x=xy\,e^x.$
5. $\frac{dy}{dx}-\frac{\tan y}{1+x}=(1+x)e^x\sec y.$
6. $\frac{dy}{dx}+\frac{y\log y}{x}=\frac{y(\log y)^2}{x^2}.$
7. $\tan y\frac{dy}{dx}+\tan x=\cos y\cos^3 x.$
8. $2\sin x\frac{dy}{dx}-y\cos x=xy^3e^x.$
9. $y'+2xy+xy^4=0.$

Integrate the following Riccati's equations :

10. $\frac{dy}{dx}=\frac{1+xy}{x^3}+y^2$ $\quad(y_1=-1/x)$
11. $y'=x^2+\frac{y}{x}-y^2$ $\quad(y_1=x)$
12. $y'=-\frac{2}{x^2}+\frac{2y}{x}-y^2$ $\quad(y_1=2/x)$
13. What are the integral curves of the equation $$xydy-(y^2-ax)dx=0\,?$$
14. Show that the equation of the previous example is invariant under the transformation $\bar{x}=t^2x,\ \bar{y}=ty$ and that $x(2ax-y^2)$ as an I.F.

(*iv*) **Exact Differential Equations.** The equation

$$M(x, y)dx+N(x, y)dy=0 \qquad ...(7)$$

is called exact if the left-hand side is the total differential of some function $u(x, y)$, *i.e.*,

$$du(x, y)=M\,dx+N\,dy.$$

The equation thus takes the form

$$du(x, y)=0.$$

Consequently, $u(x, y)=c$ where c is an arbitrary constant, is the complete primitive.

For example, $3x^2y^2dx+2x^3y^2dy=0$ is exact, because

$$d(x^3y^2)=3x^2y^2dx+2x^3y^2dy.$$

The following theorem gives a necessary and sufficient condition, first pointed out by Euler, for the exactness of (7) :

Theorem 1. The necessary and sufficient condition for the equation (7) to be exact is

$$\frac{\partial M}{\partial y}=\frac{\partial N}{\partial x} \qquad \ldots(8)$$

Proof Necessity. Let (7) be exact and $u(x, y)$ a function such that

$$du=Mdx+Ndy.$$

Since $du=\frac{\partial u}{\partial x}dx+\frac{\partial u}{\partial y}dy$, we have

$$\frac{\partial u}{\partial x}=M, \quad \frac{\partial u}{\partial y}=N.$$

Hence $$\frac{\partial M}{\partial u}=\frac{\partial^2 u}{\partial x\partial y}=\frac{\partial N}{\partial x}$$

Sufficiency. Let $\frac{\partial M}{\partial y}=\frac{\partial N}{\partial x}$.

Let V (x, y) be a function such that

$$\frac{\partial V}{\partial x}=M(x, y) \qquad \ldots(i)$$

Then $$\frac{\partial^2 V}{\partial x\partial y}=\frac{\partial M}{\partial y}=\frac{\partial N}{\partial x},$$

i.e., $$\frac{\partial}{\partial x}\left(-\frac{\partial V}{\partial y}+N\right)=0,$$

i.e., $$-\frac{\partial V}{\partial y}+N=f(y),$$

i.e., $$N=f(y)+\frac{\partial V}{\partial y}.$$

Hence M dx+N dy

$$=\frac{\partial V}{\partial x}dx+\frac{\partial V}{\partial y}dy+f(y)dy$$

$$=dV+dF(y) \qquad \text{where } dF(y)=f(y)dy$$

$$=d(V+F),$$

i.e., $$u(x, y)=V(x, y)+F(y) \qquad \ldots(ii)$$

where $$F(y)=\int f(y)dy \qquad \ldots(iii)$$

The relation $N=f(y)+\frac{\partial V}{\partial y}$ indicates that $f(y)$ is that portion of N which does not contain x. Relations (*i*), (*ii*) and (*iii*) suggest the following :

Rule to integrate Mdx+Ndy=0

If $\frac{\partial M}{\partial y}=\frac{\partial N}{\partial x}$, the equation is exact and proceed as follows :

1. Integrate M (x, y) with respect to x as if y is costant so as to get V (x, y).

2. Integrate with respect to y that part of N (x, y) which does not contain x so as to get F (y).

The complete primitive is $V(x, y)+F(y)=A$.

Integrating Factor. In certain cases when the left-hand side of the equation

$$M dx+N dy=0 \qquad \ldots(7)$$

is not a total differential, we can choose a function $\mu(x, y)$ such that after multiplying by it, the left-hand side is transformed to a total differential

$$du=\mu M\, dx+\mu N\, dy$$

Such a function μ is called an **integrating factor**. The introduction of such a factor may lead to the appearance of extraneous particular solutions making this factor zero.

It is not possible to give a general theory of finding integrating factors. In certain cases an integrating factor is suggested on inspecpection. For example $\frac{1}{xy}, \frac{1}{x^2}, \frac{1}{y^2}, \frac{1}{x^2+y^2}$ are suitable integrating factors for the equation $ydx-xdx=0$, and $\frac{1}{x^2+y^2}$ is the integrating factor for the equation $xdx+ydy+(x^2+y^2)x^2dx=0$. Roughly speaking integrating factors of a single argument (for example only of x or y or $x\pm y$ or $x^2\pm y^2$ or xy etc.) are easier to guess by inspection. For other cases the following table may prove helpful.

Equation : $M dx+N dy=0$.

S. No.	Condition	I.F.
1.	$M\, dx+N\, dy=0$ is homogeneous and $Mx+Ny\neq 0$	$\frac{1}{Mx-Ny}$
2.	$M=f(xy)y$, $N=g(xy)x$, $Mx-Ny\neq 0$	$\frac{1}{Mx-Ny}$
3.	$\frac{1}{N}\left(\frac{\partial M}{\partial y}-\frac{\partial N}{\partial x}\right)=f(x)$	$e^{\int f(x)dx}$
4.	$\frac{1}{M}\left(\frac{\partial N}{\partial x}-\frac{\partial M}{\partial y}\right)=g(y)$	$e^{\int g(y)dy}$

Example 6. Integrate the equation

$$y\,dx+x(1-3x^2y^2)dy=0.$$

Solution. Here $M=y$, $N=x(1-3x^2y^2)$.

Since $\frac{\partial M}{\partial y}=1$, $\frac{\partial N}{\partial x}=1-9x^2y^2$, $\frac{\partial M}{\partial y}\neq\frac{\partial N}{\partial x}$ and the equation is not exact.

M and N are of the forms

$$M=f(xy)y,\quad N=g(xy)x.$$

Hence $\frac{3}{Mx-Ny}=\frac{1}{x^3y^3}$ is an **integrating** factor.

The equation can be written as

$$d(xy)-x^3d(y^3)=0$$

i.e.,

$$\frac{d(xy)}{(xy)^3}-\frac{d(y^3)}{y^3}=0$$

Integrating we get

$$-\frac{1}{2}\frac{1}{(xy)^2}-\log y^3=A.$$

Finally, $y^6=ce^{-1/x^2y^2}$

Exercise 4

Integrate the following equations :

1. $xdy-ydx-(1-x^2)dx=0.$
2. $y-x\frac{dy}{dx}=x+y\frac{dy}{dx}.$
3. $y^2dx+(x^2-xy-y^2)dy=0.$
4. $(x^2y^2+xy+1)ydx+(x^2y^2-xy+1)x\,dy=0.$
5. $(x^2-2x+2y^2)dx+2xy\,dy=0.$
6. $\frac{dy}{dx}=\frac{y-xy^2-x^3}{x+x^2y+y^2}.$
7. If μ_1, μ_2 are two integrating factors of $Mdx+Ndy=0$, then prove that $\frac{\mu_1}{\mu_2}=c$ is a solution. Hence prove that $\frac{x}{y}+\frac{y}{x}=c$ is a solution of $xdy-ydx=0$.

(B) First Order but not of First Degree

The solution of (1) will depend on the following forms :

(1) **Solvable for p.** If the expression on the left of (1) can be factorised in the form

$$(p-f_1)(p-f_2)\cdots(p-f_n)=0$$

where the f_i are functions of x, y. Solving each equation $p-f_i=0$ in the form $F_i(x, y, c_i)=0$. The solution of the given equation should be of the form $F_1(x, y, c_1)\cdots F_n(x, y, c_n)=0$. Since $c_1, c_2, \ldots, c_n$ are arbitrary and the equation being of the first order admits only one arbitrary constant, we take $c=c_1=c_2=\cdots c_n$. The general solution will, therefore, be

$$F_1(x, y, c)\ldots F_n(x, y, c)=0.$$

(2) **Solvable for x or y.** If the equation can be written as

$$y=f(x, p) \qquad \ldots(i)$$

Differentiate with respect to x to get

$$p=\frac{\partial f}{\partial x}+\frac{\partial f}{\partial p}\frac{dp}{dx} \qquad \ldots(ii)$$

(i) being an equation in p and x can be solved in the form

$$F(x, p, A)=0 \qquad \ldots(iii)$$

Eliminating p from (i) and (iii) we get the required primitive.

If the equation can be solved for x as

$$x=g(y, p) \qquad \ldots(iv)$$

differentiate with respect to y to get

$$\frac{1}{p}=\frac{\partial g}{\partial y}+\frac{\partial g}{\partial p}\frac{dp}{dy} \qquad \ldots(v)$$

(v) being an equation in p and y can be solved in the form

$$G(y, p, c)=0 \qquad (vi)$$

Eliminating p from (iv) and (vi) the desired solution is obtained.

(3) **x or y is not explicity present.** Let the equation be

$$f(y, p)=0$$

Solving for p, we may get

$$\frac{dy}{dx}=\phi(y),$$

or integrating $\qquad \displaystyle\int\frac{dy}{\phi(y)}=x+A.$

A similar procedure is adopted for an equation of the form

$$g(x, p)=0.$$

The following equations with x and y of degree 1 are of particular importance :

Lagrange's Equation

$$y=x\,f(p)+g(p).$$

Differentiating with respect to x we get after simplification if $p\neq$constant,

$$\{f(p)-p)\}\frac{dx}{dp}+f'(p)\,x=-g'(p)$$

which is in linear form.

A special case of Larange's equation of frequent occurrence is **Clairaut's form** $\quad y=xp+f(p).$

Differentiating with respect to x, we get

$$p=p+x\frac{dp}{dx}+f'(p)\frac{dp}{dx}$$

i.e., $$\{x+f'(p)\}\frac{dp}{dx}=0.$$

Either $$\frac{dp}{dx}=0$$

or $$x+f'(p)=0.$$

With $\frac{dp}{dx}=0$ we get $p=c$ where c is an arbitrary constant. Making this substitution for p in the given equation we get the general solution

$$y=cx+f(c).$$

Taking together

$$y=xp+f(p)$$

and $$x+f'(p)=0$$

we can eliminate p and obtain a solution which does not involve an arbitrary constant nor can be obtained from the general solution by giving to c a particular value. We are thus led to the following definition :

Singular Solutions. A solution which does not contain any arbitrary constant, nor can be obtained from the general solution by giving to arbitrary constants particular values, is called a **singular solution.**

Geometrical Interpretation

We now give a complete geometrical interpretation of the general solution and the singular solution, if any, of a differential equation of the first order.

Consider a first order equation

$$F\left(x, y, \frac{dy}{dx}\right)=0 \quad ...(i)$$

where F is a continuously differentiable function of three arguments x, y, y' The general solution of (i) will be a relation of the form

$$G(x, y, c)=0 \quad ...(ii)$$

where c is an arbitrary constant. For varying values of c (ii) represents a one-particular family of curves, called **integral curves** of the equation (i). A particular value of c picks out a particular integral curve of this family.

Suppose that in a certain two-dimensional domain R it is possible to solve (i) for $\frac{dy}{dx}$ uniquely in the form

$$\frac{dy}{dx}=f(x, y)$$

where $f(x, y)$ is a continuously differentiable function at each front (x, y) of R.

The differential equation (iii) establishes a relation between the coordinates of a point and the slope of the tangent $\frac{dy}{dx}$ to the graph of the solution at that point. Knowledge of x, y determines $\frac{dy}{dx}$, *i.e.*, a direction at the point (x, y). The differential equation (*iii*), therefore, determines a **field of directions** on the xy-plane and integrating the equation amounts to finding the curves, the integral curves, the tangent at each point (x, y) of which coincides with the direction of the field (see Fig. 1).

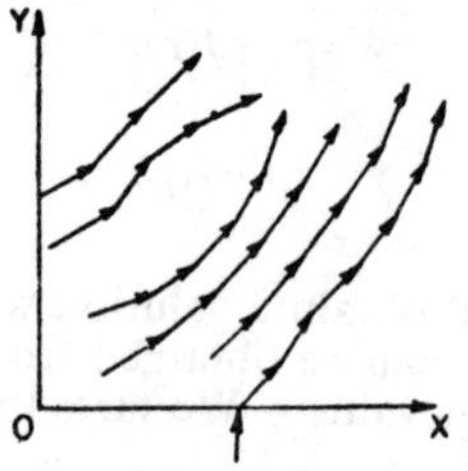

Fig. 1.

If the equation (*i*) is of degree n in $p=\frac{dy}{dx}$, through every point (x, y), there will be n directions $\left(i.e., \text{ slopes } \frac{dy}{dx}\right)$ satisfying the equation. The general solution will be of the form

$$\phi(x, y, c_1, \ldots c_n)=0,$$

but as only one arbitrary constant is admitted, these n constants $c_1 \cdots, c_n$ will be constrained by $n-1$ relations

Example 7. Interpret the equations

$$y'=\frac{y}{x} \qquad \ldots(a_1)$$

$$y'=-\frac{x}{y} \qquad \ldots(a_2)$$

geometrically.

Solution. First of all note that the slopes of the integral curves of (a_1) and (a_2) satisfy the orthogonality condition $\frac{y}{x}\left(-\frac{x}{y}\right)=-1$. Consequently the direction fields of the two equations are orthogonal.

Secondly, the integral curves of (a_1) and (a_2) are respectively,

$$y=c_1x \qquad (a_3)$$

and
$$x^2+y^2=c_2^2 \qquad \ldots(a_4)$$

(a_3) is a pencil of arrows emanating from the origin while (a_4) is a family of concentric circles centred at the origin, (see Fig. 2).

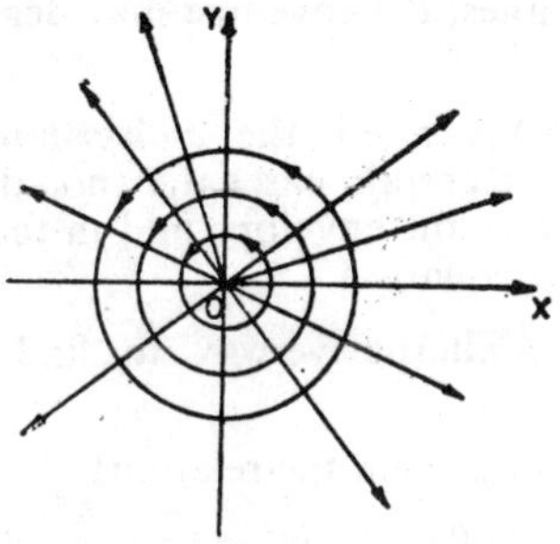

Fig. 2.

Now consider the Clairaut's equation

$$y=xp+f(p) \qquad ...(b_1)$$

again, We saw that the general solution is

$$y=cx+f(c) \qquad ...(b_2)$$

and the singular solution is given by the elimination of p from

$$y=xp+f(p)$$

and
$$x+f'(p)=0. \qquad ...(b_3)$$

The integral curve defined by (b_3) is the same as that obtained by eliminating c from the equations

$$y=cx+f(c)$$
$$x+f'(c)=0$$

as elimination takes no cognizance of the particular mode of representation of the quantities to be eliminated. What we get after this elimination is the envelope of the family of integral curves.

In general if an equation

$$F\left(x, y, \frac{dy}{dx}\right)=0$$

has the general solution

$$\phi(x, y, c)=0, \qquad ...(b_4)$$

then the eliminate of c from

$$\phi(x, y, c)=0 \qquad ...(b_5)$$
$$\frac{\partial\phi}{\partial c}=0$$

which we denote by

$$\Delta_c\, \phi=0 \qquad ...(b_6)$$

is called the singular solution and includes, besides probably other extraneous curves, the envelope of the family. The extraneous curves are either **nodal locus** or **cuspidal locus**. While the envelope represents a solution, the other two loci do not. If however, either or both of $\dfrac{\partial\varphi}{\partial x}$, $\dfrac{\partial\varphi}{\partial y}$ differ from zero, then (b_6) includes only the envelope. If (b_4) is a pencil of lines, the envelope may degenerate into a single point.

The relation (b_6) is called the **c-discriminant** relation and is also written **c-disc.** In certain cases the nodal locus and cuspidal locus may coincide with the envelope and in that case they also become solutions of the equation.

Now there is an alternative way to find the envelope of the differential equation.

The eliminant of p from the relations

$$F(x, y, p)=0$$

$$\frac{\partial F}{\partial p}=0$$

which we express by

$$\Delta p\ F=0 \qquad \ldots(b_7)$$

will be the locus of those points at which two different but consecutive members of the family of integral curves coincide. Hence the equation (b_7) will contain, besides probably other extraneous curves, the envelope of the family. The relation (b_7) is called the **p-discriminant relation** and is written **p-disc.** Besides the envelope it may contain the cuspidal locus and the **tac locus,** (*i.e.*, the locus of points at which two different but not consecutive members of the family touch). For example the straight line joining the centres of two concentric families of circles is a tac locus. The tac locus is not a solution. In certains cases the tac locus may coincide with the envelope in which case it will be a solution.

We state without proof that the c-disc includes in general the envelope locus once, node locus twice and cusp locus thrice. The p-disc includes in general the envelope once, tac locus twice and cusp locus once. This is expressed symbolically as

$$\Delta c\ \phi=\mathrm{EN}^2\ c^3$$

$$\Delta p\ F=\mathrm{ET}^2\ c$$

While determining the singular solution from $\Delta c\phi=0$ or $\Delta p F=0$, we must reject those loci which are not solutions of the given differential equation.

It may be mentioned that a first order equation of the first degree will have no singular solution. Also an equation of higher degree does not necessarily possess a singular solution. Lastly, c-disc and p-disc for the clairaut's equation are the same and only envelope may exist.

Example 8. Solve the equations and find the singular solutions if any :

(i) $xyp^2-(x^2+y^2)\ p+xy=0$

(ii) $p=\tan\left(x-\dfrac{p}{1+p^2}\right)$

(iii) $y=2xp-p^3$

(iv) $(px-y)\ (py+x)=h^2p$

(v) $4xp^2-(3x-1)^2$

Solution. (*i*) Solving the equation for p we have either $p=\frac{x}{y}$

or $$p=\frac{y}{x}.$$

The two equations have solutions

$$x^2-y^2=c_1 \quad \text{and} \quad x=c_2y.$$

Hence the general solution of the given equation is

$$(x^2-y^2-c)\,(x-cy)=0$$

(*ii*) The equation can be written as

$$x=\tan^{-1} p+\frac{p}{1+p^2}$$

Differentiating both sides with respect to y,

$$\frac{1}{p}=\frac{1}{1+p^2}\frac{dp}{dy}+\frac{1-p^2}{(1+p^2)^2}\frac{dp}{dy}$$

$$=\frac{2}{(1+p^2)^2}\frac{dp}{dy}$$

i.e., $$\frac{2p}{(1+p^2)^2}\frac{dp}{dy}=1$$

Integrating $$\frac{1}{1+p^2}=c-y$$

i.e., $$p=\sqrt{\frac{1-c+y}{c-y}}.$$

The general solution is

$$\sqrt{\frac{1-c+y}{c-y}}=\tan\left(x-\sqrt{(c-y)-(c-y)^2}\right).$$

(*iii*) This is Lagrange's equation. Differentiating we get

$$p=2p+2x\frac{dp}{dx}-3p^2\frac{dp}{dx},$$

i.e., $$\frac{dx}{dp}+\frac{2}{p}x=3p;$$

which is a linear equation. Its solution is

$$x=\frac{c}{p^2}+\frac{3}{4}p^2$$

The general solution is the eliminant of p in

$$y=2\,xp-p^3 \text{ and } x=\frac{c}{p^2}+\frac{3}{4}p^2.$$

Now while converting the given equation from $\frac{dp}{dx}$ to $\frac{dx}{dp}$ we have missed those solutions which might correspond to $\frac{dp}{dx}=0$ *i.e.* p=constant=A (say). In that case we have A=2A, *i.e.*, A=0. Hence

$p=0$, which with the given equation gives $y=0$. Thus $y=0$ is a particular solution of the given equation.

The singular solution is obtained by eliminating p from

$$y=2xp-p^3=0 \text{ and } 2x-3p^2=0.$$

This is $27y^2=32x^3$. Since this does not satisfy the given equation, there is no singular solution.

$$27y^2=32x^3 \text{ is the cusp locus.}$$

(*iv*) Putting $\quad x^2=X,\ y^2=Y$ we have

$$P=\frac{dY}{dX}=\frac{y}{x}\,p,$$

i.e.,

$$p=\sqrt{\frac{X}{Y}}\,P.$$

Making these substitutions, the equation reduces to

$$Y=PX-\frac{P}{P+1}\,h^2$$

which is in the Clairaut's form. Its general solution is

i.e.,

$$y^2=cx^2-\frac{c}{c+1}\,h^2,$$

To obtain the singular solution we eliminate c from this equation and

$$0=x^2-\frac{h^2}{(c+1)^2}$$

This gives

$$(x\pm h)^2+y^2=0$$

i.e., the points $(\pm h, 0)$. This does not satisfy the equation, hence, is not the envelope.

(*v*) By solving the equations $p=\pm\dfrac{3x-1}{2\sqrt{x}}$,

we get the complete primitive

$$(y+c)^2=x(x-1)^2$$

Eliminating c from this equation and $y+c=0$, we get $x(x-1)^2=0$.

Hence either $x=0$ or $x-1=0$ twice. $x=0$ occurs once, hence, is the envelope. This satisfies the equation as at $x=0$, (*i.e.*, y-axis) $p=\infty$ and is the singular solution.

$x=1$ does not satisfy the equation as at $x=1$, $p=\infty$ but by the equations $p=\pm 1$. Since, $x-1=0$ occurs twice in the c-disc, $x-1=0$ is the node locus.

The p-dise contains $3x-1=0$ once, hence it represents the tac locus.

Example 9. Examine the existence of a singular solution of the Lagrange's equation

$$y=2xp+p^2.$$

Solution. The p-disc of the equation obtained by eliminating p from $y=2xp+p^2$ and $2x+2p=0$ is $y=x^2$. Since it occurs once and does not satisfy the given equation, it is cusp locus. There is no singular solution.

Exercise 5

Find the complete primitives, and singular solutions, if any, of the following equations : $\left(p \equiv y' \equiv \frac{dy}{dx}\right)$

I. Equations solvable for p

1. $p^2y+2p\ x=y$
2. $p^2x^2+py(2x+y)+y^2=0$

II. Equations solvable for x or y

3. $p^2y+2px=y$
4. $y=2xp+y^2p^3$
5. $y=2xp+x^2p^4$
6. $16x^2+2p^2y-p^3x=0$

III. Clairaut's and Lagrange's equations (or equations reducible to them)

7. $y=2px+p^2y$
8. $y=2px+y^2p^3$
9. $y=2px+xp^2$
10. $(px-y)(x-py)=2p$
11. $x^2(y-xp)=yp^2$
12. $y=2px-p^2$
13. $x-y=\frac{4}{9}p^2-\frac{8}{27}p^3$

    (This is Lagrange's equation with $f(p)=1$

IV. In the following equations find the singular solutions if any :

14. $(1+p^2)y^2-4py-4x=0$
15. $p^2=4(y+x)$
16. $y'^2=y$
17. $y'^2=y^2$
18. $y(y-2px)^2=2p$

(V) Using the c-disc find the singular solutions of the following equations :

19. $y=p^2-xp+\frac{x^2}{2}$, $y=cx+c^2+\frac{x^2}{2}$
20. $(px+y)^2=py^2$, $y(c-x)=c^2$
21. $y=px+\sqrt{a^2p^2+b^2}$, $y=cx+\sqrt{a^2c^2+b^2}$
22. Show that for the equation $xp^2-2yp+4x=0$, both the c-disc and p-disc are the same and the envelopes are $y=\pm 2x$.
23. Show that $x=0$ is the tac locus of the equation

$$y'^2=4x^2$$

24. Show that for the equation $3y=2xp-\frac{2}{x}p^2$, $6y=x^3$ is the envelope and $x=0$ is the node locus.

(*c*) **Trajectories.** A curve which intersects a given family of curves according to some rule, is called a **trajectory** of the family.

If the trajectory makes an angle α with every member of the family, it is called an **α-trajectory** or an **isogonal trajectory** or an **oblique trajectory**. Of greatest importance and those which occur most frequently are the **orthogonal trajectories** for which $\alpha=90°$

Equation of an α-trajectory

(*a*) **Cartesian coordinates.** Given a family of curves

$$f(x, y, c)=0 \qquad ...(i)$$

whose differential equation obtained by eliminating c from

$$f=0, \frac{\partial f}{\partial x}=0 \text{ is}$$

$$F\left(x, y, \frac{dy}{dx}\right)=0 \qquad ...(ii)$$

If X, Y are the running coordinates of points on a trajectory cutting the family (i) at an angle α, then if

$$m_1=\frac{dY}{dX},\ m_2=\frac{dy}{dx}$$

$$\tan\alpha=\frac{m_1-m_2}{1+m_1m_2}=\frac{\dfrac{dY}{dX}-\dfrac{dy}{dx}}{1+\dfrac{dY}{dX}\dfrac{dy}{dx}},$$

i.e.,

$$\frac{dy}{dx}=\frac{\dfrac{dY}{dX}-\tan\alpha}{1+\dfrac{dY}{dX}\tan\alpha}$$

At a point of the trajectory and a curve of (i), $x=X$, $y=Y$.. Hence from (ii) we have

$$F\left(x, y, \frac{\dfrac{dY}{dX}-\tan\alpha}{1+\dfrac{dY}{dX}\tan\alpha}\right)=0$$

Since this is a definite relation between coordinates of any point (X, Y) of a trajectory, the differential equation of the family of trajectories is

$$F\left(x, y, \frac{\dfrac{dy}{dx}-\tan\alpha}{1+\dfrac{dy}{dx}\tan\alpha}\right)=0 \qquad ...(iii)$$

For an orthogonal trajectory $\alpha=90°$, *i.e.*, $\tan\alpha=\infty$.

Hence, the equation (iii) reduces to

$$F\left(x, y, -\frac{dx}{dy}\right)=0 \qquad \ldots(iv)$$

Polar coordinates. Given a family of curves

$$f(r, \theta, c)=0 \qquad \ldots(v)$$

whose differential equation is

$$F\left(r, \theta, \frac{d\theta}{dr}\right)=0. \qquad \ldots(vi)$$

Let (R, Θ) be the running coordinates of a point of an orthogonal trajectory. If the tangents T and T_1 to a curve of (*v*) and a trajectory at their common point make angles ϕ and ϕ_1 respectively with the radius vector,

then $$\tan \phi=r\frac{d\theta}{dr}, \tan \phi_1=R\frac{d\Theta}{dR}$$

Since $$|\phi-\phi_1|=\frac{\pi}{2}, \tan\phi \tan\phi_1=-1;$$

i.e., $$\left(r\frac{d\theta}{dr}\right)\left(R\frac{d\Theta}{dR}\right)=-1.$$

Making this substitution for $\frac{d\theta}{dr}$ in (*vi*)

$$F\left(r, \theta, -\frac{1}{Rr}\frac{dR}{d\Theta}\right)=0$$

Since at the point of intersection of a curve of (*v*) and an orthogonal trajectory $r=R$, $\theta=\Theta$,

$$F\left(R, \Theta, -\frac{1}{R^2}\frac{dR}{d\Theta}\right)=0.$$

Since this is a definite relation between the coordinates of any point on an orthogonal trajectory, the differential equation of the system of orthogonal trajectories is

$$F\left(r, \theta, -\frac{1}{r^2}\frac{dr}{d\theta}\right)=0 \qquad \ldots(vii)$$

Example 10. What are the orthogonal trajectories of the following families of curves :

(*i*) $\frac{x^2}{a^2+\lambda}+\frac{y^2}{b^2+\lambda}=1$

(*ii*) $r=a(1-\cos\theta)$.

Solution. (*i*) Eliminating λ, we get the differential equation of the family of confocal conics,

$$\left(x-y\frac{dx}{dy}\right)\left(x+y\frac{dy}{dx}\right)=a^2-b^2.$$

Since this equation is unaltered by replacing $\frac{dy}{dx}$ by $-\frac{dx}{dy}$, the family is self-orthogonal.

At each point of the xy-plane, there are two values of the parameter λ, one value belonging to a member of one family and there to a member of the family cutting at 90°

(*ii*) The differential equation of the family of cardiodes is

$$r\frac{d\theta}{dr}=\tan\frac{\theta}{2}.$$

Hence the differential equation of the orthogonal family is

$$-\frac{1}{r}\frac{dr}{d\theta}=\tan\frac{\theta}{2}$$

whose general solution is $r=b\,(1+\cos\theta)$.

Exercise 6

1. Find the orthogonal trajectories of the families ;

(*i*) $x^2+y^2=c^2$ (*ii*) $x^2+y^2=cx$

(*iii*) $xy=c^2$ (*iv*) $x^{4/3}+y^{4/3}=c^{4/3}$

(*v*) $r\theta=a$ (*vi*) $r^n=a^n\cos n\theta$

(*vii*) $r^n=a^n\sin n\theta$ (*viii*) $r^n\sin n\theta=a^n$

2. For what value of b the families $cy^2=x^3$ and $2x^2+by^2=A^2$ are orthogonal ?

3. For what value of p the families of curves

$x^p+y^p=c^p$ and $x^{3/2}-y^{3/2}=a^{3/2}$ are orthogonal ?

§1·2. Ordinary Differential Equations with Constant Coefficients

Consider an ordinary linear differential equation of the nth order :

$$L(y)\equiv a_0\frac{d^ny}{dx^n}+a_1\frac{d^{n-1}y}{dx^{n-1}}+\ldots+a_ny=X \qquad \ldots(N)$$

where the coefficients a_i are constants and X is a continuous function of the independent variable x on an interval I : $a\leqslant x\leqslant b$.

The equation

$$a_0\frac{d^ny}{dx^n}+a_1\frac{d^{n-1}y}{dx^{n-1}}+\ldots+a_ny=0. \qquad \ldots(H)$$

is called the **homogeneous equation** corresponding to the **non-homogeneous equation** (N).

A function $f(x)$ is said to be a solution of (N) if when it is substituted in place of y, it makes the equation an identity.

Functions $y_1,\cdots,y_n$ of x are said to be **linearly dependent** on the interval I if constants $a_1,\ldots,a_n$, not all zero, exist such that $a_1y_1+\cdots+a_ny_n\equiv 0$ on I ; otherwise they are said to be **linearly independent**.

If $y_1,\cdots,y_n$ are n linearly independent solutions of (H), then the set $s=\{y_1,y_2,\cdots y_n\}$ is called a **fundamental system** of (H).

If s is a fundamental system of (H), then

$$Y=c_1y_1+\cdots c_ny_n$$

where $c_1,\cdots,c_n$ are arbitrary constants, is also a solution of (H). This is called the **principle of superposition** or the **linearity principle of** solutions. The principle does not hold for non-homogeneous and non-linear equations. In the language of linear algebra the principle also means that the set of all solutions of (H) form a real or complex space or a subspace of the space of all differentiable functions.

The solution $Y=c_1y_1+\ldots+c_ny_n$ of (H) involving n arbitrary constants is called the **complementary function** (C.F.) of (N). The **particular integral** (P.I.) of (N) is any solution u of (N). The **general solution or complete primitive** of (N) is

$$y=\text{C.F.}+\text{P.I.}=Y+u$$

(A) To find the complementary function—The auxiliary equation

If $D\equiv\frac{d}{dx}$, the equation (N) can be written in the form

$$f(D)y=X \qquad \ldots(1)$$

where $f(D)\equiv a_0D^n+a_1D^{n+1}+\ldots+a_n, \qquad (a_0\neq 0)$

Assume $y=e^{mx}$ to be a solution of (H), then

$$(a_0m^n+a_1m^{n-1}+\ldots+a_n)e^{mx}=0.$$

Since on I $e^{mx}\not\equiv 0$, we have

$$f(m)=a_0m^n+a_1m^{n-1}+\ldots+a_n=0 \qquad \ldots(2)$$

$f(m)$ is a polynomial of degree n in m and is called the **auxiliary polynomial.**

The equation (2) is called the **auxiliary equation.** By the fundamental theorem of algebra, (2) will have n real or complex roots.

Case 1. The auxiliary equation has all roots real and simple.

If $m_1,\ldots, m_n$ are the n roots, then the complementary function is

$$Y=c_1e^{m_1x}+\ldots+c_ne^{m_nx}.$$

Case 2. The auxiliary equation has a repeated real root m=a of multiplicity r.

In this case $f(m)=(m-a)^r\, g(m)$

and $$f(a)=f'(a)=\ldots=f^{(r-1)}=0 \qquad \ldots(3)$$

Differentiating the relation

$$L(e^{mx})=f(m)e^{mx}$$

with respect to m k times, we have

$$\frac{\partial^k L(e^{mx})}{\partial m^k}=\frac{\partial^k}{\partial m^k}[f(m)e^{mx}] \qquad \ldots(4)$$

Since the coefficients a_i in L do not involve m, the operators $\frac{\partial^k}{\partial m^k}$ and L commute and hence

$$\frac{\partial^k}{\partial m^n}L(e^{mx})=L\left[\frac{\partial^k e^{mx}}{\partial m^k}\right]=L[x^ke^{mx}].$$

Applying the Leibnitz theorem, the right hand side of (4) is equal to

$$\left[\sum_{i=0}^{k} c^k{}_i\ x^{k-i}f^{(i)}(m)\right]e^{mx},$$

which because of the relations (3) vanishes as $m=a$ for $k=0$, $1,\ldots, r-1$. Thus

$$L\{x^k e^{ax}\}=0 \text{ for } k=0, 1,\ldots, r-1,$$

i.e., $e^{ax}, xe^{ax},\ldots, x^{r-1} e^{ax}$ are solutions of ⓗ.
Henee $(c_1+c_2x+\ldots+c_rx^{r-1})e^{ax}$ forms part of the complementary function corresponding to the repeated root $m=a$ of multiplicity r.

Case 3. The auxiliary equation has complex roots.

The complex roots $\alpha+i\beta$ and $\alpha-i\beta$ will occur in pairs. The part of the complementary function corresponding to the roots $\alpha\pm i\beta$ is

$$u_1=c_1e^{(\alpha+i\beta)x}+c_2e^{(\alpha-i\beta)x}$$

or, by using the Euler's formulae

$$e^{\pm i\theta}=\cos\theta\pm i\sin\theta$$

$$u_1=e^{\alpha x}(A_1\cos\beta x+A_2\sin\beta x)$$

where A_1, A_2 are arbitrary constants. The alternative forms for u_1 are $A\,c^{\alpha x}\cos(\beta x+C)$ or $A\,e^{\alpha x}\sin(\beta x+C)$.

The case of repeated complex roots is treated as that of repeated real roots. For example if $\alpha\pm i\beta$ are roots of multiplicity two, then the corresponding part of the complementary function will be $e^{\alpha x}\{A+Bx)\cos\beta x+(C+Dx)\sin Bx\}$ where A, B, C, D arc arbitrary constants.

[B] To find the particular integral—The operational rules of $D\equiv\frac{d}{dx}$.

If $$f(D)y=X \qquad \ldots(N)$$

then $y=\frac{1}{f(D)}X$ formally means any function of x without arbitrary constants which when operated on by $f(D)$ yields X. Clearly such a function $u(x)=\frac{1}{f(D)}X$ is a particular integral of the equation (N). Fortunately, the operator D satisfies several laws of ordinary algebra which affords an interpretation of $\frac{1}{f(D)}X$ in terms of known functions. The proofs of results in respect of the operator D are valid only under these algebraic rules.

O_1 **The operator D^{-1}** indicates indefinite integration with respect to x, *i.e.*,

$$D^{-1}f(x)=\int f(x)dx.$$

No arbitrary constant needs to be added, because we are interested in a particular integral.

O_2 **Iteration D^n** denotes n times differentiation while D^{-n} denotes n times indefinite integration with respect to x. Thus

$$D^n y=\frac{d^n y}{dx^n}$$

and $$D^{-n} f(x)=\int_0^x \cdots \int_0^x f(t)dt^n$$

$$=\frac{1}{(n-1)!}\int_0^x (x-t)^{n-1} f(t)dt$$

(Cauchy's formula)

O_3 **Distributivity.** If $u, v,$...are functions of x, then

$$f(D)(u+v+...)=f(D)u+f(D)v+...$$

O_4 **Commutativity.** For differential operators $f(D)$ and $g(D)$

$$f(D)+g(D)=g(D)+f(D)$$

If $f(D)$ and $g(D)$ have non-negative powers of D, then

$$f(D).g(D)=g(D).f(D)$$

O_5 **Indices.** $D^mD^n=D^{m+n}=D^nD^m,$

$$D^mD^{-n}=D^{m-n}$$

Positive and negative powers of D do not commute as can be seen from the relations :

$$D^{-1}\,Df=\int f'(x)dx+A=f(x)+A,$$

$$DD^{-1}f=\frac{d}{dx}\left[\int f(x)dx+A\right]=f(x).$$

Thus $DD^{-1}\equiv 1$, but $D^{-1}D\not\equiv 1$. However, in the context of the particular integral where we do not insist on adding an arbitrary constant, $D^{-1}\,D$ and $D\,D^{-1}$ will yield the same result.

O_6 **Partial Fractions.** If $f(D)$ can be factorised into several factors, then an operator of the form $\dfrac{g(D)}{f(D)}$ can be split up into partial fractions. Such an expression is called Heaviside's expansion.

O_7 **Inverse Operators.** In certain cases an inverse operator can be expanded both in increasing and decreasing powers of D. Such an expansion if it leads to a divergent series must be discarded. For example,

$$\frac{1}{D-a}=\sum_{k=1}^{\infty} a^{k-1}\,D^{-k} \qquad ...(i)$$

$$\frac{1}{D-a}=-\sum_{k=0}^{\infty} a^{-k-1}\,D^{k} \qquad ...(ii)$$

Expansion (i) will give

$$\frac{1}{D-a}x^2=\frac{x^3}{3}+\frac{ax^4}{3.4}+\frac{a^2x^5}{3.4.5}+...,$$

and the use of expansion (ii) leads to

$$\frac{1}{D-a}x^2=-\frac{x^2}{a}-\frac{2x}{a^2}-\frac{2}{a^3}.$$

Obviously, the latter expansion is acceptable and the former must be discarded.

Some Basic Formulae

(*i*) (The Shifting Theorem)

$$f(D)e^{\phi(x)}V(x)=e^{\phi(x)}f[D+\phi'(x)]V(x) \qquad ...(A)$$

Proof. Let $f(D)=\Sigma a_n D^n$. It is easily verified that

$$(e^{-\phi}\ De^{\phi})V\equiv(D+\phi')V$$

and

$$(e^{-\phi}\ D\ e^{\phi})^n\ V\equiv(e^{-\phi}\ D^n e^{\phi})V\equiv(D+\phi')^n V$$

Hence $e^{-\phi}\ (\Sigma\ a_n\ D^n)e^{\phi}\ V\equiv[\Sigma a_n(D+\phi')^n]V$,

i.e., $$e^{-\phi} f(D)\, e^{\phi}\ V\equiv f(D+\phi')V.$$

For the inverse operator $f^{-1}(D)$, the result follows by operating on both sides of (A) by $f^{-1}(D)$ and replacing e^{ϕ} by $e^{\phi} f^{-1}[D+\phi']$.

Corollary 1. $f(D)\, e^{ax}\ V(x)\equiv e^{ax} f(D+a)V(x)$.

For $\qquad V(x)=1$

$$f(D)\, e^{ax}=e^{ax} f(a+D).1$$
$$=e^{ax}[f(x)+f'(a)D+...].1$$
$$=e^{ax} f(a).$$

If $f(a)\neq 0$, we shall have

$$\frac{1}{f(D)}e^{ax}=\frac{e^{ax}}{f(a)}$$

Corollary 2. Let $f(D)=(D-a)^r g(D)$ where $g(a)\neq 0$.

Then $$\frac{1}{f(D)}e^{ax}=e^{ax}\frac{1}{f(D+a)}1$$
$$=e^{ax}\frac{1}{D^r}\frac{1}{g(D+a)}1$$
$$=e^{ax}\frac{1}{D^r}[g(a)+g'(a)D+...]^{-1}.1$$
$$=\frac{e^{ax}}{g(a)}\frac{1}{D^r}\left[1-\frac{g'(a)}{g(a)}D+...\right]1$$
$$=\frac{e^{ax}}{g(a)}\frac{1}{D^r}1=\frac{e^{ax}}{g(a)}\frac{x^r}{r!}$$
$$=\frac{e^{ax}x^r}{f^{(r)}(a)}$$

Remark 1. For computing the particular integral, the order of application of the operator $\frac{1}{D^r}$ and $\frac{1}{g(D+a)}$ in Cor 2 is immaterial. For, if $\frac{1}{D^r}$ operates on 1 first, then an apparently additional part of the form $(a_1+a_2x+...+a_rx^{r-1})e^{ax}$ would be included in the particular integral. Since it is already included in the complementary function corresponding to the factor $(D-a)^r$, its inclusion in the particular integral is redundant.

(*ii*) $e^{\phi(D)} f(x)V(x) \equiv f[x+\phi'(D)]e^{\phi(D)} V(x)$.

Proof. Let $f(x)=\Sigma a_n x^n$. The equivalence

$$D^n(x\, W)=xD^n W+nD^{n-1}W$$

can be generalised to

$$F(D)(xW)=xF(D)W+F'(D)W \qquad \ldots(5)$$

Put $\quad F(D)=e^{\phi(D)}$ and $W(x)=e^{-\phi(D)}\, V(x)$, then

$$[e^{\phi(D)}\, x\, e^{-\phi(D)}]V=xV+\phi'(D)V=[x+\phi'(D)]V.$$

By induction we can prove that

$$[e^{\phi(D)}\ x^n\, e^{-\phi(D)}\ ]V=[x+\phi'(D)]^n V.$$

Hence $\qquad [e^{\phi(D)}\ (\Sigma a_n x^n)\, e^{-\phi(D)}\ ]V=\Sigma a_n[x+\phi'(D)]^n V$

i.e., $\qquad [e^{\phi(D)}\ f(x)\, e^{-\phi(D)}\ ]V=f[x+\phi'(D)]V.$

Corollary 3. $e^{hD} f(x)=f(x+h)e^{hD}.1=f(x+h)$.

This is the symbolic form of the Taylor's theorem in one variable.

Corollary 4. $\dfrac{1}{f(D)}(x^n\, V)=\left[\, x-\dfrac{f'(D)}{f(D)}\right]^n \dfrac{1}{f(D)}\, V.$

The corollary follows by replacing $e^{\phi(D)}$ by $\dfrac{1}{f(D)}$ and $f(x)$ by x^n in formula (*ii*).

Corollary 5. $e^{-D^2/2}\, x^n=(x-D)^n\ 1.$

The corollary follows from formula (*ii*) or corollary (4).

Corollary 6. $(-1)^n\, e^{x^2/2}\ D^n\, e^{-x^2/2}=(x-D)^n\ 1.$

This corollary follows by putting $\phi(x)=\dfrac{x^2}{2}$ in the equivalence

$$e^{\phi(x)}\ D^n\, e^{-\phi(x)}=[D-\phi'(x)]^n\ .\ 1$$

Remark 2. The Hermite polynomials are defined by

$$He_n(x)=\sum_{k=0}^{[n/2]} \frac{(-1)^k\, n\,!\, x^{n-2k}}{2^k\, k\,!\,(n-2k)\,!}$$

It is easily verified that

$$He_n(x)=e^{-D^2/2}.x^n \qquad \ldots(6)$$

Using corollaries (5) and (6) we have the following expressions for $He_n(x)$,

$$He_n(x)=(-1)^n\ e^{x^2/2}\, D^n\, e^{-x^2/2} \qquad \ldots(7)$$

and $$He_n(x)=(x-D)^n.1 \qquad \ldots(8)$$

The relation (7) is called the **Rodrigues formula** for $He_n(x)$.

Formula (6) lends itself to direct generalisation in the form

$$H^r e_n(x) = e^{-D^r/r!}\, x^n \qquad \text{...(9)}$$

$H^r e_n(x)$ are called the **generalised Hermite polynomials. The** polynomials $He_n(x)$ and $H^r e_n(x)$ find use in combinatorial analysis.

The classical Hermite polynomials $H_n(x)$ are defined by the Rodrigues formula

$$H_n(x) = (-1)^n e^{x^2} D^n e^{-x^2}$$

and are related to the polynomials $He_n(x)$ by

$$2^{n/2} He_n(x) = H_n(x/\sqrt{2}).$$

Example 11. (a) $f(D^2) \begin{smallmatrix}\sin\\ \cos\end{smallmatrix} ax = f(-a^2) \begin{smallmatrix}\sin\\ \cos\end{smallmatrix} ax$

(b) $f(D^2) \begin{smallmatrix}\sinh\\ \cosh\end{smallmatrix} ax = f(a^2) \begin{smallmatrix}\sinh\\ \cosh\end{smallmatrix} ax.$

These results follow from the relations

$$(D^2)^n \begin{smallmatrix}\sin\\ \cos\end{smallmatrix} ax = (-a^2)^n \begin{smallmatrix}\cos\\ \sin\end{smallmatrix} ax,$$

and $$(D^2)^n \begin{smallmatrix}\sinh\\ \cosh\end{smallmatrix} ax = (a^2)^n \begin{smallmatrix}\sinh\\ \cosh\end{smallmatrix} ax.$$

Example 12. Evaluate $\dfrac{1}{(D^2+a^2)^r} \begin{smallmatrix}\sin\\ \cos\end{smallmatrix} ax.$

Solution.

$$\frac{1}{(D^2+a^2)^r} e^{iax} = \frac{1}{(D-ia)^r (D+ia)^r} e^{iax}$$

$$= \frac{1}{(D-ia)^r} \frac{e^{iax}}{(2ia)^r}$$

$$= \frac{e^{iax}}{(2ia)^r} \frac{1}{D^r} 1$$

$$= \frac{e^{iax}}{(2ia)^r} \frac{x^r}{r!}.$$

Hence $\dfrac{1}{(D^2+a^2)^r} \sin ax$ = imaginary part of $\dfrac{e^{iax}}{(2ia)^r} \dfrac{x^r}{r!}$

$$= \text{I.P. of } \frac{1}{r!}\left(\frac{2}{2a}\right)^r e^{i(ax - r\pi/2)}$$

$$= \frac{1}{r!}\left(\frac{x}{2a}\right)^r \sin\left(ax - \frac{r\pi}{2}\right)$$

Also $\dfrac{1}{(D^2+a^2)}$ **cos** ax **= real part of** $\dfrac{e^{iax}}{(2ia)^r} \dfrac{x^r}{r!}$

$$= \frac{1}{r!}\left(\frac{x}{2a}\right)^r \cos\left(ax - \frac{r\pi}{2}\right).$$

Example 13. Evaluate $\dfrac{1}{D^2+D+1} \sin x.$

Solution. $\dfrac{1}{D^2+D+1}\sin x = \text{I.P. of } \dfrac{1}{D^2+D+1}e^{ix}$

$$= \text{I.P. of } e^{ix}\ \frac{1}{i^2+i+1}$$

$$= \text{I.P. of } (-ie^{ix}) = -\cos x.$$

Example 14. Find the complete primitive of

$$(D^2+a^2)^2 y = \sin^2 \frac{ax}{2}.$$

Solution. The auxiliary equation $(m^2+a^2)^2=0$ gives $m=ia$, $ia, -ia, -ia$.

$$\text{P.I.} = \frac{1}{(D^2+a^2)^2}\sin^2\frac{ax}{2}$$

$$= \frac{1}{2}\,\frac{1}{(D^2+a^2)^2}(1-\cos ax)$$

$$= \frac{1}{2a^4} - \frac{1}{2(D^2+a^2)^2}\cos ax$$

$$= \frac{1}{2a^4} + \frac{1}{8}\left(\frac{x}{a}\right)^2 \cos ax$$

C.F. $=(c_1+c_2x)\cos ax + (c_3+c_4x)\sin ax$

Example 15. The following results are easily established :

(*a*) $\dfrac{1}{D+a}\sin ax = \dfrac{1}{2a}(\sin ax - \cos ax)$

(*b*) $\dfrac{1}{D+a}\cos ax = \dfrac{1}{2a}(\sin ax + \cos ax)$

(*c*) $\dfrac{1}{D^2+a^2}(x \sin bx) = \dfrac{x \sin bx}{a^2-b^2} - \dfrac{2\,b \cos bx}{(a^2-b^2)^2}$, $a \neq b$.

(*d*) $\dfrac{1}{D^2+a^2}(x \cos bx) = \dfrac{x \cos bx}{a^2-b^2} + \dfrac{2b \sin bx}{(a^2-b^2)^2}$, $a \neq b$.

(*e*) $\dfrac{1}{D^2+a^2}(x \sin ax) = -\dfrac{x^2 \cos ax}{4a} + \dfrac{x \sin ax}{4a^2}$.

(*f*) $\dfrac{1}{D^2+a^2}(x \cos ax) = \dfrac{x^2 \sin ax}{4a} + \dfrac{x \cos ax}{4a^2}$

(*g*) $\dfrac{1}{D^2-a^2}(x \sin ax) = -\dfrac{x \sin ax}{2a^2} - \dfrac{\cos ax}{2a^3}$.

(*h*) $\dfrac{1}{D^2-a^2}(x \cos ax) = -\dfrac{x \cos ax}{2a^2} + \dfrac{\sin ax}{2a^3}$.

[C] Cauchy-Euler Equidimensional Equation—The operator $\theta \equiv x\dfrac{d}{dx}$.

The equation

$$f(\theta)\, y \equiv x^n \frac{d^n y}{dx^n} + a_1 x^{n-1} \frac{d^{n-1} y}{dx^{n-1}} + \ldots + a_n y = X \qquad \ldots(10)$$

where $\theta \equiv x \dfrac{d}{dx}$, a_i are constants and X is a continuous function, is called a **homogeneous** or **equidimensional** linear differential equation. If we put $x=e^z$, then $\theta \equiv \dfrac{d}{dz} = x \dfrac{d}{dx}$. The operational equivalence

$$x^n \frac{d^n}{dx^n} \equiv \theta(\theta-1) \ldots (\theta-n+1)$$

is established by induction. With these replacements, the equation (10) is transformed to an equation with constant coefficients having z as the new independent variable. We can operate with θ exactly as with D. Ultimately z is to be replaced by $\log x$. For convenience a few basic results for operating with θ are mentioned below.

The Auxiliary Equation. It is easily verified that

$$f(\theta)\, x^m\, V(x) = x^m f(\theta+m)\, V(x) \qquad \ldots(11)$$

To find the complementary function of (10), assume $y=x^m$ to be a solution of the equation $f(\theta)y=0$. Then using (11) we have

$$\begin{aligned} f(\theta).\, x^m &= x^m f(\theta+m).\, 1 \\ &= x^m [f(m) + f^1(m)\, \theta + \ldots].\, 1 \\ &= x^m f(m) \\ &= 0. \end{aligned}$$

Since $x^m \not\equiv 0$, we get the auxiliary equation $f(m)=0$ in m of degree n.

Case 1. $f(m)=0$ has n real and simple roots $m_1, \ldots, m_n$. The complementary function is, $Y = c_1 x^{m_1} + \ldots + c_n x^{m_n}$.

Case 2. $f(m)=0$ has a real repeated root mk of multiplicity r. The part of the complementary function corresponding to mk will be $\{c_1 + c_2 \log x + \ldots + c_r (\log x)^{r-1}\} x^{mk}$

where $c_1, \ldots, c_r$ are arbitrary constants.

Case 3. If $f(m)=0$ has complex roots $\alpha \pm i\beta$, then the part of the complementary function corresponding to $\alpha \pm i\beta$ will be

$$x^\alpha [A_1 \cos(\beta \log x) + A_2 \sin(\beta \log x)]$$

where A_1, A_2 are arbitrary constants.

Examples 16 Solve completely the equation

$$f(\theta)y = x^2 y'' - (2r-1)\, xy' + (r^2+s^2)y = s^2 x^r \log x,$$

$$\text{where } y' \equiv \frac{dy}{dx}.$$

Solution Here

$$\begin{aligned} f(\theta) &= \theta(\theta-1) - (2r-1)\theta + r^2 + s^2 \\ &= (\theta-r)^2 + s^2. \end{aligned}$$

The auxiliary equation $(m-r)^2+s^2=0$ gives $m=r\pm i\, s$.
The complementary function is
$Y=x^r\{A_1 \cos (s \log x)+A_2 \sin (s \log x)\}$.
The particular integral is

$$u=\frac{1}{f(\theta)}\, s^2 x^r \log x=s^2 \frac{1}{f(\theta)}\, e^{rz} z$$

$$=s^2 e^{rz} \frac{1}{f(\theta+r)}\, z$$

$$=s^2 e^{rz} \frac{1}{\theta^2+s^2}\, z$$

$$=e^{rz}\left(1+\frac{\theta^2}{s^2}\right)^{-1} z$$

$$=e^{rz} z=x^r \log x.$$

Hence the complete primitive is
$y=Y+u=x^r\{A_1 \cos (s \log x)+A_2 \sin (s \log x)\}+x^r \log x$.

Exercise 7

1. Prove that
   (i) $D^n (\theta-n)^r\, y=\theta^r D^n y$
   (ii) $(D^m\, x^{m+r}\, D^r\, x^{-m}\, D^{-r})\; y=x^r D^m y$.

2. Find the complete primitive of the following differential equations :
   (i) $(D^2+D+2)y=x^3$
   (ii) $(2D^2+3D+1)y=x\, e^x$
   (iii) $(2D^2+3D+1)^2 y=x$
   (iv) $(D^4+2D^2+1)^2 y=x^2 \cos x$
   (v) $(D^2-1)y=x \sin x+(1+x^2)e^x$
   (vi) $(D^4-m^4)y=\sin mx$
   (vii) $(D^4-2D^3+5D^2-8D+4)y=e^x$
   (viii) $x^3y'''+2x^2y''+2xy'-2y=x^2+\cos (\log x)$
   (ix) $x^2y''-3xy'+y=0$
   (x) $x^3y'''-6x^2y''+18xy'-24y=48-2x^2+4x^4$

§1·3. Partial Differential Equations with Constant Coefficients

Notation. The following symbols are widely in use. For derivatives,

$$p\equiv\frac{\partial z}{\partial x},\quad q\equiv\frac{\partial z}{\partial y},\quad r\equiv\frac{\partial^2 z}{\partial x^2},\quad s\equiv\frac{\partial^2 z}{\partial x\,\partial y}=\frac{\partial^2 z}{\partial y\,\partial x},\quad t=\frac{\partial^2 z}{\partial y^2}$$

Also, $d_1\equiv\frac{\partial}{\partial x}$ and $d_2\equiv\frac{\partial}{\partial y}$. Thus, $d_1^2 z=r$,
$d_1\, d_2\, z=s$, $\quad d_2^2 z=t$.

D and D' are also used in place of d_1 and d_2 respectively.

If x_1, x_2......occur as independent variables then p_i denote the derivatives $\frac{\partial z}{\partial x_i}$ and $d_i \equiv \frac{\partial}{\partial x_i}$.

We are interested in a linear partial differential equation

$$f(d_1, d_2)\, z = g(x, y). \qquad ...(12)$$

where $f(d_1, d_2) = \sum_i \sum_j a_{ij}\, d_1^i\, d_2^j$, a_{ij}, are constants. If max $(i+j)=n$, then n is called the order of the equation (12).

The equation (12) is called a **homogeneous equation** of order n if all derivatives occuring in (12) are of order n, for example

$$f(d_1, d_2)\, z = \sum_{i=0}^{n} \sum_{j=0}^{n} a_{ij} \frac{\partial^n z}{\partial x^i\, \partial y^j} = g(x, y) \qquad ...(13)$$

Equation (12) is called **reducible** if the operator $f(d_1, d_2)$ can be factorised into linear factors of the form

$$f(d_1, d_2) \equiv \prod_{i=1}^{n} (d_1 - m_i\, d_2 - n_i) \qquad ...(14)$$

The reducible operator in (14) is called **homogeneous** if all $n_i=0$. For example $d_1^2+d_2^2$ is homogeneous reducible as $(d_1+id_2)(d_1-id_2)$, but $d_1^2+d_2$ is **irreducible**.

The Complete Integral

The **complete integral** of (12) is the most general relation in x, y, z such that when the value of z derived from it and the associated derivatives are substituted in the equation, the later becomes an identify. The complete integral may involve arbitrary constants or arbitrary functions or both.

As in the case of ordinary differential equations, the complete integral consists of two parts: the complementary function and the particular integral. The most general integral of $f(d_1, d_2)\, z=0$ is called the complementary function (C.F.) of (12) and any particular solution of (12) is called its particular integral (P.I.).

The symbol $\frac{1}{f(d_1, d_2)}\, g(x, y)$ means any function of x, y not involving arbitrary constants and arbitrary functions which when operated on by $f(d_1, d_2)$ yields $g(x, y)$. Clearly such a function $v(x, y) = \frac{1}{f(d_1, d_2)}\, g(x, y)$ is a particular integral of (12).

In what follows we shall continue to assume $f(d_1, d_2)$ to be a rational function of d_1, d_2 with constant coefficients. Considered independently of each other, d_1 and d_2 behave like $D \equiv \frac{d}{dx}$.

[A] To find the complementary function-reducible case.

Homogeneous equation—The auxiliary equation

The simplest case of the equation (12) occurs when it is homogeneous, i.e., $$f(d_1, d_2) \equiv \sum_{i=0}^{n} a_i\, d_1^{n-i}\, d_2^{i}$$

Let $\phi(y+mx)$ be a solution of $f(d_1, d_2)\, z=0$ where ϕ is an arbitrary function of one variable. Then

$$f(d_1, d_2)\, \phi(y+mx)=0$$

i.e., $$f(m, 1)\, \phi^{(n)}\, (y+mx)=0.$$

Since $\phi^{(n)} \not\equiv 0$ we have

$$f(m, 1)=0 \qquad \ldots(15)$$

The equation (15) is called **auxiliary equation.** By the fundamental theorem of algebra (15) has n real or complex roots. No separate consideration is required for real or complex roots.

Case 1. The auxiliary equation has all roots simple. If $m_1, \ldots, m_n$ are the roots of (15), then the complementary function is

$$Z=\sum_{i=0}^{n} \phi_i\, (y+m_i\, x), \qquad \ldots(16)$$

where ϕ_i are arbitrary functions of a single variable.

Case 2. The auxiliary equation has a repeated root. If $m=a$ is a root of multiplicity p, then the part of complementary function corresponding to this root is

$$\sum_{i=1}^{p} x^{i-1}\, \phi_i\, (y+ax)$$

The following theorem justifies these results and indicates a procedure to solve a non-homogeneous equation :

Theorem 1. If $(d_1-md_2-n)^p\, z=0$, then

$$z=e^{nx} \sum_{i=1}^{p} x^{i-1}\, \phi_i\, (y+mx)$$

where the ϕ_i are arbitrary functions.

Proof. Formally,

$$z=\frac{1}{(d_1-md_2-n)^p}\, 0$$

$$=e^{(md_2+n)x}\, \frac{1}{d_1^{\,p}}\, e^{-(md_2+n)x}\, 0 \quad (\S 1{\cdot}1 \text{ cor. } 1)$$

$$= e^{(md_2+n)x}\, \frac{1}{d_1^{\,p}}\, 0$$

$$= e^{(md_2+n)x} \sum_{i=1}^{p} x^{i-1}\, \phi_i(y)$$

$$= e^{nx} \sum_{i=1}^{p} x^{i-1} e^{mxd_2} \phi_i(y)$$

$$= e^{nx} \sum_{i=1}^{p} x^{i-1} \phi_i(y+mx) \qquad (\S 1{\cdot}1 \text{ cor. } 3),$$

where the ϕ_i are arbitrary functions.

Corollary 7. If $f(d_1, d_2) \equiv \prod_{i=1}^{n} (d_1 - m_i d_2 - n_i)$, then the solution of $f(d_1, d_2)\, z=0$ is given by

$$z = \sum_{i=1}^{n} e^{n_i x} \phi_i\,(y+m_i\, x)$$

If $n_i=0$, then the result (16) also follows from this corollary.

[B] To find the particular integral—Operational rules of d_1 and d_2

P_1 The operators d_1^{-1} and d_2^{-1}. The inverse operators d_1^{-1} and d_2^{-1} denote indefinite integral with respect to x and y respectively. Thus

$$d_1^{-1} f(x, y) = \int f(x, y)\, dx + F(y)$$

and

$$d_2^{-1} f(x, y) = \int f(x, y)\, dy + G(x),$$

where F and G are arbitrary functions of y and x respectively

While computing the particular integral, the arbitrary functions F and G may be overlooked.

P_2 Iteration. $d_1^m \equiv \dfrac{\partial^m}{\partial x^m}$, $d_2^n \equiv \dfrac{\partial^n}{\partial y^n}$, $d_1^m\, d_2^n \equiv \dfrac{\partial^{m+n}}{\partial x^m\, \partial y^n}$

Also, d_1^{-n} and d_2^{-n} denote n times indefinite integration with respect to x and y respectively.

P_3 Distributivity. If F, G,.........are functions of x, y and $f(d_1, d_2)$ is a rational function of d_1 and d_2, then

$$f(d_1, d_2)\,(F+G+\ldots) = f(d_1, d_2)\, F + f(d_1, d_2)\, G + \ldots$$

P_4 Commutativity. $f(d_1, d_2) + g(d_1, d_2) = g\, d_1, d_2) + f(d_1, d_2)$.

If f and g are polynomials in d_1 and d_2 with constant coefficients, then

$$f \,.\, g = g \,.\, f$$

This is not true if f and g have variable coefficients. For example,

$$(d_1 - x)\,(d_2 - x) \neq (d_2 - x)\,(d_1 - x)$$

$$(d_1 - y^2)\,(d_2 - 2y) \neq (d_2 - 2y)\,(d_1 - y^2)$$

P_5 Indices. $d_i^m\, d_i^n = d_i^{m+n}$

$$d_i^m \,.\, d_i^{-n} = d_i^{m-n} \qquad i=1, 2.$$

If $f(d_1, d_2)$ is a polynomial in d_1 and d_2 with constant coefficients, then

$$f \cdot f^{\pm n} \equiv f^{m \pm n}$$

$$f^m \cdot f^{-m} \equiv f^m \cdot \frac{1}{f^m} \equiv 1.$$

P_6 Partial fractions. If $f(d_1, d_2)$ can be factorised, then an operator of the form $\frac{g(d_1, d_2)}{f(d_1, d_2)}$ should be split into partial fractions.

P_7 Inverse operators. The operator $(d_1 - ad_2)^{-1}$ can be expanded in two ways :

$$(d_1 - ad_2)^{-1} = \frac{1}{d_1}\left(1 - a\frac{d_2}{d_1}\right)^{-1} = \frac{1}{d_1} + \frac{ad_2}{d_1} + \frac{a^2 d_2^2}{d_1^3} + \ldots$$

$$(d_1 - ad_2)^{-1} = \frac{-1}{ad_2}\left(1 - \frac{d_1}{ad_2}\right)^{-1} = \frac{-1}{ad_2} - \frac{d_1}{a^2 d_2^2} - \frac{d_1^2}{d_1^3 d_2^3} - \ldots$$

Out of these two expansions, the one leading to a conergent series is to be retained. If both expansions are finite, then the one terminating earlier should be perferred. For example if $k < l$, then

$$(d_1 - ad_2)^{-1} x^k y^l = -\left(\frac{1}{ad_2} + \frac{d_1}{a^2 d_2^2} + \ldots\right) x^k y^l$$

$$= -\left(\frac{1}{ad_2} + \frac{d_1}{a^2 d_2^2} + \ldots + \frac{d_1^k}{a^{k+1} d_2^{k+1}}\right) x^k y^l.$$

For $k=1$, $l=2$, we shall have

$$(d_1 - ad_2)^{-1} xy^2 = -\frac{x}{a}\frac{y^3}{3} - \frac{y^4}{12a^2}$$

We also have

$$(d_1 - ad_2)^{-1} xy^2 = \frac{x^2 y^2}{2} + \frac{ax^3 y}{3} + \frac{a^2 x^4}{12}.$$

Some Basic Formulae

(*i*) (The Shifting Theorem)

$$f(d_1, d_2)\, e^{\phi(x, y)}\, V(x, y) \equiv e^{\phi(x, y)}\, f\left[d_1 + \frac{\partial \phi}{\partial x}, d_2 + \frac{\partial \phi}{\partial y}\right] V(x, y).$$

Proof. Let $f(d_1, d_2) = \sum_m \sum_n a_{mn} d_1^m d_2^n$ ...(A)

By induction we can prove that

$$d_1^m\, e^{\phi(x, y)}\, V \equiv e^{\phi(x, y)} \left(d_1 + \frac{\partial \phi}{\partial x}\right)^m V,$$

and $d_2^n\, e^{\phi(x, y)}\, V \equiv e^{\phi(x, y)} \left(d_2 + \frac{\partial \phi}{\partial y}\right)^n V.$

Making these substitutions in (A) the desired formula is obtained.

Corollary 8. $f(d_1, d_2)\, e^{ax+by}\, V(x, y)$
$$\equiv e^{ax+by} f(d_1+a,\ d_2+b)\, V(x, y).$$

A particularly useful case occurs when $V\equiv 1$

$$f(d_1, d_2)\, e^{ax+by}=e^{ax+by} f(d_1+a, d_2+b)\ 1$$
$$=e^{ax+by}\left[f(a, b)+\left(d_1 \frac{\partial f}{\partial a}+d_2 \frac{\partial f}{\partial b}\right)+\ldots\right]1$$
$$=e^{ax+by} f(a, b).$$

If $f(a, b)\neq 0$, then

$$\frac{1}{f(d_1, d_2)}\, e^{ax+by}=\frac{1}{f(a, b)}\, e^{ax+by}.$$

Corollary 9. Let $f(d_1, d_2)=(d_1-a)^k\, (d_2-b)^l\, g(d_1, d_2)$,
where $g(a, b)\neq 0$. Then

$$\frac{1}{f(d_1, d_2)}\, e^{ax+by}=\frac{e^{ax+by}}{g(a, b)}\ \frac{x^k\, y^l}{k\,!\; l\,!}.$$

If $f(d_1, d_2)=(bd_1-ad_2)^r\, g(d_1, d_2)$
where $g(a, b)\neq 0$, then

either $$\frac{1}{f(d_1, d_2)}\, e^{ax+by}=\frac{1}{r\,!}\left(\frac{x}{b}\right)^r \frac{e^{ax+by}}{g\,(a, b)}$$

or $$\frac{1}{f(d_1, d_2)}\, e^{ax+by}=\frac{1}{r\,!}\left(\frac{-y}{a}\right)^r \frac{e^{ax+by}}{g\,(a, b)}$$

(*ii*) If $f\,(d_1, d_2)$ is a homogeneous polynomial of degree n in d_1 and d_2, then

$$f(d_1, d_2)\, \phi(ax+by)=f(a, b)\, [\phi^{(n)}(t)]_{t=ax+by}$$

Proof. The result follows from the relation

$$d_1^k d_2^l\, \phi\,(ax+by)=a^k b^l\, \phi^{(k+l)}\,(ax+by)$$

For the inverse operator

$$\frac{1}{f(d_1, d_2)}\phi(ax+by)=\frac{1}{f(a, b)}\int\!\!\int\ldots\ldots\int \phi(t)\, dt^n$$

where ultimately we put $t=ax+by$.

If $f(d_1, d_2)=(bd_1-ad_2)^r\, g(d_1, d_2)$,
where $g\,(a, b)\neq 0$ then

either $$\frac{1}{f(d_2, d_2)}\, \phi(ax+by)=\frac{1}{r\,!\, g(a, b)}\left(\frac{x}{b}\right)^r \phi^{-(n-r)}(ax+by)$$

or $$\frac{1}{f(d_1, d_2)}\phi(ax+by)=\frac{1}{r\,!}\ \frac{1}{g\,(a, b)}\left(\frac{-y}{a}\right)^r \phi^{\,(n-r)}\,(ax+by)$$

(*iii*) $$e^{\phi(d_1, d_2)} f(x, y)\, V(x,y)\equiv f\left[x+\frac{\partial\phi}{\partial d_1},\ y+\frac{\partial\phi}{\partial d_2}\right] e^{\phi(d_1, d_2)} V\,(x, y)$$

The proof can be completed on the same lines as in rule (*ii*) of §1-1.

Corollary 9. $e^{hd_1+kd_2} f(x, y)=f(x+h, y+k)$.

This is the symbolic form of the Taylor's theorem in two variables.

Corollary 10. Replace $e^{\phi(d_1, d_2)}$ by $\frac{1}{f(d_1, d_2)}$ and $f(x, y)$ by $x^m y^n$, we shall have

$$\frac{1}{F(d_1, d_2)} x^m y^n V(x, y)$$

$$\equiv\left[x-\frac{\frac{\partial F}{\partial d_1}}{F(d_1, d_2)}\right]^m\left[y-\frac{\frac{\partial F}{\partial d_2}}{F(d_1, d_2)}\right]^n \cdot \frac{1}{F(d_1, d_2)} V(x, y)$$

The following examples illustrate the techniques involved in the above results and will be useful in computing the particular integral.

(i) $$\frac{1}{d_1-md_2} V(x, y)=\frac{1}{d_1-md_2} e^{mxd_2} \cdot e^{-mxd_2} V(x, y)$$

$$= e^{mxd_2} \frac{1}{d_1} V(x, y-mx)$$

$$= e^{mxd_2} \int^x V(u, y-mu)\, du$$

$$=\int^x V(u, y+mx-mu)\, du$$

(ii) $$\frac{1}{(d_1-md_2)^p} V(x, y)=e^{mxd_2} \frac{1}{d_1^p} V(x, y-mx)$$

$$= e^{mxd_2} \int^x \int^u \ldots\ldots \int^u V(u, y-mu)\, du^p$$

$$=\int^x \int^u \ldots\ldots \int^u V(u, y+mx-mu)\, du^p$$

(iii) $$\frac{1}{d_1-md_2} f(y+mx)=e^{mxd_2} \frac{1}{d_1} e^{-mxd_2} f(y+mx)$$

$$= e^{mxd_2} \int f(y)\, dx$$

$$=e^{mxd_2}\, x f(y)$$

$$=x f(y+mx)$$

(iv) $$\frac{1}{(d_1-md_2)^r} x^s f(y+mx)$$

$$= e^{mxd_2} \frac{1}{d_1^r} e^{-mxd_2} x^s f(y+mx)$$

$$= e^{mxd_2} f(y) \frac{1}{d_1^r} x^s$$

$$=f(y+mx)\,,\frac{x^{r+s}}{(s+1)(s+2)\ldots(s+r)}$$

$$=\frac{s\,!}{(s+r)\,!}x^{r+s}f(y+mx)$$

(v) $$\frac{1}{d_1-md_2}f(y+mx)\,V(x,y)$$

$$=e^{mxd_2}\frac{1}{d_1}f(y)\,V(x,y-mx)$$

$$=e^{mxd_2}f(y)\int^x V(u,\,y-mu)\,du$$

$$=f(y+mx)\int^x V(u,\,y+mx-mu)\,du$$

$$=f(y+mx)\;\frac{1}{d_1-md_2}\,V(x,y) \qquad \text{[by } (i)\text{]}.$$

(vi) $$\frac{1}{d_1-md_2}x\,f(y+mx)=\log x\,f(y+mx)$$

(vii) $$\frac{1}{d_1-ad_2}e^{y+mx}=\frac{e^{y+mx}}{m-a}$$

(viii) $$\frac{1}{d_1-ad_2}\cos(y+mx)=\frac{\sin(y+mx)}{m-a}$$

$$\frac{1}{d_1-ad_2}\sin(y+mx)=-\,\frac{\cos(y+mx)}{m-a}$$

(ix) $$\frac{1}{d_1^2+d_2^2}\cos mx\cos ny=\frac{-1}{2(m^2+n^2)}\{\cos(mx+ny)+\cos(mx-ny)\}$$

$$=-\frac{\cos mx\cos ny}{m^2+n^2}$$

(x) $$\frac{1}{(d_1-d_2)(d_1-2d_2)}\cos(x+y)=-x\sin(x+y)$$

(xi) $$\frac{1}{(d_1-d_2)(d_1-3d_2)}\;(y+3x)\sin(x+y)$$

$$=\frac{1}{d_1-d_2}(y+3x)\;\frac{1}{d_1-3d_2}\sin(x+y)$$

$$=\left(\frac{1}{d_1}+\frac{d_2}{d_1^2}\right)(y+3x)\tfrac{1}{2}\cos(x+y)$$

$$=\tfrac{1}{2}(2x^2+xy)\cos(x+y).$$

(xii) $$\frac{1}{d_1^2-d_2^2}\cos(x+y)=\frac{1}{d_1-d_2}\cdot\frac{1}{d_1+d_2}\cos(x+y)$$

$$=\tfrac{1}{2}\frac{1}{d_1-d_2}\sin(x+y) \qquad =\tfrac{1}{2}\,x\sin(x+y)$$

(*xiii*) $\frac{1}{d_1^2+d_2^2}e^x \cos y$=R.P. of $\frac{1}{(d_1-id_2)(d_1+id_2)}e^{x-iy}$

$$=\text{R.P. of}\frac{1}{d_1-id_2}\ \frac{1}{2}e^{x-iy}$$

$$=\text{R.P. of } e^{ixd_2}\frac{1}{d_1}\int\frac{e}{2}^{-ixd_2}\ e^{x-iy}$$

$$=\text{R.P. of } \tfrac{1}{2}e^{ixd_2}\int e^{-iy}dx$$

$$=\text{R.P. of } \tfrac{1}{2}e^{idx_2}\,.\,e^{-iy}x$$

$$=\text{R.P. of}\frac{x}{2}e^{-i(y+ix)}$$

$$=\frac{x}{2}e^x\cos y.$$

(*xiv*) $\frac{8}{(2d_1-d_2)_2}\log(x+2y)=\frac{2}{(d_1-\frac{1}{2}d_2)^2}\log(x+2y)$

$$=2\ e^{\frac{1}{2}xd_2}\frac{1}{d_1^2}\ e^{-\frac{1}{2}xd_2}\log(x+2y)$$

$$=2\ e^{\frac{1}{2}xd_2}\frac{1}{d_1^2}\log(2y)$$

$$=e^{\frac{1}{2}xd_2}\,.\,x^2\log(2y)$$

$$=x^2\log(x+2y)$$

Irreducible Equations

If $\quad f(d_1, d_2)z=0$ ...(17)

is an irreducible equation of the nth order, e^{ax+by} will be a solution if

$$f(a, b)=0 \qquad ...(18)$$

(18) is called the auxiliary equation of (17). If $(a_1, b_1), (a_2, b_2)$... are the (a, b)-pairs satisfying (18), then

$$z=\sum_{i=1}^{\infty}A_i\ e^{a_ix+b_iy}$$

is the complete integral of (17) where A_i are arbitrary constants.

[C] Cauchy Euler Type Equidimensional Equations

Given an equation

$$f(d_1, d_2)z\equiv\Big(\sum_i\ \sum_j a_{ij}\,x^i\,y^j\,d_1{}^i d_2{}^j\Big)z=0 \qquad ...(12)$$

Make the substitutions $x=e^X$, $y=e^Y$,

$$\theta_1=\frac{\partial}{\partial X},\ \theta_2=\frac{\partial}{\partial Y},\ \text{then}$$

$$x^i\,d_1{}^i\equiv\theta_1\,(\theta_1-1)......(\theta_1-i+1)$$

$$y^j\,d_2{}^j\equiv\theta_2\,(\theta_2-1)......(\theta_2-j+1)$$

The equation (19) is then reduced to one with constant coefficients and independent variables X and Y.

Example 17. Solve the equation

$$(d_1^3-6d_1^2\,d_2+11d_1\,d_2^2-6d_2^3)z=0$$

Solution. The auxiliary equation is

$$m^3-6m^2+11m-6=0$$

Hence $m_1=1$, $m_2=2$, $m_3=3$. The complete integral is $z=\phi_1\,(y+x)+\phi_2(y+2x)+\phi_3(y+3x)$ where ϕ_1, ϕ_2, ϕ_3 are arbitrary functions.

Example 18. Solve completely the equation

$$r-3s+5t=x\,e^{x+y}.$$

Solution The equation can be written as

$$(d_1^2-2d_1d_2+5d_2^2)z=x\,e^{x+y}$$

The auxiliary equation is $m^2-2m+5=0$.

Hence $\quad m_1=1+2i$, $m_2=1-2i$.

The complementary function is

$$Z=\phi_1(y+x+2i\,x)+\phi_2(y+x-2i\,x)$$

where ϕ_1 and ϕ_2 are arbitrary functions.

The particular intergral is

$$u=\frac{1}{d_1^2-2d_1d_2+5d_2^2}\,x\,e^{x+y}$$

$$=e^{x+y}\,\frac{1}{d_1^2+5d_2^2-2d_1d_2+8d_2+4}\,x$$

$$=e^{x+y}\left[\,1+\frac{d_1^2+5d_2^2-2d_1\,d_2+8d_2}{4}\right]^{-1}\frac{x}{4}$$

$$=\frac{x}{4}\;e^{x+y}$$

The complete integral is thus

$$z=Z+u.$$

Example 19. Solve $(3d_1^2-2d_2)z=7e^{x-2y}+e^{2x+6y}$

Solution. The complementary function is

$$z_1=\sum_{i=1}^{\infty}c_i\,e^{a_ix+b_iy}\quad\text{where } 3a_i^2-2b_i=0\quad(i=1,2,\ldots)$$

$$\text{P.I.}=\frac{7\,e^{x-2y}+e^{2x+6y}}{3d_1^2-2d_2}$$

$$=e^{x-2y}+e^{2x+6y}\,\frac{1}{3(d_1+2)^2-2(d_2+6)}\,1$$

$$=e^{x-2y}+e^{2x+6y}\,\frac{1}{3d_1^2+12d_1-2d_2}\,1$$

$$=e^{x-2y}+e^{2x+6y}\left(\frac{1}{2d_2}\right)\left(1-\frac{3d_1^2+12d_1}{2d_2}\right)^{-1}1$$

$$=e^{x-2y}-\frac{y}{2}e^{2x+6y}$$

Example 20. $x^2 r+2xys+y^2 t+xp+yq=\left(\frac{y}{x}\right)^k+\log\frac{x}{y}$

Solution. The equation can be written as

$$(x^2 d_1^2+2xy\, d_1 d_2+y^2d_2^2+xd_1+yd_2)z=\left(\frac{y}{x}\right)^k+\log\frac{x}{y}$$

Put $xd_1=\theta_1$, $xd_2=\theta_2$, $x=e^X$, $y=e^Y$

Then $\theta_1=\frac{\partial}{\partial X}$, $\theta_2=\frac{\partial}{\partial Y}$

and the equation is transformed to

$$[\theta_1(\theta_1-1)+2\theta_1\theta_2+\theta_2(\theta_2-1)+\theta_1+\theta_2]z=e^{k(Y-X)}+(Y-X),$$

i.e., $$(\theta_1+\theta_2)^2z=e^{k(Y-X)}+(Y-X)$$

Hence $$\text{C.F.}=\phi_1(Y-X)+X\phi_2(Y-X)$$

$$=f_1\left(\frac{y}{x}\right)+\log x.\, f_2\left(\frac{y}{x}\right)$$

where f_1, f_2 are arbitrary functions.

$$\text{P.I.}=\frac{1}{(\theta_1+\theta_2)^2}\left\{e^{k(Y-X)}+(Y-X)\right\}$$

$$=\frac{X^2}{2}\left\{e^{k(Y-X)}+(Y-X)\right\}$$

$$=\tfrac{1}{2}\log x)^2\left\{\left(\frac{y}{x}\right)^k+\log\frac{y}{x}\right\}$$

Exercise 8

Find the complete integral of the following differential equations :

1. $r-6s+9t-4p+12q+4z=2e^{2x}\tan(y+3x)$
2. $r-s-2t=(2x^2+xy-y^2)\sin xy-\cos xy$
3. $(d_1^2+2d_1d_2+d_2^2)\,z=2\cos y-x\sin y$
4. $(d_1^2-4d_2^2)z=\frac{4x}{y^2}-\frac{y}{x^2}$
5. $r-a^2\,t=x$
6. $r+t=\cos mx\cos ny$
7. $r+3s+2t=x+y$
8. $(d_1^3-d_2^3)z=x^3\,y^3$
9. $(x^2d_1^2-y^2d_2^2)z=xy$
10. $x^2r+2xys+y^2t=x^my^n$
11. $x^2r-2xys+y^2t+3xp-yq+z=x^2y$
12. $(2d_1+d_2+5)(d_1-2d_2+1)^2\,z=0.$

§1·4. The Method of Separation of Variable (or The Product Method)

The partial equations of mathematical physics are generally solved by using the method of separation of variables. The method is explained through the following examples :

Example 21. Solve completely the P.D.E.

$$(d_1^2+d_2^2+d_1+2dz)\, z=0$$

Solution. Let $z=F(x)\,.\,G(y)$ where F and G are functions of x and y respectively. Making this substitution in the given equation we have

$$\frac{\partial^2 F}{\partial x^2}+\frac{\partial F}{\partial x}=-\frac{\partial^2 G}{\partial y^2}-2\frac{\partial G}{\partial y}$$

The two sides being functions of different variables, must be a constant, say k. This leads to

$$F=a_k\, e^{-x}+kx+b_k \text{ and } G=c_k\, e^{-2y}-ky+d_k$$

where a_k, b_k, c_k, d_k are arbitrary constants.

For a natural number n,

$$z=\sum_{k=1}^{n}(a_k\, e^{-x}+kx+b_k)\,(c_k\, e^{-2y}-ky+d_k)$$

$n=\infty$ if the series obtained is convergent.

Example 22. Solve the Laplace's equation

$$\frac{\partial^2 z}{\partial x^2}+\frac{\partial^2 z}{\partial y^2}=0$$

Subject to the conditions :

$$z(0, y)=0=z\,(\pi, y)$$

$$\lim_{y\to\infty} z(x, y)=0,\; z(x, 0)=f(x)$$

Solution. Let $z=X(x)\,.\,Y(y)$ be a solution.

Then $\dfrac{1}{Y}\,\dfrac{d^2Y}{dy^2}=-\dfrac{1}{X}\,\dfrac{d^2X}{dx^2}$.

Each side of this equality being functions of different variables, must be a constant, say k^2. Thus

$$\frac{d^2Y}{dy^2}=k^2\, Y \qquad \text{...}(a)$$

$$\frac{d^2X}{dx^2}=-k^2\, X \qquad \text{...}(b)$$

The general solutions of (a) and (b) are

$$X=A\cos kx+B\sin kx,$$

$$Y=C\, e^{-ky}+D\, e^{ky}$$

The conditions $z(0, y)=0$ and $\lim\limits_{y\to\infty} z(x, y)=0$ imply

$A=D=0, \ k>0$

The condition $z(\pi, y)=0$ implies k is an integer. The condition $z(x, 0)=f(x)$ leads to the problem of determining the a_k so that

$$f(x)=\sum_{k=1}^{\infty} a_k \sin kx \quad 0\leqslant x\leqslant \pi.$$

Exercise 9

1. Solve $r-q=0$
2. Solve $(d_1+d_2-2)\,z=0$
3. Solve the boundary value problem

$$\frac{\partial^2 z}{\partial x^2}=a^2\frac{\partial^2 z}{\partial y^2}$$

where (*i*) $z(x, 0)=z(x, 1)$

(*ii*) $z(0, y)=0$

(*iii*) $\frac{\partial z}{\partial x}=\sin\frac{\pi y}{l}$ where $x=0$

4. Solve $\nabla^2 u(x, y)=\frac{\partial^2 u}{\partial x^2}+\frac{\partial^2 u}{\partial y^2}=0$

if $u\to 0$ as $\sqrt{x^2+y^2}\to\infty$

(**Ans.** $u=e^{itx-|t|y}$)

2

The Ordinary Linear Differential Equation

§2·1. Introduction. A differential equation of the form

$$p_0(x)\, y^{(n)}+p_1(x)\, y^{(n-1)}+\ldots+p_{n-1}(x)\, y'+p_n(x)y=f(x) \qquad (A)$$

where the forcing function $f(x)$ and the coefficients $p_i(x)$ are continuous functions of x on an interval $I: a \leqslant x \leqslant b$, is said to be a **linear ordinary differential equation of order n.** The linearity of the equation is characterised by the absence of products or powers of the dependent variable (unknown) y and its derivatives.

If $p_0(x)$ does not vanish any where on I, we shall suppose the equation (A) to have been divided by $p_0(x)$ and consider the form,

$$L(y)\equiv y^{(n)}+p_1(x)\, y^{(n-1)}+\ldots+p_{n-1}(x)\, y'+p_n(x)y'=f(x) \qquad (N)$$

and for $f(x)=0$

$$L(y)\equiv y^{(n)}+p_1(x)\, y^{(n-1)}+\ldots+p_{n-1}(x)\, y'+p_n(x)y=0 \qquad (H).$$

Form (H) is called the **homogeneous** or **reduced** equation of the **non-homogeneous** equation (N).

A function $Y(x)$ is called a **solution** or **integral** or **primitive** of (A) on the interval I if for all x in I, $Y(x)$ is defined and has derivatives upto order n and when substituted in place of $y(x)$ in (A), makes it an identity.

The **general solution** or **complete primitive** of (A) is an n-parameter family of curves (or functions),

$$Y(x, y, c_1, c_2, \ldots, c_n)=0$$

where the c_i's are arbitrary constants.

A solution which may be obtained by assigning particular values to the arbitrary constants in the general solution is called a **particular solution** of the equation.

If $y=U(x)$ is the general solution of (H) and if by some means we are able to find a particular solution $y=u(x)$ of (N), then $y=U(x)+u(x)$ is the general solution of (N). $U(x)$ is the complementary function and $u(x)$ the particular integral of (N).

The principle of superposition states that if $Y_1, Y_2, \ldots Y_m$ are solutions of (H), then for arbitrary constants c_i, the expression

$\sum_{i=1}^{m} c_i Y_i(x)$ is also its solution.

A solution which is neither the general solution nor a particular solution is called a **singular solution** (S.S.)

The graph of a solution of an ordinary differential equation (*ODE*) is called an **integral curve** of the equation.

The n arbitrary constants present in the general solution of (N) may be determined by n suitably prescribed conditions. The n conditions may be prescribed values of y and its derivatives upto the $(n-1)$th order at a single point $x=a$. Such conditions are known as **initial conditions** (I.C.) A differential equation with a set of intial conditions is called an **initial-value problem** (*IVP*).

If conditions are imposed at more than one distinct points, they are known as **boundary conditions** (*BC*) and the problem then is known as a **boundary-value problem** (*BVP*). In such cases a solution may or may not exist, and may or may not be unique.

Boundary conditions are said to be unmixed if they involve values of y and its derivatives at either end of an interval. For example $Ay(0)+By'(0)=0$ is an unmixed *BC*. The are said to be **periodic** if they are of the form $y(0)=y(1)$, $y'(0)=y'(1)$.

Boundary conditions of the type $k_1y(a)+k_2y'(a)=c_1$, $l_1y(b)+l_2y'(b)=c_2$ are said to be **non-homogeneous**. If $c_1=c_2=0$, then they are said to be **homogeneous**.

The following results in linear algebraic systems will be useful in the sequel :

I (**Cramer's Rule**). If $AX=B$ where $A=[a_{ij}]_{n\times n}$, $X=(x_1,\dots, x_n)^T$, $B=(b_1,\dots, b_n)^T$, then $x_i=\dfrac{|A_i|}{|A|}$ where $|A|$ is the determinant of the metrix A and $|A_i|$ is the determinant obtained by replacing the ith column in $|A|$ by the column B.

II (*a*) The non-homogeneous system $AX=B$ is consistent, (*i.e.*, has a solution) iff $|A|\neq 0$ and in this case the solution is unique.

(*b*) The homogeneous system $AX=0$ always has the trivial solution. It also has a non-trivial solution iff $|A|=0$.

§2·2. Linear Dependence and Independence of Solutions

Definition 1. A set of functions $y_1(x)$, $y_2(x),\dots,y_n(x)$ is said to be **linearly dependent** (*LD*) over a given interval I : $a\leqslant x\leqslant b$ if n constants $c_1,\dots, c_n$, not all zero, exist such that

$$\Sigma c_i\, y_i(x)=0,\ x\in I.$$

Otherwise, the functions are called **linearly independent** (*LI*) over that interval. It is necessary to specify the interval for dependence of functions.

Definition 2. Any set of n linearly independent solutions of (H) is called its **fundamental system** of solutions.

Thus each fundamental set of (H) is a basis of the vector space of the set of all its solutions (called the **solution space**).

The following lemma stated here without proof will be useful in the sequel :

Lemma 1. The only solution $y(x)$ of (H) which satisfies the n initial conditions $y(x_0)=y'(x)_0=\dots=y^{(n-1)}(x_0)=0$ for $x_0 \varepsilon I$ is $y(x)\equiv 0$.

In other words, the solution of (H) which satisfies n given district initial conditions is unique.

Theorem 1. The equation

$$y^{(n)}+p_1(x)y^{(n-1)}+\dots+p_n(x)\,y=0 \quad \text{(H)}$$

where $p_i(x)$ are continuous on I, has n linearly independent solutions on I.

Proof. We construct a fundamental set as follows :

On account of lemma 1 a set of unique solutions $y_1(x)$, $y_2(x)$,... $y_n(x)$ can be defined such that

$$y_{k+1}{}^{(i)}(x_0)=\delta_k{}^i \ (i, k=0, 1,\dots n-1)$$

where
$$\delta_k{}^i=\begin{cases} 0 & \text{for } i\neq k \\ 1 & \text{for } i=k \end{cases}$$

Let $y(x)=\sum_{r=1}^{n} c_r y_r(x)$ and let $y(x)\equiv 0$.

Then for $i=0, 1,\dots, n-1$,

$$y^{(i)}(x_0)=\sum_{r=1}^{n} c_r\, y_r{}^{(i)}\,(x_0)=c_{i+1}=0.$$

Thus the solutions $y_i(x)$ are LI. Q.E.D.

Theorem 2. Any $n+1$ solutions of H are linearly dependent on I.

Proof. The result follows from the fact that the solution space of (H) is n-dimensional.

Corollary 1. If $\delta=\{y_1, \dots, y_n\}$ is a fundamental system of (H) then any solution $y(x)$ of (H) is a unique linear combination of $y_1,\dots, y_n$.

Corollary 2. A given fundamental system belongs to a unique linear differential equation.

Proof. If the fundamental system $\left\{y_i(x)\right\}_1^n$ belongs to (H) as well as to

$$y^{(n)}+q_1\,y^{(n-1)}+\dots+q_n\,y=0,$$

then taking the difference, we have

$$\sum_{i=1}^{n} (p_i - q_i)\, y^{(n-i)} = 0, \text{ where } p_0 = q_0 = 1.$$

This is a differential equation of order $n-1$ having n solutions $y_i(x)$. By theorem 2, the set $\left\{ y_i(x) \right\}_1^n$ will be LD which is false.

Q.E.D.

§2·3. The Wronskian

The linear dependence and independence of a set of functions can be theorctically decided by their Wronskian.

Definition 3. The **Wronskian** (or Wronskian determinant) of the set of functions $y_1, y_2, \cdots, y_n$ is the functional determinant

$$W(x) = W(y_1, \cdots, y_n) = \begin{vmatrix} y_1 & y_2 & \cdots & y_n \\ y_1' & y_2' & \cdots & y_n' \\ y_1^{(n-1)} & y_2^{(n-1)} & \cdots & y_n^{(n-1)} \end{vmatrix}$$

Definition 4. (Fundamental Matrix) If $s = \{y_1, \cdots, y_n\}$ is a fundamental system of (H), then the matrix corresponding to the Wronskian determinant $W(y_1, \cdots, y_n)$ is called a **fundamental matrix** of (H).

Clearly, a fundamental matrix is non-singular.

Theorem 3. If the set of functions $\{y_1, \cdots, y_n\}$ is linearly dependent on an interval I, then

$W(x) \equiv 0$ on I

Proof The linear dependence of functions $y_i(x)$ implies the existence of constants $c_1, c_2, \cdots, c_n$, not all zero, such that

$$\sum_{i=1}^{n} c_i y_i(x) = 0, \; x \varepsilon I$$

Differentiating successively $(n-1)$ times, this expression leades to

$$\sum_{i=1}^{n} c_i y_i'(x) = 0$$

$$\cdots\cdots\cdots\cdots\cdots\cdots$$

$$\sum_{i=1}^{n} c_i\, y_i^{(n-1)}(x) = 0$$

This is a linear system in n unknowns c_i and since it has a non-trivial solution, we shall have $W(x) = 0$ for all $x \varepsilon I$.

In other words if $W(x) \neq 0$ for all $x \in I$, then the functions y_i are LI. Q.E.D.

Remark 1. The converse of the above theorem is not true, *i.e.* the vanishing of the Wronskian is a necessary but not a sufficient condition for the linear dependence of a set of functions. In certain cases the Wronskian of a linearly independent set of functions may also vanish. For example if $y_1 = x$ and $y_2 = |x|$, then $y_1' = 1$, $y_2' = \begin{cases} 1 & x>0 \\ -1 & x<0 \end{cases}$. In the interval $x>0$, $x - |x| \equiv 0$ and in the interval $x<0$, $x + |x| = 0$. Thus in any interval which does not include the origin, y_1 and y_2 are linearly dependent.

Now consider an interval I including $x=0$ and let $c_1 y_1(x) + c_2 y_2(x) = 0$ on I. For $x=1$, $c_1 + c_2 = 0$ and for $x=-1$, $c_1 - c_2 = 0$. Thus $c_1 = c_2 = 0$ implying linear independence of y_1 and y_2 over I. However, the Wronskian $W(x) \equiv 0$ on I.

The next theorem stipulates sufficient conditions for linear dependence.

Theorem 4. If the equation

$$y^{(n)} + p_1 y^{(n-1)} + \dots + p_n y = 0 \qquad \dots(H)$$

has continuous coefficients $p_i(x)$ on I and if the Wronskian of the n solutions $y_1, y_2, \cdots, y_n$ of (H) vanishes at any point on I, then the y_i are linearly dependent on I.

Proof. Let $W(x_0) = 0$, $x_0 \in I$. Then the linear algebraic system

$$\sum_{i=1}^{n} c_i \, y_i^{(n-j)}(x_0) = 0 \qquad (j = 1, 2, \dots, n)$$

in $c_1, c_2, \cdots, c_n$ will have a non-trivial solution. With this non-trivial solution.

Let $y(x) = \sum_{i=1}^{n} c_i \, y_i(x)$.

Then $y(x_0) = y'(x_0) = \dots = y^{(n-1)}(x_0) = 0$.

Hence by lemma 1, $y(x) \equiv 0$ Q.E.D.

Remark 2. The assumption of continuity of the coefficients $p_i(x)$ of (H) in this theorem cannot be relaxed as is shown by the following example :

$y_1 = x^3$ and $y_2 = |x|^3$ are both solutions of the differential equation

$$y'' - \frac{3}{x} y' + \frac{3}{x^2} y = 0$$

on the whole real line R and are linearly independent on R. Nevertheless, their Wronskian vanishes on R. The coefficients of the differential equation are not continuous on R.

The Differential Equation Reconstructed from a Fundamental System.

Let $\{y_1,\dots,y_n\}$ be a fundamental system. It is intended to recover the related differential equation.

The y_i also satisfy the differential equation.

$$W(y, y_1,\dots,y_n)=0 \qquad \dots(B)$$

in the unknown y. The coefficient $(-1)^n W(y_1,\dots,y_n)$ of $y^{(n)}$ in B does not vanish anywhere on I. In view of corollary 2, we have the identity $y^{(n)}+p_1(x)y^{(n-1)}+\dots+p_{n-1}(x)y'+p_n(x)\,y$

$$=\frac{(-1)^n W(y,y_1,\dots,y_n)}{W(y_1,\dots,y_n)} \qquad \dots(C)$$

The coefficients $p_i(x)$ are obtained from this identity.

Liouville's Formula. Differentiating W(x) with respect to x, we get

$$W'(x)=\sum_{r=1}^{n} W_r'(x)$$

where $W_r'(x)$ is obtained from W(x) by the differentiation of its rth row. In the cases $r=1, 2,\dots, n-1$, $W_r'(x)$ vanishes as two of its rows are identical. We shall thus get

$$W'(x)=W_n'(x)=\begin{vmatrix} y_1 & y_2 & \cdots & y_n \\ y_1' & y_2' & \cdots & y_n' \\ \cdots & \cdots & \cdots & \cdots \\ y_1^{(n-2)} & & \cdots & y_n^{(n-2)} \\ y_1^{(n)} & & \cdots & y_n^{(n)} \end{vmatrix}$$

Replacing $y_i^{(n)}$ by $-\sum_{r=1}^{n} p_r y_i^{(n-r)}$, we have

$W'(x)=-p_1(x)W(x)$ and have the required formula

$$W(x)\equiv W(x_0)\exp\left(-\int_{x_0}^{x} p_1(t)\,dt\right),\ x_0\varepsilon I \qquad \dots(D).$$

Identity (D) is known as **Liouville's formula** or **Abel's identity.**

Corollary 3. The Wronskian either vanishes everywhere or nowhere.

Corollary 4. The Liouville's formula enables us to integrate a second order homogeneous linear equation

$$y''+p_1(x)y'+p_2(x)y=0 \qquad \dots(E)$$

if a non-trivial solution y_1 is known. For any solution of (E) must be a solution of

$$\begin{vmatrix} y_1 & y \\ y_1' & y' \end{vmatrix} = y'\, y_1 - y\, y_1' = c_1 e^{-\int p_1(x)dx} \qquad \ldots\text{(F)}$$

Solving the linear equation (F) we get

$$y(x) = c_1 y_1 \int \frac{e^{-\int p_1(x)dx}}{y_1^2(x)}\, dx + c_2 y_1.$$

Exercise 10

1. Find the Wronskian of the functions $e^{r_1 x}, \ldots, e^{r_n x}$ where $r_i \neq r_j$ for $i \neq j$ and show that it does not vanish anywhere.

2. Show that the following pairs of functions are linearly independent, yet their Wronskians vanish on the indicated intervals :

(*i*) $\quad y_1 = \begin{cases} x^2 & x \geqslant 0 \\ 0 & x < 0 \end{cases}, \; y_2 = \begin{cases} 0 & x \geqslant 0 \\ x^2 & x < 0 \end{cases}.$

(*ii*) $\quad y_1 = x^2,\; y_2 = x\,|\,x\,|,\; |\,x\,| \leqslant 1.$

(*iii*) $\quad y_1 = x^3,\; y_2 = x^2\,|\,x\,|,\; |\,x\,| \leqslant 1$

(*iv*) $\quad y_1 = \begin{cases} (x-1)^2 & 0 \leqslant x \leqslant 1 \\ 0 & 1 < x \leqslant 2 \end{cases}$

$\quad y_2 = \begin{cases} 0 & 0 \leqslant x \leqslant 1 \\ (x-1)^2 & 1 < x \leqslant 2 \end{cases}$

3. If y_1 and y_2 are linearly independent solutions of the homogeneous equation

$$L(y) = y'' + p_1(x)y' + p_2(x)y = 0,$$

then prove that

(*a*) $\quad p_1(x) = -\dfrac{y_1 y_2'' - y_1'' y_2}{W(x)},\; p_2(x) = \dfrac{y_1' y_2'' - y_1'' y_2'}{W(x)}.$

Hence reconstruct the equation from the linearly independent solutions e^{2x}, xe^{2x}.

(*b*) $\quad L(y) = \dfrac{W(x)}{y_1} \dfrac{d}{dx}\left[\dfrac{y_1^2}{W(x)} \dfrac{d}{dx}(y/y_1)\right].$

4. If y_1 ane y_2 are any two solutions of the equation $[p(x)y']' + q(x)y = 0$, then prove that $y_1(py_2')' - y_2(py_1')' = 0$.

Hence deduce $p(x)W(x) = A$ where A is a constant.

3

Existence and Uniqueness of Solution

In this chapter we discuss the conditions under which a unique solution to an initial value problem of a first order differential equation exists. Extension to higher order differential equations and first order differential systems is indicated.

§3·1. Definition 1

A function $f(x, y)$, defined and continuous in a domain D of the xy-plane satisfies the **Lipschitz condition** in y (or is **Lipschitz continuous** in y), written $f \in \text{Lip}_y$, if there exists a constant K such that

$$|f(x, y_1)-f(x, y_2)| \leqslant K\,|y_1-y_2| \qquad \ldots(1)$$

for every pair of points (x, y_1), (x, y_2) in D. The least value of K for which (1) holds is called the **Lipschitz constant** of $f(x, y)$ in D.

For example $f(x, y)=xy^2$ which is continuous in x, y in any finite part of the plane, is also Lipschitz continuous. In the domain D : $|x| \leqslant 1$, $|y| \leqslant 1$, its Lipschitz constant is 2.

Remark 1. In case $\left|\dfrac{\partial f}{\partial y}\right|$ is bounded in D, then $f(x, y)$ is Lipschitz continuous in y in D. For, if $\left|\dfrac{\partial f}{\partial y}\right| \leqslant K$, then by the mean value theorem

$$|f(x, y_1)-f(x, y_2)| \leqslant \left|\frac{\partial f(x, y)}{\partial y}\right| |y_1-y_2|$$
$$\leqslant K\,|y_1-y_2|$$

In this case Lipschitz constant

$$= \sup_{(x,\,y)\in D} \left|\frac{\partial f}{\partial y}\right|.$$

In a closed bounded domain D, continuity of $\dfrac{\partial f}{\partial y}$ will imply boundedness of $\left|\dfrac{\partial f}{\partial y}\right|$, and hence Lipschitz continuity in y.

Remark 2. Lipschitz continuity in y, implies continuity in y but the converse is not true. This is clear from the definition (1) as $y_1 \to y_2$ implies

$$f(x, y_1) \to f(x, y_2).$$

On the other hand $f(x, y)=y^{1/3}$ is continuous in D : $|x| \leqslant 1$, $|y| \leqslant 1$. But the difference quotient $\frac{f(0, y)-f(0, 0)}{y-0}=1/y^{2/3}$ is unbounded in every neighbourhood of the origin. Hence $f(x, y)$ is not Lipschitz continuous in D.

3·2. Picard's Method of Successive Approximations

Consider the initial value problem

$$y'=f(x, y), \; y(x_0)=y_0 \qquad \ldots(2)$$

where $f(x, y)$ is continuous in some neighbourhood of the point (x_0, y_0).

The procedure is to replace the IVP (2) by the equivalent integral equation

$$y(x)=y_0+\int_{x_0}^{x} f(t, y(t))\, dt \qquad \ldots(3)$$

The problem is now directed to solving (3) by a process of iteration. We begin with a rough approximation (called the **starting solution**) and improve it step by step with the hope to reach the exact solution ultimately. The conditions under which this goal is achieved is the subject matter of this chapter.

The main advantage of considering (3) intead of (2) is that it produces the successive approximations more easily and the procedure is analytically justified.

At the first step take $y_0(x)=y_0$ as the starting solution. Insert this approximation in the right hand side of (3) to get the better approximation *viz.*

$$y_1(x)=y_0+\int_{x_0}^{x} f(t, y_0(t))\, dt.$$

The next step will give

$$y_2(x)=y_0+\int_{x_0}^{x} f(t, y_1(t))\, dt.$$

At the $(n+1)$th step we get

$$y_n(x)=y_0+\int_{x_0}^{x} f(t, y_{n-1}(t))\, dt.$$

The functions $y_n(x)$ are called **Picard's approximations** and the procedure is called Picard's method of successive approximatious.

We discuss the sufficient conditions under which the $y_n(x)$ tend to the required solution as $n \to \infty$. The function $y_n(x)$ will be called the nth Picard's approximation.

3·3. Theorem 1

Let $f(x, y)$ be a continuous function in a domain D : $|x-x_0| \leqslant a$, $|y-y_0| \leqslant b$, satisfying in D the Lipschitz condition in y,

$$|f(x, y_1)-f(x, y_2)| \leqslant K |y_1-y_2|$$

and let sup $|f(x, y_2)| \leqslant M$ in D. If

$h=\min\left(a, \frac{b}{M}\right)$, then the initial value problem

$$y'=f(x, y), \ y(x_0)=y_0$$

has a unique solution on the interval $|x-x_0| \leqslant h$.

Proof. Construct the sequence of approximations $y_0(x)=y_0$

$$y_n(x)=y_0+\int_{x_0}^{x} f(t, y_{n-1}(t))\, dt \quad (n=1, 2,...)$$

to the solution of the equivalent problem

$$y(x)=y_0+\int_{x_0}^{x} f(t, y(t))\, dt.$$

The proof that $y_n(x)$ converges to the required solution as $n\to\infty$ is taken up in several steps.

(*i*) On the interval $|x-x_0| \leqslant h$, the curve $y=y_n(x)$ lies in D, *i.e.*, $|y_n-y_0| \leqslant b$.

For $n=1$,

$$|y_1-y_0| \leqslant \left|\int_{x_0}^{x} |f(t, y_0)|\, dt\right| \leqslant M|x-x_0| \leqslant Mh \leqslant b.$$

Let $y=y_{n-1}(x)$ lie in D. Then

$$|y_n-y_0| =\left|\int_{x_0}^{x} f(t, y_{n-1}(t))\, dt\right| \leqslant M|x-x_0| \leqslant Mh \leqslant b$$

By induction the result is true for all n.

(*ii*) $|y_n(x)-y_{n-1}(x)| \leqslant \frac{MK^{n-1}}{n!} |x-x_0|^n$

For $n=1$, $|y_1-y_0| \leqslant M|x-x_0|$. Let the result be true for n. Then

$$|y_{n+1}-y_n| \leqslant \left|\int_{x_0}^{x} |f(t, y_n)-f(t, y_{n-1})|\, dt\right|$$

$$\leqslant K\left|\int_{x_0}^{x} |y_n-y_{n-1}|\, dt\right|$$

$$\leqslant \frac{MK^n}{n!}\left|\int_{x_0}^{x} |t-x_0|^n\, dt\right|$$

$$=\frac{MK^n}{(n+1)!} |x-x_0|^{n+1}.$$

(*iii*) **The sequence $\{y_n(x)\}$ is uniformly convergent in the closed interval $|x-x_0|\leqslant h$.**

As $y_n=y_0+\sum_{r=1}^{n}(y_r-y_{r-1})$, we have

$$|y_n|\leqslant|y_0|+\sum_{r=1}^{n}|y_r-y_{r-1}|$$

$$\leqslant|y_0|+\sum_{r=1}^{n}\frac{MK^{r-1}}{r!}|x-x_0)^r$$

$$\leqslant|y_0|+\frac{M}{K}\sum_{r=1}^{n}\frac{K^r h^r}{r!}$$

$$\leqslant|y_0|+\frac{M}{K}(eKh-1)$$

The sequence $y_n(x)$ thus converges uniformly to a function $y(x)$ on the interval $|x-x_0|\leqslant h$. Each $y_n(x)$ being an indefinite integral is continuous; hence $y(x)$ is also continuous and $y(x_0)=y_0$.

(*iv*) **$y(x)$ is the required solution.**

Since, $|f(x, y)-f(x, y_n)|\leqslant K|y-y_n|$ and $y_n\to y$ uniformly, $f(x, y_n)\to f(x, y)$ uniformly on $|x-x_0|\leqslant h$,

Also $y(x)=\lim_{n\to\infty} y_{n+1}(x)=y_0+\lim_{n\to\infty}\int_{x_0}^{x} f(t, y_n(t)\, dt$

$$=y_0+\int_{x_0}^{x}\lim_{n\to\infty} f(t, y_n(t)\, dt$$

$$=y_0+\int_{x_0}^{x} f(t, y(t)\, dt.$$

(*iv*) **Uniqueness of the solution.**

Let $z(x)$ be another continuous solution on $|x-x_0|\leqslant h$. Assume $|y(x)-z(x)|\leqslant N$ on $|x-x_0|\leqslant h$. Then

$$|y(x)-z(x)|\leqslant\left|\int_{x_0}^{x}|f(t, y(t))-f(t, z(t))|\,dt\right|$$

$$\leqslant K\left|\int_{x_0}^{x}|y-z|\,dt\right|\leqslant KN|x-x_0|.$$

By induction we can prove that

$$|y(x)-z(x)|\leqslant\frac{NK^n}{n!}|x-x_0|^n\leqslant\frac{NK^n h^n}{n!},$$

which tends to zero as $n\to\infty$. Hence $y(x)\equiv z(x)$

Remark 4. If Lipschitz continuity is replaced by the stronger condition of continuity of $\frac{\partial f}{\partial y}$, theorem 1 is known as the **Picard's theorem.**

If only continuity, and not Lipschitz continuity, is assumed fof $f(x, y)$ is assumed, then the existence, but not the uniqueness, or the solution is guaranteed. This existence theorem is known as **Peano's theorem.**

Remark 4. Theorem 1 is essentially a local existence and uniqueness theorem. It guarantees the existence of a unique solution on the neighbourhood $|x-x_0| \leqslant h$ of the point $x=x_0$ where h may be quite small. In a certain case when h is very small, in reality a unique solution may exist on a much wider interval. Such conditions are called **local** in contrast to **global** ones which guarantee existence over a finite region (instead of a neighbourhood of a point).

Example 1. Find Picard's first three approximations y_1, y_2, y_3 for the initial value problem

$$y'=x^2+y,\ y(0)=1,\ D: |x| \leqslant 1,\ |y-1| \leqslant 1.$$

What is the local interval of existence ?

Answer. With the starting solution $y_0(x)=1$, Picard's method yields :

$$y_1(x)=1+\int_0^x (t^2+1)\ dt=1+x+\frac{x^3}{3}$$

$$y_2(x)=1+\int_0^x \left(t^2+1+t+\frac{t^3}{3}\right) dt$$

$$=1+x+\frac{x^2}{2}+\frac{x^3}{3}+\frac{x^4}{12},$$

$$y_3(x)=1+\int_0^x \left(t^2+1+t+\frac{t^2}{2}+\frac{t^3}{3}+\frac{t^4}{12}\right) dt$$

$$=1+x+\frac{x^2}{2}+\frac{x^3}{6}+\frac{x^4}{12}+\frac{x^5}{60}.$$

In D, $M=\max |f(x, y)| =\max |x^2+y| =3$.

Since $h=\min(1, \frac{1}{3})=\frac{1}{3}$, the existence and uniqueness of the solution is guaranteed in the interval $|x| \leqslant \frac{1}{3}$.

The exact solution of the problem is $y(x)=3e^x-x^2-2x-2$, which is valid for all x.

Exercise 11

1. Define the terms Lipschitz continuity and Lipschitz constant of a function in a region.

Show that Lipschitz continuity of a function implies continuity, but the converse is not true.

2. Show that Lipschitz continuity of f and g implies that of $f \pm g$ but not of $f.g$.

3. Find the Lipschitz constant in y for the following functions in the assigned regions :

(a) $4x^2+y^2$, $|x| \leqslant 1$, $|y| \leqslant 1$.
(b) $x^2 \cos^2 y + y \sin^2 x$, $|x| \leqslant 1$, $|y| < \infty$.
(c) $x-y$, $|x| < \infty$, $|y| < \infty$.
(d) x^2y, $|x| \leqslant 1$, $|y| \leqslant 1$.

4. Show that $f(x, y)=xy^2 \varepsilon \text{ Lip}_y$ in the rectangle $a \leqslant x \leqslant b$, $c \leqslant y \leqslant d$, but not in any strip $a \leqslant x \leqslant b$, $|y| \infty$.

5. Show that $xy \varepsilon \text{ Lip}_y$ in any rechange $a \leqslant x \leqslant b$, $c \leqslant y \leqslant d$ and any strip $a \leqslant x \leqslant b$, $|y| < \infty$, but not on the entire plane.

6. Taking the initial value as the starting solution, find the Picard's approximations y_1, y_2, y_3 for the following IVP's mentioning in each case the local existence interval. Where possible state the exact solution also.

(a) $y'=y$, $y(0)=1$; $|x| \leqslant 1$, $|y=1| \leqslant 1$.
(b) $y'=x+y^2$, $y(0)=0$, $|x| \leqslant 1$, $|y| \leqslant 1$.
(c) $x'=1+xy^2$, $y(0)=0$; $|x| \leqslant 1$, $|y| \leqslant 1$.
(d) $y'=2y/x$, $y(1)=1$; $|x-1| \leqslant \frac{1}{2}$, $|y-1| \leqslant \frac{1}{2}$.
(e) $y'=x^2+y^2$, $y(0)=0$; $|x| \leqslant 1$, $|y| \leqslant 1$.
(f) $y'=x^2-y^2$, $y(-1)=0$; $|x+1| \leqslant 1$, $|y| \leqslant 1$.
(g) $y'=x+y$, $y(0)=1$.

7. Show that $(f(x, y)=x+|\sin y|$ satisfies the Lipschitz condition in y on the xy—plane, but $\dfrac{\partial f}{\partial y}$ does not exist in the neighbourhood of the points $(x, 0)$.

§3·4. Change in Initial Condition and Perturbation

Theorem 2. With the hypotheses of theorem 1 if $y=Y(x)$ is the solution with $Y(x_0)=y_0+\delta$, then for $|x-x_0| \leqslant h$ we shall have $|Y(x)-y(x)| \leqslant \delta e^{kh}$.

Proof. Construct the sequence

$$Y_0(x)=y_0+\delta$$

$$Y_i(x)=y_0+\delta+\int_{x_0}^{x} f(t, Y_{i-1}(t))\, dt \qquad (i=1, 2, \ldots)$$

By induction we can prove that

$$|Y_n(x)-y_n(x)| \leqslant \delta e^{k|x-x_0|} \leqslant \delta e^{kh}.$$

The theorem follows by letting $n \to \infty$.

Remark 5. The theorem implies that a small change in the initial condition results in a small change in the solution. Such problems are called well-posed.

On the above lines it can be shown that the solutions of the problem

$$y'=f(x, y ; \lambda), y(x_0)=y_0$$

vary continuously with λ, called the **perturbation parameter** of the problem.

3·5. Linear Systems

A system of differential equations is called **linear** if it is linear in all unknowns and their derivatives. A linear system of the first order in n unknown has the **normal form** (or is called a **normal system**) if

$$\frac{dx_i}{dt}=\sum_{i=1}^{n} a_{ij}(t)x_j+f_i(t) \quad (i=1, 2,..., x) \qquad ...(4)$$

Or in the vector form

$$\frac{dX}{dt}=AX+F \qquad ...(5)$$

where $A=[a_{ij}(t)]_{n\times n}$ and X and F are single-column matrices.

The system (4) is called **autonomous** if the right hand side does not involve t explicitly, otherwise it is **non-autonomous.**

Two systems are said to be equivalent if they have the same solution set.

If a linear operator L is defined by $L[X]=\frac{dX}{dt}-AX$, then (5) may be written as $L(X)=F$. If $F=0$, it is called a **homogeneous system.**

Vectors $X_1,\cdots, X_n$ are **linearly dependent** on the interval $a\leqslant t\leqslant b$ if there exist constants $\alpha_1,\ldots, \alpha_n$ such that $\alpha_1X_1+\ldots+\alpha_nX_n \equiv 0$ for $a\leqslant t\leqslant b$ and at least one $\alpha_i\neq 0$, otherwise they are **linearly independent.** The vector identify $\Sigma\alpha_iX_i\equiv 0$ can be replaced by the condition,

$$W(t)= | [X_1X_2\cdots X_n] | = \begin{vmatrix} x_{11} & x_{12} & x_{1n} \\ x_{22} & \cdots & x_{2n} \\ x_{n1} & \cdots & x_{nn} \end{vmatrix} =0.$$

This determinent $W(t)$ is called the **Wronskian** of the vector set $\{X_1, \ldots X_n\}$. If $X_1, \ldots, X_n$ are linearly independent solution vectors of the system $X'=AX$, then the set $S=\{X_1, \ldots, X_n\}$ is called a **fundamental system** and the matrix $[X_1, \ldots X_n]$ with columns $X_1, \ldots, X_n$ is called a **fundamental matrix** of the system.

It is possible to reduce an nth order ordinary differential equation to a linear system. For example let $y^{(n)}=f(t, y, y', \ldots, y^{(n-1)})$ with IC's : $y(t_0)=y_0, y'(t_0)=y_0', \ldots, y^{(n-1)}(t_0)=y_0^{(n-1)}$.

Put $\quad y=x_1, y'=x_2, y''=x_3\ldots y^{(n-1)}=x_n.$

Then $y^{(n)}=f(t, x_1, x_2, ..., x_n)$

and $x_1(0)=y_0, x_2(0)=y_0', ..., x_n(0)=y_0^{(n-1)}$.

The given equation reduces to a linear system.

Example 1. Change the sysstem $x''-y=0$, $t^3y'-2x=0$ to a normal system.

Solution. Put $x=x_1, x'=x_2, y=x_3$.

Then

$$x_1'=x_2$$
$$x_2'=x_3$$
$$x_3'=\frac{2}{t^3}\cdot x_1$$

Theorem 2. (Existence and Uniqueness of solution).

Given the system

$$\frac{dx}{dt}=x'=f(t, x, y), x(t_0)=x_0$$

$$\frac{dy}{dt}=y'=g(t, x, y), y(t_0)=y_0.$$

where f and g are continuous and satisfy the Lipschitz condition in x and y in a given region D : $|t-t_0| \leqslant a$, $|x-x_0| \leqslant b$, $|y-y_0|=c$, i.e., $|f(t, x_1, y_1)-f(t, x_2, y_2)| \leqslant K_1|x_1-x_2|+K_2|y_1-y_2|$ for some positive constants K_1, K_2. Then the system has a unique solution in a neighbourhood of the point (x_0, y_0, z_0).

The proof follows Picard's procedure of theorem 1. The nth Picard's approximations in this case are :

$$x_n(t)=x_0+\int_{t_0}^{t} f(t, x_{n-1}(t), y_{n-1}(t))\, dt$$

$$y_n(t)=y_0+\int_{t_0}^{t} g(t, x_{n-1}(t), y_{n-1}(t))\, dt$$

The theorem is easily extended to a linear system in n unknowns.

Corollary 1. The only solution of

$$y^{(n)}+p_1(x)y^{(n-1)}+\dots+p_n(x)y=0$$

where the $p_i(x)$ are continuous functions and

$$y(a)=y'(a)=\dots=y^{(n-1)}(a)=0 \text{ is } y\equiv 0.$$

Corollary 2. The solution set $S=\{X_1, ..., X_n\}$ of the homogeneous system $X'=AX$ with continuous coefficients $a_{ij}(t)$ on the interval, I : $a \leqslant t \leqslant b$ is linearly dependent on I if and only if $W(t)\equiv 0$ on I.

Proof. Let $W(t_0)=0$ for some $t_0 \in I$. Then a non-trivial system $(c_1, ... c_n$ exists for which $\sum_{i=1}^{n} c_i X_i(t_0)=0$ i.e., $\sum_{i=1}^{n} c_i x_{ji}(t_0)=0$ $(j=1, ... n)$.

Let $X = \sum_{i=1}^{n} c_i X_i(t)$. Then by the uniqueness of the solution $\sum_{j=1}^{n} c_i X_i(t) \equiv 0$, *i.e.*, S is linearly dependent. The converse is obvious.

Remark 6. The theorem cannot be extended to an arbitray vector set. For example $X_1 = \begin{bmatrix} t \\ t \end{bmatrix}$, $X_2 = \begin{bmatrix} t^2 \\ t^2 \end{bmatrix}$ are linearly independent yet their Wronskian vanishes for all t. X_1, X_2 are not solutions a homogeneous linear systems with continuous coefficients.

4

Methods of Solving Linear Equations

§4·1. Depression of Order

If a particular solution $y_1(x)$ of an nth order homogeneous equation is known, the substitution $y=y_1v$ reduces it to that of order $n-1$ in the new dependent variable v.

Example 1. Find the general solution of the equation $xy''+2y'+xy=1$, if $y_1=\dfrac{\sin x}{x}$ is a particular solution of the corresponding homogeneous equation.

Solution. Put $y=\dfrac{\sin x}{x}v$ where v is the new dependent variable then substituting $y'=y_1'v+y_1v'$, $y''=y_1''v+2y_1'v'+y_1v''$ in the given equation we get

$$xy_1v''+2(xy_1'+(y_1)v'+xy_1''+2y_1'+(xy_1)v=1.$$

Since $xy_1''+2y_1'+xy_1=0$, after taking

$$y_1=\frac{\sin x}{x} \text{ we get}$$

$$v''+2\cot xv'=\frac{1}{\sin x}.$$

This gives

$$v'=-\mathrm{A}\operatorname{cosec}^2 x-\frac{\cos x}{\sin^2 x}$$

$$v=\mathrm{A}\cot x+\mathrm{B}+\frac{1}{\sin x},$$

and finally.

$$y=\mathrm{A}\,\frac{\cos x}{x}+\mathrm{B}\,\frac{\sin x}{x}+\frac{1}{x}.$$

Exercise 12

Solve the following differential equations by reducing the order :

(1) $x^3y'''+xy'-y=\dfrac{1}{x}$, $y_1=x$.

(2) $\sin xy''' + \cos xy'' + \sin xy' + \cos xy = 0$

$y_1 = \sin x,\ y_2 = \cos x.$

(3) $(x^2-1)y'' - 2xy' + 2y = (x^2-1)^2$; $y_1 = x^2+1.$

(4) $x^2y'' - x(x+2)y' + (x+2)y = x^3$; $y_1 = x.$

(5) $x^2(\log x - 1)y'' - xy' + y = 0$; $y_1 = x.$

(6) $(3x+2x^2)y'' - 6(1+x)y' + 6y = 6$; $y_1 = 1+x.$

(7) $(1-x)y'' + xy' - y = (1-x)^2$; $y_1 = x.$

§4·2. Exact Differential Equations and Adjoints

Definition 1. An nth order differential equation is called **exact** if it can be obtained by differentiating once an $(n-1)$th order differential equation (called a **first integral**) without any other process like multiplication, elimination etc.

Condition of Exactness. Let the given linear differential equation of order n be

$$P_0 y^{(n)} + P_1 y^{(n-1)} + \ldots + P_n y = X \qquad \ldots(1)$$

where the $P_i(x)$ and X are functions of x on a given interval. If we integrate both sides of (1) and use the formula

$$\int P_{n-r} y^{(r)} dx = (-1)^r \int \overset{(r)}{P}_{n-r} \cdot y dx + (-1)^{r-1} \overset{(r-1)}{P}_{n-r} \cdot y + \ldots$$
$$+ P_{n-r}\, y^{(r-1)} \quad (r = 0, 1, 2, \ldots, n).$$

we shall get

$$\int Q_n\, y dx + Q_{n-1}\, y + Q_{n-2}\, y' + \ldots + Q_0 y^{(n-1)} = \int X dx + C$$

where $\quad Q_0 = P_0,\ Q_1 = P_1 - P_0',\ Q_2 = P_2 - P_1' + P_0'', \ldots,$

and $\quad Q_n = P_n - P'_{n-1} + P''_{n-2} - \ldots + (-1)^n P_0^{(n)}.$

The necessary and sufficient condition of exactness of (1), therefore, is

$$Q_n = P_n - P_{n-1}' + \ldots + (-1)^n P_0^{(n)} = 0 \qquad \ldots(2)$$

The Adjoint Operator

Let $L(y) = P_0 y^{(n)} + P_1 y^{(n-1)} + \ldots + P_n y \qquad \ldots(3)$

If we multiply (3) by z and use the formula

$$zy^{(r)} = \frac{d}{dx}\left\{ \sum_{i=1}^{r} (-1)^{i-1} y^{(r-i)} z^{(i-1)} \right\} + (-1)^r y z^{(r)}$$

We shall get an expression of the form

$$z\, L(y) = y\, \bar{L}(z) + \frac{d}{dx}\{P(y, z)\}$$

i.e., $$z\, L(y) - y\, \bar{L}(z) = \frac{d}{dx}\{P(y, z)\} \qquad \ldots(4)$$

where $\quad \bar{L}(z) \equiv (-1)^n (P_0\, z)^{(n)} + (-1)^{n-1} (P_1\, z)^{n-1} + \ldots + P_n\, z.$

The operators $L \equiv P_0 \dfrac{d^n}{dx^n} + P_1 \dfrac{d^{n-1}}{dx^{n-1}} + \ldots + P_n$

$$\bar{L} \equiv (-1)^n \left(\frac{d^n}{dx^n}\right) P_0 + (-1)^{n-1} \left(\frac{d^{n-1}}{dx^{n-1}}\right) P_1 + \ldots P_n$$

are called formal adjoints and the differential equations $L(y)=0$ and $\bar{L}(z)=0$ are called **adjoint differential equations.** The relation (4) is called **Lagrange's identity** and the expression $P(y, z)$ is called the **bilinear concomitant** or **conjunct** of y and z.

From (4) we shall get

$$\int_{x_0}^{x} \{z\, L(y) - y\, \bar{L}(z)\}\, dx = \Big[P(y, z) \Big]_{x_0}^{x} \qquad \ldots(5)$$

which is called the **Green's formula.** If $L \equiv \bar{L}$ the operator is called a **self-adjoint operator.**

Taking the case of a second order linear equation

$$L(y) = p_0\, y'' + p_1\, y' + p_2\, y = 0 \qquad \ldots(6)$$

the Lagrange's identify will be

$$z\, L(y) - y\, \bar{L}(z) = \frac{d}{dx} \{p_0\, (y'\, z - yz') + (p_1 - p_0')\, yz\}.$$

It can be easily verified that the equation (6) is self-adjoint if and only if $p_1 = p_0'$ and the Lagrange's identity is

$$z\, L(y) - y\, \bar{L}(z) = \frac{d}{dy} \{p_0\, (y'\, z - yz')\}.$$

In case y and z are solutions of the self-adjoint equation $L(y)=0$, then

$$p_0(y'\, z - yz') = C \qquad \ldots(7)$$

where C is a constant. Relation (7) is called **Abel's formula.** For different solution pairs (y_1, y_2), value of C will differ.

A second order equation of the form

$$[p(x)\, y']' + q(x)\, y = 0 \qquad \ldots(8)$$

is in the self-adjoint form. A differential equation

$$p_0\, y'' + p_1\, y' + p_2\, y = 0$$

where p_0, p_1, p_2 are defined and continuous on a closed interval $I : a \leqslant x \leqslant b$ and $p_0(x) > 0$ on I, can be put in the form (8) where $p(x) = p_0(x)\, r(x)$, $q(x) = p_2(x) \,.\, r(x)$

and $$r(x) = \frac{1}{p_0(x)} \exp \left[\int_{x_0}^{x} \frac{p_1(t)}{p_0(t)}\, dt \right] \qquad x_0 \in I,$$

A self-adjoint differential equation of the form

$$[p(x)\, y']' + [q(x) + \lambda\, r(x)]\, y = 0 \qquad \ldots(9)$$
$$r(x) > 0,$$

where λ is a parameter, real or complex, is called a **Sturm-Liouville differential equation.** The equation (9) with appropriate homogeneous boundary conditions is called a **Sturm-Liouville eigenvalue problem.** Those values of λ for which non-trivial solutions for such problems exist, are called eigenvalues and the corresponding solutions are called eigen functions. For example, for the problem

$$y''+\lambda y=0\ ;\ y(0)=y(\pi)=0$$

$\lambda=1^2, 2^2,\ldots, n^2,\ldots$ are eigenvalues and the corresponding solutions $\sin x$, $\sin 2x,\ldots, \sin nx,\ldots$ are eigenfunctions. The eigenfunctions satisfy an important orthogonal property as proved in the following theorem :

Theorem 1. The eigenfunctions of a Sturm-Liouville problem form an orthogonal set.

Proof. Let the problem be

$(p\,y')'+(q+\lambda r)\,y=0,\quad p(x)>0,\quad r(x)>0$ on $[a, b]$

BC : $y(a)=y(b)=0.$

If λ_m, λ_n are any two eigenvalues and y_m, y_n the corresponding eigenfunctions,

then
$$(p\,y_m')'+(q+\lambda_m r)\,y_m=0 \quad \ldots(i)$$
$$(p\,y_n')'+(q+\lambda_n r)\,y_n=0 \quad \ldots(ii)$$

Multiplying (*i*) by y_m, (*ii*) by y_n and subtracting we get

$$\Big[\,p(y_m'\,y_n-y_m\,y_n')\,\Big]_a^b+(\lambda_m-\lambda_n)\int_a^b r(x)\,y_m(x)\,y_n(x)\,dx=0.$$

Owing to boundary conditions, the expression in the square bracket vanishes and consequently if $m\neq n$, we get the orthogonality relation with **weight function** $r(x)$:

$$\int_a^b r(x)\,y_m(x)\,y_n(x)\,dx=0 \qquad \text{Q.E.D.}$$

Corollary 1. All eigenvalues of a Sturm-Liouville problem are real.

Proof. Consider the problem of theorem 1. If $\alpha+i\beta$ is an eigenvalue, then $\alpha-i\beta$ is also an eigenvalue. If $y_m=u+iv$ and $y_n=u-iv$ are the corresponding eigenfunctions, then

$$\int_a^b r(x)\,(u^2+v^2)\,dx=0$$

As $r(x)>0$, this is possible only if $u=0=v$. This is false because eigenfunctions are non-zero. Hence the eigenvalues are not complex.

Corollary 2. (Eigenfunction expansion). If $\{y_i(x)\}$ is the set of eigenfunctions of the preceding problem and $f(x)$ is a function on $[a, b]$ such that $f(a)=f(b)=0$, then

$$f(x)=\sum_{i=1}^{\infty} c_i\,y_i(x) \qquad \ldots(10),$$

where
$$c_i=\frac{1}{\mu_i}\int_a^b r(x)\,f(x)\,y_i(x)\,dx$$

$$\mu_i=\int_a^b r(x)\, y_i^2(x)\, dx.$$

Expansions as in (10) are called eigenfunction expansions.

Example 2. Solve $x^3 y''+(4x^2-3x)\, y'+(2x-3)\, y=0$.

Solution. Here $P_0=x^3$, $P_1=4x^2-3x$, $P_2=2x-3$. Hence $Q_2=P_2-P_1'+P_0''=0$, *i.e.*, the equation is exact.

Also $Q_1=P_1-P_0'=x^2-3x$, $Q_0=P_0=x^3$.

A first integral is $x^3y'+(x^2-3x)y=A$

Finally, $xy=Ae^{-3/x}+B$.

Example 3. Solve $2x^2\, (x+1)y''+x(7x+3)y'-3y=x^2$.

Solution. As the condition of integrability is not satisfied, the equation is not exact in this form. Multiplying the cquation by x^m, we shall have

$$P_0=2x^{m+2}(x+1),\quad P_1=x^{m+1}\,(7x+3),\quad P_2=-3x^m.$$

Hence $Q_2=P_2-P_1'+P_0''=0$ if $m=-2$. The integrating factor is thus x^{-2}, *i.e.*, the given equation becomes exact after multiplication by x^{-2}. Hence the equation.

$$2(x+1)y''+\left(7+\frac{3}{x}\right)y'-3x^{-2}\, y=1$$

is exact. Here $P_0=2x+2$, $P_1=7+3/x$, $P_2=-3x^{-2}$.

Hence $Q_0=P_0=2x+2$, $Q_1=P_1-P_0'=5+3/x$, $Q_2=0$.

A first integral is

$$(2x+2)y'+\left(5+\frac{3}{x}\right)y=x+A$$

Integrating further we get finally,

$$x^{3/2}\,(x+1)\;\; y=c_1+c_2x^{5/2}+\frac{1}{7}x^{7/2}$$

Remark 1. It is important to remember that the value of C in the Abel's formula : $C=p\,(y_1\, y_2'-y_2\, y_1')$ depends upon the particular ordered pair of solutions (y_1, y_2) of the given equation. For example, for the equation $y''+y=0$ if $y_1=\sin x$. $y_2=\cos x$ then $C=-1$ while for the pair $(\sin x+\cos x, \sin x-\cos x)$ it is equal 2. For the equation $y''=0$ for $(1, x)$ C is 1 while for $(1, 1-x)$ it is -1.

Problem. Solve the BVP : $(1+x^2)y''+4xy+3y=0$; $y(0)=1$, $y(2)=0{\cdot}2$.

Exercise 13

1. Prove that in theorem 1 if $p\,(a)=0$, then the condition $y(a)=0$ can be dropped. If $p(b)=0$, then the condition $y\,(b)=0$ can be dropped. If $p(a)=p(b)$, then the boundary conditions $y\,(a)=y(b)=0$ can be replaced by the periodic conditions

$$y(a)=y(b),\; y'\,(a)=y'\,(b)$$

2. **Solve the following differential equations :**

(*i*) $\sin^2 x y''=2y$, (*ii*) $\cos^2 x\, y''=2y$,

(*iii*) $x^{1/2} y''+2xy'+3y=x$, (*iv*) $y''-\cot x y'+2y=\cos x$.

3. **For the equation $y''=0$ determine C for the solution pairs :**

(*i*) $(1, x)$, (*ii*) $(x, 1-z)$, (*iii*) $(1, 1-x)$.

4·3. The Normal Form

We saw that if a particular solution y_1 of a homogeneous equation is known, then a change to a new dependent variable z by the substitution $y=y_1 z$ reduces the order by one.

An alternative change of dependent variable can be adopted to eliminate the second term in the equation. Let the equation be

$$y''+Py'+Qy=0 \qquad ...(11)$$

Replacement of y by y_1v will give

$$v''+\left(\frac{2y_1'}{y_1}+P\right)v'+\frac{1}{y_1}(y_1''+Py_1'+Qy_1)\,v=0$$

If $\dfrac{2y_1'}{y_1}+P=0$, then $y_1=e^{-\frac{1}{2}\int P dx}$

and we have

$$v''+Iv=0 \qquad ...(12)$$

where $$I=Q-\tfrac{1}{2}P'-\tfrac{1}{4}P^2$$

Equation (12) is called the **normal form** of (11). In many cases the normal form is simpler to solve.

The coefficient I in (12) is invariant under transformations of the type $y=y_1 z$, where y_1 is known. Putting $y=y_1z$ in (1).

$$z''+P_1z'+Q_1z=0 \qquad ...(13)$$

where $$P_1=P+\frac{2}{y_1}y_1',$$

$$Q_1=\frac{1}{y_1}(y_1''+Py_1'+Qy_1)$$

Let the normal form of (13) be $v''+I_1v=0$;

then $$I_1=Q_1-\tfrac{1}{2}P_1'-\tfrac{1}{4}P_1^2$$

$$=\frac{1}{y_1}(y_1''+Py_1'+Qy_1)-\tfrac{1}{2}\left(P'+\frac{2(y_1y_1''-y_1'^2)}{y_1^2}\right)-\tfrac{1}{4}\left(P+\frac{2y_1'}{y_1}\right)^2$$

$$=Q-\tfrac{1}{2}P'-\tfrac{1}{4}P^2=I$$

Second order equations having the same normal form are equivalent and can be transformed into one another by a change of the dependent variable.

Example 4. Solve the equation

$$x^2y''-2(x^2+x)y'+x^2+2x+2)y=0$$

by reducing it to normal form.

Solution. Here $P=-2-\frac{2}{x}$,

$$Q=1+\frac{2}{x}+\frac{2}{x^2}.$$

Hence $\quad I=Q-\frac{1}{2}P'-\frac{1}{4}P^2=0,$

$$e^{-\frac{1}{2}\int P dx}=xe^x.$$

The substitution $y=x\,e^x\,v$ reduces the given equation to $v''=0$. Thus $v=Ax+B$ and finally $y=x\,e^x\,(Ax+B)$.

Exercise 14

Solve the following equations by reduction to normal form :

(1) $y''-4xy'+(4x^2-3)y=e^{x^2}$

(2) $y''-\frac{1}{\sqrt{x}}y'+\frac{1}{4x^2}(-8+\sqrt{x}+x)y=0$

(3) $y''-\frac{2}{x}y'+\left(a^2+\frac{2}{x^2}\right)y=0$

(4) $y''+2n\cot nx\,y'+(m^2-n^2)y=0$

(5) If u and su are two solutions of the equation $v''+Iv=0$, then prove that

$$\frac{s''}{s'}+2\frac{u'}{u}=0$$

and
$$\frac{s'''}{s'}-\frac{3}{2}\left(\frac{s''}{s'}\right)^2=2I \qquad \text{...(A)}$$

Prove that if s is a solution of (A), then $\frac{1}{\sqrt{s'}}$ and $\frac{s}{\sqrt{s'}}$ are also its solutions.

(The expression on the left hand side of (A) is called the **Schwarzian derivative** and is denoted by $\{s, x\}$).

6. Verify the following properties of the Schwarzian derivative :

(*i*) $\{s, x\}=-s'^2\{x, s\}$

(*ii*) $\left\{\frac{as+b}{cs+d}, x\right\}=\{s, x\}$

(*iii*) $\{s, x\}=y'^2\,[\{s, y\}-\{x, y\}]$

(*iv*) $\{s, x\}=y'^2\,\{s, y\}-v'^2\{x, v\}+v'^2\{y, v\}$

4·4. Change of the Independent Variable

The following theorem answers the question whether a second order linear equation should be solved by a change of the independent variable and if so, to suggest the appropriate substitution.

Theorem 2[1]. The necessary and sufficient condition that the differential equation

$$y''+Py'+Qy=R \qquad (i)$$

where P, Q, R are functions of x, can be converted to one with constant coefficients by a change of the independent variable is

$$Q'+2PQ=KQ^{3/2} \qquad (ii)$$

where K is some constant. The new indendendent variable z is given either by

$$z=\int e^{-\int P dx} dx \quad \text{or by} \quad z=\int \sqrt{\frac{Q}{a}}\,dx$$

where a is a suitable constant.

Proof. Necessity. Let z be the new independent variable given by $z=f(x)$. Equation (i) will be changed to

$$f'^2 \frac{d^2y}{dz^2}+(f''+Pf')\frac{dy}{dz}+Qy=R \qquad \text{...(A)}$$

Choose $f(x)$ in such a way that

$$f'^2=\lambda^2 (a+bz)^2\, Q \qquad \text{...(B)}$$

and

$$f''+Pf'=\mu\,(a+bz)\, Q \qquad \text{...(C)}$$

for some constants a and b. In case $b=0$ (A) reduces to an equation with constant coefficients. If $b\neq 0$, then it reduces to the Cauchy-Euler form, which again reduces to one with constant cocfficients by the change $a+bz=e^t$.

Eliminating f from (B) and (C), we get

$$Q'+2PQ=KQ^{3/2}$$

where

$$K=2\left(\frac{\mu}{\lambda}-\lambda b\right)$$

Sufficiency. Let the condition (ii) hold. The change of x to another variable z in (i) leads to

$$\frac{d^2y}{dz^2}+L\frac{dy}{dz}+My=N \qquad \text{...(D)}$$

where

$$L=\frac{z''+Pz'}{z'^2}, \quad M=\frac{Q}{z'^2}, \quad N=\frac{R}{z'^2}$$

In (ii) put $V=2/Q^{1/2}$, then

$$V'-PV=-K \qquad \text{...(E)}$$

Solving (E), $$V e^{-\int P dx}=-A-K\int e^{-\int P dx} dx \qquad \text{...(F)}$$

If we choose the substitution $z=\int e^{-\int P dx}\, dx$,

---

1. The author is thankful to Prof. Prem Chand of Vikram University for suggesting this theorem.

then $$z'=e^{-\int P dx},\ L=0$$

and $$V\,e^{-\int P dx}=\frac{2}{\sqrt{Q}}=z'$$

Thus using (F) we get $M=\dfrac{4}{(A+Kz)^2}$

Since $z''+Pz'=0$, the equation (D) now takes the form

$$4\ (A+Kz)^2\frac{d^2y}{dz^2}+Q=G\ (z)$$

which is at once reducible to an equation with constant cofficients by putting $A+Kz=e^t$.

Alternatively, if we choose the substitution

$$z=\int\sqrt{\frac{Q}{a}}\,dx,\ \text{then } M=a \ \text{ and } L=\frac{K}{2\sqrt{a}}=b$$

(say), thus leading to on equation with constant coefficients.

Q.E.D.

Remark 1. When $K=0$, both the substitutions are the same. When $K\neq 0$, the simplest substitution should be applied.

Example 5. Solve the differential equation

$$(1+x^2)^2y''+2x(1+x^2)y'+y=(\tan^{-1} x)^2.$$

Solution Here $P=\dfrac{2x}{1+x^2}$, $Q=\dfrac{1}{(1+x^2)^2}$.

Hence,

$$Q'+2PQ=0.$$

A change of independent variable is thus feasible. Put

$$z=\int\sqrt{Q}\,dx=\tan^{-1} x.$$

With this substitution the given equation will reduce to

$$\frac{d^2y}{dz^2}+y=z^2$$

Hence,

$$y=A\cos z+B\sin z+z^2-2$$

and finally

$$y=\frac{A}{\sqrt{1+x^2}}+\frac{Bx}{\sqrt{1+x^2}}+(\tan^{-1}x)^2-2.$$

Example 6. Solve the equation

$$y''+(3\sin x-\cot x)\,y'+2\sin^2 x y=e^{-\cos x}\sin^2 x.$$

Solution Here $P=3\sin x-\cot x$, $Q=2\sin^2 x$.

Hence $Q'+2PQ=12\sin^3 x=3\sqrt{2}\,Q^{3/2}$.

A change in the independent variable is thus warranted. Of the two choices

$$z=\int e^{-\int P\,dx}\,dx \quad \text{and} \quad z=\int\sqrt{\frac{Q}{a}}\,dx,$$

the later is simpler. Put $z=-\cos x$.

Then the given equation is changed to

$$\frac{d^2y}{dy^2}+3\frac{dy}{dz}+2y=e^z.$$

The complete solution is

$$y=A\,e^{-2z}+Be^{-z}+\frac{1}{6}e^z$$

and finally

$$y=Ae^{2\cos x}+B\,e^{\cos x}+\frac{1}{6}\,e^{-\cos x}$$

Exercise 15

Solve the following differential equations by changing the independent variable :

(1) $xy''-y'-4x^3\,y=8x^3\sin x^2$.

(2) $\cos xy''+\sin xy'-2\cos^3 x\,.\,y=2\cos^5 x$.

(3) $y''+(\tan x-3\cos x)\,y'+2\cos^2 xy=\cos^4 x$.

(4) $y''+(\tan x-1)^2\,y'-n\,(n-1)\sec^4 x\cdot y=0$.

(5) $xy''+(4x^2-1)\,y'+4x^3\,y=2x^3$.

(6) $xy''-y'+4x^3\,y=x^5$.

4·5. Lagrange's Method of Variation of Parameters

In cases where a particular integral of a non-homogeneous equation is difficult to obtain, if a fundamental system of the corresponding homogeneous equation, (*i.e.*, the C.F.) is known, then the method of variation of parameters can be used to find the general solution.

Let $\quad L(y)=X \quad$ ...(N)

be and nth order linear differential equation where

$$L\equiv D^n+p_1\,D^{n-1}+\ldots+p_n.$$

Let $y_1, y_2,\ldots,y_n$ be the n linearly independent solutions of the homogeneous equation

$$L(y)=0 \qquad \ldots(H)$$

Assume $\quad y=\sum_{i=1}^{n} A_i y_i \quad$ ...(i)

to be a solution of (N) where the coefficients A_i instead of being constants, are functions of x to be determined.

Differentiating (i) we get

$$y' = \sum_{i=1}^{n} A_i y_i' + \sum_{i=1}^{n} A_i' y_i.$$

Put $$\sum_{i=1}^{n} A_i' y_i = 0$$

Differentiating further,

$$y'' = \sum_{i=1}^{n} A_i y_i'' + \sum_{i=1}^{n} A_i' y_i'.$$

Put $$\sum_{i=1}^{n} A_i' y_i' = 0.$$

Iterating the procers we shall get

$$y^{(n)} = \sum_{i=1}^{n} A_i \, y_i^{(n)} + \sum_{i=1}^{n} A_i' \; y_i^{(n-1)}.$$

Now put $A_1' y_1^{(n-1)} + \ldots + A_n' y_n^{(n-1)} = X$.

The resultant expressions for $y, y', \ldots, y^{(n)}$ when substituted in (N) make it an identify. The n unknowns $A_1', \ldots, A_n'$ are found from the linear algebraic system

$$A_1' y_1 + A_2' y_2 + \cdots + A_n' y_n = 0$$
$$A_1' y_1' + A_2' y_2' + \ldots + A_n' y_n' = 0$$
$$\ldots\ldots\ldots\ldots\ldots\ldots\ldots\ldots\ldots\ldots$$
$$A_1' y_1^{(n-1)} + \ldots + A_n' y_n^{(n-1)} = X.$$

The determinant of the coefficients of this system is the Wronskian $W(y_1, \ldots, y_n)$ of a fundamental system, hence, $W \neq 0$. Hence, the system will have a unique solution.

Solving we shall get

$$A_i' = \frac{W_i}{W} \quad (i = 1, \ldots n),$$

where W_i is obtained from W by replacing its ith column by $(0, 0, \ldots, X)^T$. Finally,

$$A_i = \int \frac{W_i}{W} dx + B_i$$

where the B_i are arbitrary constants. Complete primitive of (N) is thus obtained.

Example 7. Find the general solution of the equation

$$(x^2 - 1)y'' - 2xy' + 2y = (x^2 - 1)^2 \qquad \ldots(i)$$

Solution. We first solve the corresponding homogeneous equation

$$(x^2-1)y''-2xy'+2y=0 \qquad ...(ii)$$

$y=x$ is a solution of (*ii*). Put $y=xv$ in (*ii*) to get

$$x(x^2-1)\ v''-2v'=0.$$

Hence $v=A\left(x+\dfrac{1}{x}\right)+B,$

and $$y=A(x^2+1)+Bx \qquad ...(iii)$$

Let A and B be functions of x so that the expression on the right side of (*iii*) is a solution of (*i*). Differentiating (*iii*)

$$y'=2xA+B+A'(x^2+1)+B'x.$$

Put $A'(x^2+1)+B'x=0.$ ...(*iv*)

Differentiating further,

$$y''=2A+2xA'+B'$$

Now put $2xA'+B'=x^2-1$ ...(*v*)

Solving (*iv*) and (*v*) for A′ and B′, we get

$$A'=x,\ B'=-x^2-1,$$

i.e., $$A=\frac{x^2}{2}+c_1,$$

$$B=-\frac{x^3}{3}-x+c_2.$$

The required solution thus is

$$y=c_1(x^2+1)+c_2x-\frac{x^2}{2}+\frac{x^4}{6}.$$

Exercise 16

Solve the following differential equations by the method of variation of parameters :

(1) $y''+y=\operatorname{cosec} x$

(2) $(1-x)\ y''+xy'-y=(1-x)^2$

(3) $xy''+2y'+xy=1/x$

(4) $y'''+y'=\tan x$

(5) $(x^2+x)\ y''+(2-x^2)y'-(2+x)\ y=x\ (x+1)^2$

(6) If $y''+\lambda^2y=p(x)\ y(x)$, than prove that

$$y(x)=A\cos\lambda x+B\sin\lambda x+\frac{1}{\lambda}\int_{x_0}^{x}\sin(x-t)\lambda.\ p(t)\ y(t)\ dt$$

5

Solution in Series and Special Functions

Introduction. In this chapter we describe the method of Frobenius for finding the solution of a second order linear ordinary differential equation in the form of power series. The method will be applied to find the solutions of the Gauss's hypergeometric equation, the Legendre's equation, the Bessel's equation, Kummer's equation and the Hermite's equation. Important properties of these solutions, popularly known as special functions of mathematical physics, are studied.

§5·1. Ordinary and Singular Points

A function is called *analytic* (or *regular* or *holomorphic*) at a point if it is possible to expand it in a Taylor's series, valid in some neighbourhood of the point ; in other words, the function is single-valued in some neighbourhood of the point and is infinitely differentiable at that point.

Given a differential equation :

$$y''+p(x)\,y'+q(x)\,y=0 \qquad \ldots(1)$$

A point $x=a$ is an *ordinary point* of the differential equation (1) if the coefficients $p(x)$ and $q(x)$ are analytic at that point ; otherwise it is called a *singular point* or a *singularity* of the equation.

A singularity $x=a$ of (1) is called *regular* if $(x-a)\,p(x)$ and $(x-a)^2\,q(x)$ are analytic at the point $x=a$; otherwise it is called an *irregular singularity*. A change of the variable by $x-a=t$ transfers the point $x=a$ to $t=0$

A series of the form

$$x^{\rho}\sum_{r=0}^{\infty} a_r\,x^r, \qquad (a_0\neq 0) \qquad \ldots(2)$$

where ρ is a real or complex number and the power series $\sum_{r=0}^{\infty} a_r\,x^r$ is convergent is some interval $|\,x\,|<R$, is called a ***generalised*** **or *quasi power series*** **(or *Frobenius series*). If ρ is a non-negative integer, then (2) becomes an ordinary power series.**

The Point at Infinity :

To find the nature of the point at infinity (*i.e.*, $x=\infty$), we put $x=\frac{1}{t}$ in (1), thereby mapping $x=\infty$ into $t=0$. The nature of $t=0$ is the nature of $x=\infty$.

By putting $x=\frac{1}{t}$, equation (1) is transformed to

$$\frac{d^2y}{dt^2}+\left\{\frac{2}{t}-\frac{1}{t^2}\,p\left(\frac{1}{t}\right)\right\}\frac{dy}{dt}+\frac{1}{t^4}\,q\left(\frac{1}{t}\right)y=0 \qquad \ldots(3)$$

The following conclusions can be drawn on the basis of the equation (3) :

(*i*) $x=\infty$, an ordinary point if for $x \to \infty$,

$$p(x)=\frac{2}{x}+0(x^{-2}), \quad q(x)=0(x^{-4}).$$

i.e., $xp(x)$ and $x^4q(x)$ are bounded for $x \to \infty$.

(*ii*) $x=\infty$, a regular singularity if for $x \to \infty$,

$$p(x)=0(x^{-1}) \text{ and } q(x)=0(x^{-2}).$$

An equation is called *Fuchsian* if all its singularities are regular.

Example 1. (*a*) For the Legendre's equation

$$(1-x^2)\,y''-2xy'+n\,(n+1)\,y=0.$$

$x=0$ is an ordinary point ;

$x=-1,\ 1,\ \infty$ are regular singularities.

(*b*) For the Bessel's equation

$$x^2y''+xy'+(x^2-n^2)\,y=0.$$

$x=0$ is a regular singularity and $x=\infty$ is an irregular singularity.

(*c*) For the hypergeometric equation

$$x(1-x)\,y''+\{c-(a+b+1)\,x\}\,y'-aby=0$$

the only singularities are 0, 1, ∞ which are regular.

(*d*) For the Kummer's equation

$$xy''+(c-x)\,y'-ay=0$$

$x=0$ is a regular singularity while $x=\infty$ is an irregular singularity.

(*e*) For the Hermite's equation

$$y''-2xy'+2ny=0.$$

$x=0$ is an ordinary point while $x=\infty$ is an irregular singularity.

(*f*) For the equation $x^2y''+\sin x y'+\cos x y=0$

$x=0$ and $x=\infty$ are regular singularities.

(*g*) For the equation $x^2(x-4)^2\,y''+3xy'-(x-4)\,y=0$.

$x=0$ and $x=\infty$ are regular singularities, while $x=4$ is an irregular singularity.

§5·2. Series Solutions—The Method of Frobenius

Given the differential equation

$$L(y)=x^2y''+xp(x)\,y'+q(x)\,y=0 \qquad ...(4)$$

$x=0$ is a regular singularity if $p(x)$ and $q(x)$ are analytic at $x=0$.

Let $$p(x)=p_0+p_1x+p_2x^2+\dots\dots\dots=\sum_{r=0}^{\infty} p_r\,x^r$$

$$q(x)=q_0+q_1x+q_2x^2+\dots\dots\dots=\sum_{r=0}^{\infty} q_r\,x^r$$

In the special case $p_0=0$, $q_0=q_1=0$, the point $x=0$ is an ordinary point.

Assume a solution,

$$y=x^{\rho}\sum_{r=0}^{\infty} a_r\,x^r, \qquad (a_0\neq 0) \qquad ...(5)$$

for the equation (4). Substituting the series (5) in (4) and collecting like powers of x, we shall get

$$L(y)=\sum_{r=0}^{\infty} g_r\,x^{\rho+r}=0 \qquad ...(6)$$

where $$g_r=\{(\rho+r)^2+(p_0-1)(\rho+r)+q_0\}\,a_r$$

$$+\sum_{k=0}^{r-1}\{(\rho+k)\,p_{r-k}+q_{r-k}\}\,a_k \qquad ...(7)$$

For (5) to be a solution of (4), we should have

$g_r=0$ for $r=0, 1, 2, \dots$.

The coefficient of the lowest degree term (also called the *dominant term*) in (6) gives $g_0=0$,

i.e., $$F(\rho)=\rho^2+(p_0-1)\,\rho+q_0=0. \qquad ...(8)$$

The equation (8) is called the *indicial equation* (or *governing equation*). Its roots ρ_1, ρ_2 are called the *exponents* at $x=0$.

In the special case when $x=0$ is an ordinary point, the exponents will be $\rho=0$, 1 as can be seen by putting $p_0=q_0=0$ in (8).

As will be seen later, both the exponents 0 and 1 lead to the same series solution. It is thus enough to assume an ordinary series solution

$$y=\sum_{r=0}^{\infty} a_r\,x^r \quad \text{at the ordinary point } x=0.$$

For given ρ, $g_r=0$ is a *recurrence relation* in consecutive coefficients. Upon simplification we shall get

$$a_r=\frac{a_0\, h_r(\rho)}{[F(\rho+r)\,F(\rho+r-1)\ldots F(\rho+1)]},\ r\geqslant 1 \qquad \ldots(9)$$

For exponents ρ_1 and ρ_2 we shall have the relations

$$\begin{aligned} F(\rho)&=(\rho-\rho_1)(\rho-\rho_2)\\ F(\rho_1+r)&=r(r+\rho_1-\rho_2)\\ F(\rho_2+r)&=r(r+\rho_2-\rho_1)\end{aligned} \qquad \ldots(10)$$

If after substituting the series (5) in (4) we put $g_r=0$ for $r\geqslant 1$ without putting $F(\rho)=0$, then we shall get

$$L(y)=a_0(\rho-\rho_1)(\rho-\rho_2)\,x^\rho \qquad \ldots(11)$$

The following cases now arise :

Case 1. $\rho_1-\rho_2$ **is neither zero nor an integer.** Two linearly independent solutions, each for $\rho=\rho_1$ and $\rho=\rho_2$ are obtained.

Case 2. $\rho_1=\rho_2=\alpha$**(say).** We have

$L(y)=a_0(\rho-\alpha)^2\, x^\rho$

$$\frac{\partial L(y)}{\partial\rho}=L\left(\frac{\partial y}{\partial\rho}\right)=a_0\left\{2(\rho-\alpha)x^\rho+(\rho-\alpha)^2\, x^\rho \log x\right\}.$$

Hence, $\left[\dfrac{\partial y}{\partial\rho}\right]_{\rho=\alpha}$ is also a solution. We thus have two linearly independent solutions :

$$y_1=\sum_{r=0}^{\infty} a_r\, x^{\alpha+r},\quad y_2=y_1\log x+x^\alpha\sum_{r=1}^{\infty}\left[\frac{\partial a_r}{\partial\rho}\right]_{\rho=\alpha} x^r.$$

$\dfrac{\partial a_r}{\partial\rho}$ should be calculated by logarithmic differentiation.

Case 3. $\rho_1-\rho_2=n$**, a positive integer.** In this case there are two possibilities :

(*a*) $h_n(\rho_2)\neq 0$. Since $F(\rho_1)=0=F(\rho_2+n)$, the coefficient a_n becomes infinite if $\rho=\rho_2$. To overcome this difficulty we put $a_0=\rho-\rho_2$. Then

$$L(y)=(\rho-\rho_1)(\rho-\rho_2)^2\, x^\rho$$

As in case 2, it can be seen that $\left[\dfrac{\partial y}{\partial\rho}\right]_{\rho=\rho_2}$ will also be a solution. In this case we shall have two linearly independent solutions

$$y_1=\left[\,y\,\right]_{\rho=\rho_1}\quad\text{and}\quad y_2=\left[\frac{\partial y}{\partial\rho}\right]_{\rho=\rho_2}$$

A third solution y_3 will be obtained by using the exponent $\rho=\rho_2$ in the expression for a_r after putting $a_0=\rho_2+n$. But this being non-logarithmic, will be a constant multiple of y_1.

y_2 will be of the form

$$y_2 = A\, y_1 \log x + x^{\rho_2} \sum_{r=0}^{\infty} b_r\, x^r$$

where A is some constant. In the particular case when A=0, both solutions are free of a logarithmic term.

No set procedure can be prescribed to compute A.

Problem. By referring to relation (9) explain why the difficulty is removed by taking $a_0 = \rho - \rho_2$.

(*b*) $h_n(\rho_2) = 0$. In this case since $F(\rho_2 + n) = 0$, a_n becomes indeterminate. The general solution will be of the form

$$y = a_0\, y_1 + A y_2$$

where y_1 is a terminating series and y_2 is free of logarithm, a_0, A are arbitrary.

Example 2. Solve the equation $2xy'' - y' - y = 0$

Solution. Assume a solution

$$y = \sum_{r=0}^{\infty} a_r\, x^{\rho+r} \qquad a_0 \neq 0.$$

Its substitution in the given equation gives

$$\sum_{r=0}^{\infty} a_r(\rho+r)(2\rho+2r-3)x^{\rho+r-1} - \sum_{r=0}^{\infty} a_r\, x^{\rho+r} = 0 \quad ...(i)$$

Equating to zero the coefficient of the lowest degree term, we get the indicial equation $\rho(2\rho-3)=0$. The exponents are $\rho_1 = 0$, $\rho_2 = 3/2$.

Collecting like powers of x in (*i*), we get

$$\sum_{r=0}^{\infty} [a_{r+1}(\rho+r+1)(2\rho+2r-1) - a_r]\, x^{\rho+r} = 0 \quad ...(ii)$$

Equating to zero the coefficients of all powers of x in (*ii*), we get the recurrence relation

$$a_{r+1} = \frac{a_r}{(\rho+r+1)(2\rho+2r-1)}$$

For $\rho = 0$

$$a_r = \frac{-a_0}{r\,!\; 1\,.\,3 \ldots (2r-1)}, \text{ and}$$

$$y_1 = 1 - x - \frac{x^2}{2\,!} - \frac{x^3}{3\,!\;1\,.\,3} - \frac{x^4}{4\,!\;3\,.\,5} - \ldots$$

For $\rho = 3/2$

$$a_r = \frac{2^r a_0}{5\,.\,7 \ldots (3+2r)\; 2\,.\,4 \ldots (2r)}, \text{ and}$$

$$y_2 = x^{3/2}\left\{1 + \frac{x}{5} + \frac{x^2}{2\,!\;5.7} + \frac{x^3}{3\,!\;5.7.9} + \ldots\right\}$$

The general solution is

$$y=Ay_1+By_2$$

Example 3. Solve $xy''+y=0$.

Solution.

Let $y=x^{\rho}\sum_{0} a_r x^r$ be a solution. The indicial equation is $\rho(\rho-1)=0$. Also

$$a_r=\frac{(-1)^r a_0}{\rho[(\rho+1)(\rho+2)\ldots(\rho+r-1)]^2(\rho+r)}$$

The root $\rho=1$ gives a solution

$$y_1=\sum_{r=0}^{\infty}\frac{(-1)^r x^{r+1}}{r!(r+1)!}$$

For $\rho=0$, a_1 becomes infinite. Hence put $a_0=\rho$. Then

$$\left[\frac{\partial ar}{\partial\rho}\right]_{\rho=0}=\frac{(-1)^r}{(r-1)!\,r!}(H_r+H_{r-1})$$

where

$$H_r=1+\frac{1}{2}+\frac{1}{3}+\ldots+\frac{1}{r}.$$

Since $y=\rho x^{\rho}+\sum_{r=1}^{\infty} a_r x^{\rho+r}$, we get the second linearly independent solution.

$$y_2=\left[\frac{\partial y}{\partial\rho}\right]_{\rho=0}=y_1\log x+1-\sum_{r=1}^{\infty}\frac{(-1)^r(H_r+H_{r-1})x^r}{(r-1)!\;r!}$$

It is easily verified that $[y]_{\rho=0}=-y_1$.

Example 4. Consider $x^2y''+x^2y'-2y=0$.

Solution.

Assuming the solution $y=\sum_{r=0}^{\infty} a_r x^{\rho+r}$, we get the indicial equation $(\rho-2)(\rho+1)=0$. Hence the exponents are $\rho_1=2$, $\rho_2=-1$. The recurrence relation in coefficients is

$$a_r=\frac{-(\rho+r-1)}{(\rho+r-2)(\rho+r+1)}a_{r-1} \qquad \ldots(i)$$

For the exponent $\rho_2=-1$, we get $a_1=-\frac{1}{2}a_0$, $a_2=0$ and a_3 is Indeterminate. For $r\geqslant 4$,

$$a_r=\frac{6(-1)^{r-3}(r-2)}{r!}a^3$$

The general solution is

$$y=a_0\ x^{-1}\left(1-\frac{x}{2}\right)+a_3\ x^{-1}\sum_{r=3}^{\infty}\frac{(-1)^{r-3}\ (r-2)\ x^r}{r\ !}$$

where a_0 and a_3 are arbitrary.

The non-terminating solution can be expressed as

$$y_2=\sum_{r=0}^{\infty}\frac{(-1)^r\ (r+1)}{(r+3)\ !}x^{r+2}$$

Use of $\rho=2$ in (*i*) also leads to the solution y_2.

Example 5. **Solve $x\ (x-1)\ y''+(3x-1)y'+y=4x$.**

Solution. **To find the C.F., consider**

$$L(y)=x(x-1)\ y''+(3x-1)y'+y=0 \qquad \text{...(ii)}$$

Use of $y=\sum_{r=0}^{\infty} a_r\ x^{\rho+r}$ in (*ii*) gives

$$\sum_{r=1}^{\infty}(c+r)^2\ (a_{r-1}-a_r)x^{\rho+r-1}-\rho^2\ a_0\ x^{\rho-1}=0.$$

The exponents are $\rho_1=\rho_2=0$.

For $\rho=0$, $\quad a_r=a_{r-1},\ r\geqslant 1$. One solution is

$$y_1=1+x+x^2+...=(1-x)^{-1}.$$

Since $y=x^\rho+x^\rho\sum_{r=1}^{\infty}a_r\ x^r$ and each a_r is the same constant,

$$\frac{\partial y}{\partial \rho}=\log\ x\ x^\rho\sum_{r=0}^{\infty}x^r.$$

Hence
$$y_2=\left[\frac{\partial y}{\partial \rho}\right]_{\rho=0}=\frac{\log x}{1-x}$$

Thus
$$\text{C.F.}=\frac{A+B\log x}{1-x}$$

To find the particular integral, put $y=x^\rho\sum_{r=0}^{\infty}a_r\ x^r$ in the given non-homogeneous equation, retain the recurrence relation $a_r=a_{r-1}$ and put $-a_0\rho^2\ x^{\rho-1}=4x$.

This is satisfied if $a_0=-1$ and $\rho=2$. Thus

$$y_p=-x^2\sum_{r=0}^{\infty} x^r=\frac{-x^2}{1-x}.$$

The required general solution is thus

$$y=\frac{A+B\log x-x^2}{1-x}$$

Exercise 17

A. Find the series solutions of the following linear differential equations at $x=0$:

(1) $4xy''+2y'+y=0$

(2) $2x(1-x)y''+(1-x)y'+3y=0$

(3) $3xy''+(2-x)y'-2y=0$

(4) $x^2y''-x(1+x)y'+y=0$

(5) $xy''+(1-2x)y'+(x-1)y=0$

(6) $xy''+y'-y=0$

(7) $xy''+y=x$

(8) $xy''+2(1-x)y'+(x-2)y=0$

(9) $xy''+(1-x)y'+ny=0$, (n a natural number).

(10) $x^2y''+x(x-1)y'+(1-x)y=x^2$.

B. Prove the following:

(*i*) (Cauchy Product)

$$\sum_{n=0}^{\infty}\sum_{k=0}^{\infty} a_n b_k\ x^{n+k}=\sum_{n=0}^{\infty}\sum_{k=0}^{n} a_{n-k}\,b_k\,x^n$$

(*ii*) $$\sum_{n=0}^{\infty}\sum_{k=0}^{\infty} a_n b_k\ x^{n+2k}=\sum_{n=0}^{\infty}\sum_{k=0}^{[n/2]} a_{n-2k}\,b_n\ x^n$$

where $[m]$=greatest integer in m.

(*iii*) $$\sum_{n=0}^{\infty}\sum_{k=0}^{n} a_n b_k\ x^{n+k}=\sum_{n=0}^{\infty}\sum_{k=0}^{\infty} a_{n+k}\,b_k\,x^{n+2k}$$

$$=\sum_{n=0}^{\infty}\sum_{k=0}^{[n/2]} a_{n-k}b_k\,x^n$$

§5·3. The Gauss's Hypergeometric Equation

$$x(1-x)y''+\{c-(a+b+1)x\}y'-aby=0 \qquad ...(12)$$

This is a Fuchsian equation. The only singularities are regular at 0, 1, ∞. On applying the method of Frobenius, it is found that the exponents are as follows :

(*i*) $x=0$, exponents 0, $1-c$;

(*ii*) $x=1$, exponents 0, $c-a-b$;

(*iii*) $x=\infty$, exponents a, b.

All this information is symbolically represented by the scheme

$$y=P\begin{Bmatrix} 0 & 1 & \infty & \\ 0 & 0 & a & x \\ 1-c & c-a-b & b & \end{Bmatrix} \quad \ldots(13)$$

The symbol on the right of (13) is called the *Riemann-P-function* or the *Papperitz symbol* of the equation. The main use of this symbol is that it is capable of representing all solutions of the equation.

We now consider the solutions at $x=0$. The exponents are 0, $1-c$ and the recurrence relation is

$$a_{n+1}=\frac{(\rho+a+n)\,(\rho+b+n)}{(\rho+n+1)\,(\rho+c+n)}a_n$$

For $\rho=0$, if we let

$$a_0=1,$$

then $$a_r=\frac{(a)_r\,(b)_r}{r\,!\,(c)_r},$$

where $(a)_r$ is the *Barne's symbol* (also called the *Pochhammer symbol* or the *factorial function* defined by

$$(a)_r=a(a+1)\ldots(a+r-1)=\frac{\Gamma(a+r)}{\Gamma(a)} \quad \ldots(14)$$

The relevant solution is

$$y_1=\sum_{r=0}^{\infty}\frac{(a)_r\,(b)_r}{r\,!\,(c)_r}x^r=1+\frac{a\,.\,b}{1\,.\,c}x+\frac{a(a+1)\;b(b+1)}{2\,!\;c(c+1)}x^2+\ldots \quad \ldots(15)$$

The series on the right of (15) is called the *hypergeometric function* and is denoted by ${}_2F_1\left[\begin{matrix}a,\,b\,;\,x\\ c\end{matrix}\right]$ or ${}_2F_1[a, b; c ; x]$ or simply $F[a, b; c ; x]$, It is symmetric in the numerator parameters a, b.c is called the denominator parameter.

By the ratio test, the series in (15) is convergent if $|x|<1$, as is also obvious from the fact that the nearest singularity to $x=0$ of the equation is $x=1$.

If $x=1$, by using Raabe's test, the series is convergent if $\text{Re}\,(c-a-b)>0$.

Many familiar functions are special cases of the hypergeometric function. For example

$$(1-x)^n=F[-n, 1 ; 1 ; x]$$

$$(1+x)^n+(1-x)^n=2F[-\tfrac{1}{2} n, -\tfrac{1}{2} n+\tfrac{1}{2} ; \tfrac{1}{2} ; x^2]$$

$$\log (1+x)=xF[1, 1 ; 2, -x]$$

$$\log\frac{1+x}{1-x}=2\, xF\left[\frac{1}{2}, 1 ; \frac{3}{2} ; -x^2\right]$$

$$\sin^{-1}x=xF\left[\frac{1}{2}, \frac{1}{2} ; \frac{3}{2} ; x^2\right]$$

$$\tan^{-1}x=xF\left[\frac{1}{2}, 1 ; \frac{3}{2} ; -x^2\right].$$

The second solution near $x=0$ is

$$y_2=x^{1-c}\, F[a-c+1, b-c+1 ; 2-c ; x] \qquad \text{...(16)}$$

provided c is not an integer and if we follow the procedure for obtaining $F[a, b ; c ; x]$.

We can make a short-cut to obtain the second solution (16) by a transformation of the P-symbol. It is easy to see that

$$x^{c-1}\, P\begin{Bmatrix} 0 & 1 & \infty & \\ 0 & 0 & a & x \\ 1-c & c-a-b & b & \end{Bmatrix}$$

$$=P\begin{Bmatrix} 0 & 1 & \infty & \\ c-1 & 0 & a+1-c & x \\ 0 & c-a-b & b+1-c & \end{Bmatrix} \qquad \text{...(17)}$$

The solution corresponding to the exponent 0 at $x=0$ for the equation of the right side of (17) is by comparison with (15)

$$F[a+1-c, b+1-c ; 2-c ; x].$$

Since $P\begin{Bmatrix} 0 & 1 & \infty & \\ 0 & 0 & a & x \\ 1-c & c-a-b & b & \end{Bmatrix}$

$$=x^{1-c}\, P\begin{Bmatrix} 0 & 1 & \infty & \\ c-1 & 0 & a+1-c & x \\ 0 & c-a-b & b+1-c & \end{Bmatrix} \qquad \text{...(18)}$$

$$y_2=x^{1-c}\, F[a+1-c, b+1-c ; 2-c ; x]$$

is a solution of (12).

To test the linear independence of y_1 and y_2, we examine their Wronskian,

$$W(x)=W(y_1, y_2)=\begin{vmatrix} y_1 & y_2 \\ y_1' & y_2' \end{vmatrix}.$$

By using the Liouville's formula, we have

$$W(x)=W(o)e^{-\int_0^x p(t)\,dt}$$

where $$p(x)=\frac{c-(a+b+1)x}{x(1-x)}.$$

Hence $$W(x)=W(o)x^{-c}(1-x)^{c-a-b-1}$$

i.e., $$\begin{vmatrix} 1+\frac{ab}{1.c}x+\ldots & x^{1-c}(1+\ldots) \\ \frac{ab}{c}+\ldots & (1-c)x^{-c}(1+\ldots) \end{vmatrix}=W(o)x^{-c}(1-x)^{c-a-b-1}$$

A comparison of the two sides suggests that

$$W(o)=1-c. \quad \text{Hence if } c\neq 1,$$

the two solutions y_1 and y_2 are linearly independent. For $c=1$, they are identical.

In this case the second solution will have a logarithmic term.

As an illustration of the use of the P-symbol in generating solutions of, (12) we note

$$(1-x)^{a+b-c}\,P\left\{\begin{matrix} 0 & 1 & \infty & \\ 0 & 0 & a & x \\ 1-c & c-a-b & b & \end{matrix}\right\}$$

$$=P\left\{\begin{matrix} 0 & 1 & \infty & \\ 0 & a+b-c & c-b & x \\ 1-c & 0 & c-a & \end{matrix}\right\}.$$

Hence $$y_3=(1-x)^{c-a-b}\,F[c-b, c-a\,;\,c\,;\,x]$$

is also a solution of (12).

Let $$y_3=Ay_1+By_2$$

At $x=0$, y_1 and y_3 are analytic, but y_2 has a branch point. Hence $B=0$. Equating the constant terms or letting $x\to 0$, we find $A=1$. Thus finally $y_3=y_1$.

It should be noted that if either a or b is a negative integer, then $F[-n, b\,;\,c\,;\,x]$ is a polynomial.

The Generalised Hypergeometric Function. The notation

$${}_2F_1\left[\begin{matrix} a:b\,;\,x \\ c \end{matrix}\right]=\sum_{r=0}^{\infty}\frac{(a)_r(b)_r}{r\,!\,(c)_r}x^r$$

is extended in the form

$$
{}_pF_q\left[\begin{matrix} a_1, a_2, \dots, a_p\,;\, x \\ b_1, b_2, \dots, b_q \end{matrix}\right]=\sum_{r=0}^{\infty}\frac{(a_1)_r\dots(a_p)_r}{r\,!\,(b_1)_r\dots(b_q)_r}x^r \quad \dots(19)
$$

The series on the right of (19) is called the generalised hypergeometric function. It is convergent for all x if $p\leqslant q$, for $p=q+1$ if $|x|<1$ and for $\left(\sum_1^q b_i-\sum_1^p a_i\right)>0$ if $|x|=1$. If $p>q+1$, the series diverges for all x except $x=0$. If any one of the numerator parameters is a negative integer, the series terminates and the question of convergence does not arise.

The function ${}_pF_q$ remains unaltered by the interchange of any two numerator parameters a_i, a_j or any two denominator parameters b_i, b_j. If a numerator parameter a_i equals a denominator parameter, then the ${}_pF_q$ reduces to a ${}_{p-1}F_{q-1}$.

In this notation,

$$
e^x={}_0F_0[- \,;\, x]={}_1F_1[\alpha\,;\,\alpha\,;\,x]
$$

$$
(1-x)^{-a}={}_1F_0[a\,;\,-\,;\,x]=\sum_{r=0}^{\infty}\frac{(a)_r}{r\,!}x^r
$$

Note. In the ${}_pF_q$ notation p denotes the number of numerator parameters and q the number of denominator parameters.

Theorem 1. If $Re\ c > Re\ b>0$, then

$$
F[a, b\,;\,c\,;\,x]=\frac{\Gamma(c)}{\Gamma(b)\Gamma(c-b)}\int_0^1 t^{b-1}(1-t)^{c-b-1}(1-xt)^{-a}\,dt \quad \dots(20)
$$

Proof. The formula (20) is obtained by expanding the binomial as

$$
(1-xt)^{-a}=\sum_{r=0}^{\infty}\frac{(a)_r}{r\,!}x^r t^r, \quad |xt|<1
$$

integrating term-by-term and using the result

$$
\int_0^1 x^{m-1}(1-x)^{n-1}\,dx=\frac{\Gamma(m)\Gamma(n)}{\Gamma(m+n)}. \qquad \text{Q.E.D.}
$$

Corollary 1. (Gauss's theorem). If $Re\ (c-a-b)>0$,

$$
F[a, b\,;\,c\,;\,1]=\frac{\Gamma(c)\Gamma(c-a-b)}{\Gamma(c-a)\Gamma(c-b)}.
$$

Corollary 2. (Vandermonde's theorem).

$$
F[-n, b\,;\,c\,;\,1]=\frac{(c-b)_n}{(c)_n}.
$$

In particular $F[-n, n+1\,;\,1\,;\,1]=(-1)^n$.

Confluence of Singularities. If we replace x by x/b in (12), the new equation will have regular singularities at 0, b, ∞. If $b \to \infty$, the two regular singularities b and ∞ coalesce into an irregular singularity $x=\infty$ of the new equation

$$xy''+(c-x)y'-ay=0 \qquad ...(21)$$

which is called the *confluent hypergeometric equation* or the *Kummer's equation*. It has only one regular singularity $x=0$. If c is neither zero nor an integer, (21) has two linearly independent solutions

$$y_1={}_1F_1[a : c ; x],$$
$$y_2=x^{1-c}\,{}_1F_1[a-c+1 ; 2-c ; x].$$

The confluent hypergeometric equation

$$xy''+(1+\alpha-x)y'+ny=0 \qquad ...(22)$$

is called Laguerre's equation. Its polynomial solution for n a natural number is

$$L_n^{(\alpha)}(x)=\frac{(1+\alpha)_n}{n!}\,{}_1F_1[-n ; 1+\alpha ; x]$$

which is called the generalised Laguerre polynomial. For α 0,

$$L_n(x)={}_1F_1[-n ; 1 ; x]$$

is called the (ordinary) Laguerre polynomial.

Exercise 18

1. Prove the following relations :

   (*i*) $(a-1)(a)_{n-1}=(a-1)_n=(a-1)_{n-1}(a+n-2)$

   (*ii*) $(a)_{n+k}=(a)_n(a+n)_k$

   (*iii*) $(a)_{-n}=\dfrac{(-1)^n}{(1-a)_n}$

   (*iv*) $(-a)_n=\dfrac{(-1)^n}{(1+a)_{-n}}$

   (*v*) $(1)_n=n!$

   (*vi*) $(-n)_k=(-1)^k n!/(n-k)!$; $(-n)_n=(-1)^n n!$

   (*vii*) $(a-n)_n=(-1)^n(1-a)_n$

   (*viii*) $(a+k)_{n-k}=(-1)^{n-k}(1-a-n)_{n-k}$

   (*ix*) $(a)_{n-k}=\dfrac{(-1)^k (a)_n}{(1-a-n)_k}=\dfrac{(-1)^n(a-n)_k}{(1-a)_n}$

   (*x*) $(1+n)_n=4^n(\tfrac{1}{2})_n$

2. Prove that $\dfrac{d}{dx}$ F$[a, b ; c ; x]=\dfrac{ab}{c}F[a+1, b+1 ; (c+1); x]$.

3. If $\binom{n}{k}$ or $({}^nC_k)$ denotes the $(k+1)$th binomial coefficient, tnen prove that

   (*a*) $\binom{n}{k}=\dfrac{(-1)^k(-n)_k}{k!}$ (*b*) $\binom{n+2a-1}{n}=\dfrac{(2a)_n}{n!}$

4. Prove the duplication formulae

(i) $\Gamma(2a)=\dfrac{2^{2a-1}}{\sqrt{\pi}}\,\Gamma(a)\Gamma(a+\tfrac{1}{2})$

(ii) $(a)_{2n}=2^{2n}\left(\dfrac{a}{2}\right)_n\left(\dfrac{a}{2}+\dfrac{1}{2}\right)_n$

5. Prove that

$$x^{c-1}(1-x)^{a+b-c}\,P\begin{Bmatrix} 0 & 1 & \infty & \\ 0 & 0 & a & x \\ 1-c & c-a-b & b & \end{Bmatrix}$$

$$=P\begin{Bmatrix} 0 & 1 & \infty & \\ 0 & 0 & 1-a & x \\ c-1 & a+b-c & 1-b & \end{Bmatrix}.$$

Deduce that

$$(1-x)^{c-a-b}F\,[1-a,\,1-b\,;\,2-c\,;\,x]$$
$$=F(a-c+1,\,b-c+1\,;\,2-c\,;\,x]$$

§ 5·4. The Legendre's Equation

We are concerned with the derivation and properties of solutions of the Legendre's equation

$$(1-x^2)y''-2xy'+n(n+1)y=0 \qquad \ldots(23)$$

where n is a positive integer.

It is a Fuchsian equation with regular singularities at $1, -1, \infty$. The substitution $t=\frac{1}{2}(1-x)$ sends the point $x=1$ to $t=0$ and the equation (23) is transformed to

$$t(1-t)y''+(1+2t)y'+n(n+1)y=0 \qquad \ldots(24)$$

The equation (24) is in the hypergeometric form with $a=-n$, $b=n+1$, $c=1$. All solutions of the equation (24) are represented by the P-symbol

$$y=P\begin{Bmatrix} 0 & 1 & \infty & \\ 0 & 0 & n+1 & t \\ 0 & 0 & -n & \end{Bmatrix} \qquad \ldots(25)$$

Hence all solutions of (23) are given by

$$y=P\begin{Bmatrix} 1 & -1 & \infty & \\ 0 & 0 & n+1 & x \\ 0 & 0 & -n & \end{Bmatrix}.$$

One of the solutions of (24) is the polynomial

$F[-n, n+1\,;\,1\,;\,t]$. Hence

$$P_n(x)=F\left[-n, n+1; 1; \frac{1-x}{2}\right] \quad ...(26)$$

is the polynomial solution of (23). By (26) $P_n(x)=P_{-n-1}(x)$.

$P_n(x)$ is called the nth *Legendre's polynomial*. It is obvious that $P_n(1)=1$.

Relation (26) is called the *Murphy's formula* for $P_n(x)$.

Since (23) remains unaltered by changing x to $-x$, $P_n(-x)$ is also a polynomial solution.

Hence, $P_n(-x)=A\,P_n(x)$. Thus

$$\begin{aligned} A\,P_n(1)=A&=P_n(-1)\\ &=F[-n, n+1; 1; 1]\\ &=(-1)^n. \end{aligned}$$

Hence $\qquad P_n(-x)=(-1)^n P_n(x),$

and $$P_n(x)=(-1)^n F\left[-n, n+1; 1; \frac{1+x}{2}\right].$$

We now fully solve the equation (23).

Solution in the Neighbourhood of $x=\infty$.

We can change equation (23) by putting $x=\dfrac{1}{t}$ to the form

$$t^2(t^2-1)\frac{d^2y}{dt^2}+2t^3\frac{dy}{dt}+n(n+1)\,y=0$$

and assume a solution $y=t^{\rho}\sum_{r=0}^{\infty} a_r t^r$ in the neighbourhood of $t=0$,

or alternatively, assume a solution $y=x^{-\rho}\sum_{r=0}^{\infty} a_r x^{-r}$ of equ. (23).

In either case, the exponents obtained are $-n$ and $n+1$ and $a_1=a_3=a_5=\ldots\ldots\ldots=0$. Also,

$$a_{2r}=\frac{\rho(\rho+1)(\rho+2)\ldots(\rho+2r-1)\,a_0}{[(\rho-n+1)(\rho-n+3)\ldots(\rho-n+2r-1)][(\rho+n+2)(\rho+n+4)\ldots(\rho+n+2r)]} \quad r\geqslant 1 \quad ...(27)$$

In case $2n+1$ is neither zero nor an integer, if $a_0=1$, then

$$\Big[a_{2r}\Big]_{\rho=-n}=\frac{\left(-\frac{n}{2}\right)_r\left(\frac{1}{2}-\frac{n}{2}\right)_r}{r!\,(\frac{1}{2}-n)_r},$$

$$\Big[a_{2r}\Big]_{\rho=n+1}=\frac{\left(\frac{n+1}{2}\right)_r\left(\frac{n+2}{2}\right)_r}{r!\,(n+\frac{3}{2})_r}$$

We thus obtain two linearly independent solutions of (23), *viz.*,

$$y_1=x^n\,F\left[-\frac{n}{2},\;-\frac{n}{2}+\frac{1}{2}\;:\;-n+\frac{1}{2}\;;\;\frac{1}{x^2}\right]$$

$$y_2=x^{-n-1}\,F\left[\frac{n+1}{2},\;\frac{n+2}{2}\;;\;n+\frac{3}{2}\;;\;\frac{1}{x^2}\right].$$

For a positive integer n, either $-\frac{n}{2}$ or $-\frac{n}{2}+\frac{1}{2}$ is a negative integer, hence y_1 is a polynomial. On using the Vandermonde's theorem, it is seen that

$$P_n(x)=\frac{(2n)\,!}{2^n(n\,!)^2}\,x^n\,F\left[-\frac{n}{2},\;-\frac{n}{2}+\frac{1}{2}\;;\;-n+\frac{1}{2}\;;\;\frac{1}{x^2}\right] \qquad \ldots(28)$$

If $$Q_n(x)=\frac{2^n(n\,!)^2}{(2n+1)\,!}\,x^{-n-1}\,F\left[\frac{n+1}{2},\;\frac{n+2}{2}\;;\;n+\frac{3}{2}\;;\;\frac{1}{x^2}\right] \qquad \ldots(29)$$

Then $Q_n(x)$ is called the *Legendre's function of the second kind.*

From the forms of y_1, y_2 if is seen that if one solution of (23) is a polynomial, the second independent solution is an infinite series. By putting $y=v\,P_n(x)$, the general solution of (23) can be computed as

$$y=A\,P_n(x)+B\,P_n(x)\int_x^\infty \frac{dx}{(x^2-1)\,P_n^2(x)}.$$

A comparison of the dominant terms shows that

$$Q_n(x)=P_n(x)\int_x^\infty \frac{dx}{(x^2-1)\,P_n^2(x)} \qquad \ldots(30)$$

The solution $P_n(x)$ is of importance for $|x|<1$ and $Q_n(x)$ is of importance for $|x|>1$.

In the expanded form, (28) can be written as

$$P_n(x)=\frac{1}{2^n}\sum_{k=0}^{[n/2]}\frac{(-1)^k\,(2n-2k)\,!\,x^{n-2k}}{k\,!\,(n-k)\,!\,(n-2k)\,!} \qquad \ldots(31)$$

where $[x]$=largest integer in x

A few Legendre's polynomials are listed below,

$P_0(x)=1,\qquad P_1(x)=x,\qquad P_2(x)=\frac{1}{2}(3x^2-1)$

$P_3(x)=\frac{1}{2}(5x^3-3x),\qquad P_4(x)=\frac{1}{8}(35x^4-30x^2+3)$

$P_5(x)=\frac{1}{8}(63x^5-70x^2+15x).$

Problem. Solve the equation (23) in the neighbourhood of the ordinary point $x=0$ by assuming a solution $y=\sum_{r=0}^{\infty} a_r x^r$ and show that the general solution is $y=A\,y_1+B\,y_2$

where $y_1=1-\frac{n(n+1)}{2!}x^2+\frac{n(n-2)(n+1)(n+3)}{4!}x^4-\ldots$

$$y_2=x-\frac{(n-1)(n+2)}{3!}x^3+\frac{(n-1)(n-3)(n+2)(n+4)}{5!}x^5-\ldots$$

The Wronskian of $Q_n(x)$ and $P_n(x)$. We have

$(1-x^2)P_n''-2xP_n'+n(n+1)P_n=0 \qquad (i),$

$(1-x^2)Q_n''-2xQ_n'+n(n+1)Q_n=0 \qquad (ii).$

Multiplying (i) by Q_n, (ii) by P_n and subtracting we get

$(x^2-1)(Q_nP_n''-Q_n''P_n)+2x(Q_nP_n'-Q_n'P_n)=0,$

i.e., $\qquad (x^2-1)(Q_nP_n'-Q_n'P_n)=A,$

where A is a constant to be determined.

From the formulae (28) and (29), we find for large x

$$Q_nP_n'\sim\frac{n}{2n+1}\frac{1}{x^2} \text{ and } Q_n'P_n\sim\frac{-n-1}{2n+1}\frac{1}{x^2}$$

Hence $W(x)=W(Q_n, P_n)=Q_nP_n'-Q_x'P_n\sim\frac{1}{x^2}$

Also $(x^2-1)W(x)\sim\frac{x^2-1}{x^2}\to 1$ as $|x|\to\infty$.

Hence $A=1$. Thus finally

$$W(Q_n, P_n)=\frac{1}{x^2-1} \qquad \ldots(32)$$

(32) is equivalent to the relation

$$Q_n(x)=P_n(x)\int_x^\infty\frac{dx}{(x^2-1)P_n^2(x)} \qquad \ldots(30)$$

Problem. Derive (32) by using the Liouville's formula.

Further Properties of $P_n(x)$

(A) **Generating Function.** For $|2xt-t^2|<1$, using the binomial theorem, we have

$$(1-2xt+t^2)^{-1/2}=\sum_{r=0}^{\infty}\frac{(\frac{1}{2})_r(2xt-t^2)^r}{r!}$$

$$=\sum_{r=0}^{\infty}\sum_{s=0}^{r}\frac{(\frac{1}{2})_r(-1)^s(2x)^{r-s}t^{r+s}}{s!(r-s)!}$$

$$=\sum_{r=0}^{\infty}\sum_{s=0}^{r}\frac{(-1)^s(2r)!x^{r-s}t^{r+s}}{2^{r+s}r!s!(r-s)!}$$

$$=\sum_{n=0}^{\infty}t^n\sum_{s=0}^{[n/2]}\frac{(-1)^s(2n-2s)!x^{n-2s}}{2^n s!(n-s)!(n-2s)!}$$

$(r+s=n),$

where $\left[\frac{n}{2}\right]$ is the greatest integer in $\frac{n}{2}$. Using relation (31) we have finally

$$(1-2xt+t^2)^{-1/2}=\sum_{n=0}^{\infty} t^n P_n(x) \qquad \ldots(33)$$

$(1-2xt+t^2)^{-1/2}$ is called the *generating funetion* and (33) the generating function relation of $P_n(x)$.

The relation $P_n(-x)=(-1)^n P_n(x)$ also follows from (33). The following particular values of $P_n(x)$ can be derived from (33):

$$P_n(1)=1, \quad P_n(-1)=(-1)^n, \quad P_n'(1)=\tfrac{1}{2} n(n+1),$$

$$P_{2n}(0)=\frac{(-1)^n (\frac{1}{2})_n}{n!}=\frac{(-1)^n (2n)!}{2^{2n} (n!)^2}, \quad P_{2n+1}(0)=0,$$

$$P_{2n}'(0)=0, \quad P_{2n+1}'(0)=\frac{(-1)^n(\frac{3}{2})_n}{n!}=\frac{(-1)^n(2n+1)!}{2^{2n}(n!)^2},$$

$$P_{2n}''(0)=\frac{(-1)^n(2n+)!}{2^{2n-1}\, n!\, (n-1)!}, \quad P_{2n+1}''(0)=0$$

Corollary 1. Put $x=\cos\theta$ in (33). We get

$$\sum_{n=0}^{\infty} t^n P_n(\cos\theta)=(1-t\,e^{i\theta})^{-1/2}(1-t\,e^{-i\theta})^{-1/2}$$

$$=\sum_{r=0}^{\infty}\sum_{s=0}^{\infty} \frac{(\frac{1}{2})_r\, (\frac{1}{2})_s}{r!\, s!}\, e^{i(r-s)\theta}\, t^{r+s}$$

$$=\sum_{n=0}^{\infty}\sum_{r=0}^{n} \frac{(\frac{1}{2})_r\, (\frac{1}{2})_{n-r}}{r!\,(n-r)!}\, e^{i(2r-n)\theta}\, t^n \qquad (r+s=n)$$

Thus $$P_{2n}(\cos\theta)=\frac{(\frac{1}{2})_n}{n!}+2\sum_{r=0}^{n-1} \frac{(\frac{1}{2})_r\, (\frac{1}{2})_{2n-r}}{r!\,(2n-2r)!}\cos(2n-2r)\theta,$$

$$P_{2n+1}(\cos\theta)=2\sum_{r=0}^{n} \frac{(\frac{1}{2})_r\, (\frac{1}{2})_{2n+1-r}}{r!\,(2n+1-r)!}\cos(2n+1-r)\,\partial.$$

These expansions can be condensed in the form

$$P_n(\cos\theta)=\sum_{r=0}^{[n/2]} C_r \cos(n-2r)\,\theta, \quad C_r>0$$

For $\theta=0$, we shall get $1=\sum_{r=0}^{[n/2]} C_r$.

Hence $|P_n(\cos\theta)| \leqslant \Sigma |C_r| = \Sigma C_r \doteq 1$.

Thus for $|x| \leqslant 1$, $|P_n(x)| \leqslant 1$. ...(34)

Corollary 2. Put $x=\cos\theta$, $t=e^{i\phi}$, then for $0 \leqslant \phi < \theta < \pi$

$$\sum_{k=0}^{\infty} P_k(\cos\theta)\cos k\phi = \cos\frac{\phi}{2}\left[2(\cos\phi-\cos\theta)\right]^{-1/2}$$

$$\sum_{k=0}^{\infty} P_k(\cos\theta)\sin k\phi = -\sin\frac{\phi}{2}\left[2(\cos\phi-\cos\theta)\right]^{-1/2}$$

Problem. Putting $x=\cos\theta$ in the Murphy's formula (26) and using the result ${}_2F_1[-n, n+1\,;\,1\,;\,1]=(-1)^n$, prove

$$|P_n(\cos\theta)| \leqslant 1.$$

(B) Recurrence relations. Differentiating both sides of (33) with respect to t and x we get respectively,

$$(x-t)(1-2xt+t^2)^{-3/2} = \sum_{n=0}^{\infty} nt^{n-1} P_n(x) \qquad \text{...}(i)$$

$$t(1-2xt+t^2)^{-3/2} = \sum_{n=0}^{\infty} t^n P_n'(x) \qquad \text{...}(ii)$$

Relation (i) can be written as

$$(x-t)\sum_{n=0}^{\infty} t^n P_n(x) = (1-2xt+t^2)\sum_{n=0}^{\infty} nt^{n-1} P_n(x) \qquad \text{...}(iii)$$

Combining (i) and (ii) we get

$$(x-t)\sum_{n=0}^{\infty} t^n P_n'(x) = t\sum_{n=0}^{\infty} nt^{n-1} P_n(x) \qquad \text{...}(iv)$$

Equating the coefficients of t^n on both sides of (iii) and (iv) we get respectively,

$$(n+1)P_{n+1}(x)-(2n+1)xP_n+nP_{n-1}(x)=0 \qquad \text{...I}$$

$$xP_n'-P_{n-1}'=nP_n \qquad \text{...II}$$

I is the classical *three-term-relation* of P_n. It is a pure recurrence relation. II is a differential recurrence relation.

Differentiate I and use II to get

$$P_{n+1}'-P_{n-1}'=(2n+1)P_n \qquad \text{...III}$$

Substitute P_{n-1}' obtained from II into III, we get

$$xP_n'-P_{n+1}'=-(n+1)P_n \qquad \text{...IV}$$

Replace n by $n-1$ in IV, multiply II throughout by x and add to get

$$(x^2-1)P_n'=n(xP_n-P_{n-1}) \qquad \text{...V}$$

V with I leads to

$$(x^2-1)\, P_n' = (n+1)\,(P_{n+1} - x\, P_n) \qquad \text{...VI}$$

Multiply V by $(n+1)$, VI by n and add, we get the ***Beltrami's result,***

$$(2n+1)\,(x^2-1)\, P_n'(x) = n(n+1)\,(P_{n+1}) - (P_{n-1}) \text{VII}$$

From III we have the ***Christoffel's expansion***

$$P_n'(x) = \sum_{r=0}^{p} (2n-4r-1)\, P_{n-2r-1},$$

$$p = \frac{n-1}{2}$$

or $\frac{1}{2}n-1$ **according as n is odd or even.**

(c) The Rodrigues' Formula

$Y=(x^2-1)^n$ **satisfies the equation**

$$(x^2-1)\, Y'' + 2\,(1-n)\, x\, Y' - 2nY = 0.$$

Differentiating this equation n times applying the Leibnitz rule, we get after putting $y = \dfrac{d^nY}{dx^n}$,

$$(1-x^2)\, y'' - 2xy' + n\,(n+1)\, y = 0$$

Hence $$P_n(x) = A\, \frac{d^n\,(x^2-1)^n}{dx^n}$$

Putting $x=1$, **we get**

$$1 = A\,.\,2^n\, n\,!$$

i.e., $$A = \frac{1}{2^n\, n\,!},$$

Thus we get the Rodrigues' formula,

$$P_n(x) = \frac{1}{2^n\, n\,!}\, \frac{d^n\,(x^2-1)^n}{dx^n} \qquad \text{...(35)}$$

Problem. **Derive (35) directly by expanding $(x^2-1)^n$ by the binomial theorem, differentiating term-by-term and using (31).**

Corollary 3. **Writing $(x^2-1)^n = (x-1)^n\, (x+1)^n$ in (35) and applying the Leibnitz rule, we shall get**

$$P_n(x) = \frac{1}{2^n\, n\,!} \sum_{k=0}^{n} {}^nC_k\, \frac{n\,!}{(n-k)\,!}\, \frac{n\,!}{k\,!} (x-1)^{n-k}\, (x+1)^k$$

$$= \sum_{k=0}^{n} ({}^nC_k)^2 \left(\frac{x-1}{2}\right)^{n-k} \left(\frac{x+1}{2}\right)^k \qquad \text{...(36)}$$

Put $$x=\frac{1+t}{1-t} \text{ in (36) to get}$$

$$(1-t)^n P_n\left(\frac{1+t}{1-t}\right)=\sum_{k=0}^{n} ({}^nC_k)^2 t^k \qquad ...(37)$$

Again put $t=\tan h^2 \frac{u}{2}$, then $t \geqslant 0$

and $\frac{1+t}{1-t}=\cosh u \geqslant 1$. Hence

$$P_n (\cosh u)=(1+t)^{-n} \sum_{k=0}^{n} ({}^nC_k)^2 t^k.$$

The first term on the right of this equality is 1 and all subsequent terms are positive. Hence

$$P_n (\cosh u) \geqslant 1 \qquad (38)$$

If $x=\cosh u$, then $x \geqslant 1$ and since

$$P_n(-x)=(-1)^n P_n(x),$$

$$| P_n(x) | \geqslant 1 \text{ if } | x | \geqslant 1 \qquad (39)$$

Corollary 4. The repeated use of Rolle's theorem shows that $P_n(x)=\frac{1}{2^n n!} \frac{d^n (x^2-1)^n}{dx^n}$ has n simple zeros lying between -1 and 1.

(D) Orthogonality. The Legendre's equation can be written in the self-adjoint form

$$[(1-x^2) y']'+n(n+1) y=0 \qquad ...(40)$$

Comparing (40) with the form (9) of chapter IV, $p(x)=1-x^2$, $q(x)=0$, $r(x)=1$, $\lambda=n(n+1)$. Since $p(x)=0$ for $x=-1, 1$, (40) represents a Sturm–Liouville problem without explicit boundary conditions (see theorem 1 and exercise 7 (1) of chapter IV). Its eigenfunctions are $P_n(x)$ with the related eigenvalues $n(n+1)$ $(n=0, 1, 2,...)$. Hence $\{P_n(x)\}$ is an orthogonal set of polyomials over the interval $-1 \leqslant x \leqslant 1$ with weight function equal to 1, *i.e.*,

$$\int_{-1}^{1} P_m(x) P_n(x) dx=0, \qquad m \neq n.$$

There are other approaches for establishing orthogonality of the Legendre sequence. The following is the complete statement

$$\int_{-1}^{1} P_m(x) P_n(x) dx=\begin{cases} 0 & m \neq n \\ \frac{2}{2n+1} & m=n \end{cases} \qquad ...(41)$$

One of the proofs runs as follows : Let $m \neq n$. we have

$$[(1-x^2) P_m']'+m(m+1) P_m=0 \qquad ...(i)$$

$$[(1-x^2) P_n']'+n(n+1)P_n=0 \qquad ...(ii)$$

Multiply (*i*) by P_n, (*ii*) by P_m and subtract. After simplification we get

$$\{m(m+1)-n(n+1)\}\int_{-1}^{1} P_m(x)\, P_n(x)\, dx$$
$$=\Big[(1-x^2)\,(P_m'(x)P_n(x)-P_m(x)\, P_n'(x))\Big]_{-1}^{1}$$
$$=0$$

Hence $\int_{-1}^{1} P_m\, P_n\, dx=0 \qquad m\neq n$

For $m=n$, we proceed as follows : Squaring both sides of the generating relation (33) and integrating with respect to x between -1 and 1

$$\int_{-1}^{1}\frac{dx}{1-2xt+t^2}=\sum_{m=0}^{\infty}\sum_{k=0}^{m}\int_{-1}^{1} P_{m-k}^{(x)}\, P_k^{(x)}\, .\, t^m\, dx$$

The integrals on the right vanish when $m-k\neq k$ and are non-zero when $m-k=k$, *i.e.*, m is even. Let $m=2n$, then

$$\sum_{n=0}^{\infty} t^{2n}\int_{-1}^{1} P_n^2\, dx=\Big[-\frac{1}{2t}\log\,(1-2\,xt+t^2)\Big]_{-1}^{1}$$

$$=\frac{1}{t}\log\frac{1+t}{1-t}=2\sum_{n=0}^{\infty}\frac{t^{2n}}{2n+1}$$

Hence $\int_{-1}^{1} P_n^2\, dx=\dfrac{2}{2n+1}$.

An alternative way to prove (41) is to integrate by parts n times the integral

$$\int_{-1}^{1} f(x)\frac{d^n\,(x^2-1)^n}{dx^n} \text{ to get } (-1)^n\int_{-1}^{1} f^{(n)}(x)\, .\, (x^2-1)^n\, dx \qquad ...(iii).$$

If $f(x)=P_m(x)$ and $m<n$, then $f^{(n)}(x)=0$,

If $f(x)=P_n(x)$, $f^{(n)}(x)=\dfrac{(2n)\,!}{2^n\, n\,!}$

$\int_{-1}^{1} P^2{}_n\, dx$ is finally evaluated by using the result

$$(-1)^n\int_{-1}^{1}(x^2-1)^n\, dx=\frac{\Gamma(\tfrac{1}{2})\,\Gamma(n+1)}{\Gamma(n+3/2)}$$

and using the duplication formula.

Since $P^2{}_n$ is an even function,

$$\int_{0}^{1} P^2{}_n\, dx=\frac{1}{2n+1},$$

Corollary 5. We have

$$(n-m)\,(n+m+1)\int_0^1 P_m(x)P_n(x)\,dx = P_m(0)\,P_n'(0) - P'_m(0)\,P_n(0) \qquad \ldots(42)$$

Put $m=2r+1$, $n=2s$ and use the results.

$$P_{2r+1}(0)=0,\ P'_{2r+1}(0)=\frac{(-1)^r\,(3/2)_r}{r\,!},$$

$$P_{2s}(0)=\frac{(-1)^s\,(\frac{1}{2})_s}{s\,!},\ P'_{2s}(0)=0$$

We shall get

$$\int_0^1 P_{2r+1}(x)\,P_{2s}(x)\,dx=\frac{(-1)^{r+s}\,(\frac{1}{2})_{r+1}\,(\frac{1}{2})_s}{(r+s+1)\,(2r+1-2s)\,r\,!\,s\,!}$$

Corollary 6. If $m>n$ and $m-n=2s$, the using (*iii*)

$$\int_{-1}^{1} x^m\,P_n(x)\,dx=\frac{m\,!\,\Gamma(s+\frac{1}{2})}{2^n(2s)\,!\,\Gamma(n+s+3/2)} \qquad \ldots(43)$$

The integral vanishes if $m<n$ or $m-n$ is odd.

For $m=n$,

$$\int_{-1}^{1} x^n\,P_n(x)\,dx=\frac{n\,!}{2^n(\frac{1}{2})_{n+1}}=\frac{2^{n+1}\,(n\,!)^2}{(2n+1)\,!} \qquad \ldots(44)$$

Lemma 1. Given $f(x)=\sum_{n=0}^{\infty}\frac{a_n\,x^n}{n\,!}$. If

$$f(x)=\sum_{n=0}^{\infty} b_n\,P_n(x), \text{ then}$$

$$b_n=\sum_{k=0}^{\infty}\frac{(2n+1)\,a_{n+2k}}{2^{n+2k}\,k\,!\,(3/2)_{n+k}}$$

Proof. Since $\sum_{n=0}^{\infty} b_n\,P_n(x)=\sum_{k=0}^{\infty}\frac{a_k\,x^k}{k\,!}$

$$\frac{2}{2n+1}b_n=\sum_{k=0}^{\infty}\int_{-1}^{1}\frac{a_k}{k\,!}\,x^k\,P_n(x)dx$$

$$=\sum_{k=n}^{\infty}\frac{a_k}{k!}\int_{-1}^{1} x^k\,P_n(x)\,dx$$

$$=\sum_{k=0}^{\infty}\frac{a_{k+n}}{(k+n)\,!}\int_{-1}^{1} x^{k+n}\,P_n(x)\,dx$$

$$=\sum_{k=0}^{\infty}\frac{a_{n+2k}}{(n+2k)!}\int_{-1}^{1} x^{n+2k}\, P_n(x)\, dx \qquad \text{(Cor. 6)}$$

$$=\sum_{k=0}^{\infty}\frac{a_{n+2k}}{2^n\,(2k)!}\;\frac{\Gamma(k+3/2)}{\Gamma(n+k+3/2)}$$

After using the duplication formula

$(2k)! = \Gamma(2k+1) = \dfrac{2^{2k}\,\Gamma(k+\frac{1}{2})\,\Gamma(k+1)}{\Gamma(1/2)}$, we have the required result.

A series of the form $\Sigma b_k P_k(x)$ is called a Legendre series.

Theorem 2. If $|x| < y$, then

$$\frac{1}{y-x}=\sum_{n=0}^{\infty}(2n+1)\, P_n(x)\, Q_n(y) \qquad \text{...(45)}$$

Proof. $\dfrac{1}{y-x}=\displaystyle\sum_{n=0}^{\infty}\frac{x^n}{y^{n+1}}$

Let $\dfrac{1}{y-x}=\displaystyle\sum_{n=0}^{\infty} b_n\, P_n(x)$, using lemma 1,

$$b_n=(2n+1)\sum_{k=0}^{\infty}\frac{(n+2k)!}{2^{n+2k}\,k!\,(\frac{3}{2})_{n+k}}\;\frac{1}{y^{n+2k+1}}$$

$$=(2n+1)\, Q_n(y).$$

Note. If $\{p_n(x)\}$ is an orthogonal sequence on $[a, b]$ with weight function $w(x)$, then the sequence of functions

$$q_n(x)=\frac{1}{2}\int_a^b \frac{w(t)\, p_n(t)}{x-t}\, dt$$

is a *sequence of the second Kind*.

Corollary 7. $Q_n(x)=\frac{1}{2}\displaystyle\int_{-1}^{1}\frac{P_n(t)}{x-t}\, dt \qquad \text{...(46)}$

This follows from (45) and the orthogonal property (46) is called the *Neumann's formula*.

Corollary 8. $Q_n(x)=\dfrac{1}{2}P_n(x)\log\dfrac{x+1}{x-1}-W_{n-1} \qquad \text{...(47)}$

where $W_{n-1}(x)$ is a polynomial of degree $n-1$.

Proof. Using formula (46),

$$Q_n(x)=\frac{1}{2}\int_{-1}^{1}\frac{P_n(t)-P_n(x)+P_n(x)}{x-t}dt\,.$$

$$=\tfrac{1}{2}P_n(x)\int_{-1}^{1}\frac{dt}{x-t}-\tfrac{1}{2}\int_{-1}^{1}\frac{P_n(x)-P_n(t)}{x-t}dt.$$

$\dfrac{P_n(x)-P_n(t)}{x-t}$ will involve expressions of the type

$$\frac{x^r-t^r}{x-t}=x^{r-1}+a_1x^{r-2}\,t+\ldots+t^{r-1}$$

for $\qquad r=1, 2, \ldots, n.$

We shall thus have formula (47).

If we substitute $Q_n(x)$ from (47) in the Legendre's equation, we shall get

$$(1-x^2)\,W_{n-1}-2x\,W'_{n-1}+n(n+1)\,W_{n-1}=2\,P_n'(x) \qquad \ldots(i)$$

Let $\quad W''_{n-1}=\displaystyle\sum_{r=0}^{n-1} C_r\,P_r(x)$

Using the relation $P_n'(x)=\displaystyle\sum_{r=0}^{[n-1/2} (2n-4r-1)\,P_{n-2r-1}(x),$

we shall find $C_{n-2s}=0$ and

$$C_{n-2s-1}=\frac{2n-4s-1}{(2s+1)\,(n-s)}$$

Thus finally

$$Q_n(x)=\tfrac{1}{2}P_n(x)\,.\,\log\frac{x+1}{x-1}-\sum_{s=0}^{p}\frac{2n-4s-1}{(2s+1)\,(n-s)}P_{n-2s-1}(x)$$

where $p=\tfrac{1}{2}(n-1)$ or $\quad \tfrac{1}{2}n-1$ according as n is odd or even,

(E) Integral Formulae

(i) Schlafli's Integral formula

Put $f(z)=(z^2-1)^n$ in the Cauchy's formula for the nth derivative of an analytic function *viz.*,

$$\frac{d^nf(x)}{dx^n}=\frac{n\,!}{2\pi_i}\int_C\frac{f(z)}{(z-x)^{n+1}}dz$$

where C is a simple closed contour enclosing $z=x$, then using the Rodrigues' formula,

$$P_n(x)=\frac{1}{2\pi_i\,2^n}\int_C\frac{(z^2-1)^n}{(z-x)^{n+1}}dz \qquad \ldots(48)$$

(ii) **Laplace's Integral Formulae.** Let C be the circle with centre $z=x$ and radius $|\sqrt{x^2-1}|$. Then $z=x+\sqrt{x^2-1}\,e^{i\phi}$ and

$$z^2-1=2(x^2-1)^{1/2}\, e^{i\phi}\, \{x+1\sqrt{x^2-1}\,\cos\phi\},$$

$$z-x=(x^2-1)^{1/2}\, e^{i\phi}.$$

Substitute in (48) to get

$$P_n(x)=\frac{1}{2\pi}\int_{-\pi}^{\pi}\{x+\sqrt{x^2-1}\,\cos\phi\}^n\, d\phi.$$

Since the integrand is an even function of ϕ, we get the *Laplace's first integral for $P_n(x)$*:

$$P_n(x)=\frac{1}{\pi}\int_0^{\pi}\{x+\sqrt{x^2-1}\,\cos\phi\}^n\, d\phi. \qquad ...(49)$$

Since $P_n(x)=P_{-n-1}(x)$, we get *Laplace's second integral.*

$$P_n(x)=\frac{1}{\pi}\int_0^{\pi}\frac{d\phi}{(x+\sqrt{x^2-1}\,\cos\phi)^{n+1}} \qquad ...(50)$$

Corollary 9. For $|x|\leqslant 1$,

$$|x+\sqrt{x^2-1}\,\cos\phi|^2$$
$$=x^2+(1-x^2)\cos^2\phi$$
$$\leqslant x^2+(1-x^2)=1.$$

Hence using (49) we have $|P_n(x)|\leqslant 1$.

Remark 1. Formula (49) can also be established by putting $a=1-xt$, $b=t\sqrt{x^2-1}$ in the well-known result

$$\int_0^{\pi}\frac{d\phi}{a+b\cos\phi}=\frac{\pi}{\sqrt{a^2-b^2}},$$

expanding both sides in powers of t and equating the coefficients of t^n.

Problem. Prove that $P_n(x)$ tends uniformly to zero as $n\to\infty$ in any interval $-1+\delta\leqslant x\leqslant 1-\delta$, $\delta>0$. [In the interval $\varepsilon\leqslant\phi\leqslant\pi-\varepsilon$, $\varepsilon>0$

$|x+\sqrt{x^2-1}\,\cos\phi|<1$, hence $|x+\sqrt{x^2-1}\,\cos\phi|^n\to 0$ as $n\to\infty$. Now use (49)].

Corollary 10. $\displaystyle\sum_{k=0}^{n} {}^nC_k\, P_k(\cos\theta)=\left(2\cos\frac{\theta}{2}\right)^n P_n\left(\cos\frac{\theta}{2}\right).$

Proof. Using (49) with $x=\cos\theta$,

$$\sum_{k=0}^{n} {}^nC_k\, P_k(\cos\theta)=\frac{1}{\pi}\int_0^{\pi}\sum_{k=0}^{n} {}^nC_k\,(\cos\theta+i\sin\theta\cos\phi)^k\, d\phi$$

$$=\frac{1}{\pi}\int_0^{\pi}(1+\cos\theta+i\sin\theta\cos\phi)^n\,d\phi$$

$$=\frac{\left(2\cos\frac{\theta}{2}\right)^n}{\pi}\int_0^{\pi}\left(\cos\frac{\theta}{2}+i\sin\frac{\theta}{2}\cos\phi\right)^n d\phi$$

$$=\frac{\left(2\cos\frac{\theta}{2}\right)^n}{\pi}\int_0^{\pi}\left(\cos\frac{\theta}{2}+\sqrt{\cos^2\frac{\theta}{2}-1}\cos\phi\right)^n d\phi$$

$$=\left(2\cos\frac{\theta}{2}\right)^n P_n\left(\cos\frac{\theta}{2}\right) \qquad \text{...(51)}$$

$\theta=0$ in (51) gives the familiar result

$$\sum_{k=0}^{n}{}^nC_k=2^n.$$

Put $\cos\theta=x$ in (51) to get

$$2^{-n/2}\sum_{k=0}^{n}{}^nC_k\;P_k(x)=(1+x)^{1/2n}\;P_n(\sqrt{\tfrac{1}{2}(1+x)}\cdot \qquad \text{...(52)}$$

If $m \leqslant n$, then using orthogonal property, (52) will give

$$\int_{-1}^{1}\left(\frac{1+x}{2}\right)^{1/2n}P_n(\sqrt{\tfrac{1}{2}(1+x)}.\;P_m(x)\,dx=\frac{2^{-n+1}}{2m+1}\;{}^nC_m \qquad \text{...(53)}$$

The substitution $1+x=2t^2$ in (53) gives

$$\int_0^1 t^{n+1}\,P\;(t)\,P_m(2t^2-1)\;dt=\frac{2^{-n-1}}{2m+1}\;{}^nC_m \qquad \text{...(54)}$$

Corollary 11. (*Mehler—Dirichlet Formula*).

$$P_n(\cos\theta)=\frac{2}{\pi}\int_0^{\theta}\frac{\cos(n+\frac{1}{2})\psi}{\sqrt{2(\cos\psi-\cos\theta)}}\,d\psi,\ 0<\theta<\pi,\ n=0,1,2.$$

Proof. In the Laplace's first integral formula (49)

put $x=\cos\theta$ and change the variable of integration by $\cos\theta+i\sin\theta\cos\phi=e^{i\psi}$, we shall get

$$P_n(\cos\theta)=\frac{1}{\pi}\int_{-\theta}^{\theta}\frac{e^{(n+1)i\psi}\,d\psi}{\sqrt{1-2\cos\theta\,.\,e^{i\psi}+e^{2i\psi}}}$$

which reduces to the desired formula.

Exercise 19

1. If $u_n=\int_{-1}^{1} \frac{P_n(x)\, P_{n-1}(x)}{x}\, dx$, then prove that

$(n+1)\, u_{n+1}+n\, u_n=2$. Hence evaluate u_n.

2. For $x \triangleright 1$ show by induction that $P_n(x) \triangleleft P_{n+1}(x)$.

3. Prove the Christoffel's expansion

$$P_n'=(2n-1)\, P_{n-1}+(2n-5)\, P_{n-3}+(2n-9)\, P_{n-5}+\ldots\ldots,$$

the last term being $3x$ or 1 according as x is even or odd.

Hence deduce (*i*) $|\, P_n'(x)\,| \leqslant \frac{1}{2}\, n(n+1)$ for $|\, x\,| \leqslant 1$,

(*ii*) $\int_{0}^{1} P_n'^2(x)\, dx=\frac{1}{2}\, n(n+1)$.

4. For $|\, x\,| \leqslant 1$, prove the inequalities.

(*i*) $|\, P_n(x)\,| \leqslant 2^{2n-2} \left(\left[\frac{n}{2}\right]+1\right)$,

(*ii*) $|\, P_n'(x)\,| \leqslant 2^{2n-2}\, n^2$,

(*iii*) $P_n^2(x)-P_{n-1}(x)\, P_{n+1}(x) \geqslant 0$. (Turan's inequality)

5. Establish the *Christoffel summation formula*,

$$\sum_{r=0}^{n} (2r+1)\, P_r(x)\, P_r(y)=(n+1)\, \frac{P_n(x)\, P_{n+1}(y)-P_n(y)\, P_{n+1}(x)}{y-x}$$

Hence deduce

(*i*) $$\sum_{r=0}^{n} (2r+1)\, P_r^2(x)=(n+1)\{P_n(x)\, P'_{n+1}(x)-P_n'(x)\, P_{n+1}(x)\}$$

$$=(n+1)^2\, P_n^2(x)-(x^2-1)\, P_n'^2(x),$$

(*ii*) $$\int_{-1}^{1} \frac{P_n(x)}{x-x_k}\, dx=\frac{2}{n\, P_{n-1}(x_k)}$$

where x_k is a zero of $P_n(x)$.

6. Prove $x^n=\frac{n\,!}{2^n} \sum_{k=0}^{[n/2]} \frac{(2n-4k+1)\, P_{n-2k}(x)}{k\,!\,(3/2)_{n-k}}$.

7. Prove $P_{2n}(0)=(n^{-1/2})$.

8. From $(1-2xt+t^2)^{-1/2}=\Sigma t^n P_n(x)$ deduce the following

(*i*) $P^{(r)}{}_n(1)=\frac{(-1)^r\,(-n)_r\,(n+1)_r}{2^r\, r\,!}$, $r \leqslant n$, (*ii*) $P_n(x)=F\left[-n, n+1\ ;\ 1; \frac{1-x}{2}\right]$.

[Expand $P_n(x)$ in powers of $x-1$ by Taylor's theorem and use (*i*)].

9. Prove that $P_n(x)$ has n real and distinct zeros all lying between -1 and 1.

10. Prove that $Q_n(x)=\frac{1}{2} P_n(x) . \log\frac{1+x}{1-x} - W_{n-1}(x)$

where $$W_{n-1}(x)=\sum_{m=1}^{n}\frac{1}{m}P_{m-1}(x)\,P_{n-m}(x),\ W_{-1}=0.$$

11. Prove the following :

(*i*) $Q_n(x)=(-1)^{n+1}\,Q_n(-x)$,

(*ii*) $$\int_1^{\infty} P_m(x)Q_n(x)\,dx=\frac{1}{(n-m)(n+m+1)}$$

(*iii*) $$(y-x)\sum_{m=0}^{n}(2m+1)P_m(x)Q_m(y)$$

$$=1-(n+1)\{P_{n+1}(x)\,Q_n(y)-P_n(x)\,Q_{n+1}(y)\}$$

12. Prove the following results :

(*i*) $$\sum_{m=0}^{n} P_m(\cos\theta)\,P_{n-m}(\cos\theta)=\frac{\sin(n+1)\,\theta}{\sin\theta}$$

(*ii*) $$(2\cos\theta)^{-n}\sum_{m=0}^{n}{}^nC_m\,P_m(\cos\theta)=P_n(\cos\theta),\ \cos\theta\neq0$$

(*iii*) $$\sum_{m=0}^{n}{}^nC_m\,(-\cos\theta)^m\,P_m(\cos\theta)=\sin\theta^n\,P_n(\sin\theta)$$

(*iv*) $$\sum_{m=0}^{n-1}P_m(\cos\theta)\cos(n-m)\theta=n\,P_n(\cos\theta)$$

(*v*) $$\sum_{m=0}^{2n}(-1)^m\,P_m(\cos\theta)\,P_{2n-m}(\cos\theta)=P_n(\cos\theta)$$

(*vi*) $$\sum_{n=0}^{\infty}\frac{P_n(\cos\theta)}{n+1}=\log\left(1+\operatorname{cosec}\frac{\theta}{2}\right)$$

(vii) $\sum_{n=0}^{\infty} P_n(\cos\theta)=\frac{1}{2}\ \text{cosec}\frac{\theta}{2}$

(viii) $\sum_{k=0}^{n}(2k+1)\ P_k(x)=P'_{n+1}(x)+P_n'(x)$

13. Prove $(2n+1)\ (x^2-1)P_n'(x)=n(n+1)\ (xP_{n+1}-P_n)$.

14. Prove the following :

(i) $\int_{-1}^{1} x^2\ P_n^2\ (x)\ \ dx=\frac{1}{8}\ \ \frac{1}{2n-1}+\frac{3}{4(2n+1)}+\frac{1}{8(2n+3)}$

(ii) $\int_{-1}^{1} x^2 P_{n-1}(x)\ P_{n+1}(x)\ dx=\frac{2n\ (n+1)}{(2n-1)\ (2n+1)\ (2n+3)}$

(iii) $\int_{-1}^{1}(1-x^2)\ P_n'^2\ \ (x)\ dx=\frac{2n\ (n+1)}{2n+1}$

(iv) $\int_{-1}^{1}\ (x^2-1)P_{n+1}\ \ P_n'\ dx=\frac{2n\ (n+1)}{(2n+1)\ (2n+3)}$

(v) $\int_{0}^{1}\ P_{2m+1}(x)\ \ dx=\frac{(-1)^m\ (2m)\ !}{2^{2m}\ (m+1)\ !\ m\ !}$

(vi) $\int_{0}^{1}\ x\ (P_{2m}(x)\ dx=\frac{(-1)^{m-1}\ (2m-2)\ !}{2^{2m}\ (m+1)\ !\ (m-1)\ !}$

(vii) $\int_{-1}^{1}(1+x)^{m+n}\ \ P_n(x)\ dx=\frac{2^{m+n+1}\ \{(m+n)\ !\}^2}{m\ !\ (m+2n+1)\ !}$

14. Evaluate $\lim_{n\to\infty} P_n\left(\cos\ \frac{x}{n}\right)$.

15. (a) If $m\geqslant n$, then prove that :

$$x^m Q_n(x)=\frac{1}{2}\int_{-1}^{1}\ \frac{t^m\ P_n\ (t)}{x-t}\ dt$$

$$P_m(x)\ Q_n\ (x)=\frac{1}{2}\int_{-1}^{1}\frac{P_m\ (t)\ P_n\ (t)}{x-t}\ dt$$

(b) For $m=n+1$, prove

$$x^{n+1}\ Q_n\ (x)=\frac{1}{2}\int_{-1}^{1}\frac{t^{n+1}\ P_n\ (t)}{x-t}+\frac{2^n\ (n\ !)^2}{(2n+1)\ !}$$

$$P_{n+1}(x)\ \ Q_n(x)=\frac{1}{2}\int_{-1}^{1}\frac{P_{n+1}\ (t)\ P_n\ (t)}{x-t}+\frac{1}{n+1}$$

(c) $P_{n+1}(x)\ Q_n\ (x)-P_n(x)\ Q_{n+1}\ (x)=\frac{1}{n+1}$

16. Prove the recurrence relations,

(i) $(n+1)Q_{n+1}(x)-(2n+1)xQ_n(x)+n\ Q_{n-1}(x)=0$

(ii) $(2n+1)\ Q_n\ (x)=Q_n'{}_{+1}(x)-Q_n'{}_{-1}(x)$

17. Prove that if $x>1$

$$Q_n(x)=\frac{1}{2^{n+1}}\int_{-1}^{1}\frac{(1-t^2)^n}{(x-t)^{n+1}}dt$$

18. If $I_n=(2n+1)\int_{-1}^{1}P_n^2(x)\,dx$, then prove that $I_n=I_{n-1}$. Hence deduce

$$I_n=\frac{2}{2n+1}.$$

(Use I).

19. Prove $\frac{1-x^2}{n^2}\{P'^2_n+P_n^2\}=\frac{1-x^2}{n^2}\{P'^2_{n-1}+P^2_{n-1}\}$

20. Prove $\frac{1-x^2}{n^2}\{P'^2_n+P_n^2\}\leqslant 1$; $n\geqslant 1$, $|x|\leqslant 1$

§6·5. Bessel's Equation

$$x^2y''+xy'+(x^2-n^2)y=0 \qquad \text{...(55)}$$

$x=0$ is the regular singular point. It is not a Fuchsian equation.

Assuming a solution $y=x^\rho\sum_{r=0}^{\infty}a_r x^r$, $a_0\neq 0$, we shall have

$$a_0(\rho^2-n^2)=0$$

$$a_1\{(\rho+1)^2-n^2\}=0$$

$$a_r\{(\rho+r)^2-n^2\}+a_{r-2}=0 \qquad (r\geqslant 2)$$

The exponents are $\rho=\pm n$ and $a_1=a_3=\ldots\ldots=0$.

For $\rho=n$, we have

$$a_{2r}=\frac{(-1)^r a_0}{2^{2r}\,r!\,(n+1)_r}$$

Taking $a_0=\frac{1}{2^n\,\Gamma(n+1)}$, one of the solutions is

$$J_n(x)=\sum_{r=0}^{\infty}\frac{(-1)^r\left(\frac{x}{2}\right)^{n+2r}}{r!\,\Gamma(n+r+1)} \qquad \text{...(56)}$$

$J_n(x)$ is called the *Bessel's function* of the first kind of order n (or index n).

To find a second linearly independent solution we consider some particular cases.

Case 1. 2n is not an integer. A second linearly independent solution $J_{-n}(x)$ is

$$J_{-n}(x)=\sum_{r=0}^{\infty}\frac{(-1)^r\left(\frac{x}{2}\right)^{-n+2r}}{r\,!\,\Gamma(-n+r+1)}$$

Case 2. If n is an integer, $J_{-n}(x)=(-1)^n J_n(x)$. To seek a second linearly independent solution put

$$Y_n(x)=\frac{J_n(x)\cos n\pi-J_{-n}(x)}{\sin n\pi}$$

For n an integer, the limit of the right hand side is to be obtained and in this case Y_n involves a logarithmic term.

$Y_n(x)$ is called the *Bessel's function of the second kind* of order n or the *Neumann's function.*

Case 3. n=½. In this case there is no logarithmic term. Two linearly independent solutions are $x^{-1/2}\cos x$, $x^{-1/2}\sin x$.

Case 4. n=0. One of the solutions is

$$[y]_{\rho=0}=J_0(x)=\sum_{r=0}^{\infty}\frac{(-1)^r}{(r\,!)^2}\left(\frac{x}{2}\right)^{2r},$$

where $\quad y=x^{\rho}\displaystyle\sum_{r=0}^{\infty}\frac{(-1)^r x^{2r}}{[(\rho+2)^2(\rho+4)^2\ldots(\rho+2r)^2]}$

The other solution is

$$\left[\frac{\partial y}{\partial \rho}\right]_{\rho=0}=Y_0(x)$$

$$=J_0(x)\log x+\sum_{r=1}^{\infty}\frac{(-1)^{r-1}\left(\frac{x}{2}\right)^{2r}}{(r\,!)^2}H_r$$

where $\quad H_r=1+\frac{1}{2}+\frac{1}{3}+\ldots+\frac{1}{r}.$

$Y_0(x)$ is the Neumann's function of order zero.

Further Properties of the Bessel function $J_n(x)$

As found earlier

$$J_n(x)=\sum_{r=0}^{\infty}\frac{(-1)^r\left(\frac{x}{2}\right)^{n+2r}}{r\,!\,\Gamma(n+r+1)} \quad \ldots(56)$$

$J_{\pm\frac{1}{2}}(x)$ are elementary. Indeed

$$J_{1/2}(x)=\sqrt{\frac{2}{\pi x}}\sin x,\ J_{-1/2}(x)\sqrt{\frac{2}{\pi x}}\cos x.$$ (56) can be written in the form

$$J_n(x)=\frac{x^n}{2^n\,n\,!}\sum_{r=0}^{\infty}\frac{1}{r\,!\,(n+1)_r}\left(-\frac{x^2}{4}\right)^r$$

$$=\frac{x^n}{2^n\,n\,!}\,{}_0F_1\left[-\,;\,n+1\,;\,-\frac{x^2}{4}\right]$$

(A) **Generating Function :**

$$e^{x/2(t-1/t)}=\sum_{r=0}^{\infty}\sum_{s=0}^{\infty}\frac{(-1)^s\left(\frac{x}{2}\right)^{r+s}}{r\,!\,s\,!}\,t^{r-s}$$

$$=\sum_{n=-\infty}^{\infty}\sum_{s=0}^{\infty}\frac{(-1)^s\left(\frac{x}{2}\right)^{n+2s}}{s\,!\,\Gamma(n+s+1)}\,t^n,\ (r-s=n).$$

Hence finally,

$$e^{x/2(t-1/t)}=\sum_{n=-\infty}^{\infty}J_n(x)\,t^n. \qquad \text{...(57)}$$

By changing the sign of x and t together in (57) it follows that,

$$J_n(-x)=(-1)^n J_n(x)=J_{-n}(x).$$

Corollary 11. Change t to $\frac{1}{t}$ in (57) and multiply the resulting equality to (57), we get

$$1=J_0^2(x)+2\sum_{n=1}^{\infty}J_n^2(x)$$

Hence $\quad |J_0(x)|\leqslant 1, \qquad |J_n(x)|\leqslant\frac{1}{\sqrt{2}}.$

Corollary 12. Put $t=e^{i\theta}$ in (57) and equate the real and imaginary parts to get

$$\cos(x\sin\theta)=J_0(x)+2\sum_{m=1}^{\infty}J_{2m}(x)\cos 2m\,\theta \qquad \text{...(58)}$$

$$\sin(x\sin\theta)=2\sum_{m=0}^{\infty}J_{2m+1}(x)\sin(2m+1)\,\theta \qquad \text{...(59)}$$

Put $\theta=0$,

$$1=J_0(x)+2\sum_{m=1}^{\infty} J_{2m}(x) \qquad \text{...(60)}$$

Put $\theta=\frac{\pi}{2}$,

$$\cos x=J_0(x)+2\sum_{m=1}^{\infty}(-1)^m J_{2m}(x) \qquad \text{...(61)}$$

$$\sin x=2\sum_{m=0}^{\infty}(-1)^m J_{2m+1}(x) \qquad \text{...(62)}$$

Multiply (58) by cos $2m\ \theta$, (59) by sin $(2m+1)\ \theta$ and integrate, we shall get

$$J_{2m}(x)=\frac{1}{\pi}\int_0^{\pi} \cos (x \sin \theta) \cos 2m\ \theta\ d\theta,$$

$$J_{2m+1}(x)=\frac{1}{\pi}\int_0^{\pi} \sin (x \sin \theta) \sin (2m+1)\ d\theta.$$

These results can be lumped into

$$J_n(x)=\frac{1}{\pi}\int_0^{\pi} \cos (n\theta-x \sin \theta)\ d\theta \qquad \text{...(63)}$$

for integers n. In particular

$$J_0(x)=\frac{1}{\pi}\int_0^{\pi} \cos (x \sin \theta)\ d\theta. \qquad \text{...(64)}$$

The integral in (63) is called the *Bessel's integral* for J_n.

(B) Recurrence Relations :

Differentiating the generating relation (57) with respect to x and t we get respectively,

$$\frac{1}{2}\left(t-\frac{1}{t}\right)e^{x/2(t-1/t)}=\sum_{n=-\infty}^{\infty} J_n'(x)\ t^n$$

$$\frac{x}{2}\left(1+\frac{1}{t^2}\right)e^{x/2(t-1/t)}=\sum_{n=-\infty}^{\infty} n\ J_n(x)\ t^{n-1}.$$

These are equivalent to

$$\frac{1}{2}\sum_{n=-\infty}^{\infty}(J_{n-1}-J_{n+1})\ t^n=\sum_{n=-\infty}^{\infty} J_n'(x)\ t^n \qquad \text{...}(i)$$

$$\frac{x}{2} \sum_{n=-\infty}^{\infty} (J_{n-1}+J_{n+1})\, t^{n-1} = \sum_{n=-\infty}^{\infty} n\, J_n\, t^{n-1} \qquad \text{...(ii)}$$

Equating coefficients of like powers of t on both sides of (*i*) and (*ii*) we get respectively,

$$2\, J_n'(x) = J_{n-1}(x) - J_{n+1}(x) \qquad \text{...I}$$

$$\frac{2n}{x} J_n(x) = J_{n-1}(x) + J_{n+1}(x) \qquad \text{...II}$$

Sum and difference of I and II gives

$$x\, J_n' = x\, J_{n-1} - n\, J_n \qquad \text{...III}$$

$$x\, J_n' = n\, J_n - x\, J_{n+1} \qquad \text{...IV}$$

A special case for $n=0$ is $J_0' = -J_1$.

From the relation (56) we shall get

$$\frac{d}{dx}(x^n J_n) = x^n J_{n-1} \qquad \text{...(65)}$$

$$\frac{d}{dx}(x^{-n} J_n) = -x^{-n} J_{n+1} \qquad \text{...(66)}$$

More generally for a positive integer m,

$$\left(\frac{1}{x}\frac{d}{dx}\right)^m \{x^n J_n\} = x^{n-m} J_{n-m} \qquad \text{...(67)}$$

$$\left(\frac{1}{x}\frac{d}{dx}\right)^m \{x^{-n} J_n\} = (-1)^m x^{-n-m} J_{n+m} \qquad \text{...(68)}$$

A particular case of (67) and (68) is

$$(-1)^n \left(\frac{1}{x}\frac{d}{dx}\right)^n J_0(x) = x^{-n} J_n \qquad \text{...(69)}$$

(69) suggests that J_n may be obtained from J_0.

From (56) we derive $\lim\limits_{x \to 0} x^{-n} J_n(x) = \dfrac{1}{2^n\, n!}$

i.e., for small x $J_n(x) \sim \dfrac{x^n}{2^n\, n!}$

The following special cases may be evaluated from the recurrence relation II

$$J_{3/2}(x) = \frac{1}{x} J_{1/2} - J_{-1/2} = \sqrt{\frac{2}{\pi x}} \left(\frac{\sin x}{x} - \cos x \right)$$

$$J_{-3/2}(x) = -\frac{1}{x} J_{-1/2} - J_{1/2} = \sqrt{\frac{2}{\pi x}} \left(-\frac{\cos x}{x} - \sin x \right)$$

$$J_{5/2}(x) = \sqrt{\frac{2}{\pi x}} \left(\frac{3 \sin x}{x^2} - \frac{3 \cos x}{x} - \sin x \right)$$

$$J_{-5/2}(x)=\sqrt{\frac{2}{\pi x}}\left(\frac{3\cos x}{x^2}+\frac{3\sin x}{x}-\cos x\right).$$

In general for a positive integer n,

$$J_{n+1/2}(x)=A(x)\cos x+B(x)\sin x$$

where $A(x)$ and $B(x)$ are polynomials in $x^{-1/2}$.

Bessels functions of index half an odd integer are called *Spherical Bessel functions* and occur in physical problems.

(C) Addition Formula for $J_n(x)$:

$$\sum_{n=-\infty}^{\infty} J_n(x+y)\, t^n=e^{x/2(t-1/t)}\,.\,e^{y/2(t-1/t)}$$

$$=\sum_{r=-\infty}^{\infty}\sum_{s=-\infty}^{\infty} J_r(x)\, J_s(y)\, t^{r+s}$$

$$=\sum_{n=-\infty}^{\infty}\sum_{r=-\infty}^{\infty} J_r(x)\, J_{n-s}(x)\,.\,t^n.$$

Hence $$J_n(x+y)=\sum_{r=-\infty}^{\infty} J_r(x)\, J_{n-r}(y) \qquad \text{...(70)}$$

This is the addition formula for J_n.

For $n=0$, $y=-x$, since $J_0(0)=1$,

$$1=\sum_{r=-\infty}^{\infty}(-1)^r J_r^2(x)$$

$$=J_0^2-2J_1^2+2J_2^2-\ldots \qquad \text{...(71)}$$

(D) Multiplication Formula for $J_0(x)$

By Taylor's theorem we have

$$J_0(\sqrt{x+h})=\sum_{n=0}^{\infty}\frac{h^n}{n\,!}\frac{d^n}{dx^n}J_0(\sqrt{x}).$$

Replacing x by x^2 yields

$$J_0(\sqrt{x^2+h})=\sum_{n=0}^{\infty}\frac{h^n}{2^n\, n\,!}\left(\frac{d}{x\,dx}\right)^n J_0(x)$$

$$=\sum_{n=0}^{\infty}\frac{1}{n!}\left(\frac{-h}{2x}\right)^n J_n(x).$$

Replace h by $\lambda^2 x^2-x^2$, to obtain

$$J_0(\lambda x)=\sum_{n=0}^{\infty}\frac{1}{n!}\left\{(1-\lambda^2)\frac{x}{2}\right\}^n J_n(x) \qquad \ldots(72)$$

(72) is called the *multiplication formula* for J_0.

The modified Bessel function of the first kind of index zero is defined as

$$I_0(x)=J_0(ix)=\sum_{r=0}^{\infty}\frac{1}{(r!)^2}\left(\frac{x}{2}\right)^{2r} \qquad \ldots(73)$$

Putting $\lambda=i$ in (72) yields

$$I_0(x)=\sum_{n=0}^{\infty}\frac{x^n}{n!}J_n(x) \qquad \ldots(74)$$

$\lambda=0$ yields

$$1=\sum_{n=0}^{\infty}\frac{(x/2)^n}{n!}J_n(x) \qquad \ldots(75)$$

(E) The Wronskian of J_n and J_{-n} :

We have

$$J_n''+\frac{1}{x}J_n'+\left(1-\frac{n^2}{x^2}\right)J_n=0 \qquad \ldots(i)$$

$$J''_{-n}+\frac{1}{x}J'_{-n}+\left(1-\frac{n^2}{x^2}\right)J_{-n}=0 \qquad \ldots(ii)$$

Multiply (*i*) by J_{-n}, (*ii*) by J_n and subtract, we get,

$$\frac{d}{dx}(J_n' J_{-n}-J_n J_{-n}')+\frac{1}{x}(J_n' J_{-n}-J_n J_{-n}')=0$$

Hence $x(J_n J_{-n}'-J_n' J_{-n})=A.$

Equating the constant terms on both sides,

$$A=\frac{1}{\Gamma(n+1)\,\Gamma(-n)}-\frac{1}{\Gamma(n)\,\Gamma(1-n)}=-\frac{2\sin n\pi}{\pi}$$

where we have used the result $\Gamma(n)\,\Gamma(1-n)=\dfrac{\pi}{\sin n\pi}$.

Thus $$W(J_n, J_{-n})=-\frac{2\sin n\pi}{\pi x} \qquad \ldots(76)$$

The result (76) can also be expressed in the form

$$\frac{d}{dx}\left(\frac{J_{-n}}{J_n}\right)=-\frac{2\sin n\pi}{\pi\, J_n^2(x)\, x} \qquad ...(77)$$

(F) **Orthogonality.** $J_m(\lambda x)$ and $J_n(\mu x)$ satisfy the relations

$$x^2\frac{d^2 J_m(\lambda x)}{dx^2}+x\frac{d J_m(\lambda x)}{dx}+(\lambda^2 x^2-m^2)\, J_m(\lambda x)=0 \qquad ...(i)$$

$$x^2\frac{d^2 J_n(\mu x)}{dx^2}+x\frac{d J_n(\mu x)}{dx}+(\mu^2 x^2-n^2)\, J_n(\mu x)=0 \qquad ...(ii)$$

Multiply (*i*) by $\frac{1}{x} J_n(\mu x)$, (*ii*) by $\frac{1}{x} J_m(\lambda x)$, integrate between 0 to a and subtract to obtain, if $m \not< -1,\ n \not< -1$,

$$(\mu^2-\lambda^2)\int_0^a x\, J_m(\lambda x)\, J_n(\mu x)\, dx+(m^2-n^2)\int_0^a J_m(\lambda x)\, J_n(\mu x)\frac{dx}{x}$$
$$=a[\lambda\, J_m'(\lambda a)\, J_n(\mu a)-\mu\, J_m(\lambda a)\, J_n'(\mu a)].$$

If $m=n$, for $\lambda\neq\mu$ we get

$$\int_0^a x\, J_n(\lambda x)\, J_n(\mu x)\, dx=\frac{a}{\mu^2-\lambda^2}[\lambda\, J_n'(\lambda a)\, J_n(\mu a) -\mu\, J_n(\lambda a)\, J_n'(\mu a)] \qquad ...(78)$$

For $\lambda=\mu$, put $\mu=\lambda+h$, use Taylor's theorem, let $h\to 0$ and use the relation

$$J_n''(\lambda a)=-\frac{1}{\lambda a}J_n'(\lambda a)-\left(1-\frac{n^2}{\lambda^2 a^2}\right)J_n(\lambda a).$$

We shall get

$$\int_0^a x\, J_n^2(\lambda x)\, dx=\frac{a^2}{2}\left[\{J_n'(\lambda a)\}^2+\left(1-\frac{n^2}{\lambda^2 a^2}\right)J_n^2(\lambda a)\right] \qquad ...(79)$$

If $a=1$ and λ, μ are roots of $J_n(x)$, then (78) and (79) lead to

$$\int_0^1 x\, J_n(\lambda x)\, J_n(\mu x)\, dx=\begin{cases}0 & \text{if } \lambda\neq\mu\\ \frac{1}{2}\{J_n'(\lambda)\}^2 & \text{if } \lambda=\mu\end{cases} \qquad ...(80)$$

(G) **Integral Relations :**

(*i*) **Sonine's integral.** Expanding $J_0(x\sin\theta)$ and integrating term-by term we get

$$J_n(x)=\frac{x^n}{2^{n-1}\,\Gamma(n)}\int_0^{\pi/2}\sin\theta\cos^{2n-1}\theta\, J_0(x\sin\theta)\, d\theta \qquad ...(81)$$

where we have used $\int_0^{\pi/2}\sin^{2m-1}\theta\cos^{2n-1}\theta\, d\theta=\frac{\Gamma(m)\,\Gamma(n)}{2\Gamma(m+n)}$.

(*ii*) $\int_0^\infty e^{-px} J_0(ax)\, dx = \frac{1}{\sqrt{a^2+p^2}}$.

Several such relations are obtained adopting term-by-term integration.

(*iii*) $\frac{1}{\pi}\int_0^\pi e^{ix\cos\phi}\, d\phi = J_0(x)$.

Example 6. Prove $\sum_{n=0}^{\infty} \frac{x^n P_n(\cos\theta)}{n!} = e^{x\cos\theta} J_0(x\sin\theta)$.

Solution. Using the Laplace's first integral for P_n,

$$\sum_{n=0}^{\infty} \frac{x^n P_n(\cos\theta)}{n!} = \frac{1}{\pi}\int_0^\pi \left\{\sum_{n=0}^{\infty} \frac{x^n}{n!}(\cos\theta + i\sin\theta\cos\phi)^n\right\} d\phi$$

$$= \frac{1}{\pi}\int_0^\pi e^{x\cos\theta + ix\sin\theta\cos\phi}\, d\phi$$

$$= e^{x\cos\theta} J_0(x\sin\theta).$$

Modified Bessel functions. Modified Bessel function $I_n(x)$ of the first kind of index n is defined as

$$I_n(x) = i^{-n} J_n(ix) = \frac{(x/2)^n}{\Gamma(n+1)}\, {}_0F_1\left[-;\, n+1;\, \frac{x^2}{4}\right]$$

If satisfies the equation,

$$x^2 y'' + xy' - (x^2+n^2)\, y = 0.$$

Exercise 14

1. Prove $W(J_n, Y_n) = \frac{2}{\pi x}$.

2. Prove (*i*) $4 J_n'' = J_{n-2} - 2 J_n + J_{n+2}$,
   (*ii*) $2 J_0'' = J_2 - J_0$
   (*iii*) $8 J_n''' = J_{n-3} - 3 J_{n-1} + 3 J_{n+1} - J_{n+3}$
   (*iv*) $4 J_0''' + 3 J_0' + J_3 = 0$.

3. Prove $J_n(2\sqrt{x}) = (-1)^n x^{n/2} \frac{d^n}{dx^n} J_0(2\sqrt{x})$.

4. Prove $\lim_{n\to\infty} P_n\left(\cos\frac{x}{n}\right) = J_0(x)$.

5. Prove (*a*) $J_{n+1/2}(x) = (-1)^n x^{n+1/2} \sqrt{\frac{2}{\pi}} \left(\frac{d}{x\,dx}\right)^n \left(\frac{\sin x}{x}\right)$

   (*b*) $J_{-n+1/2} = x^{n-1/2} \sqrt{\frac{2}{\pi}} \left(\frac{d}{x\,dx}\right)^{n-1} \left(\frac{\cos x}{x}\right)$

6. Prove $\int_0^\infty e^{-at} J_0(bt)\, dt = \dfrac{1}{\sqrt{a^2+b^2}}$.

7. Prove the following :

$$\int_0^\infty t^{m-1} J_n(t)\, dt = \frac{2^{m-1}\Gamma\left(\dfrac{m+n}{2}\right)}{\Gamma(\frac{1}{2}n-\frac{1}{2}m+1)} \qquad m+n>0,\ n>1$$

8. Prove the following :

(i) $J_n(x) = \dfrac{1}{2^{n-1}\Gamma(n)\, x^n} \displaystyle\int_0^x (x^2-t^2)^{n-1} J_0(t)\, dt, \qquad n>0.$

(ii) $J_n(x) = \dfrac{x^n}{2^{n-1}\Gamma(n)} \displaystyle\int_0^{\pi/2} \sin\theta \cos^{2n-1}\theta\, J_0(x\sin\theta)\, d\theta, \; n>0.$

(iii) $J_n(x) = \dfrac{1}{\sqrt{\pi}\,\Gamma(n+\frac{1}{2})}\left(\dfrac{x}{2}\right)^n \displaystyle\int_0^\pi \cos(x\sin\phi)\cos^{2n}\phi\, d\phi,$

$n>-\frac{1}{2}.$

(iv) $J_{n+1/2}(x) = \dfrac{\left(\dfrac{x}{2}\right)^{n+1/2}}{\sqrt{\pi}\; n!} \displaystyle\int_{-1}^{1} (1-t^2)^n e^{ixt}\, dt, \qquad n>0.$

9. Prove $J_0(x) = \dfrac{2}{\pi}\displaystyle\int_0^\infty \dfrac{\sin(x+y)}{x+y} J_0(y)\, dy$

10. Prove $\displaystyle\sum_{n=0}^{\infty} \dfrac{J_n(x)t^n}{n!} = J_0(\sqrt{x^2-2xt})$ (Use 72)

11. Prove $e^{x/2(t+1/t)} = \displaystyle\sum_{n=-\infty}^{\infty} I_n(x)\,.\,t^n.$

12. Prove (i) $2I_n' = I_{n-1} + I_{n+1}.$

(ii) $\dfrac{2n}{x} I_n = I_{n-1} + I_{n+1}$

(iii) $x I_n' = x I_{n-1} - n I_n.$

(iv) $x I_n' = n I_n + x I_{n+1}.$

(v) $\dfrac{d}{dx}(x^n I_n) = x^n I_{n-1}.$

(vi) $\dfrac{d}{dx}(x^{-n} I_n) = x^{-n} I_{n+1}.$

13. Prove that $J_n(x)$ has no complex zeros.

§5·6. Laguerre's Differential Equation

$$xy''+(1+\alpha-x)\,y'+ny=0, \qquad 1+\alpha > 0 \qquad \ldots(82)$$

There is a polynomial solution at the regular singularity $x=0$, called the *generalised Laguerre polynomials*, given by

$$L_n^{(\alpha)}(x)=\frac{(1+\alpha)_n}{n\,!}\ {}_1F_1\,[-n\,;\,1+\alpha\,;\,x] \qquad \ldots(83)$$

For $\alpha=0$, it is denoted by $L_n(x)$ and is called the ordinary Laguerre polynomial.

(A) Generating Functions :

(*i*) $$(1-t)^{-1-\alpha}\ \exp.\left(\frac{-xt}{1-t}\right)=\sum_{n=0}^{\infty} L_n^{(\alpha)}(x)\,t^n.$$

(*ii*) $$e^t\ {}_0F_1\,[-\,;\,1+\alpha\,;\,-xt]=\sum_{n=0}^{\infty}\frac{L_n^{(\alpha)}(x)\,t^n}{(1+\alpha)_n}.$$

The relations are proved by expanding the left hand sides in powers of t.

(B) The Rodrigues Formula :

$$L_n^{(\alpha)}(x)=\frac{x^{-\alpha}\,e^x}{n\,!}\,D^n\,[e^{-x}\,x^{n+\alpha}]$$

(C) Recurrence Relations

(*i*) $DL_n^{(\alpha)}(x)=-L_{n-1}^{(\alpha+1)}(x)$.

(*ii*) $L_n^{(\alpha)}(x)=L_{n-1}^{(\alpha)}(x)+L_n^{(\alpha-1)}(x)$

(*iii*) $(n-x)\,L_n^{(\alpha)}(x)=(\alpha+n)\,L_{n-1}^{(\alpha)}(x)-x\,L_n^{(\alpha+1)}(x)$.

(*iv*) $(1+\alpha+n)\,L_n^{(\alpha)}(x)=(n+1)\,L_{n+1}^{(\alpha)}(x)+x\,L_n^{(\alpha+1)}(x)$.

(D) Orthogonality. Equation (82) can be put in the self-adjoint form

$$[x^{1+\alpha}\,e^{-x}\,y']'+nx^{\alpha}\,e^{-x}\,y=0.$$

Here $p(x)=x^{1+\alpha}\,e^{-x}$, $q(x)=0$, $r(x)=x^{\alpha}\,e^{-x}$, $\lambda=n$.

Since for $1+\alpha > 0$, $p(0)=p(\infty)=0$, on the interval $(0, \infty)$, the sequence $\{L_n^{(\alpha)}(x)\}$ is orthogonal with weight function $x^{\alpha}\,e^{-x}$. The eigenfunctions are $L_n^{(\alpha)}(x)$ with eigenvalues $n=0, 1, 2,\ldots\ldots$

Using the Rodrigues' formula, it can be proved

$$\int_0^{\infty} x^{\alpha}\,e^{-x}\,\{L_n^{(\alpha)}\}^2\,dx=\frac{\Gamma(1+\alpha+n)}{n\,!} \qquad \ldots(84)$$

§5·7. Hermite's Equation

$$y''-2xy'+2ny=0 \qquad \ldots(85)$$

$x=0$ is an ordinary point. The polynomial solution at $x=0$ is called the *Hermite polynomial* given by

$$H_n(x)=(2x)^n\ {}_2F_0\left[-\frac{n}{2},\ -\frac{n}{2}+\frac{1}{2}\ ;-;\ -\frac{1}{x^2}\right]$$

$$=\sum_{k=0}^{[n/2]}\frac{(-1)^k\,n\,!\,(2x)^{n-2k}}{k\,!\,(n-2k)\,!}$$

(A) Generating Function :

$$e^{2xt-t^2}=\sum_{n=0}^{\infty}\frac{H_n(x)}{n\,!}\,t^n \qquad \text{...(86)}$$

(B) Rodrigues' Formula :

By Taylor's theorem,

$$H_n(x)=e^{x^2}\left[\frac{\partial^n}{\partial t^n}\ e^{-(x+t)^2}\right]_{t=0}=(-1)^n\ e^{x^2}\ \frac{d^n}{dx^n}\ (e^{-x^2}) \qquad \text{...(87)}$$

(C) Recurrence Relations :

Differentiating the generating relation with respect to x and t, we get

(*i*) $H_n'(x)=2n\ H_{n-1}$.

(*ii*) $2x\ H_n=2n\ H_{n-1}+H_{n+1}$.

(*iii*) $H_n'(x)=2x\ H_n-H_{n+1}$.

(*iv*) $x\ H_n'=n\ H'_{n-1}+n\ H_n$.

(D) Orthogonality :

The Hermite's equation can be put in the self-adjoint form

$$(e^{-x^2}\,y')'+2n\ e^{-x^2}\ y=0.$$

Here $p(x)=e^{-x^2}$, $q(x)=0$, $r(x)=e^{-x^2}$, $\lambda=2n$.

Since $p(-\infty)=p(\infty)=0$, the sequence $\{H_n(x)\}$ is orthogonal on $(-\infty,\ \infty)$ with weight function e^{-x^2}

We have $$\int_{-\infty}^{\infty}e^{-x^2}\ H_m(x)\ H_n(x)\ dx=\begin{cases}0, & m\neq n\\ 2^n\ n\,!\ \sqrt{\pi}, & m=n\end{cases} \qquad \text{...(88)}$$

Problem. Prove : (*i*) $L_n^{(\alpha)}(x)=\dfrac{-x)^n}{n!}\ {}_2F_0\left[-n,\ -n-\alpha\ ;-\dfrac{1}{x}\right]$

(*ii*) $H_{2n}^{(x)}=(-1)^n\ \dfrac{(2n)\,!}{n\,!}\ {}_1F_1\left[-n\ ;\ \dfrac{1}{2}\ ;\ x^2\right]$

(*iii*) $H_{n_2+1}^{(x)}=(-1)^n\ \dfrac{(2n+1)\,!}{n\,!}\ {}_1F_1\left[-n\ ;\ \dfrac{3}{2}\ ;\ \dfrac{2}{x}\right]$.

Exercise 20

1. Prove $(n+1)\ L_{n+1}^{(\alpha)}(x)-(2n+\alpha+1-x)\ L_n^{(\alpha)}(x)$
$+(n+\alpha)\ L_{n-1}^{(\alpha)}(n)=0.$

2. Verify $L_n^{(\alpha)}(0)=\dfrac{(1+\alpha)_n}{n\,!}$.

3. Prove $x\dfrac{d}{dx}\ L_n^{(\alpha)}(x)=n\ L_n^{(\alpha)}(x)-(n+\alpha)\ L_{n-1}^{(\alpha)}(x).$

4. Prove $H_{2n}(0)=(-)^n\dfrac{(2n)\,!}{n\,!}$; $H_{2n+1}{}^{(0)}=0.$

$H_{2n}{}'^{(0)}=0$;

$H'_{2n+1}{}^{(0)}=2(-1)^{n+1}\dfrac{2n+1)\,!}{n\,!}.$

5. Prove $\displaystyle\int_0^\infty e^{-n}\,x^k\,L_n(x)=\begin{cases}0, & k=0,1,\ldots n-1\\ (-1)^n\,n\,!, & k=n\end{cases}$

6. Prove $L_n'(x)=-\displaystyle\sum_{k=0}^{n-1} L_k(x).$

7. Prove (*a*) $L_n^{(\alpha+\beta+1)}(x+y)=\displaystyle\sum_{k=0}^{n} L_n^{(\alpha)}(x)\,.\,L_{n-k}^{(\beta)}(y).$

(*b*) $2^{n/2}\,H_n\left(\dfrac{x+y}{\sqrt{2}}\right)=\displaystyle\sum_{r=0}^{n} {}^nC_r\ H_r(x)\ H_{n-r}(y).$

8. Prove the Christoffel's summation formula

$$\sum_{k=0}^{n}\frac{H_k(x)\ H_k(y)}{2^k\,k\,!}=\frac{H_{n+1}(x)\ H_n(y)-H_{n+1}(y)\ H_n(x)}{2^n\,n\,!\,(y-x)}$$

Hence deduce $\displaystyle\sum_{k=0}^{n} H_k^2(x)=\frac{nH_{n+1}\ H_{n-1}-(n+1)\ H_n^2}{2^n\,n\,!}$.

9. Prove $s\displaystyle\int_0^\infty e^{-st}\ L_n(t)\ dt=\left(1-\frac{1}{s}\right)^n$

10. Prove that $L_n^{(\alpha)}(x)=\displaystyle\sum_{k=0}^{n}\binom{\alpha+n}{n-k}\frac{x^k}{k\,!}$.

11. Let $y_1(x)=x^\alpha\displaystyle\sum_{r=0}^{\infty} a_r\,x^r,\quad y_2(x)=x^\beta\sum_{r=0}^{\infty} b_r\,x$

or $y_2(x)=y_1(x)\log x$ be linearly independent solutions of the equation $y''+p(x)\,y'+q(x)\,y=0$, then prove that

$x\,p(x)<\infty,\ x^2\,q(x)<\infty$ as $x\to\infty$ (see ex. 3. p 48)

6

Ordinary Equations in More Than Two Variables

§6·1. Total Differential Equations

Differential equations of the form

$$P\,dx+Q\,dy+R\,dz=0 \qquad ...(1)$$

where P, Q, R are differentiable functions of x, y, z in a given region are called *total differential equations* or *Pfaffian equations*.

Functional Dependence

Lemma 1. Given a function $v=g(x, y)$. y can be expressed as a function of x and v if and only if $\dfrac{\partial v}{\partial y}\neq 0$.

Proof. Let $y=G(x, v)$. Then $1=\dfrac{\partial G}{dv}\cdot\dfrac{\partial v}{\partial y}$. Since $\dfrac{\partial G}{\partial v}\neq 0$, we have $\dfrac{\partial v}{\partial y}\neq 0$.

Conversely, $\dfrac{\partial v}{\partial y}\neq 0$ implies v is a function of y. Hence $v=g(x, y)$.

Lemma 2. Let $u=f(x, y)$, $v=g(x, y)$ be functions of two independent variables x and y. Then u and v have a functional relation not explicity involving x and y if and only if the Jacobian

$$\frac{\partial(u, v)}{\partial(x, y)}=\begin{vmatrix} \dfrac{\partial u}{\partial x} & \dfrac{\partial v}{\partial x} \\ \dfrac{\partial u}{\partial y} & \dfrac{\partial v}{\partial y} \end{vmatrix}=0.$$

Proof. Necessity. Let u, v have a functional relation

$$H(u, v)=0. \qquad ...(i)$$

Differentiating (i) with respect to x and y we get respectively

$$\frac{\partial H}{\partial u}\frac{\partial u}{\partial x}+\frac{\partial H}{\partial v}\frac{\partial v}{\partial x}=0$$

$$\frac{\partial H}{\partial u}\frac{\partial u}{\partial y}+\frac{\partial H}{\partial u}\frac{\partial v}{\partial y}=0.$$

Eliminating $\frac{\partial H}{\partial u}\Big/\frac{\partial H}{\partial v}$ from these two relations, the required condition is obtained.

Sufficiency. Let $\frac{\partial(u, v)}{\partial(x, y)}=\frac{\partial u}{\partial x}\frac{\partial v}{\partial y}-\frac{\partial u}{\partial y}\cdot\frac{\partial v}{\partial x}=0.$

If $\frac{\partial v}{\partial y}\neq 0$, then by lemma 1, $v=g(x, y)$ defines y as a function of x and v, *i.e.*, $y=G(x, v)$.

Hence $\quad u=f(x, y)=f[x, G(x, v)]$
$\quad\quad\quad\quad=F(x, v)\quad$ (say).

$$\text{Now}\quad 0=\frac{\partial u}{\partial x}\frac{\partial v}{\partial y}-\frac{\partial u}{\partial y}\frac{\partial v}{\partial x}$$

$$=\left(\frac{\partial F}{\partial x}+\frac{\partial F}{\partial v}\frac{\partial v}{\partial x}\right)\frac{\partial v}{\partial y}-\frac{\partial F}{\partial v}\frac{\partial v}{\partial y}\frac{\partial v}{\partial x}$$

$$=\frac{\partial F}{\partial x}\frac{\partial v}{\partial y}$$

Hence $\frac{\partial F}{\partial x}=0$, *i.e.*, F does not involve x. Hence $u=F(v)$.

Lemma 3. P, Q, R are differentiable functions of x, y, z in a given region and λ is an arbitrary function of x, y, z defined in the same region. If,

$$P\left(\frac{\partial Q}{\partial z}-\frac{\partial R}{\partial y}\right)+Q\left(\frac{\partial R}{\partial x}-\frac{\partial P}{\partial z}\right)+R\left(\frac{\partial P}{\partial y}-\frac{\partial Q}{\partial x}\right)=0\quad ...(ii)$$

then the relation (*ii*) remains unaltered if P, Q, R are replaced by λP, λQ, λR throughout.

Proof. The result is obtained by the stated replacements.

Condition of Integrability

Theorem 1. The necessary and sufficient condition that the equation (1) should be integrable is

$$P(Q_z-R_y)+Q(R_x-P_z)+R(P_y-Q_x)=0\quad ...(2)$$

where $P_x=\frac{\partial P}{\partial x}$ etc.

Proof. Necessity. Let equation (1) be integrable and $U(x, y, z)=C$ be its integral. Then equation (1) and

$$U_x\,dx+U_y\,dy+U_z\,dz=0\quad ...(3)$$

are the same. This will be so if there exists an integrating factor $\mu(x, y, z)$ such that

$$\mu P=U_x,\quad \mu Q=U_y,\quad \mu R=U_z.$$

$$\text{Consequently}\quad \frac{\partial(\mu Q)}{\partial z}=\frac{\partial(\mu R)}{\partial y},\quad \frac{\partial(\mu R)}{\partial x}=\frac{\partial(\mu P)}{\partial z},$$

$$\frac{\partial(\mu P)}{\partial y}=\frac{\partial(\mu Q)}{\partial x}$$

or
$$Q_z-R_y=\frac{1}{\mu}(R\,\mu_y-Q\mu_z)$$

$$R_x-P_z=\frac{1}{\mu}(P\,\mu_z-R\,\mu_x)$$

$$P_y-Q_x=\frac{1}{\mu}(Q\,\mu_x-P\,\mu_y).$$

Multiplying the first, second and third of these relations respectively by P, Q, R and adding, we obtain condition (2). The condition ensures the existence of an integrating factor μ which makes the left hand side of (1) an exact differentiate.

Sufficiency. Let the condition (2) be true. Temporarily consider z (or some other variable x or y) fixed and integrate the ordinary equation

$$P\,dx+Q\,dy=0 \qquad \ldots(4)$$

in which z appears as a parameter. Let the integral of (4) be

$$U(x, y, z)=C(z). \qquad \ldots(5)$$

in which the arbitrary constant C may be a function of z. Now choose $C(z)$ in such a manner that (5) satisfies equation (1).

Differentiating (5) we get

$$U_x\,dx+U_y\,dy+U_z\,dz-C'(z)\,dz=0. \qquad \ldots(6)$$

Equations (1) and (6) must be the same, hence

$$\frac{U_x}{P}=\frac{U_y}{Q}=\frac{U_y-C'}{R}$$

i.e., $U_x=\lambda P,\ U_y=\lambda Q,\ U_z-C'(z)=\lambda R$

where λ is some function of x, y, z.

$C(z)$ can be determined if $U_z-\lambda R$ is a function of z and $C(=U)$ alone, *i.e.*, a functional relation exists between $U_z-\lambda R$ and U which does not involve x, y explicitly. By lemma 2, this will be so if the Jacobian of U and $U_z-\lambda R$ with respect to x, y is zero,

i.e., $$U_x\frac{\partial}{\partial y}(U_z-\lambda R)-U_y\frac{\partial}{\partial x}(U_z-\lambda R)=0. \qquad \ldots(7)$$

Now replace P, Q, R in (1) by λP, λQ, λR respectively and put $\lambda P=U_x$, $\lambda Q=U_y$ we shall get the condition (7).

A Recipe to Integrate the Equation (1)

(*i*) Regarding one of the variables, say z, as a temporary constant, solve the equation $P\,dx+Q\,dy=0$.

Suppose this equation is U=costant.

(*ii*) Find λ such that $\lambda P=\dfrac{\partial U}{\partial x}\cdot$

(*iii*) Express the equation $\lambda(P\ dx+Q\ dy+R\ dz)=0$ in the form $dU+\Delta dz=0$ where Δ is a function of z and solve it.

Geometrical Interpretation of the Pfaffian Equation

Let $r=(x, y, z)$, $X=(P, Q, R)$. Since P, Q, R are differentiable, X represents a continuous vector field. The equation (1) can be written in the vector notation in the form

$$X \cdot dr=0. \qquad \ldots(A)$$

The condition of integrability (2) can be written as

$$X \cdot \text{Curl}\ X=0. \qquad \ldots(B)$$

Let the integral surfaces of (1) be $U(x, y, z)=C$.

The vector elements dr are tangential to the integral surfaces and the vector lines X are orthogonal to those surfaces at the points $r=(x, y, z)$. The solution of (1), therefore, entails determination of the family of surfaces which are orthogonal to the vectors (P, Q, R) at the points (x, y, z). A comparison of (A and B) shown that $dr=(dx, dy, dz)$ is proportional to cari X_2 giving auxiliary equations for solution.

If the vector field $X=(P, Q, R)$ is *potential, i.e.*, $X=\text{grad}\ U$, then the desired surfaces are *level* (or *equipotetial*) surfaces $U(x, y, z)=$constant of the potential function $U(x, y, z)$.

If X is not a potential, it can be made so by the multiplication of a scalar factor $\mu(x, y, z)$ called the integrating factor.

Example 1. Integrate the equation

$$(yz+xyz)\ dx+(zx+xyz)\ dy+(xy+xyz)\ dz=0.$$

The condition of integrability is satisfied. After dividing the equation by xyz and rearranging, we get

$$\frac{dx}{x}+\frac{dy}{y}+\frac{dz}{z}+dx+dy+dz=0.$$

This is a complete differential. Integrating $e^{x+y+z}=A\ xyz$ where A is an arbitrary constant.

Example 2. Integrate $(y+z)\ z\ dx-zx\ dy+xy\ dz=0$. The condition of integrability is satisfied. Regard x as a constant. Then

$$\frac{dy}{y}-\frac{dz}{z}=0,$$

i.e., $\dfrac{y}{z}=\phi(x)$ where ϕ is an arbitrary function of x to be determined.

Differentiating and rearraging we get

$$x\phi'(x)\ z^2\ dx-zx\ dy+xy\ dz=0.$$

Comparing this with the given equation,

$$x\,\phi'(x)=\frac{y+z}{z}=\phi(x)+1.$$

Solving this first order linear equation $\phi(x)=Cx-1$ where C is an arbitrary constant. The required solution is

$$y+z=C\,zx.$$

Homogeneous equations. If P, Q, R are homogeneous in x, y, z, the change $x=uz$, $y=vz$ may prove useful.

Exercise 21

Integrate the equations :

1. $(y^2+yz)\,dx+(z^2+zx)\,dy+(y^2-yx)\,dz=0.$
2. $yz\,dx+(x^2\,y-zx)\,dy+(x^2\,z-xy)\,dz=0.$
3. $(2x^3\,y+1)\,dx+x^4\,dy+x^2\tan z\,dz=0.$
4. $(y^2+yz+z^2)\,dx+(z^2+zx+x^2)\,dy+(x^2+xy+y^2)\,dz=0.$
5. $(e^x\,y+e^z)\,dx+(e^y\,z+e^x)\,dy+(e^y-e^x\,y-e^y\,z)\,dz=0.$

§6·2. Systems of Differential Equations

The Matrix Notation For a Linear System.

Consider the linear system in n unknowns x_i and the independent variable t,

$$x_i'(t)=\sum_{j=1}^{n} a_{ij}(t)\,.\,x_j+b_i(t)\ (i=1, \ldots, n) \qquad \ldots(N)$$

where the coefficient $a_{ij}(t)$ are continuous functions on an interval I : $a\leqslant t\leqslant b$.

The system (N) can be expressed in the matrix-vector notation as

$$X'=A(t)\,.\,X+B(t) \qquad \ldots(8)$$

where $A(t)=[a_{ij}(t)]_{n\times n}$, $X=(x_1, \ldots, x_n)^T$,

$$X'=\left(\frac{dx_1}{dt}, \frac{dx_2}{dt}, \ldots, \frac{dx_n}{dt}\right)^T \text{ and } B(t)=[b_1(t), \ldots, b_n(t))^T.$$

$(x_1, \ldots, x_n)^T$ denotes a column vector (matrix).

The homogeneous system corresponding to (N) is

$$X'=A(t)\,.\,X. \qquad \ldots(H)$$

If the right side of (N) does not explicity involve t, it is called an *autonomous system* ; otherwise it is a *non-autonomus system.*

The general solution of (N) consists of the general solution of (H) plus a particular integral of (N).

A *solution* of (N) is an n-dimensional vector $X_i=(x_{i1}, x_{i2}, \ldots, x_{in})^T$ such that when substituted in place of X, it makes (8) an identity. A set consisting of solution vectors of the system is called a *solution set*.

When t is regarded as a parameter, a solution $X=X(t)$ defines in the Euclidean n-space a certain curve called an *integral curve* of the system.

If t is regarded as time, in the Euclidean space with rectangular coordinates $x_1(t), \ldots, x_n(t)$, $\frac{dX}{dt}$ is a velocity vector and $X=X(t)$ denotes a *trajectory*. With this interpretation of the variables, a system

$$\frac{dX}{dt}=F(t, X)$$

is called a *dynamical system*, the space of coordinates $x_1, \ldots, x_n$ the *phase space* and the integal curve $X=X(t)$ the *phase trajectory*.

Linear Dependence of Solutions

A linear system

$$\frac{dX}{dt}=A\,X+B$$

contains all unknowns and their derivatives in the first degree. In such a case an operator L defined by

$$\dot{L}(x)=\frac{dX}{dt}-AX$$

is a linear operator.

Definition 1. If $X_1, \ldots, X_n$ are solution vectors of the homogeneous system (H), then the solution set $\{X_1, \ldots, X_n\}$ is said to be *linearly dependent* on an interval $I : a \leqslant t \leqslant b$ if constants $C_1, \ldots, C_n$, not all zero, exist such that

$$\sum_{i=0}^{n} C_i\, X_i(t)=0, \qquad t \in I \qquad \ldots(9)$$

or in terms of coordinates

$$\sum_{j=1}^{n} C_j\, x_{ij}(t)=0, \qquad (i=1, \ldots, n)$$

A solution set which is not linearly dependent is called *linearly independent*.

Theorem 2. If $A(t)=[a_{ij}(t)]_{n\times n}$ is a matrix with elements $a_{ij}(t)$ continuous on an interval $I : a \leqslant t \leqslant b$, then the solution set S of the linear system

$$X'=A(t)\,.\,X \qquad \ldots(H)$$

is an n-dimensional vector space.

Proof. It is straight forward to prove that if X_i are solutions of (H) then a linear combination $\Sigma\, C_i\, X_i$ is also a solution. This is called the *principle of superposition* (or the *linearity principle*). S is thus a linear space.

The system satisfies conditions of existence and uniqueness of an initial value problem discussed in chapter III. n solutions $X_1, \ldots, X_n$ satisfying the initial conditions

$$X_i(t_0)=(0, 0, \ldots, 1, 0, \ldots, 0)^T, \qquad t_0 \in I$$

exist. It is simple to prove that $X_1, \ldots, X_n$ are linearly independent.

Definition 2. A basis of the solution space of (H) is called a *fundamental system.*

If $\{X_1, \ldots, X_n\}$ is a fundamental system of (H), then the matrix $[X_1, \ldots, X_n]$ where each X_i is a column, is called a *fundamental matrix* of (H).

Definition 3 (Wronskian). **The Wronskian (or Wronskian determinant)** of a vector set $\{X_1, \cdots X_n\}$ where each $X_i=(x_{i1}, \ldots, x_{in})^T$, is defined to be the determinant.

$$W(t) \simeq W(X_1, \ldots X_n) = \begin{vmatrix} x_{11} & x_{12} & \cdots & x_{1n} \\ x_{21} & \cdots & \cdots & x_{2n} \\ \cdots & \cdots & \cdots & \cdots \\ x_{n1} & \cdots & \cdots & x_{nn} \end{vmatrix}$$

Theorem 3. A solution set $S=\{X_1, \ldots, X_n\}$ is a fundamental system of (H) if and only if the Wronskian $W(X_1, \ldots, X_n) \neq 0$ on I.

Proof. Necessity. Let $W(t_0)=0$ for some $t_0 \in I$. Then there exists a non-null constant vector $(c_1, \ldots, c_n)$ such that

$$c_1\, x_{11}\,(t_0)+\ldots \cdots c_n\, x_{1n}(t_0)=0$$

$$\ldots\ldots\ldots\ldots\ldots\ldots\ldots\ldots$$

$$c_1 x_{n1}(t_0)+\ldots+c_n\, x_{nn}\,(t_0)=0$$

i.e., $\qquad c_1 X_1\,(t_0)+\ldots+c_n x_{nn}\,(t_0)=0 \qquad \ldots(i)$

Since $X_1, \ldots, X_n$ are solutions of (H),

$$X(t)=c_1 X_1(t)+\ldots+c_n\, X_n\,(t)$$

is also a solution satisfying the initial condition (i), *i.e.*, $X(t_0)=0$. Owing to the uniqueness of the solution $X \equiv 0$ is the only solution. Hence $c_1 X_1\,(t)+\ldots+c_n X_n(t)=0$ for all $t \in I$, *i.e.*, S is linearly dependent. We also get $W(t)=0$ for all $t \in I$.

Sufficiency. Let $W(t) \neq 0$ for all $t \in I$ and $\sum_{i=0}^{n} c_i\; X_i(t)=0$. We shall have

$$c_1x_{11}(t)+\ldots+c_nx_{1n}(t)=0$$

$$\ldots\ldots\ldots\ldots\ldots\ldots$$

$$c_1x_{n1}(t)+\ldots+c_nx_{nn}(t)=0\cdot$$

Since the coefficient determinant $W(t)\neq 0$, this system has only the trivial solution $c_1=c_2=\ldots=c_n=0$. Hence S is linearly independent.

Remark 1. The theorem does not apply to an arbitrary set of vectors which are not solutions of a homogeneous linear system with continuous coefficients. For example vectors

$$X_1=\begin{bmatrix} t \\ t \end{bmatrix},\quad X_2=\begin{bmatrix} t^2 \\ t^2 \end{bmatrix}$$

are linearly independent on *h*. But

$$W(X_1, X_2)=\begin{vmatrix} t & t^2 \\ t & t^2 \end{vmatrix}\equiv 0.$$

The catch of this discrepancy is that X_1, X_2 are not solutions of a linear homogeneous system with continuous coefficients.

Liouville's Formula. Let $W(t)$ be the Wronskian of the solution set $\{X_1, \ldots, X_n\}$ of the homogeneous system

$$X'=A(t)\,.\,X,\ t\in I \qquad \text{(H)}$$

$$W(t)=W(t_0)\ \exp\left\{\int_{t_0}^{t}(t_r\ A)\,dt\right\},\ t_0\in I \qquad \ldots(10)$$

Proof. Let $X_i(t)=(x_{i1}(t), \ldots, x_{in}(t))^T$ and

$$\left.\begin{aligned} x'_{i1}&=\sum_{k=1}^{n} a_{ik}^{(t)}\,x_{k1} \\ x'_{i2}&=\sum_{k=1}^{n} a_{ik}^{(t)}\,x_{k2} \\ \ldots\quad &\ldots\quad \ldots\quad \ldots \\ x'_{in}&=\sum_{k=1}^{n} a_{ik}^{(t)}\,x_{kn} \end{aligned}\right\}\ (i=1,\ \ldots\ n), \qquad \ldots(11)$$

Then $X_1, \ldots, X_n$ will be solutions of she system

$$X'=A(t)\,X,$$

Also,

$$W(t)=\begin{vmatrix} x_{11} & x_{12} & \ldots & x_{1n} \\ x_{21} & x_{22} & \ldots & x_{2n} \\ \ldots & \ldots & \ldots & \ldots \\ x_{n1} & x_{n2} & \ldots & x_{nn} \end{vmatrix}$$

Differentiating with respect to t,

$$W'(t)=W_1(t)+W_2(t)+\dots+W_n(t),$$

where $W_r(t)$ is the determinant obtained from $W(t)$ by differentiating its rth row, *i.e.*

$$W_r(t)=\begin{vmatrix} x_{11} & x_{12} & \dots & x_{1n} \\ \dots & \dots & \dots & \dots \\ x'_{r1} & x'_{r2} & \dots & x'_{rn} \\ \dots & \dots & \dots & \dots \\ x_{n1} & x_{n2} & \dots & x_{nn} \end{vmatrix}$$

Substituting $x'_{rj}=\sum_{k=1}^{n} a_{rk}\, x_{kj}$, we shall get

$$W_r(t)=a_{rr}(t)\ W(t).$$

Hence $\quad W'(t)=\left(\sum_{r=1}^{n} a_{rr}(t)\right)W(t)=t_r\ A)\ W(t).$

Integrating both sides

$$W(t)=K \exp\left\{\int_{t_0}^{t} (t_r\ A)\ dt\right\}$$

Putting $t=t_0$ we obtain $K=W(t_0)$.

Formula (10) is also called *Abel's identity*.

Remark 2. If $W(t_0)=$ for some $t_0 \in I$, then by (10),

$$W(t)\equiv 0 \text{ on } I.$$

§6·3. First Order Systems with Constant Coefficients

We are concerned with a linear system

$$X'=AX \qquad \dots(12)$$

where $X=(x_1, \dots, x_n)^T$ and A is a cônstant matrix $[a_{ij}]_{n\times n}$

Assume a solution $X=e^{\lambda t}\,K$ of (12) where K is a non-zero constant column vector of dimension n and λ is a constant. Substituting in (12) we get,

$$e^{\lambda t}\,AK=\lambda\, e^{\lambda t}\,K,$$

i.e., $\quad e^{\lambda t}\,(A-\lambda\, I)K=0,$

where I is the $n\times n$ unit matrix. As $e^{\lambda t}\neq 0$.

$$(A-\lambda I)K=0,$$

i.e.,

$$\begin{aligned} (a_{11}-\lambda)K_1+a_{12}\,K_2+\dots+a_1 K_n&=0 \\ a_{21}K_1+(a_{22}-\lambda)K_2+\dots+a_{2n}K_n&=0 \\ \dots\dots\dots\dots\dots\dots& \\ a_{n1}K_1+\ \dots\dots+(a_{nn}-\lambda)K_n&=0 \end{aligned} \qquad \dots(13)$$

where $K=(k_1, \ldots, k_n)^T$. For a non-trivial solution of (13) in $k_1, \ldots,$ k_n, we have

$$p(\lambda)=\begin{vmatrix} a_{11}-\lambda & a_{12} & \cdots & a_{1n} \\ a_{21} & a_{22}-\lambda & \cdots & a_{2n} \\ \cdots & \cdots & \cdots & \cdots \\ a_{n1} & & & a_{nn}-\lambda \end{vmatrix}=0 \qquad \ldots(14)$$

(14) is called the *characteristic* (or *auxiliary*) equation of the system (12). Equation (14) has n roots $\lambda_1, \ldots, \lambda_n$ called eigenvalues of the matrix A. The corresponding eigenvectors $(k_1, \ldots, k_n)^T$ are obtained from the algebraic system (13).

The following cases arise

Case 1. Characteristic roots are real and simple.

Since the characteristic roots $\lambda_1, \ldots, \lambda_n$ are distinct, the corresponding eigenvectors $K_1, \ldots, K_n$ are linearly independent. We thus have n linearly independent solutions $K_1\ e^{\lambda_1 t}, \ldots, K_n\ e^{\lambda_n t}$ of the system.

Case 2. Characteristic Roots are Complex and Simple.

Complex conjugate roots occur in pairs. Let $\lambda_1=\alpha+i\beta$, $\lambda_2=\alpha-i\beta$ be two characteristic roots and K_1, K_2 the corresponding eigenvectors. Let $K_1=R+iS$, $K_2=R-iS$. Then we have solutions.

$$X_1=e^{\lambda_1 t}\, K_1=U(t)+i\, V(t)$$

$$X_2=e^{\lambda_2 t}\, K_2=U(t)-i\, V(t)$$

where

$$U(t)=e^{\alpha t}\, (R \cos \beta t - S \sin \beta t)$$

$$V(t)=e^{\alpha t}\, (R \sin \beta t + S \cos \beta t).$$

$U(t)$ and $V(t)$ are two linearly independent solution of the system corresponding to λ_1 and λ_2.

Case 3. Multiple roots. The method will be explained by an example.

The Meaning of e^{tA}.

Let A be an n-square matrix with *distinct eigenvalues* $\lambda_1, \ldots, \lambda_n$. If $X_1, \ldots, X_n$ are the associated eigenvectors, then $X_1, \ldots, X_n$ are linearly independent and the matrix $P=[X_1 \ldots X_n]$ where X_i are columns, is non-singular. P is called the diagonalising matrix of A and

$$P^{-1}AP=\begin{bmatrix} \lambda_1 & 0 & \cdots & 0 \\ 0 & \lambda_2 & \cdots & 0 \\ \cdots & \cdots & \cdots & \cdots \\ 0 & 0 & \cdots & \lambda_n \end{bmatrix}=D=\text{diag}\,(\lambda_1, \ldots, \lambda_n)$$

Thus $A = PDP^{-1}$. For an integer k, we have

$$A^k = PD^kP^{-1} = P \text{ diag } (\lambda_1{}^k, \ldots, \lambda_n{}^k) \; P^{-1},$$

In general $f(A) = P \text{ diag } [f(\lambda_1), \cdots, f(\lambda_n)] P^{-1}$.

For a real or complex number t,

$$e^{tA} = P \text{ diag } (e^{t\lambda_1}, \ldots, e^{t\lambda_n}) P^{-1} \qquad \ldots(15)$$

To illustrate the application of this concept in the solution of a linear system with distinct eigenvalues we consider the following examples :

(*i*) Solve the system $X' = AX$ where

$$A = \begin{bmatrix} -3 & -1 \\ 2 & -1 \end{bmatrix}, \; X' = \begin{bmatrix} \dfrac{dx_1}{dt} \\ \dfrac{dx_2}{dt} \end{bmatrix}$$

Solution. Integrating $X' = AX$ with respect to t,

$$X = e^{tA} \, K,$$

where K is an arbitrary constant column vector of dimension 2.

The eigenvalues of A are given by

$$\begin{vmatrix} -3-\lambda & -1 \\ 2 & -1-\lambda \end{vmatrix} = 0, \; i.e., \; \lambda_1 = -2+i, \; \lambda_2 = -2-i$$

The associated eigenvectors are

$$Y_1 = \begin{bmatrix} -1 \\ 1+j \end{bmatrix}, \; Y_2 = \begin{bmatrix} -1 \\ 1-i \end{bmatrix}.$$

Hence the diagonalising matrix of A is

$$P = \begin{bmatrix} -1 & -1 \\ 1+i & 1-i \end{bmatrix}$$

Also,

$$P^{-1} = \frac{1}{2i} \begin{bmatrix} 1-i & 1 \\ -1-i & -1 \end{bmatrix}$$

Hence,

$$D = P^{-1} AP = \begin{bmatrix} -2+i & 0 \\ 0 & -2-i \end{bmatrix}$$

$$e^{tA} = Pe^{tD} P^{-1} = \frac{1}{2i} \begin{bmatrix} -1 & -1 \\ 1+i & 1-i \end{bmatrix} \begin{bmatrix} e^{(-2+i)t} & 0 \\ 0 & e^{(-2-i)t} \end{bmatrix} \begin{bmatrix} 1-i & 1 \\ -1-i & -1 \end{bmatrix}$$

$$=e^{-2t}\begin{bmatrix}\cos t-\sin t & -\sin t\\ 2\sin t & \cos t+\sin t\end{bmatrix}$$

$X=e^{tA}$ is the solution of the system.

Hence two linearly independent solutions of the system are

$$X_1=e^{-2t}\begin{bmatrix}\cos t-\sin t\\ 2\sin t\end{bmatrix},\ X_2=e^{-2t}\begin{bmatrix}-\sin t\\ \cos t+\sin t\end{bmatrix}$$

If $K=\begin{bmatrix}k_1\\ k_2\end{bmatrix}$ is an arbitrary constant vector, the general solution of the system is given by

$$X=e^{tA}\,K=k_1\,X_1+k_2X_2$$

i.e.,

$$x_1=e^{-2t}\{k_1\,(\cos t-\sin t)-k_2\sin t\}$$

$$x_2=e^{-2t}\{2k_1\sin t+k_2\,(\cos t+\sin t)\}.$$

(*ii*) Solve the initial value problem

$$X'=AX,\ X(0)=\begin{bmatrix}1\\ 2\end{bmatrix}$$

where

$$A=\begin{bmatrix}3 & 1\\ -5 & -3\end{bmatrix}$$

Solution. Integrating $X'=AX$, we have $X(t)=e^{tA}\,K$. Also

$$X(0)=\begin{bmatrix}1\\ 2\end{bmatrix}=K$$

We now compute e^{tA}. The characteristic values of A are $\lambda_1=2$, $\lambda_2=-2$ and the associated eigenvectors are

$$Y_1=\begin{bmatrix}1\\ -1\end{bmatrix},\ Y_2=\begin{bmatrix}2\\ -5\end{bmatrix}$$

The diagonalising matrix of A is

$$P=\begin{bmatrix}1 & 1\\ -1 & -5\end{bmatrix}.$$

Also

$$P^{-1}=\tfrac{1}{4}\begin{bmatrix}5 & 1\\ -1 & -1\end{bmatrix}.$$

$$D=P^{-1}\,AP=\begin{bmatrix}2 & 0\\ 0 & -2\end{bmatrix}$$

Hence $$e^{tA}=\mathrm{P}\,e^{tD}\,\mathrm{P}^{-1}=\frac{1}{4}\begin{bmatrix} 5\,e^{2t}-e^{-2t} & e^{2t}-e^{-2t} \\ -5e^{2t}+5\,e^{-2t} & e^{2t}+5\,e^{-2t} \end{bmatrix}$$

which represents the general solution. The unique solution of the given IVP is

$$\mathrm{X}=e^{tA}\mathrm{K}=e^{tA}\begin{bmatrix} 1 \\ 2 \end{bmatrix}=\frac{1}{4}\begin{bmatrix} 7e^{2t}-3e^{-2t} \\ -7e^{2t}+15e^{-2t} \end{bmatrix}$$

$$=\frac{7}{4}e^{2t}\begin{bmatrix} 1 \\ -1 \end{bmatrix}+\frac{1}{4}e^{-2t}\begin{bmatrix} -3 \\ 15 \end{bmatrix}.$$

The following examples illustrate the procedure to solve linear systems without computing matrix e^{tA}.

Example 3. Solve the system

$$\frac{dx}{dt}=2x+5y$$

$$\frac{dy}{dt}=x-2y.$$

Solution. In the matrix-vector form the system can be expressed as

$$\mathrm{X}'=\mathrm{AX} \quad \text{where } \mathrm{A}=\begin{bmatrix} 2 & 5 \\ 1 & -2 \end{bmatrix}$$

The characteristic equation

$$\begin{vmatrix} 2-\lambda & 5 \\ 1 & -2-\lambda \end{vmatrix}=0 \text{ gives } \lambda_1=3,\ \lambda_2=-3$$

Replacing X by $e^{\lambda t}$ K where $\mathrm{K}=\begin{bmatrix} k_1 \\ k_2 \end{bmatrix}$ we get after simplification

$$\begin{bmatrix} 2 & 5 \\ 1 & -2 \end{bmatrix}\begin{bmatrix} k_1 \\ k_2 \end{bmatrix}=\lambda\begin{bmatrix} k_1 \\ k_2 \end{bmatrix}$$

i.e., $(2-\lambda)k_1+5k_2=0$ ...(*i*)

$k_1-(\lambda+2)k_2=0$ ...(*ii*)

For $\lambda=3$ both (*i*) and (*ii*) give $k_1=5k_2$

Hence one solution is

$$\mathrm{X}_1=e^{3t}\begin{bmatrix} 5 \\ 1 \end{bmatrix}.$$

For $\lambda=-3$, $k_1=-k_2$ and another linearly independent solution is

$$X_2=e^{-3t}\begin{bmatrix}-1\\1\end{bmatrix}.$$

The general solution of the system is

$$X=c_1\,e^{3t}\begin{bmatrix}5\\1\end{bmatrix}+c_2\,e^{-3t}\begin{bmatrix}-1\\1\end{bmatrix}$$

or $$x=5c_1\,e^{3t}-c_2\,e^{-3t},\ y=c_1\,e^{3t}+c_2\,e^{-3t}.$$

Example 4. Solve the system $X'=AX$ where $A=\begin{bmatrix}1 & -1\\1 & 3\end{bmatrix}$.

Solution. The characteristic roots are $\lambda_1=\lambda_2=2$.

Let $$X=\begin{bmatrix}a+bt\\c+dt\end{bmatrix}e^{2t}$$

be a solution. Then substitution in the system gives

$$2a+b+2bt=a-c+(b-d)\,t$$

i.e., $$c=-a-b,\ d=-b.$$

Keeping a and b arbitrary we get the general solution

$$X=a\,e^{2t}\begin{bmatrix}1\\-1\end{bmatrix}+b\,e^{2t}\begin{bmatrix}t\\-1-t\end{bmatrix}.$$

The solution can also be expressed in the form

$$x=(c_1+c_2t)e^{2t},\ y=-(c_1+c_2+c_2t)\,e^{2t}$$

where c_1, c_2 are arbitrary.

Example 5. Find the complete solution of the non-homogeneous system.

$$x'=x-2y+e^t$$
$$y'=x+4y+e^{2t}$$

Solution. The system can be written as

$$X'=\begin{bmatrix}1 & -2\\1 & 4\end{bmatrix}X+\begin{bmatrix}e^t\\e^{2t}\end{bmatrix}. \quad ...(i)$$

The general solution of the corresponding homogeoeous system is

$$X=c_1\,e^{2t}\begin{bmatrix}-2\\1\end{bmatrix}+c_2\,e^{3t}\begin{bmatrix}-1\\1\end{bmatrix} \quad ...(ii)$$

For (*ii*) to be a solution of the given system (*i*) regard c_1 and c_2 as functions of t. We shall have

$$c_1'\, e^{2t}\begin{bmatrix} -2 \\ 1 \end{bmatrix} + c_2'\, e^{3t}\begin{bmatrix} -1 \\ 1 \end{bmatrix} = \begin{bmatrix} e^t \\ e^{2t} \end{bmatrix}.$$

Hence

$$-2c_1'\, e^{2t} - c_2'\, e^{3t} = e^t$$

$$c_1'\, e^{2t} + c_2'\, e^{3t} = e^{2t}$$

Solving for c_1', c_2' we have

$$c_1' = -1 - e^{-t},\ c_2' = 2e^{-t} + e^{-2t}$$

i.e., $c_1 = -t + e^{-t} + A,\ c_2 = -2e^{-t} - \frac{1}{2}e^{-2t} + B$

Hence the general solution required is

$$X = A\, e^{2t}\begin{bmatrix} -2 \\ 1 \end{bmatrix} + B\, e^{3t}\begin{bmatrix} -1 \\ 1 \end{bmatrix} + \tfrac{1}{2}\, e^t\begin{bmatrix} -3 \\ 1 \end{bmatrix} + e^{2t}\begin{bmatrix} 2t \\ -t-1 \end{bmatrix}$$

Example 6. Solve the initial value problem

$$X' = \begin{bmatrix} -2 & 4 \\ 1 & 1 \end{bmatrix} X,\ X(0) = \begin{bmatrix} -2 \\ 3 \end{bmatrix}.$$

Solution. The general solution is

$$X(t) = k_1\, e^{2t}\begin{bmatrix} 1 \\ 1 \end{bmatrix} + k_2\, e^{-3t}\begin{bmatrix} 4 \\ -1 \end{bmatrix}$$

$$= F(t)\ K$$

where $F(t) = \begin{bmatrix} e^{2t} & 4e^{-3t} \\ e^{2t} & -e^{-3t} \end{bmatrix}$ and $K = \begin{bmatrix} k_1 \\ k_2 \end{bmatrix}$.

Hence $F(0) = \begin{bmatrix} 1 & 4 \\ 1 & -1 \end{bmatrix}$ and $F^{-1}(0) = \frac{1}{5}\begin{bmatrix} 1 & 4 \\ 1 & -1 \end{bmatrix}$.

$$K = F^{-1}(0)X(0) = \tfrac{1}{5}\begin{bmatrix} 1 & 4 \\ 1 & -1 \end{bmatrix}\begin{bmatrix} -2 \\ 3 \end{bmatrix} = \begin{bmatrix} 2 \\ -1 \end{bmatrix}.$$

Hence the required solution is

$$X(t) = \begin{bmatrix} e^{2t} & 4e^{-3t} \\ e^{2t} & -e^{-3t} \end{bmatrix}\begin{bmatrix} 2 \\ -1 \end{bmatrix}$$

$$= 2\, e^{2t}\begin{bmatrix} 1 \\ 1 \end{bmatrix} - e^{-3t}\begin{bmatrix} 4 \\ -1 \end{bmatrix}.$$

§6·4. Systems Reducible to Integrable Combinations

We consider linear and non-linear systems which may be expressed in the form

$$\frac{dx}{P}=\frac{dy}{Q}=\frac{dz}{R} \qquad ...(16)$$

where P, Q, R are continuous functions of x, y, z in a given region. An integrable combination of (16) is a differential equation deducible from it by algebraic operations or change of variables which is readily integrable.

If $u(x, y, z)=c_1$, $v(x, y, z)=c_2$ are two linearly independent solutions of (16), then $\phi(u, v)=0$ is called its *general solution* where ϕ is an arbitrary function.

A solution of the spstem which is not included in its general solution is called a *special integral.*

Geometrically interpreted, the general solution $\phi(u, v)=0$ denotes a system of surfaces containing the two-parameter family of curves $u(x, y, z)=a$, $v(x, y, z)=b$. This family of curves is orthogonal to the integral surfaces represented by the total differential equation

$$Pdx+Qdy+Rdz=0$$

Example 7. Solve the system

$$\frac{dx}{x^2-y^2-z^2}=\frac{dy}{2xy}=\frac{dz}{2xz}.$$

Solution. The equation $\dfrac{dy}{2xy}=\dfrac{dz}{2dz}$ gives a solution

$$u(x, y, z)=\frac{y}{z}=c_1.$$

Multiplying the numerators and denominators of the first, second and third ratios by x, y, z respectively, and adding we get

$$\frac{xdx+ydy+zdz}{x(x^2+y^2+z^2)}=\frac{dy}{2xy}$$

After integration we get

$$v(x, y, z)=\frac{x^2+y^2+z^2}{y}=c_2.$$

The general solution is

$$\phi\left(\frac{y}{z}, \frac{x^2+y^2+z^2}{y}\right)=0$$

where ϕ is an arbitrary function.

§6·5. The Method of Elimination

A general differential system in n unknowns $x_1, \cdots, x_n$ and independent variable t with constant coefficients can be represented in the form

$$\sum_{j=1}^{u} P_{ij}(D)\, x_j=b_i(t) \qquad (i=1, ..., n) \qquad ...(17)$$

where $D=\frac{d}{dt}$ and $P_{ij}(D)$ are polynomials in D with constant coefficients. The system is called *non-degenerate* if the determinent $\Delta(D)=|P_{ij}(D)|\not\equiv 0$; otherwise it is called *degenerate*. The number of arbitrary constants in the general solution is equal to the degree of D in $\Delta(D)$.

Example 8. Solve the system

$$-(D+2)x_1+(D^2-4)x_2+4t \qquad \text{...(i)}$$

$$(D+3)x_1+(D+y)x_2=0 \qquad \text{...(ii)}$$

Solution. Here $\Delta(D)=\begin{vmatrix} -(D+2) & D^2-y \\ D+3 & D+y \end{vmatrix}$

$$=-D^3-4D^2-5D-2\not\equiv 0$$

The system is non-degenerate. Since $\Delta(D)$ has degree 3 in D, the general solution will have three arbitrary constants.

We now eliminate x_1. Adding (*i*) and (*ii*) we get

$$x_1+(D^2+D+3)x_2=4t \qquad \text{...(iii)}$$

Operate on (*iii*) by D+2 and add to (*i*),

$$(D+1)^2(D+2)x_2=12t+4$$

Hence $\quad x_2=(c_1+c_2t)e^{-t}+c_3e^{-2t}+6t-13.$

Substituting this value of x_2 in (*iii*) we get

$$x_1=-3c_1e^{-t}+c_2(1-3t)e^{-t}-5c_3-14t+33.$$

Exercise 22

1. Find the general solution of the following systems :

(*i*) $X'=\frac{1}{2}\begin{bmatrix} 1 & 1 \\ -3 & 5 \end{bmatrix}X$

(*ii*) $X'=\begin{bmatrix} -3 & -1 \\ 2 & -1 \end{bmatrix}X$

(*iii*) $X'=\begin{bmatrix} 3 & -1 \\ 1 & 1 \end{bmatrix}X$

(*iv*) $X'=\begin{bmatrix} 1 & 2 \\ 4 & 3 \end{bmatrix}X$

(*v*) $X'=\begin{bmatrix} 1 & 1 & -2 \\ -1 & 2 & 1 \\ 0 & 1 & -1 \end{bmatrix}X$

2. Solve the initial value problems :

(*i*) $X' = \begin{bmatrix} 2 & 2 \\ 1 & 3 \end{bmatrix} X, \quad X(0) = \begin{bmatrix} 1 \\ 4 \end{bmatrix}$

(*ii*) $X' = \begin{bmatrix} 2 & 4 \\ -2 & -2 \end{bmatrix} X, \quad X(0) = \begin{bmatrix} 1 \\ 3 \end{bmatrix}$

(*iii*) $\begin{matrix} x' = y \\ y' = -x \end{matrix}, \quad x(0) = 0, \; y(0) = 1$

3. Solve completely the following systems :

(*i*) $X' = \begin{bmatrix} 1 & 1 \\ 4 & -2 \end{bmatrix} X + \begin{bmatrix} -5t+2 \\ -8t-8 \end{bmatrix}$

(*ii*) $(D+1)x = y + e^t$
$(D+1)y = x + e^t$.

(*iii*) $X' = \begin{bmatrix} 3 & -1 \\ 4 & -1 \end{bmatrix} X + \begin{bmatrix} 1 \\ t \end{bmatrix}, \quad X(0) = \begin{bmatrix} 1 \\ 0 \end{bmatrix}$

4. Find the general solutions of the following systems :

(*i*) $\dfrac{dx}{x(y-z)} = \dfrac{dy}{y(z-x)} = \dfrac{dz}{z(x-y)}$

(*ii*) $\dfrac{dx}{xz-y} = \dfrac{dy}{yz-x} = \dfrac{dz}{1-z^2}$

(*iii*) $\dfrac{dx}{y^2+yz+z^2} = \dfrac{dy}{z^2+zx+x^2} = \dfrac{dz}{x^2+xy+y^2}$

(*iv*) $\dfrac{dx}{1} = \dfrac{dy}{-2} = \dfrac{dz}{3x^2 \sin(y+2x)}$

5. Solve the systems :

(*i*) $\dfrac{2d^2x}{dt^2} - \dfrac{dy}{dt} - 4x = 2t, \quad \dfrac{2dx}{dt} + \dfrac{4dy}{dt} - 3y = 0$

(*ii*) $(D^2-3)x - 4y = 0, \quad x + (D^2+1)y = 0$

7

Partial Differential Equations

Differential equations in which the unknown functions are functions of more than one variable, are called partial differential equations (PDE's). They arise from geometrical concepts and from the mathematical formulation of physical processes. Formally they can be obtained either from the elimination of arbitrary functions or arbitrary constants.

In the case of three variables x, y, z, the following abbreviations are used :

$$\frac{\partial z}{\partial x}=p, \quad \frac{\partial z}{\partial y}=q, \quad \frac{\partial^2 z}{\partial x^2}=r,$$

$$\frac{\partial^2 z}{\partial x \partial y}=\frac{\partial^2 z}{\partial y \partial x}=s, \quad \frac{\partial^2 z}{\partial y^2}=t$$

If more than three variables are in use, the dependent variable is denoted by z and the independent variables by $x_1, \cdots, x_n$. For derivatives, $p_i \equiv \frac{\partial z}{\partial x_i}$. In what follows we shall abbreviate 'partial differential equations' as *pde* or PDE.

7·1. Construction of Partial Differential Equations of First Order

(A) Elimination of Arbitrary Constants. Given a functional relation

$$f(x, y, z, a, b)=0 \qquad \ldots(i)$$

where a and b are arbitrary constants.

Differentiating (i) with respect to x, y we get

$$\frac{\partial f}{\partial x}+\frac{\partial f}{\partial z} p=0 \qquad \ldots(ii)$$

$$\frac{\partial f}{\partial y}+\frac{\partial f}{\partial z} q=0 \qquad \ldots(iii)$$

Eliminating a, b from $(i), (ii), (iii)$ we get an equation of the form

$$F(x, y, z, p, q)=0 \qquad \ldots(1)$$

(1) is a *pde* of the first order.

As an illustration the elimination of a, b from the relation

$z=ax+by+ab$ leads to the *pde* $\quad z=px+qy+pq$.

(B) Elimination of Arbitrary Functions

Let two given linearly independent differentiable functions $u(x, y, z)$ and $v(x, y, z)$ be related by an arbitrary functional relation

$$\phi(u, v)=0 \qquad \text{...}(iv)$$

Differentiation of (*iv*) with respect to x and y respectively gives

$$\frac{\partial \phi}{\partial u}\left(\frac{\partial u}{\partial x}+\frac{\partial u}{\partial z}p\right)+\frac{\partial \phi}{\partial v}\left(\frac{\partial v}{\partial x}+\frac{\partial v}{\partial z}p\right)=0 \qquad \text{...}(v)$$

$$\frac{\partial \phi}{\partial u}\left(\frac{\partial u}{\partial y}+\frac{\partial u}{\partial z}q\right)+\frac{\partial \phi}{\partial v}\left(\frac{\partial v}{\partial y}+\frac{\partial v}{\partial z}q\right)=0 \qquad \text{...}(vi)$$

Elimination of $\frac{\partial \phi}{\partial u} \Big/ \frac{\partial \phi}{\partial v}$ from these relations will give

$$\mathrm{P}p+\mathrm{Q}q=\mathrm{R} \qquad \text{...}(2)$$

where P, Q, R are continuous functions of x, y, z in the region of definition of u, v.

An equation of the form (2) is called a 'Lagrange's linear equation'.

In general if P, Q, R are functions of x, y, z an equation of the form (2) is called a *quasilinear equation*. If P, Q are functions of x, y only, it is called a *semi-linear equation*. If in addition to P, Q, R also does not involve z, it is called linear. By convention the term Lagrange's linear equation includes all the three types.

7·2. Types of Integrals

For an equation of the first order $\mathrm{F}(x, y, z, p, q)=0$, the following types of solutions are possible :

I. Complete integral which has two arbitrary constants.

II. General solution which consists of an arbitrary function. If u=constant, v=constant are two linearly independent solutions of the equation, then its general solution can be represented as $\phi(u, v)=0$, where ϕ is an arbitrary function.

A general solution can also be generated by the complete integral. For example, if

$$f(x, y, z, a, b)=0 \qquad \text{...}(i)$$

is the complete integral, then assume an arbitrary functional relation in a and b, *i.e.*, $b=\phi(a)$. Thus

$$f(x, y, z, a, \phi(a))=0 \qquad \text{...}(ii)$$

Differentiating (*ii*) partially with respect to a

$$\frac{\partial f}{\partial a}=0 \qquad \text{...}(ii)$$

Eliminating a from (*ii*) and (*iii*) we theoretically get a relation involving an arbitrary function ϕ. This represents the general solution.

III. Singular solution. Let the complete integral be

$$f(x, y, z, a, b)=0$$

where a and b have degree more than one. Eliminating a and b from the three relations $f=0$, $\frac{\partial f}{\partial a}=0$, $\frac{\partial f}{\partial b}=0$ we get the singular solution. The singular solution besides containing the envelop of the surfaces represented by the complete integral may contain extraneous loci *viz.*, conical points or double lines. These extraneous loci do not represent solutions to the given equation.

IV. Special integrals are those which are not included in any of the above three categories.

A Cauchy problem. In practical problems it is ordinarily required to find solutions which satisfy certain initial conditions. For example we may require an integral surface which contains a given space curve, called the *initial curve.* Such a problem is called a *Cauchy problem.*

7·3. Lagrange's Linear Equation

The equation

$$Pp+Qq=R \qquad ...(2)$$

where P, Q, R are continuous functions of x, y, z in a given domain, is called the Lagrange's linear equation. In vector form (2) can be written

$$N.F=0 \qquad ...(3)$$

where $N=(p, q, -1)$ and $F=(P, Q, R)$

(3) indicates that if $z=f(x, y)$ is an integral surface of (2), then a vector N is directed normally to the surface and the vector lines F are collinear with the tangent vector (dx, dy, dz) of the surface. The condition of collinearity determines the following *subsidiary (or characteristic)* equations of (2),

$$\frac{dx}{P}=\frac{dy}{Q}=\frac{dz}{R} \qquad ...(4)$$

The integral surface of (2) are made up of the integral curves of the equations (4).

To integrate (2) we therefore, find the solution curves of (4) *viz.*,

$$u(x, y, z)=a, \qquad v(x, y, z)=b$$

called the *characteristic curves.*

If $R=0$ we shall get the projections of the characteristic curves on the xy plane. These projected curves are called *ground characteristics* or *base characteristics.*

If from this two-parameter family of curves we isolate in some manner a one-parameter family by choosing some continuous functional relation $\phi(a, b)=0$ in a and b, we get the equation of the integral surfaces of (2) as

$$\phi(u, v)=0 \qquad ...(5)$$

where ϕ is an arbitrary function. (5) is called the *general solution* of (2).

In a Cauchy problem we require not an arbitrary surface, but one which passes through a given (initial) curve

$$F(x, y, z)=0, \quad G(x, y, z)=0.$$

In this case ϕ will not be arbitrary, but will be determined by eliminating x, y, z from the four relations $u=a$, $v=b$, $F=0$, $G=0$.

If $H(a, b)=0$ is the eliminant, then the required surface (*i.e.*, solution to the Cauchy problem) is $H(u, v)=0$. The condition for a unique solution $H(u, v)=0$ to exist is that the given initial curve should itself not be a characteristic curve

It is often convenient to represent space curves parametrically. Let the initial curve be

$$\Gamma : \begin{cases} x=\phi(s) \\ y=\psi(s) \\ z=\chi(s) \end{cases} \qquad \ldots(6)$$

The integral surface will be the locus of those characteristic curves which emanate from Γ. As parametrically such an integral surface will be represented by

$$x=x(t, s), \quad y=y(t, s), \quad z=z(t, s) \qquad \ldots(7)$$

for which

$$x(0, s)=\phi(s)$$

$$y(0, s)=\psi(s)$$

$$z(0, s)=\chi(s).$$

A given parameter $s=s_0$ fixes a point on Γ. From this point will emanate a characteristic curve. On this characteristic curve, t will be the parameter.

For the elimination of t, s in the relations (7) to be possible, the Jacobians $\dfrac{\partial(x, y)}{\partial(t, s)}, \dfrac{\partial(y, z)}{\partial(t, s)}, \dfrac{\partial(z, x)}{\partial(t, s)}$ should not vanish.

In case $\dfrac{\partial(x, y)}{\partial(t, s)}=0$, then the direction of projection of Γ on the xy-plane is parallel to the direction of projection of a tangent to a characteristic curve.

Equation (2) can be generalised to

$$P_1 p_1+\ldots+P_n p_n=R$$

where the P_i and R are continuous functions in an $(n+1)$-dimensional manifold of $x_1, \ldots, x_n, z$ and $p_i=\dfrac{\partial z}{\partial x_i}$ $(i=1, \ldots, n)$.

If $u_i(x_1, \ldots, x_n, z)=C_i$ are solutions of the subsidiary equations,

$$\frac{dx_1}{P_1}=\frac{dx_2}{P_2}=\ldots=\frac{dx_n}{P_n}=\frac{dz}{R},$$

then the general solution is $\phi(u_1, \cdots, u_n)=0$ where ϕ is an arbitrary function.

Example 1. Solve the equation

$$z(x+2y)\,p-z(y+2x)\,q=y^2-x^2.$$

Solution. The characteristic equations are

$$\frac{dx}{z(x+2y)}=\frac{dy}{-z(y+2x)}=\frac{dz}{y^2-x^2}.$$

Each ratio is equal to $\dfrac{x\,dx+y\,dy+z\,dz}{0}$

whence $$u=x^2+y^2+z^2=a.$$

The first two ratios yield

$$\frac{dy}{dx}=-\frac{y+2x}{x+2y} \quad \text{which gives}$$

$$v=x^2+y^2+xy=b.$$

The general solution is $\phi(x^2+y^2+z^2,\ x^2+y^2+xy)=0.$

Example 2. (A Cauchy problem). Find the integral surface of the equation $(y-z)\,p+(z-x)\,q=x-y$ which passes through the straight line $z=0$, $y=2x$.

Solution. The subsidiary equations are

$$\frac{dx}{y-z}=\frac{dy}{z-x}=\frac{dz}{x-y}.$$

Each ratio is equal to

$$\frac{dx+dy+dz}{0}=\frac{x\,dx+y\,dy+z\,dz}{0}.$$

This gives $x+y+z=C_1$ and $x^2+y^2+z^2=C_2$.

Since $z=0$, $y=2x$, we shall have

$$C_1=3x,\quad C_2=5x^2=\frac{5}{9}C_1^2.$$

Eliminating C_1, C_2 from the three relations

$$x+y+z=C_1,\quad x^2+y^2+z^2=C_2,\quad 5C_1^2=9\,C_2$$

we get the required integral surface

$$5(x+y+z)^2=9(x^2+y^2+z^2).$$

Example 3. Solve the Cauchy problem

$$xp+yq=kz$$

$$\text{IC}:\ z(x, 1)=F(x).$$

Solution. The initial curve Γ is given by

$$y=1,\quad z=F(x).$$

Γ can be parametrically represented by the equations $x=s$, $y=1$, $z=F(s)$. (In fact x itself could be taken as a parameter).

The subsidiary equations are

$$\frac{dx}{x}=\frac{dy}{y}=\frac{dz}{kz}(=dt \text{ say}).$$

Here t is another parameter. Solving we get

$$x=\mathrm{A}\,e^t,\quad y=\mathrm{B}\,e^t,\quad z=\mathrm{C}\,e^{kt}$$

The integral surfaces are thus parametrically given by

$$x(t, s)=\mathrm{A}\,e^t,\quad y(t, s)=\mathrm{B}\,e^t,\quad z(t, s)=\mathrm{C}\,e^{kt}.$$

Since on Γ $t=0$, we have

$$x(0, s)=\mathrm{A}=s,\quad y(0, s)=\mathrm{B}=1,\quad z(0, s)=\mathrm{C}=\mathrm{F}(s).$$

The required integral surface is, therefore,

$$x=se^t,\quad y=e^t,\quad z=\mathrm{F}(s)\,e^{kt}.$$

Eliminating s and t we get the Cartesian equation of the surface as

$$z=y^k\,\mathrm{F}\left(\frac{y}{x}\right)$$

Example 4. Examine the solvability of the Cauchy problem

$$p+zq=1$$

$$\text{IC}:\quad z=x+1,\quad y=\tfrac{1}{2}x^2+x,$$

Solution. From the characteristic equations

$$\frac{dx}{1}=\frac{dy}{z}=\frac{dz}{1}$$

we get the characteristic curves $z-x=\mathrm{C}_1$, $y=\frac{1}{2}z^2+\mathrm{C}_2$.

We can write
$$y=\tfrac{1}{2}(x+\mathrm{C}_1)^2+\mathrm{C}_2$$
$$=\tfrac{1}{2}x^2+\mathrm{C}_1x+\tfrac{1}{2}\mathrm{C}_1^2+\mathrm{C}_2.$$

For $\mathrm{C}_1=1$, $\mathrm{C}_2=-\frac{1}{2}$, we get a characteristic curve

$$z=x+1,\quad y=\tfrac{1}{2}x^2+x.$$

Since the initial curve is a characteristic curve, there is no unique solution. An infinity of integral surfaces pass through the initial curve.

Exercise 23

1. Find the general solution of the following pde's :
   (*i*) $(y^2+z^2)\,p-xyq=-xz.$
   (*ii*) $z-xp-yq=a\sqrt{x^2+y^2+z^2}.$
   (*iii*) $(y+zx)\,p-(x+yz)\,q=x^2-y^2$
   (*iv*) $2xyp+(y^2-x^2+1)\,q=x^2\,y.$
   (*v*) $xp+2yq=2\left(z-\dfrac{x^2}{y}\right).$

2. Solve the Cauchy problems :
   (*i*) $p-q=1$, IC : $y=x^2$, $z=x^3$.
   (*ii*) $x(y^2+z)\,p-y(x^2+z)\,q=(x^2-y^2)\,z.$
   IC : $x+y=0$, $z=1$.
   (*iii*) $yp-xq=0$, IC : $z=1$, $x^2+y^2=4$.

(*iv*) $z(xp-ydy)=y^2-x^2$, IC : $z=0$, $x^2+y^2=1$.

(*v*) $(x+2)\,p+2yq=2z$, IC : $z(-1, y)=\sqrt{y}$.

3. Integrate the following equations :

(*i*) $(y^2+z^2-x^2)\,p-2xy\,q+2xz=0$.

(*ii*) $(y^3\,x-2x^4)\,p+(2y^4-x^3\,y)\,q=9z\,(x^3-y^3)$.

(*iii*) $(x_3-x_2)\,p_1+x_2\,p_2-x_3\,p_3=x_2(x_1+x_3)-x_2^2$.

(*iv*) $p_1\,x_1+p_2\,x_2+p_3\,x_3=z+\dfrac{x_1\,x_2}{x_3}$.

4. Prove that the elimination of the functions f, g from

$$z=f(2x+y)+g(x-y)-xy$$

leads to the second order PDE $r-s-2t=1$

7·4. Non-linear Equations of the First Order Standard Forms

Complete integrals, as is verified by direct substitution, can be found in the following canonical cases :

Standard I. $F(p, q)=0$

$z=ax+by+C$ is a complete integral where a, b, c are arbitrary constants and a, b are related by $F(a, b)=0$.

No singular solution exists.

Standard II. $f(x, p)=g(y, q)$.

Put $f(x, p)=g(y, q)=a$ where a is an arbitrary constant. Solve for p, q if it is possible. Let $p=\phi_1(x, a)$, $q=\phi_2(y, a)$. Then

$$dz=\phi_1(x, a)\,dx+\phi_2(y, a)\,dy.$$

Hence $z=\int\phi_1(x, a)\,dx+\int\phi_2(y, a)\,dy+b$ is the required solution.

No singular solution exists.

Standard III. $F(z, p, q)=0$.

Put $x=x+ay$ where a is an arbitrary constant. The given equation will change to the ordinary differential equation

$$F\left(z, \frac{dz}{du}, a\,\frac{dz}{du}\right)=0.$$

A complete integral will be of the form

$$z=\phi(x+ay, a, b).$$

A singular solution may exist.

Standard IV. (Clairaut's Form).

$$z=px+qy+f(p, q).$$

The complete integral is

$$z=ax+by=f(a, b).$$

A singular solution may exist.

Transformation of Variables

Sometimes a suitable change of variables transforms a non-linear *pde* to some standard form. Though comprehensive rules of change cannot be given, the following devices are suggestive :

(*i*) If the equation is $F(x^m p, y^n q)=0$, the substitutions $X=x^{1-m}$, $Y=y^{1-n}$ change it to $F(P, Q)=0$

where $$P\equiv\frac{\partial z}{\partial X},\ Q\equiv\frac{\partial z}{\partial Y}.$$

(*ii*) If $k\neq -1$, the substitution $Z=z^{k+1}$ changes the *p.d.e.* $F(z^k p, z^k q)=0$ to $F(P, Q)=0$,

where $$P=\frac{\partial Z}{\partial x}=(k+1)\, z^k p,\quad Q=\frac{\partial Z}{\partial y}=(k+1)\, z^k q.$$

In case $k=-1$, substitute $Z=\log z$.

(*iii*) A combination of (*i*) and (*ii*) will change $(F(x^m z^k p, y^n z^k q)=0$ to $F(P, Q)=0$.

(*iv*) The use of devices (*i*) and (*ii*) will change $F(z, x^m p, y^n q)$ 0 to the form $F(z, P, Q)=0$.

Example 5. Solve completely the equation

$$q^2=z^2 p^2(1-p^2).$$

Solution. If $x+ay=u$, where a is an arbitrary constant, the equation changes to

$$a^2\left(\frac{dz}{du}\right)^2-z^2\left(\frac{dz}{du}\right)^2\left\{1-\left(\frac{dz}{du}\right)^2\right\}=0. \quad ...(i)$$

Hence either $\frac{dz}{du}=0$ giving $z=b_1$,

or $$a^2-z^2\left\{1-\left(\frac{dz}{du}\right)^2\right\}=0. \quad ...(ii)$$

(*ii*) yields the complete integral

$$z^2-a^2=(x+ay+b)^2.$$

To find the general solution put $b=\phi(a)$ where ϕ is an arbitrary function. Differentiating with respect to a, we get

$$-a=\{(x+ay+\phi(g)\}\,\{y+\phi'(a)\}. \quad ...(iii)$$

The eliminant of a from (*ii*) and (*iii*) is the general solution.

To find the singular solution write the complete integral in the form $$f(x, y, z, a, b)=z^2-a^2-(x+ay+b)^2=0. \quad ...(iv)$$

Then,

$$\frac{\partial f}{\partial a}=-2a-2y(x+ay+b)=0 \quad ...(v)$$

$$\frac{\partial f}{\partial b}=-2(x+ay+b)=0. \quad ...(vi)$$

Elimination of a and b from (*iv*), (*v*), (*vi*) leads to the singular solution $z=0$.

Example 6. Solve the equation

$$p^m \sec^{2m} x+z^l q^n \operatorname{cosec}^{2n} y=z^{\frac{lm}{m-n}}.$$

Solution. The equation can be written as

$$\left(z^{\frac{-l}{m-n}} \frac{\partial z}{\partial x} \sec^2 x\right)^m+\left(z^{\frac{-l}{m-n}} \frac{\partial z}{\partial y} \operatorname{cosec}^2 x\right)^n=1.$$

Put $dZ=z^{\frac{-l}{m-n}}\, dz$, $dX=\cos^2 x\, dx$, $dy=\sin^2 y\,dy$,

i.e., $$Z=\frac{m-n}{m-l-n}\, z^{\frac{m-l-n}{m-n}},\quad X=\tfrac{1}{4} \sin 2x+\tfrac{1}{2} x,$$

$$Y=-\tfrac{1}{4} \sin 2y+\tfrac{1}{2} y.$$

Also put $P=\dfrac{dZ}{dX}$, $Q=\dfrac{dZ}{dY}$

Then the given equation is transformed to $P^m+Q^n=1$

The complete integral is $Z=aX+bY+c$

where a, b, c are arbitrary constants bound by the relation $a^m+b^n=1$.

Substituting for X, Y, Z, we get the desired integral.

There is no singular solution.

Exercise 24

Solve completely the equations :

1. $pq=xy$
2. $pq=x^m y^n z^{2l}$
3. $x^2 p^2+y^2q^2=z^2$
4. $z^2 (p^2+q^2)=x+y$
5. $px+qy=q(px+z)^2$ (Put $Z=xz$)
6. $9(p^2z+q^2)=4$
7. $z=px+qy-2p^2-3q^2$
8. $z=px+qy$
9. $px-qy=(z-xy)^2$ (Put $z-xy=Z$)
10. $xp+2yq=2\left(z-\dfrac{x^2}{y}\right)^2$ $\left(\text{Put } z-\dfrac{x^2}{y}=Z\right)$

§7·5. Charpit's General Method of Solution

Given a general equation of the first order

$$F(x, y, z, p, q)=0 \qquad ...(i)$$

The integral of (i) is a relation, as yet unknown, in x, y, z such that the values of z, p, q derived from it when substituted in (i) make it an identity. Suppose there is aoother relation

$$G(x, y, z, p, q)=0 \qquad ...(ii)$$

which also is unknown, such that the values of p, q derived from (ii) are the same as those in (i). We can solve (i) and (ii) for p and q provided the Jacobian $\dfrac{\partial(F, G)}{\partial(p, q)}=\dfrac{\partial F}{\partial p}\dfrac{\partial G}{\partial q}-\dfrac{\partial F}{\partial q}\dfrac{\partial G}{\partial p}\neq 0$. We assume this condition in what follows. If these values of p, q in terms of x, y, z make the Pfaffian equation

$$dz=pdx+qdq \qquad ...(iii)$$

exact, then we can integrate it and obtain an interal of (i).

The condition of integrability (§6·1) of (iii) is

$$p\frac{\partial q}{\partial z}-q\frac{\partial p}{\partial z}+\frac{\partial q}{\partial x}-\frac{\partial p}{\partial y}=0 \qquad ...(iv)$$

The four derivatives in (iv) are obtained from (i) and (ii). Differentiate (i) and (ii) with respect to x, we get.

$$\frac{\partial F}{\partial x}+\frac{\partial F}{\partial p}\frac{\partial p}{\partial x}+\frac{\partial F}{\partial q}\frac{\partial q}{\partial x}=0$$

$$\frac{\partial G}{\partial x}+\frac{\partial G}{\partial q}\frac{\partial p}{\partial x}+\frac{\partial G}{\partial q}\frac{\partial q}{\partial x}=0,$$

whence $$\frac{\partial q}{\partial x}=-\frac{\partial(F, G)}{\partial(p, x)}\bigg/\frac{\partial(F, G)}{\partial(p, q)}$$

Now differentiate (i) and (ii) with respect to y and solve for $\dfrac{\partial p}{\partial y}$ to get

$$\frac{\partial p}{\partial y}=-\frac{\partial(F, G)}{\partial(y, q)}\bigg/\frac{\partial(F, G)}{\partial(p, q)}.$$

Differentiation with respect to z leads to

$$\frac{\partial p}{\partial z}=-\frac{\partial(F, G)}{\partial(z, q)}\bigg/\frac{\partial(F, G)}{\partial(p, q)},$$

$$\frac{\partial q}{\partial z}=-\frac{\partial(F, G)}{\partial(p, z)}\bigg/\frac{\partial(F, G)}{\partial(p, q)}.$$

Substitute these computed derivatives into the integrability condition (iv) and multiply throughout by the Jacobian $\dfrac{\partial(F, G)}{\partial(p, q)}(\neq 0)$. After rearrangement we shall get

$$\frac{\partial F}{\partial p}\frac{\partial G}{\partial x}+\frac{\partial F}{\partial q}\frac{\partial G}{\partial y}+\left(p\frac{\partial F}{\partial p}+q\frac{\partial F}{\partial q}\right)\frac{\partial G}{\partial z}$$

$$-\left(\frac{\partial F}{\partial x}+p\frac{\partial F}{\partial z}\right)\frac{\partial G}{\partial p}-\left(\frac{\partial F}{\partial y}+q\frac{\partial F}{\partial z}\right)\frac{\partial G}{\partial q}=0 \qquad ...(v)$$

The subsidiary equations of (v) are

$$\frac{dx}{\frac{\partial F}{\partial p}}=\frac{dy}{\frac{\partial F}{\partial q}}=\frac{dz}{p\frac{\partial F}{\partial q}+q\frac{\partial F}{\partial q}}=\frac{dp}{-\frac{\partial F}{\partial x}-p\frac{\partial F}{\partial z}}=\frac{dq}{-\frac{\partial F}{\partial y}-q\frac{\partial F}{\partial z}} \quad ...(8)$$

One integral of these equations (say U$=a$) which involves p or q or both and which is functionally distinct from F$=0$, *i.e.* $\frac{\partial (F, U)}{\partial (p, q)}\neq 0$, can be taken as the equation G $(x, y, z, p, q)=0$.

Determine $p=p(x, y, z, b)$, $q=q(x, y, z, Q)$ from the system

$$F(x, y, z, p, q)=0$$
$$U(x, y, z, p, q)=$$

Substituting into $dz=p(x, y, z, a)dx+q\ (x, y, z, a)\ dy$,

We get an equation integrable by a single relation

$$\phi\ (x, y, z, a, b)=0 \quad ...(vi)$$

which is a complete integral of the given equation.

The subsidiary equations (8) are often called *Charpits equations* and the method of integration *Charpit's method.*

Remark 1. Charpit's method is involved and should be resorted to only when a standard form is not possible.

Example 7. Solve by Charpit's method the equation $xzq^2=p$.

Solution. $F=xzq^2-p$.

The subsidiary equations are

$$\frac{dx}{-1}=\frac{dy}{2xzq}=\frac{dp}{-zq^2-xpq^2}=\frac{dq}{-xq^3}$$

$$=\frac{dz}{2xzq^2-p}$$

Since $2\ xz\ q^2-p=xzq^2$,

$$\frac{dz}{xzq^2}=\frac{dq}{-xq^3},$$

i.e.,

$$\frac{dz}{z}=-\frac{dq}{q}.$$

Hence $qz=a$. The function $G=qz-a=0$.

$$p=xz\frac{a^2}{z^2}=a^2\frac{x}{z}.$$

Hence

$$dz=a^2\frac{x}{z}\,dx+\frac{a}{z}\,dy.$$

Integrating we get the complete integral

$$z^2=a^2x^2+2ay+b.$$

There is no sigular integral.

[The problem can be solved by reducing the equation to the form $2P=zq^2$ by putting $x^2=X$, $P=\frac{\partial z}{\partial Z}$].

Exercise 25

Solve by Charpit's method the equations (in case possible solve by reducing to standard form) :

1. $yzp^2-q=0$.
2. $p^2+qy-z=0$.
3. $p^2+q^2=z$.
4. $z^2=pq\,xy$.
5. $2zq^2-y^2p+y^2q=0$.

Cauchy Problem for F(x, y, z, p, q)=0.

Let us recall the various types of first-order equations ;

(*i*) *Linear.* $P(x, y)p+Q(x, y)q=R(x, y)$.

(*ii*) *Semilinear.* $P(x, y)p+Q(x, y)q=R(x, y, z)$,

(*iii*) *Quasilinear.* $P(x, y, z)p+Q(x, y, z)q=R(x, y, z)$.

(*iv*) *Non-linear.* $F(x, y, z, p, q)=0$.

where p and/or q appear in more than 1 degree.

In the linear, semilinear and quasilinear cases the equation is of the form

$$Pp+Qq=R \qquad \ldots(i)$$

Let $z=f(x, y)$ be an integral surface of (*i*). Through every point of an integral surface there is a unique normal $(p, q, -1)$ perpendicular to the tangential direction (P, Q, R). An integral surface is uniquely fixed if it is made to pass through a prescribed non-characteristic (initial) curve C.

Given a fully non-linear equation

$$F(x, y, z, p, q)=0 \qquad \ldots(ii)$$

In this case at a given point $P(x, y, z)$ of an integral suface of (*ii*), a normal $N(p, q, -1)$ is not uniquely determined, rather a single infinity of normals $(p, q, -1)$ pass through (x, y, z) where p, q satisfy (*ii*). At each point of an integral surface we thus have a *cone of normals* at P. The geometrial meaning of the equation (*ii*) is that at each point P of any of its integral surface, the normals would generate a cone N(P), (see Fig. 7·1). Equation to the cone at (x, y, z) will be

$$F\left(x, y, z, \frac{X-x}{z-Z}, \frac{Y-y}{z-Z}\right)=0.$$

In the case of a Cauchy problem connected with a non-linear equation (*ii*), the determination of a unique integral surface is a bit more complicated than in case (*i*). To determine the unique integral surface which passes through a given initial curve C, not only should we determine the normal cone at a point of C, but

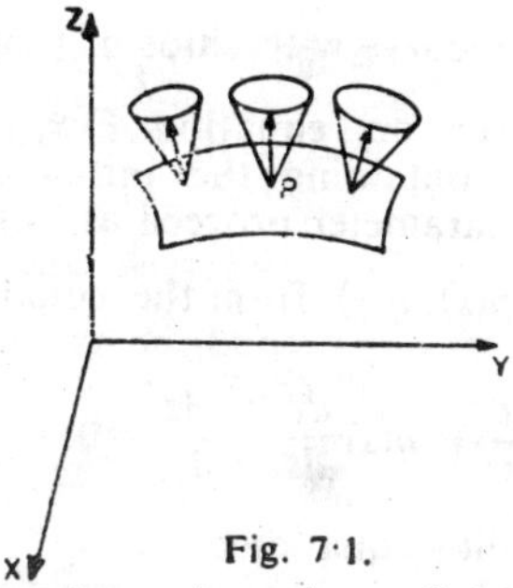

Fig. 7·1.

we must also determine the orientation of the tangent plane at all points of C. If the curve C is given by parameter t, then the p, q in the given equation are such that

$$\frac{dz}{dt}=p\frac{dx}{dt}+q\frac{dy}{dt} \qquad \text{...}(iii)$$

The equation (*iii*) is called the *strip condition*. The Cauchy problem in a genuinely non-linear case cannot be posed for an initial curve only, but rather with an initial curve together with the strip condition. An initial curve C with the strip condition is called an *initial strip*. By virtue of the strip condition the tangential chips are smoothey joined together to form the integral surface along C (See Fig. 7·2).

The tangent planes at P (x, y, z) with running coordinates X, Y, Z.

$$p(X-x)+q(Y-y)=Z-z \qquad \text{...}(iv)$$

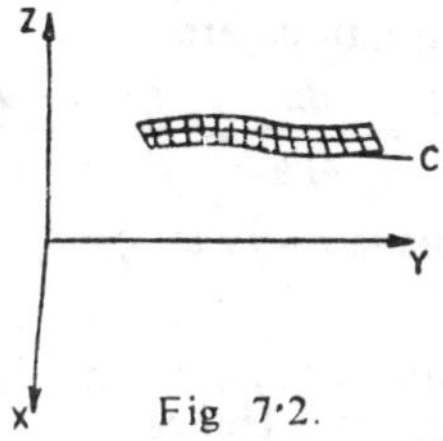

Fig 7·2.

by virtue of relation (*ii*) form a one-parameter family and envelope another cone T(P) at P. This cone is called the *Monge cone* at P. In the quasi-linear case T(P) degenerates into the tangent line at P and N (P) degeneretes into a plane (Prove).

In the quasilinear case there is only one characteristic direction (P, Q, R)at each point C. In the general case there is whole Monge cone of characteristic directions. Each integral surface through a point P is tangent to the monge cone T (P) at P and they touch along a characteristic curve. A characteristic curve together with p, q, satisfying the strip condition is called a *characteristic strip*. If by charpit's equations x, y, z, p, q are determined as functions of a parameter t, then $x=p(t)$, $y=y(t)$ $z=z(t)$ is a characteristic curve which together with the tangent plane $p(X-x)+q(Y-y)=Z-C)$ is a characteristic strip.

Definition 1. A *strip* is a set of five functions $\{x(t), y(t), z(t). p(t), z(t)\}$ which are bound by the *strip condition* $z'=px'+qy'$,

Intuitively, a strip is a curve with chips of tangent planes attached.

Summary. Given an equation $F(x, y, z, p, q)=0$. To find an integral surface containing the initial curve $x=x(s)$, $y=y(s)$ $z=z(s)$ where s is a parameter proceed as follows :

We determine $p(s)$, $q's)$ from the equations $F[x(s), y(s), z(s), p(s), q(s)]=0$.

$$p(s)\frac{dx}{ds}+q(s)\frac{dy}{ds}-\frac{dz}{ds}=0$$

Solve the Charpit's equations

and
$$\frac{dx}{F_p}=\frac{dy}{F_q}=\frac{dz}{pF_p+qF_q}=\frac{dp}{-F_x-pF_z}=\frac{dq}{-F_y-qF_z}=dt$$

where the initial conditions for $t=0$ are $x(0, s)=x(s)$, $y(0, s)=y(s)$, $z(0, s)=z(s)$, $p(0, s)=p(s)$, $q(0, s)=q(s)$.

The solution $x=x(t, s)$, $y=y(t, s)$ $z=z(t, s)$ thus obtained represents the integral surface of the Cauchy problem.

Example 8. Find the integral surface of the equation $z=pq$ passing through the curve C. $x=1$, $y=z$.

Solution. The parametric equations of the initial curve are $x=1$, $y=s$, $z=s$.

From the strip condition $z'=px'+qy'$ we get $1=q$.

Solving $z=pq$, $1=q$ for p and q we get $p=s$, $q=1$.

The Charpit's equations are

$$\frac{dx}{q}=\frac{dy}{p}=\frac{dz}{2pq}=\frac{dp}{p}=\frac{dq}{q}=dt.$$

Integrating we have $p=c_1 e^t$, $q=c_2 e^t$, $x=c_2 e^t+c_3$, $y=c_1 e^t+c_4$ $z=c_1 c_2 e^{2t}+c_5$.

For $t=0$, we have $p=s$, $q=1$, $x=1$, $y=s$, $z=s$. Henee $c_1=s$, $c_2=1$, $c_3=0$, $c_4=0$, $c_5=0$. Hence

$$p=s\,e^t,\ q=e^t,\ x=e^t,\ y=s\,e^t,\ z=s\,e^{2t}.$$

The required integral surface is

$$x=e^t,\ y=s\,e^t,\ z=s\,e^{2t},$$

i.e.,
$$z=xy.$$

Example 9. Solve the Cauchy problem

$$pq=xy$$

$$\text{IC}: x=1,\ y^3=z^2$$

Solution. The parametric equations of the initial curve are $x=1$, $y=s^2$, $z=s^3$.

The strip condition $z'=px'+qy'$ gives $q=\frac{3}{2}s$. Solving $pq=xy$ and $q=\frac{3}{2}s$ for p and q we get $p=\frac{2}{3}s$, $q=\frac{3}{2}s$

The Charpit's equations are

$$\frac{dx}{q}=\frac{dy}{p}=\frac{dz}{2pq}=\frac{dp}{y}=\frac{dq}{x}=dt \qquad \ldots(i)$$

Solving we get $p^2=y^2+c_1$, $q^2=x^2+c_2$. Also $\frac{dx}{\sqrt{x^2+c_2}}=\frac{dy}{\sqrt{y^2+c_1}}=dt$ yield

$$x=\sqrt{c_2}\sin h\,(t+c_3),$$
$$y=\sqrt{c_1}\sin h\,(t+c_4).$$

For $t=0$ we shall have

$$p=\frac{2}{3}s,\quad q=\frac{3}{2}s,$$
$$x=1,\ y=s^2,\ z=s^3.$$

Hence
$$c_1=\frac{4}{9}s^2-s^4,\qquad c_2=\frac{9}{4}s^2-1,$$
$$1=\sqrt{c_2}\sinh c_3$$
$$s^2=\sqrt{c_1}\sinh c_4.$$

Hence
$$\cosh c_3=\sqrt{\frac{1+c_2}{c_2}}=\frac{\frac{3}{2}s}{\sqrt{c_2}},\ \sinh c_3=\frac{1}{\sqrt{c_2}}$$
$$\cosh c_4=\frac{\sqrt{s^4+c_1}}{\sqrt{c_1}},\ \sinh c_4=\frac{s^2}{\sqrt{c_1}}.$$

Also
$$x=\sqrt{c_2}\,[\sinh t\cosh c_3+\cosh t\sinh c_3]$$
$$=\cosh t+\frac{3}{2}s\sinh t \qquad \ldots(ii)$$
$$y=\sqrt{c_1}[\sinh t\cosh c_4+\cosh t\sinh c_4]$$
$$=s^2\cosh t+\frac{2}{3}s\sinh t \qquad \ldots(iii)$$

Hence
$$p^2=y^2+c_1=(s^2\sinh t+\tfrac{2}{3}s\cosh t)^2,$$

i.e.,
$$p=s^2\sinh t+\tfrac{2}{3}s\cosh t.$$

Similarly $q=\sinh t+\frac{3}{2}s\cosh t$

On the characteristic curves represented by the Charpit's equations (*i*), t is the parameter.

$$\frac{dz}{dt}=2p\frac{dx}{dt}$$
$$=2(s^2\sinh t+\tfrac{2}{3}s\cosh t)\left(\sinh t+\frac{3}{2}s\cosh t\right)$$
$$z=s^2\ \sinh 2t+\left(\frac{s}{3}+\frac{3}{4}s^3\right)\cosh 2t+A$$

For $t=0,\ z=s^3$,

hence $$A=\frac{s^3}{4}-\frac{s}{3}$$

Thus $$z=\frac{s^3}{4}-\frac{s}{3}+s^2\sinh 2t+\left(\frac{s}{3}+\frac{3}{4}s^3\right)\cosh 2t \quad \ldots(iv)$$

Relations (*ii*), (*iii*) and (*iv*) represent the parametric equations of the integral surface of the problem.

Exercise 26

Solve the Cauchy problems :

1. $p^2+q^2=2$, IC : $x=0,\ y=z$.
2. $p^2+q^2=4z$, IC : $x=0,\ z=y^2$.
3. $z=pq+1$, IC : $y=2,\ z=2x+1$.

§7·6. Partial Differential Equations of the Second Order : Monge's Equation

A general second order PDE can be expressed as

$$F(x, y, z, p, q, r, s, t)=0 \quad \ldots(1)$$

If one or more of the second order derivatives r, s, t appear in more than one degree, it is called *non-linear*.

An equation of the form

$$Rr+Ss+Tt+Pp+Qq+Zz=G \quad \ldots(2)$$

where R, S, T are functions of x, y, z, p, q is called a *quasi-linear equation*. It is called *semi-linear* if R, S, T are functions of x, y, z only. If all coefficients R, S, T, P, Q, Z and G are functions of x, y only, then (2) is called a *linear equation*.

A *complete integral* of the equation (1) is the most general relation possible between x, y, z such that, when the value of z derived from it and the associated differential coefficients are substituted in the equation, the later becomes an identity. The complete integral may involve either arbitrary constants or arbitrary functions or both.

An *intermediate integral* is a relation in the form of a first order PDE from which the given equation can be deduced.

For example, let

$$u=f(v) \quad \ldots(3)$$

where u, v are given functions of x, y, z, p, q and f is an arbitrary function.

Differentiating relation (3) with respect to x, y we get respectively.

$$\frac{\partial u}{\partial x}+p\frac{\partial u}{\partial z}+r\frac{\partial u}{\partial p}+s\frac{\partial u}{\partial q}=\frac{\partial f}{\partial v}\left(\frac{\partial v}{\partial x}+p\frac{\partial v}{\partial z}+r\frac{\partial v}{\partial p}+s\frac{\partial v}{\partial q}\right)$$

$$\frac{\partial u}{\partial y}+q\frac{\partial u}{\partial z}+s\frac{\partial u}{\partial p}+t\frac{\partial u}{\partial q}$$
$$=\frac{\partial f}{\partial v}\left(\frac{\partial v}{\partial y}+q\frac{\partial v}{\partial z}+s\frac{\partial v}{\partial p}+t\frac{\partial v}{\partial q}\right)$$

Eliminating $\frac{\partial f}{dv}$ we shall have an equation of the form

$$Rr+Ss+Tt+U(rt-s^2)=V \qquad ...(4)$$

Equation (4) is called *Monge's non-linear equation.*

$u=f(v)$ is an *intermediate integral* for equation (G).

In the special case $U=0$, Monge's equation is

$$Rr+Ss+Tt=V \qquad ...(5)$$

which we shall discuss in this section.

Problem. Let $z=f(u)+g(v)$ where u, v are given functions of x, y and f, g are arbitrary functions. Then prove that z satisfies a linear equation of the form

$$Rr+Ss+Tt+Pp+Qq=V$$

Find the PDE for the relation $z=f(x+ay)+g(x-ay)$ f, g arbitrary.

Special Forms Reducible to ODE or Lagrange's Form

Consider the equations

(*a*) $r+M(x)\,p=N(x)$
(*b*) $s+M(x)\,p=N(x)$
(*c*) $s+M(y)\,q=N(y)$
(*d*) $t+M(y)\,q=N(y)$
(*e*) $Ar+2Bs+Cp=G$
(*f*) $2Bs+Ct+Eq=G$

Equations (*a*), (*b*), (*c*) can be reduced to ODE and (*e*), (*f*) can be reduced to a first order PDE of Lagrange's form.

Example 10. $xys-px=y^2$.

Solution. The equation can be written as

$$\frac{\partial p}{\partial y}-\frac{1}{y}\,p=\frac{y}{x}$$

which is a linear ordinary equation with p as dependent and y independent variable.

$$\text{I.F.}=e^{-\int\frac{dy}{y}}=\frac{1}{y}$$

Hence $\qquad \dfrac{p}{y}=\dfrac{y}{x}+\psi(x)$

i.e., $\qquad p=\dfrac{y^2}{x}+y\psi(x)$

where ψ is an arbitrary function.

Integrating further,

$$z=y^2 \log x+y\,\phi_1(x)+\phi_2(y)$$

where ϕ_1, ϕ_2 are arbitrary functions.

Example 11. $ys-2xr-2p=6xy$.

Solution. The equation can be written as

$$2x\frac{\partial p}{\partial x}-y\frac{\partial p}{\partial y}=-6xy-2p,$$

which is the Lagrange's form in p.

The auxiliary equations are

$$\frac{dx}{2x}=\frac{dy}{-y}=\frac{dp}{-6xy-2p}$$

$$\frac{dx}{2x}=\frac{dy}{-y} \Rightarrow xy^2=c_1$$

$$\frac{dy}{y}=\frac{dp}{6xy+2p} \Rightarrow \frac{dy}{y}=\frac{ydp}{6c_1+2py}$$

$$\Rightarrow \frac{dp}{dy}-\frac{2}{y^2}p=\frac{6c_1}{y}$$

$$\Rightarrow \frac{p}{y^2}=\frac{-2c_1}{y^3}+\phi_1(x)$$

$$\Rightarrow p=-\frac{2c_1}{y}+\phi_1(x).y^2$$

$$\Rightarrow \frac{\partial z}{\partial x}=-2xy+\phi_1(x)\,y^2.$$

Hence finally

$$z=-x^2y+\phi(x)\,y^2+\psi(y).$$

Problem. Solve the equations :

1. $2xr-ys+2p=4xy^2$.
2. $xs+yt+q=10x^3y$
3. $xyr+x^2s-yp=(x^2-y)\,e^y$

4. $xs-2x^2t=1$.
5. $xr-ys+p=y^2$.

Answers

1. $z=x^2y^2+\phi_1(xy^2)+\phi_2(y)$.
2. $z=x^3y^2+\phi_1(y/x)+\phi_2(x)$.
3. $z=\phi_1(x^2-y^2)+\phi_2(y)+xe^y$.
4. $z=\phi_1(x^2+y)+\phi_2(x)+xe^y$.
5. $z=-x^2y+\phi_1(xy)+\phi_2(y)$.

Monge's Method

Given the equation

$$Rr+Ss+Tt=V \qquad \ldots(5)$$

where R, S, T, V are functions of x, y, z, p, q.

We first determine an intermediate integral

$$u=f(v) \qquad \ldots(6)$$

for the equation (5).

Now r and t can be obtained in terms of s from the relations

$$dp=rdx+sdy$$

$$dq=sdx+tdy$$

and after substitution in (5), we get the *subsidary equations*.

$$Rdy^2-Sdx\,dy+Tdx^2=0 \qquad \ldots(7)$$

$$Rdydx+Tdx\,dy-Vdx\,dy=0 \qquad \ldots(8)$$

A solution of the subsidiary equations is also a solution of (5).

Let $\frac{dy}{dx}=\lambda_1$ and $\frac{dy}{dx}=\lambda_2$ be solutions of (7) where λ_1, λ_2 are functions of x, y, z, p, q. Making the substitutions in (8) we shall have the pair of equations

$$dy=\lambda_1 dx,\ \lambda_1 Rdp+Tdq-\lambda_1 Vdx=0 \qquad \ldots(9)$$

$$dy=\lambda_2 dx,\ \lambda_2 Rdy+Tdq-\lambda_2 Vdx=0 \qquad \ldots(10)$$

Let $u_1=c_1$, $v_1=c_2$ be two integrals of the total differential equation (9) and $u_2=c_3$, $v_2=c_4$ integrals of (10) where u_1, v_1, u_2, v_2 are functions of x, y, z, p, q.

Hence we get two intermediate integrals of (5).

$$u_1=\phi_1(v_1) \qquad \ldots(11)$$

$$u_2=\phi_2(v_2) \qquad \ldots(12)$$

Solving equations (11), (12) for p, q and substituting in

$$dz=pdx+qdy$$

we get after integration the required solution. In case $\lambda_1=\lambda_2$, we get only one intermediate integral.

Example 12. $q^2r-2pqs+p^2t=pq^2z.$

Solution. $R=q^2$, $S=-2pq$, $T=p^2$, $V=p^2qz$.

The subsidiary equations are

$$Rdy^2-Sdxdy+Tdx^2=0$$

$$Rdy\ dp+Tdxdq-Vdxdy=0$$

i.e., $$q^2dy^2+2pqdydx+p^2dx=0 \quad \ldots(i)$$

$$q^2dydp+p^2dxdq-p^2qzdxdy=0 \quad \ldots(ii)$$

From (i) $$(pdx+qdy)^2=0 \quad \ldots(iii)$$

i.e. $$dz=0$$

$$z=A$$

Here $\lambda_1=\lambda_2=-\dfrac{p}{q}$.

From (ii) and (iii), we get

$$-pqdp+p^2dq+Ap^3dx=0$$

i.e., $$\frac{qdp-pdq}{p^2}=Adx.$$

Integrating further

$$\frac{q}{p}=Ax+\phi(z)=zx+\phi(z).$$

Hence an intermediate integral is

$$[zx+\phi(z)]p-q=0 \quad \ldots(iv)$$

The characteristic equations for (iv) are

$$\frac{dx}{zx+\phi(z)}=\frac{dy}{-1}=\frac{dz}{0}$$

Hence $z=A$ as before.

Also $\dfrac{dx}{Ax+\phi(A)}=\dfrac{dy}{-1}$ gives

$$\log[Ax+\phi(A)]=-Ay+B(z)$$

i.e., $$zx+\phi(z)=-\psi(z)\,e^{-yz}.$$

Example 13. $qr-ps=0$.

Solution. $R=q$, $S=-p$, $T=0$, $V=0$.

Subsidiary equations are

$$Rdy^2 - Sdxdy + Tdx^2 = 0$$

$$Rdy\,dp + Tdx\,dq - Vdx\,dy = 0$$

i.e., $$qdy^2 + pdx\,dy = 0 \quad \text{...}(i)$$

and $$qdy\,dp = 0 \quad \text{...}(ii)$$

$dy=0$ holds for both (i) and (ii).

If $pdx+qdy=dz=0$ then $z=c$

In this case from (ii)

$$dp=0$$

i.e., $$p=B$$

i.e., $$\frac{\partial z}{\partial x}=\phi(z)$$

i.e., $$\psi(z)=x+F(y)$$

Finally the complete integral can be expressed as

$$x=f(y)+g(z)$$

where f, g are arbitrary functions.

Example 14. $(q+1)s=(p+1)t$.

Solution. $R=0,\ S=q+1,\ T=-(p+1),\ V=0$.

Subsidiary equations are

$$Rdy^2 - Sdx\,dy + Tdx^2 = 0$$

$$Rdy\,dp + Tdx\,dq - Vd\,xdy = 0$$

i.e., $$-(q+1)\,dxdy - (p+1)\,dx^2 = 0 \quad \text{...}(i)$$

and $$-(p+1)\,dxdq = 0 \quad \text{...}(ii)$$

$dx=0$ satisfies both (i) and (ii).

If from (i), we take

$$pdx+qdy+dx+dy=0$$

then from (ii), $$dq=0$$

Hence we have

$$dz+dx+dy=0$$

and $$dq=0$$

Hence $$x+y+z=C$$

$$q=D$$

i.e., $$\frac{\partial z}{\partial y}=\phi(x+y+z)$$

i.e., $$z=\phi_1(x+y+z)+\phi_2(x).$$

Example 15. $r-a^2t=0$.

Solution. $R=1,\ S=0,\ T=-a^2,\ V=0$.

Subsidiary equations are

$$R dy^2 - S dx\, dy + T dx^2 = 0$$

$$R dy\, dp + T dx dq - V dx\, dy - 0$$

i.e., $\quad dy^2 - a^2 dx^2 = (dy + a dx)(dy - a dx) = 0 \qquad \ldots(i)$

and $\quad dy\, dp - a^2 dx dq = 0 \qquad \ldots(ii)$

From (*i*) $\quad y + ax = A,\ y - ax = B.$

Equation (*ii*) with $dy + a dx = 0$ gives

$$dp + a dq = 0$$

i.e., $\quad p + aq = C = \phi_1(y + ax) \qquad \ldots(iii)$

Also $\quad dy - a dx = 0 \quad$ gives

$$p - aq = D = \phi_2(y - ax) \qquad \ldots(iv)$$

Solving (*iii*), (*iv*) for p, q.

Substituting in

$$dz = p dx + q dy$$

we have

$$dz = \frac{1}{2}(\varphi_1 + \varphi_2)\, dx + \frac{1}{2a}(\varphi_1 - \varphi_2)\, dy$$

$$= \frac{1}{2a}\phi_1(y + ax)(dy + a dx) - \frac{1}{2a}\phi_1(y - ax)(dy - a dx)$$

Integrating we have finally

$$z = \phi(y + az) + \psi(y - ax)$$

the arbitrary constant of integration being absorbed in either of the arbitary functions ϕ or ψ.

Problem. Solve the following equations by Monge's method :

1. $pt - qs = q^3$.
2. $q^2 r + 2pqs + p^2 t = 0$.
3. $q^2 r - 2pqs + p^2 t = pq^2$.
4. $q^2 r - 2pqs + p^2 t = pt - qs$
5. $q^2 r - 2pqs + p^2 t = -\frac{2}{x} pq^2$.
6. $z(qs - pt) = pq^2$.
7. $q^2 r - 2pqs + p^2 t = 0$.

Answers

1. Intermediate integral (I.I.) : $x - \dfrac{1}{q} = \phi(z)$

   General solution (G.S.) : $y = xz + \phi_1(z) + \phi_2(x)$.
2. I.I. $p = q\phi(z)$; G.S. : $y + x f(z) = F(z)$.
3. I.I. $e^x q - p\phi(z) = 0$; G.S. $e^y + \phi(z)\, y = \psi(z)$.

4. I.I. $p=1+q f(z)$; G.S. $y=e^x\phi_1(z)+\phi_2(z)$.
5. I.I. $x^2p=q\phi(z)$; G.S. $xy=\phi_1(z)+x\phi_2(z)$.
5. G.S. $y=\phi(z)+z\psi(x)$.
6. I.I. $p=q f(z)$; G.S. $y+x f(z)=g(z)$.

§7·7. Classification of Second Order Linear Equations

A general second order partial differential equation can be expressed as

$$F(x, y, z, p, q, r, s, t)=0. \qquad ...(i)$$

If one or more of the second order derivatives r, s, t appear in more than one degree, it is called a *non-linear equation.*

An equation of of the type

$$A.r+2B.s+C.t=f(x, y, z, p, q) \qquad ...(ii)$$

where A, B, C are functions of x, y, z, p, q is called *quasilinear.*

A *linear equation* is of the form

$$A.r+2B.s+C.t+D.p+E.q+F.z=G \qquad ...(iii)$$

where A, B, C, D, E, F, G are functions of x, y only. It is the aim of this article to classify equations of form (*iii*).

The classification depends on the part $A.r+2B.s+C.t$ which is called the *principal part* of (*iii*). If we write $\frac{\partial}{\partial x}\equiv\xi$, $\frac{\partial}{\partial y}\equiv\eta$, the principal part can be written as

$$Q(\xi, \eta)\, z=(A\xi^2+2B\xi\,\eta+C\eta^2)\, z. \qquad ...(iv)$$

The quadratic form $Q(\xi, \eta)$ is called the *characteristic form* of (*iii*) and the equation

$$A\, dy^2+2B\, dx\, dy+C\, dx^2=0 \qquad ...(v)$$

is called its *characteristic differential equation.*

If we put $\zeta=-\frac{\xi}{\eta}$, then $Q(\xi, \eta)=\eta^2\, q(\zeta)$,

where $q(\zeta)=A\zeta^2-2B\zeta+C$.

If $\zeta_+(x, y)$ and $\zeta_-(x, y)$ are the roots of the quadratic equation $q(\zeta)=0$, then two families of curves in the xy-plane are obtained by solving the ordinary differential equations

$$\frac{dy}{dx}=\zeta_+(x, y) \quad \text{and} \quad \frac{dy}{dx}=\zeta_-(x, y).$$

These curves are called the *characteristic curves* of (*iii*).

Definition 2. The P.D.E. (*iii*) is called *elliptic* at a point (x, y) if $B^2-AC<0$ there, *parabolic* if $B^2-AC=0$ and *hyperbolic* if $B^2-AC>0$ at the point.

These classifications are said to hold in a plane region D if they hold at every point of D.

If A, B, C are constants then the type of (*iii*) is the same everywhere,

In the non-linear case the situation is more complicated. As the expression B^2-AC will involve z also, the type of the equation may depend on the particular solution considered. For some solutions it may be elliptic, for some it may be parabolic and for some hyperbolic. Since a solution is not known in advance, the type also cannot be known except in special cases. For example $z^2 r+3pq\, s-z^2 t=0$ is hyperbolic for all solutions z. On the other hand $(1-p^2)\, r-2pq\, s+(1-q^2)\, t=0$ is hyperbolic for those solutions z for which $\sqrt{p^2+q^2}>1$ and elliptic for those for which $\sqrt{p^2+q^2}<1$.

The following are typical examples of these forms :

(*i*) The Laplace's (or potential) equation

$\dfrac{\partial^2 z}{\partial x^2}+\dfrac{\partial^2 z}{\partial y^2}=0$ is elliptic everywhere in the (x, y)-plane.

(*ii*) The heat (or diffusion) equation

$\dfrac{\partial^2 z}{\partial x^2}=\dfrac{1}{C}\dfrac{\partial z}{\partial t}$ is parabolic everywhere in the (x, t)-plane.

(*iii*) The wave equation $\dfrac{\partial^2 z}{\partial x^2}-\dfrac{\partial^2 z}{\partial t^2}=0$ is hyperbolic everywhere in the (x, t)-plane.

For Laplace's equation the characteristic equation is $dy^2+dx^2=0$; hence the characteristic curves are imaginary at all points.

The heat equation has coincident characteristics while the wave equation has real and distinct characteristics.

Reduction to Canonical Forms

By a suitable change of independent variables, it is possible to reduce the given equation to a canonical form.

Let $\xi=\xi(x, y)$, $\eta=\eta(x, y)$. If the Jacobian $\dfrac{\partial(\xi, \eta)}{\partial(x, y)}\neq 0$, we can obtain x, y in terms of ξ, η. The linear equation $A.r+2B.s+C.t=f(x, y, z, p, q)$ will change to

$$A_1\, z_{\xi\xi}+2B_1\, z_{\xi\eta}+C_1\, z_{\xi\eta}=f_1(\xi, \eta, z, z_\xi, z_\eta)$$

where
$$A_1=A\,\xi_x^2+2B\,\xi_x\,\xi_y+C\,\xi_y^2$$
$$B_1=A\,\xi_x\,\eta_x+B(\xi_x\,\eta_y+\xi_y\,\eta_x)+C\,\xi_y\,\eta_y$$
$$C_1=A\,\eta_x^2+2B\,\eta_x\,\eta_y+C\,\eta_y^2.$$

Case 1. $B^2>AC$. In this case $A\lambda^2-2B\lambda+C=0$ has two real and distinct roots λ_1 and λ_2. Hence $A_1=A(\xi_x+\lambda_1\,\xi_y)(\xi_x+\lambda_2\,\xi_y)$,

$$C_1=A(\eta_x+\lambda_1\,\eta_y)\,(\eta_x+\lambda_2\,\eta_y).$$

Choose ξ and η so that $\xi_x+\lambda_1 \xi_y=0$ and $\eta_x+\lambda \eta_y=0$. Then $A_1=C_1=0$. We shall get $\xi=\phi_1(x, y)$, $\eta=\phi_2(x, y)$ as the desired transformations and the equation will change to

$$\frac{\partial^2 z}{\partial \xi \, \partial \eta}=\phi\left(\xi, \eta, z, \frac{\partial z}{\partial \xi}, \frac{\partial z}{\partial \eta}\right).$$

which is the canonical form of a hyperbolic equation.

Case 2. $B^2=AC$. $A \lambda^2-2B \lambda+C$ is a complete square

and
$$A_1=A(\xi_x+\lambda \xi_y)^2$$
$$B_1=A(\xi_x+\lambda \xi_y)(\eta_x+\lambda \eta_y)$$
$$C_1=A(\eta_x+\lambda \eta_y)^2.$$

If we choose η so that $\eta_x+\lambda \eta_y=0$ and let ξ be an arbitrary function independent of η, then $B_1=C_1=0$ and $A_1\neq 0$. We get the cononical form in parabolic case

$$\frac{\partial^2 z}{\partial \xi^2}=\phi_1(\xi, \eta, z, z_\xi, z_\eta).$$

Case 3. $B^2 < AC$. In this case $A \lambda^2-2B \lambda+C=0$ has two conjugate complex roots $\xi=\alpha+i \beta$, $\eta=\alpha-i \beta$.

Proceeding as in case 1 we shall get

$$\frac{\partial^2 z}{\partial \xi \, \partial \eta}=\phi(\xi, \eta, z, z_\xi, z_\eta),$$

and finally we get the canonical form of an elliptic equation

$$\frac{\partial^2 z}{\partial \alpha^2}+\frac{\partial^2 z}{\partial \beta^2}=\phi_2(\alpha, \beta, z, z_\alpha, z_\beta).$$

Example 10. Classify and solve the equation

$$y^2 r-2xy\, s+x^2 t-\frac{y^2}{x} p-\frac{x^2}{x} q=0.$$

Solution. Here $A=y^2$, $B=-xy$, $C=x^2$.
Since $B^2=AC$, the equation is parabolic at all points.
The characteristic differential equation is

$$y^2\, dy^2+2xy\, dx\, dy+x^2\, dx^2=0.$$

This leads to the characteristic curves

$$x^2+y^2=C_1, \quad x^2-y^2=C_2.$$

If we put $\xi=x^2+y^2$, $\eta=x^2-y^2$, the given equation reduces to

$$\frac{\partial^2 z}{\partial \xi^2}=0.$$

This gives $z=\phi_1(\eta)+\xi\, \phi_2(\eta)$
$$=\phi_1(x^2-y^2)+(x^2+y^2)\, \phi_2(x^2-y^2)$$

where ϕ_1, ϕ_2 are arbitrary functions.

Example 11. In the hyperbolic equation

$$\frac{\partial^2 z}{\partial x^2}-\frac{\partial^2 z}{\partial y^2}=0$$

$\lambda^2-1=0$, *i.e.*, $\lambda=1, -1$.

Put $\xi=x+y$, $\eta=x-y$ in the equation. We get $\frac{\partial^2 z}{\partial\xi\,\partial\eta}=0.$

Hence the general solution is $z=\phi_1(x+y)+\phi_2(x-y)$

where ϕ_1, ϕ_2 are arbitrary functions.

Example 12. Classify the following equation and solve it :

$$y(x+y)(r-s)-xp-yq-z=4xy+y^2.$$

Solution. Here $A=y(x+y)$, $B=\frac{-y}{2}(x+y)$, $C=0$

$B^2=y^2(x+y)^2$, $AC=0$. Hence $B^2>0$ at all points. The equation is elliptic.

Since $y(x+y)\,\lambda^2+y(x+y)\,\lambda=0$,

$\lambda=-1, 0.$ Hence $\frac{\partial\xi}{\partial x}-\frac{\partial\xi}{\partial y}=0,$

$$\frac{\partial\eta}{\partial x}=0.$$

Thus $\xi=x-y$, $\eta=y$.

With these substitutions the given equation is changed to

$$\frac{\partial^2 z}{\partial\xi\,\partial\eta}+\frac{1}{\eta}\frac{\partial z}{\partial\xi}+\frac{1}{\xi}\frac{\partial z}{\partial\eta}+\frac{1}{\xi\eta}z=\frac{3\eta}{\xi}-4,$$

i.e., $$\frac{\partial}{\partial\xi}\left(\frac{\partial z}{\partial\eta}+\frac{1}{\eta}z\right)+\frac{1}{\eta}\left(\frac{\partial z}{\partial\eta}+\frac{1}{\eta}z\right)=\frac{3\eta}{\xi}-4.$$

Put $$\frac{\partial z}{\partial\eta}+\frac{1}{\eta}z=u.$$

Then $$\frac{\partial u}{\partial\xi}+\frac{1}{\xi}u=\frac{3\eta}{\xi}-4.$$

Hence $$\xi u=3\eta\,\xi-2\xi^2+\phi(\eta).$$

Also $$\frac{\partial z}{\partial\eta}+\frac{1}{\eta}z=3\eta-2\xi+\frac{1}{\xi}\phi(\eta)$$

Integrating we get

$$\eta z=\eta^3-\xi\eta^2+\frac{1}{\xi}\phi_1(\eta)+\phi_2(\xi)$$

or $$z=-xy+\frac{\phi_1(y)}{x+y}+\frac{\phi_2(x+y)}{y}$$

where ϕ_1, ϕ_2 are arbitrary functions.

Note. The method of solving a P.D.E. adopted in this example is known as *Laplace's method* of change of the independent variables.

Exercise 27

Classify the following P.D.E.'s, reduce to canonical forms and solve where possible using the Laplace's method :

1. $r-x^2t=0.$
2. $r+x^2t=0.$
3. $r+2s+t=0.$

4. $x^2 r + xs - yt = 0.$

5. $(n-1)^2 r - y^{2n} t = ny^{2n-1} q.$

6. $x^2 r - 2xy s + y^2 t + (2x - x^2 y^2) p + xy^3 q - 2x^2 y^4 z = y^2 e^{xy}.$

7. $x^2 r - y^2 t = xp - yq.$

8. $(r-s) x = (t-s) y.$

9. $(r-s) y + (s+t) x + q - p.$

10. $r + \dfrac{2}{x} p - a^2 t = 0.$

11. $\dfrac{\partial^2 z}{\partial x^2} - 2 \sin x \dfrac{\partial^2 z}{\partial x \, \partial y} - \cos^2 x \dfrac{\partial^2 z}{\partial y^2} - \cos x \dfrac{\partial z}{\partial y} = 0.$

12. $y^2 r + x^2 t = 0.$

13. $y^2 r - x^2 t = 0.$

14. $(1-x^2) r - t = 0$

$$\left[\text{Ans. } \frac{2\partial^2 z}{\partial \xi \partial y} = \tan (\xi - \eta) \left\{ \frac{\partial z}{\partial \xi} - \frac{\partial z}{\partial \eta} \right\} \right]$$

15. $y^2 r + 2xys + x^2 t = 0.$

$$\left[\text{Ans. } (\eta^2 - \xi^2) \frac{\partial^2 z}{\partial \eta^2} = \xi \frac{\partial z}{\partial \xi} = 0 \right]$$

16. $xyr - (x^2 - y^2)s - xyt + py - qx = z(x^2 - y^2).$

[Put $x^2 + y^2 = c_1$ $y/x = c_2$]

8

Green's Function : ODE

§8·1. Introduction

The aim of this chapter is to describe the theory of one-dimensional Green's function for a second order ordinary linear differential equation with homogeneous boundary conditions. The use of the Green's function leads to the integral equation formulation of a boundary-value problem.

We are concerned with a self-adjoint equation of the form

$$L(y)=[p(x)\, y']'+q(x)\, y=f(x) \qquad \text{...(1)}$$

where $p(x)$ and $q(x)$ are continuous functions of x on a given interval I : $a \leqslant x \leqslant b$ and $p(x) > 0$ on I.

A second order equation

$$p_0(x)\, y''+p_1(x)\, y'+p_2(x)\, y=g(x) \qquad \text{...(2)}$$

$$p_0(x) > 0 \text{ on I},$$

can be reduced to the form (1) by multiplying both sides by the factor

$$\frac{1}{p_0(x)}\; exp \left(\int_a^x \frac{p_1(t)}{p_0(t)}\, dt \right) \qquad a \in \text{I}.$$

We shall suppose this to have been done and consider form (1) for a second order equation.

Together with equation (1) we shall also be concerned with homogeneous boundary conditions of the following type.

(*i*) **Unmixed boundary conditions.** Boundary conditions are said to be unmixed if they involve the function and its derivative at either $x=a$ or $x=b$, but not at both. For example,

$$B_1 : a_1\, y(a)+a_2\, y'(a)=0, \qquad |\, a_1 \,|+|\, a_2 \,| \neq 0$$

$$B_2 : b_1\, y(b)+b_2 y'(b)=0, \qquad |\, b_1 \,|+|\, b_2 \,| \neq 0$$

are unmixed homogeneous boundary conditions.

(*ii*) **Periodic boundary conditions.** Boundary conditions are said to be periodic if they have the form

$$y(a)=y(b), \quad y'(a)=y'(b).$$

In what follows by a boundary value problem we shall mean a self-adjoint differential operator L together with boundary conditions of type (*i*) or (*ii*).

Remark 1. By a change of dependent variable, unmixed conditions of the type B_1, B_2 can by reduced to non-homogeneous conditions of the type $y(a)=A$, $y(b)=B$.

Further, if we substitute

$$y(x)=z(x)+\frac{x-b}{a-b}A+\frac{x-a}{b-a}B,$$

then the B V P

$$\begin{cases} L(y)\equiv(py')'+qy=0 \\ BC: y(a)=A,\ y(b)=B. \end{cases}$$

where at least one of A and B is non-zero is reduced to

$$\begin{cases} L(z)=(pz')'+qz=-F(x) \\ BC: z(a)=0,\ z(b)=0 \end{cases}$$

where $F(x)=p'(x)\dfrac{A-B}{a-b}+\dfrac{q(x)}{a-b}\{A(x-b)-B(x-a)\}.$

Thus a homogeneous equation with non-homogeneous boundary conditions can be changed to a non-homogeneous equation with homogeneous boundary conditions.

Domain and range of the operators. $L^2[a, b]$ is the Hilbert space of all square integrable functions on $[a, b]$ in the Lebesgue sense, *i.e.*, $f\in L^2(a, b])$ if $\int_a^b |f|^2\,dx<\infty.$

Let S be the linear manifold of $L^2[a, b]$ such that for $y(x)\in S$, $L(y)\in L^2[a, b]$ and $y(x)$ satisfies given boundary conditions. Then S and $L^2[a, b]$ will be the *domain* and *range* spaces of the differential operators L involved in the problems. S will be called the space of *testing functions* of the operator L.

Adjoint problem. For a linear differential operator L, where

$$L(y)\equiv p(x)y''+q(x)y'+r(x)y \qquad \ldots(a)$$

the *formal adjoint* $\bar{L}$ of L is defined by

$$\bar{L}(z)\equiv(pz)''-(qz)'+rz \qquad \ldots(b)$$

Consider the BVP

$$L(y)\equiv py''+qy'+ry=f(x) \qquad \ldots(i)$$

$$B_1: \quad a_1y(a)+a_2y'(a)=0$$

$$B_2: \quad b_1y(b)+b_2y'(b)=0 \qquad \ldots(ii)$$

The boundary conditions adjoint to the conditions B_1 and B_2 are given by the equality

$$0=\int_b^a zL(y)dx-\int_a^b y\bar{L}(z)dx$$

$$=\Big[p(x)(zy'-yz')+yz(q-p')\Big]_a^b \qquad \ldots(iii)$$

For (*iii*) to hold, the expression in the bracket must vanish at $x=a$ and $x=b$. Using conditions (*ii*) we get the adjoint boundary conditions, namely

$$A_1: \quad \{a_1p(a)+a_2p'(a)-a_2q(a)\}z(a)+a_2p(a)z'(a)=0 \qquad \ldots(iv)$$

A_2 : $\{b_1p(b)+b_2p'(b)-b_2q(b)\}z(b)+b_2p(b)z'(b)=0$

The adjoint to the problem (*i*), (*ii*) is given by $\bar{L}(z)=0$ and boundary conditions (*iv*).

The adjoint operation L together with the adjoint boundary conditions is called the *exact adjoint* of the operator L.

The case of periodic conditions is similar. If $p'(x)=q(x)$ *i.e.* the operator L is self-adjoint, then the boundary conditions A_1, A_2 reduce to B_1, B_2 respectively. Such a problem is called a self-adjoint problem.

§8·2. One-dimensional Green's Functions : BVPs

It is possible to theoretically perceive the existence of a Green's function as the operator inverse to a self-adjoint problem. Given

$$L(y)=[p\,y']'+q\,y=f(x) \qquad \ldots(1)$$

$$B_1 : \quad a_1y(a)+a_2y'(a)=0$$

$$B_2 : \quad b_1y(b)+b_2y'(b)=0$$

If L^{-1} denotes the operator inverse to the problem (1)$-B_1$. B_2, then

$$y(x)=L^{-1}f(x).$$

Since L is a differential operator. we expect L^{-1} to be an integral operator. Thus y will be of the form

$$y(x)=\int_a^b G(x,t)f(t)\,dt \qquad \ldots(3)$$

The bivariate function $G(x, t)$ defined on the region $a\leqslant x\leqslant b$, $a\leqslant t\leqslant b$ is called the Green's function of the differential problem. It is called one-dimensional, because one of the two arguments x and t is regarded as the independent variable and the other as a parameter.

We shall later see that the use of the Green's function enables the conversion of a self-adjoint eigenvalue problem (Sturm-Liouville problems) to a Fredholm integral equation of the second kind.

A heuristic approach. Consider the self-adjoint problem (1), B_1, B_2 cited above. If the corresponding homogeneous problem with $L(y)=0$ has only the trivial solution $y\equiv 0$, then the non-homogeneous problem will have a unique solution.

Let $y_1(x)$ be a solution of the homogeneous equation satisfying condition B_1, but not B_2 and $y_2(x)$ a solution satisfying condition B_2, but not B_1. $y_1(x)$, $y_2(x)$ will thus be linearly independent. We shall call $y_1(x)$ and $y_2(x)$ respectively the *left hand* and *right hand* solutions. Their Wronskian is non-vanishing.

i.e, $$W(x)=W(y_1, y_2)\neq 0$$

where $$W(x)=y_1(x)y_2'(x)-y_2(x)y_1'(x)$$

We now solve the non-homogeneous problem by the method of variation of parameters. Let

$$y(x)=A(x)y_1(x)+B(x)y_2(x) \qquad \ldots(i)$$

be the required solution of (1). Then

$$A'(x)y_1(x)+B'(x)y_2(x)=0 \qquad \ldots(ii)$$

$$A'(x)y_1'(x)+B'(x)y_2'(x)=\frac{f(x)}{p(x)} \qquad \ldots(iii)$$

Solving (*ii*) and (*iii*) for $A'(x)$ and $B'(x)$ we have

$$A'(x)=-\frac{y_2(x)\,f(x)}{p(x)W(x)},$$

$$B'(x)=\frac{y_1(x)\,f(x)}{p(x)W(x)} \qquad \ldots(iv)$$

We now show that $A(b)=0$, $B(a)=0$

From (*i*) $y'(x)=A(x)y_1'(x)+B(x)y_2'(x)$

Hence $\qquad 0=a_1y(a)+a_2y'(a)$

$$=a_1\{A(a)y_1(a)+B(a)y_2(a)\}+a_2\{A(a)y_1'(a)+B(a)y_2'(a)\}$$

$$=B(a)\{a_1y_2(a)+a_2y_2'(a)\}$$

Since $a_1y_2(a)+a_2y'(a)\neq 0$, we have $B(a)=0$. Similarly using condition B_2 we can prove that $A(b)=0$. Using the Abel's formula $p(x)W(x)=\text{constant}=C$ (say) we shall get

$$A(x)=\int_x^b \frac{y_2(t)\,f(t)}{C}\,dt. \qquad C=p(t)W(t)$$

$$B(x)=\int_a^x \frac{y_1(t)\,f(t)}{C}\,dt$$

Making these substitutions in (*i*) we get

$$y(x)=\int_x^b \frac{y_1(x)y_2(t)}{C}\,f(t)dt+\int_a^x \frac{y_1(t)y_2(x)}{C}\,f(t)dt$$

$$=\int_a^b G(x, t)\,f(t)dt \qquad \ldots(v)$$

t x t
a|-----|-----|b

where
$$G(x, t)=\begin{cases} \dfrac{y_1(x)y_2(t)}{C} & a\leqslant x\leqslant t\leqslant b \\[2ex] \dfrac{y_1(t)y_2(x)}{C} & a\leqslant t<x\leqslant b \end{cases} \qquad \ldots(vi)$$

Owing to the symmetry of x and t in $G(x, t)$ we can also express the relation (*v*) in the form

$$y(t)=\int_a^b G(x, t)\,f(x)dx \qquad \ldots(vii)$$

x t x
a|-----|-----|b

The function $G(x, t)$ defined in (*vi*) is called the Green's function of the problem. One of the two arguments is regarded as the independent variable and the other as a parameter. For example in (*v*) t is the independent variable while in (*vii*) x is the independent variable. It is very important to keep this dichotomy in mind while operating with a Green's function $G(x, t)$. $G(x, t)$ is also sometimes called the *influence function* of the differential problem.

Properties of the Green's Function

The Green's function $G(x, t)$ of a self-adjoint homogeneous boundary-value problem.

$$L(y)=0 \qquad a\leqslant x\leqslant b$$

$$B_1: \quad a_1 y(a)+a_2 y'(a)=0$$

$$B_2: \quad b_1 y(b)+b_2 y'(b)=0$$

is characterised by the following properties :

(*i*) $G(x, t)$ is continuous in the domain $a\leqslant x, t\leqslant b$.

(*ii*) $\dfrac{\partial G(x, t)}{\partial x}$ is discontinuous at $x=t$

with the jump given by

$$\frac{\partial G(t-0, t)}{\partial x}-\frac{\partial G(t+0, t)}{\partial x}=\frac{-1}{p(t)} \qquad \ldots(viii)$$

where $\dfrac{\partial G(t-0, t)}{\partial x}=\left[\dfrac{\partial G(x, t)}{\partial x}\right]_{x=t}$ when $x\leqslant t$

and $\dfrac{\partial G(t+0, t)}{\partial x}=\left[\dfrac{\partial G(x, t)}{\partial x}\right]_{x=t}$ when $x\geqslant t$

(*iii*) With x as the independent variable, $G(x, t)$ satisfies the differential equation $L(y)=0$ except at $x=t$.

(*iv*) $G(x, t)$ as a function of x satisfies the boundary conditions B_1, B_2.

(*v*) $G(x, t)=G(t, x)$.

The above properties are said to define $G(x, t)$.

Remark 2. If $G(x, t)$ is regarded as a function of t, then the jump (*viii*) could be expressed in the form

$$\frac{\partial G(x, x+0)}{\partial t}-\frac{\partial G(x, x-0)}{\partial t}=\frac{1}{p(x)}$$

Remark 3. A bivariate function satisfying properties (*i*), (*ii*) and (*iii*), but not necessarily (*iv*) is called the *fundamental solution* or the free-space Green's function of the equation $L(y)=0$.

Remark 4. If there is a non-trivial solution $y_0(x)$ of the homogeneous problem satisfying both the boundary conditions B_1, B_2, then the non-homogeneous problem with the same boundary conditions does not have a unique solution and the Green's function of the problem is not unique. Such a non-trivial solution $y_0(x)$ is associated with the eigenvalue $\lambda=0$ of the equation

$$L(y)+\lambda y=0 \qquad \text{with } B_1,\ B_2.$$

Remark 5. (LG(x, t) as a distribution)

Consider the Green's function for a problem over an infinite interval, *i.e.*, $a=-\infty$, $b=\infty$. The problem $L(y)=f(x)$ together with assigned boundary conditions can be inverted in the form

$$y(x)=\int_{-\infty}^{\infty} G(x, t) f(t) dt$$

If we express this symbolically in the form

$$y(x)=L^{-1} f(x)$$

then $$L(y)=(LL^{-1}) f(x)=f(x),$$

i.e., $$\int_{-\infty}^{\infty} LG(x, t). f(t)\, dt=f(x) \qquad \text{...(A)}$$

where L acts on G regarded as a function of x.

Recall the definition of the *Dirac's delta function* $\delta(x)$ on the interval $-\infty<x<\infty$, namely $\delta(x)=0$ every where except at $x=0$ where it becomes so great that

$$\int_{-\infty}^{\infty} \delta(x)\, dx=1$$

Dirac's function has the following sifting property

$$\int_{-\infty}^{\infty} \delta(x-t) f(t) dt=f(x) \qquad \text{...(B)}$$

It can be verified that no ordinary function satisfies property (B) and hence $\delta(x)$ is to be considered as a generalised function or a distribution.

Comparing (A) and (B) we have

$$LG(x, t)=\delta(x-t)$$

A solution of $L(y)=\delta(x-t)$ satisfies properties (*i*), (*ii*) and (*iii*) of a Green's function and hence is a fundamental solution of the equation $L(y)=0$.

Generalisation to Equation of nth order

Consider the BVP

$$L(y)=p_0(x)y^{(n)}+p_1(x)\, y^{(n-1)}+...+p_n(x)y=0 \qquad \text{...(M)}$$

$$\text{BC} : \sum_{i=1}^{n} a_{ik}\, y^{(n-i)}\ (a)+\sum_{i=1}^{n} b_{ik}\, y^{(n-i)}\ (b)=0\ (k=1,\dots, n) \quad \text{...(N)}$$

where the $p_i(x)$ are continuous and $p_0>0$ on the interval $I=a\leqslant x\leqslant b$.

The Green's function of the BVP (M), (N) is a bivariate funcion $G(x, t)$, $a\leqslant x$, $t\leqslant b$ characterised by the following properties :

(*i*) $G(x, t)$ is continuous in x, t and has continuous derivatives with respect to x or t upto order $n-2$ on the interval I.

(*ii*) $\dfrac{\partial^{n-1}G\,(x, t)}{\partial\, x^{n-1}}$ is discontinuous at $x=t$ with a jump given by

$$\frac{\partial^{n-1} G(t-0, t)}{\partial x^{n-1}} - \frac{\partial^{n-1} G(t+0, t)}{\partial x^{n-1}} = -1/p(t).$$

(*iii*) $G(x, t)$ satisfies the equation $L(y)=0$ except at $x=t$.

(*iv*) $G(x, t)$ as a function of x satisfies the boundary conditions (N).

In addition if L is a self-adjoint differential operator, then $G(x, t)$ is symmetric,

i.e., $\qquad G(x, t)=G(t, x)$.

If $y\equiv 0$ is the only solution of the homogeneous problem, then the $2n$ conditions given in the properties ensure the unique determination of the Green's function $G(x, t)$.

Note : Heavisile's unit function $H(x)=\begin{cases} 1 & x>0 \\ 0 & x<0 \end{cases}$

$$H'(x)=\delta(x)$$

Problem. Find the Green's function of the BVP

$$y''=0\ ;\ y(0)=y(1)=0 \text{ by integrating } y''=\delta(x-t)$$

Problem. Find the Green's function of the IVP;

$$y''=\delta(x-t)\ y(0)=y''(0)=0.$$

A Recipe fo Find the Green's Function

Given a boundary-value problem

$$L(y)\equiv [p(x)y']'+q(x)y=0$$

$$B_1 : a_1 y(a)+a_2 y'(a)=0$$

$$B_2 : b_1 y(b)+b_2 y'(b)=0 \qquad a\leqslant x\leqslant b.$$

Let $y_1(x)$ and $y_2(x)$ be any two linearly independent solutions of the equation on the interval I : $a\leqslant x\leqslant b$. The Green's function also being a solution on I except at $x=t$, must be a linear combination of y_1 and y_2. Let

$$G(x, t)=\begin{cases} c_1 y_1(x)+c_2 y_2(x) & a\leqslant x\leqslant t\leqslant b \\ c_3 y_1(x)+c_4 y_1(x) & a\leqslant t \angle x\leqslant b, \end{cases}$$

where c_1, c_2, c_3, c_4 are functions of t. Application of properties (*i*), (*ii*) of the Green's function and the two boundary conditions determine c_1, c_2, c_3, c_4, and hence $G(x, t)$ uniquely.

The procedure fails if a non-trival solution $y_0(x)$ of the equation satisfying both the boundary conditions B_1, B_2 exists. This necessitates generalisation of the present definition which we consider in the next section The procedure besides showing the existence and uniqueness of the Green's function, also provides a method of its construction.

The following theorem affords the solution of a two-point

boundary-value problem in the form of an integral and eventually leads to the conversion of such a problem to a Fredholm integral equations.

Theorem 1. $G(x, t)$ is the Green's function of the homogeneous problem.

$$L(y)=(py')'+qy=0 \qquad ...(i)$$

$$B_1 : a_1y(a)+a_2y'(a)=0, \qquad B_2 : b_1y(b)+b_2y'(b)=0 ;$$
$$p>0, \; a\leqslant x\leqslant b.$$

Then the solution to the non-homogeneous problem

$$L(y)=f(x) \qquad ...(ii)$$

with conditions B_1, B_2 is given by

$$y(x)=\int_a^b G(x, t)f(t)dt \qquad ...(iii)$$

The converse holds.

Proof. Put $u(x)=y(x)$, $v(x)=G(x, t)$ in the Green's formula

$$\int_a^b [vL(u)-uL(v)]\,dx=\Big[p(x)(v\,u'-u\,v')\Big]_a^b$$

Then $$\int_a^b [G(x, t)L(y)-y(x)L(G)]dx=\int_a^{t-0}[GL(y)-y\,L(G)]\,dx$$
$$+\int_{t+0}^b [GL(y)-y\,L(G)]dx$$

$=I_1+I_2$ where I_1 and I_2 are given below.

Since $L(G)=0$ and $L(y)=f(x)$, the left-hand side is equal to

$$\int_a^b G(x, t)f(x)\,dx.$$

Also $$I_1=p(t)\left[\frac{\partial G(t+0, t)}{\partial x}-\frac{\partial G(t-0, t)}{\partial x}\right]y(t)$$

$$=p(t)\frac{1}{p(t)}y(t)=y(t).$$

$$I_2=p(b)\left[G(b, t)\,y'(b)-y(b)\frac{\partial G(b, t)}{\partial x}\right]$$
$$-p(a)\left[G(a, t)y'(a)-y(a)\frac{\partial G(a, t)}{\partial x}\right]$$

Since $y(x)$ and $G(x, t)$ satisfy the boundary conditions B_1, B_2

$$a_1y(a)=-a_2y'(a)$$

and $$a_2\frac{\partial G(a, t)}{\partial x}=-a_1G(a, t),$$

i.e. $$G(a, t)\,y'(a)-\frac{\partial G(a, t)}{\partial x}y(a)=0.$$

Similarly $$G(b, t)y'(b)-\frac{\partial G(b, t)}{\partial x}y(b)=0.$$

Hence $\qquad I_2=0.$

$$y(t)=\int_a^b G(x, t) f(x)\, dx.$$

Owing to the symmetry of x, t in $G(x, t)$, we have finally

Thus $\quad y(x)=\int_a^b G(x, t) f(t)\, dt$

This proves the necessity. We now prove the converse.

Assume the relation (*iii*). Then we shall prove that $y(x)$ satisfies the equation :

$L(y)=f(x)$ and the boundary conditions B_1, B_2,

By the relations (*ii*), we have

$$\begin{aligned} y(x) &= \int_a^b G(x, t) f(t)\, dt \\ &= \int_a^{x-0} G(x, t)\, f(t) dt+\int_{x+0}^b G(x, t) f(t)\, dt \\ &= \frac{y_2(x)}{c}\int_a^x y_1(t) f(t)+\frac{y_1(x)}{c}\int_x^b y_2(t) f(t)\, dt \qquad \ldots(iv) \end{aligned}$$

where $\qquad c=pW$ (constant).

Differentiating (*vi*) with respect to x

$$\begin{aligned} y'(x) &= \frac{y_2'(x)}{c}\int_a^x y_1(t) f(t) dt+\frac{y_1'(x)}{c}\int_x^b y_2(t) f(t)\, dt \\ &\quad +\frac{y_2(x)\, y_1(x) f(x)}{c}-\frac{y_1(x) y_2(x) f(x)}{c} \\ &= \int_a^b \frac{\partial G(x, t)}{\partial x} f(t)\, dt \qquad \ldots(v) \end{aligned}$$

and

$$\begin{aligned} y''(x) &= \frac{y_1''(x)}{c}\int_a^x y_1(t) f(t)+\frac{y_1''(x)}{c} \\ &\quad \int_x^b y_2(t) f(t)\, dt+(y_1y_2'-y_2y_1')\frac{f}{c} \\ &= \int_a^b \frac{\partial^2 G(x, t)}{\partial x^2} f(t)\, dt+\frac{f(x)}{p(x)} \qquad \ldots(vi) \end{aligned}$$

Hence $\quad L(y)=py''+p'y'+qy$

$$\begin{aligned} &= \int_a^b \left[p\frac{\partial^2 G}{\partial x^2}+p'\frac{\partial G}{\partial x}+qG \right] f(t)\, dt+f(x) \\ &= \int_a^b L(G) f(t)\, dt+f(x)-f(x). \end{aligned}$$

Remark 6. (*a*) Result (*vi*) could be derived from (*v*) by direct differentiation as follows :

$$y'(x)=\int_a^{x-0}\frac{\partial G}{\partial x}f(t)\,dt+\int_{x+0}^{b}\frac{\partial G}{\partial x}f(t)\,dt$$

Hence $$y''(x)=\int_a^{x-0}\frac{\partial^2 G}{\partial x^2}f(t)\,dt+\int_{x+0}^{b}\frac{\partial^2 G}{\partial x^2}f(t)\,dt$$

$$+\left[\frac{\partial G(x,x-0)}{\partial x}-\frac{\partial G(x,x+0)}{\partial x}\right]f(x)$$

$$=\int_a^b\frac{\partial^2 G(x,t)}{\partial x^2}f(t)\,dt+\left[\frac{\partial G(x+0,x)}{\partial x}-\frac{\partial G(x-0,x)}{\partial x}\right]f(x)$$

$$=\int_a^b\frac{\partial^2 G(x,t)}{\partial x^2}f(t)\,dt+\frac{f(x)}{p(x)}$$

(*b*) (*iii*) is plausible by the sifting property of the delta function.

Since $G(x, t)$ is a fundamental solution,

$$LG(x,t)=\delta(x-t).$$

Hence using relation (*iv*),

$$L(y)=\int_a^b LG(x,t)f(t)\,dt$$

$$=\int_a^b \delta(x-t)f(t)\,dt=f(x).$$

Conversion of a BVP to an Integral Equation

Using therorem 1, it is possible to convert a Sturm Lionville eigen value problem to a Predholm integral equation of the second kind.

Given a Sturm-Lionville equation

$$L(y)+\lambda r(x)y\equiv(py')'+\{q(x)+\lambda r(x)\}y=f(x)$$

together with boundary conditions B_1, B_2. Let $G(x, t)$ be the Green's function of the problem $L(y)=0$ with B_1, B_2. Then replacing $f(x)$ in theorem 1 by $f(x)-\lambda r(x)\,y(x)$, we shall get the Fredholm equation

$$y(x)=F(x)-\lambda\int_a^b G(x,t)\,r(t)\,y(t)\,dt,$$

where $$F(x)=\int_a^b G(x,t)f(t)\,dt.$$

§8·3. Adjoint Problems

Hitherto we have considered the construction of Green's function of a second order self-adjoint problem. In this section we consider the construction of a general linear second order boundry value problem and its adjoint problem. The results are sammed up in the following

Theorem 2. Given a boundry value problem

$$L(y)\equiv p(x)y''+q(x)y'+r(x)y=0 \qquad \ldots(A)$$

$$\begin{aligned} B_1 &: a_1 y(a)+a_2 y'(a)=0 \\ B_2 &: b_1 y(b)+b_2 y'(b)=0 \\ I &: a\leqslant x\leqslant b,\ p(x)>0 \text{ on I.} \end{aligned} \qquad \ldots(B)$$

Let $y_1(x)$ and $y_2(x)$ be the left-hand and right-hand solutions of the problem. The Green's function G (x, t) of the problem is defined by

$$G(x,t)=\begin{cases} \dfrac{y_1(x)\,y_2(t)}{p(t)\,W(t)} & x\leqslant t \\[2ex] \dfrac{y_1(t)\,y_2(x)}{p(t)\,W(t)} & x\geqslant t \end{cases}$$

In this general case pW may not be a constant on I).

Let $\bar{y}_1, \bar{y}_2$, $\overline{W}$ and $\overline{G}(, t)$ be the left-hand, right-hand solutions their Wronskian and the Green's function of the related adjoint problem, then the following relations hold :

(*i*) $$\bar{y}_1=\frac{y_1(x)}{p(x)\,W(x)},\quad \bar{y}_2=\frac{y_2(x)}{p(x)\,W(x)}$$

(*ii*) $$p(x)\,\overline{W}(x)=\frac{1}{p(x)\,W(x)}$$

(*iii*) $$\overline{G}(x,t)=\begin{cases} \dfrac{\bar{y}_1(x)\,\bar{y}_2(t)}{p(t)\,\overline{W}(t)} & x\leqslant t \\[2ex] \dfrac{\bar{y}_1(t)\,\bar{y}_2(x)}{p(t)\,\overline{W}(t)} & x\geqslant t \end{cases}$$

(*iv*) $\overline{G}(x, t)=G(t, x)$.

Proof. The adjoint problem to the given problem (A), (B) is given by

$$\overline{L}(z)\equiv(pz)''-(qz)''+rz=0 \qquad (\overline{A})$$

$$\overline{B_1}: [a_1\, p(a)+a_2 p'(a)-a_2 q(a)]\, z(a)+a_2\, p(a)\, z'(a)=0 \quad \ldots(\overline{B})$$

$$\overline{B}_2: [b_1\, p(b)+b_2\, p'(b)-b_2 q(b)]\, z(b)+a_2 p(b)\, z'(b)=0$$

(*i*) We shall verify that $\bar{y}_1(x)$ and $\bar{y}_2(x)$ satisfy equation $(\overline{A})$ and conditions $\overline{B}_1$, $\overline{B}_2$ respectively.

By Liouville's formula

$$W(x)=K\, e^{-\int_{x_0}^{x}\frac{q(t)}{p(t)}dt}$$

where $K=W(x_0),\ x_0 \in I$

Assuming $\bar{y}_1, \bar{y}_2$ as in (*i*), we have

$$p\bar{y}_1=\frac{y_1}{K}\, e^{\int_{x_0}^{x}\frac{q}{p}dt},\quad q\bar{y}_1=\frac{q\,y_1}{p\,K}\, e^{\int_{x_0}^{x}\frac{q}{p}dt}$$

$$r\bar{y}_1 = \frac{r\,y_1}{p\,K}\, e^{\int_{x_0}^{x} \frac{q}{p}\,dt}$$

$$(p\bar{y}_1)' - q\bar{y}_1 = \frac{y_1'}{K}\, e^{\int_{x_0}^{x} \frac{q}{p}\,dt} = \frac{y_1'(x)}{W(x)} \qquad \cdots(\overline{C})$$

Differentiating further we get

$$(p\bar{y}_1)'' - (q\bar{y}_1)' = \frac{y_1'}{K}\frac{q}{p}\, e^{\int_{x_0}^{x} \frac{q}{p}\,dt} + \frac{y_1''}{K}\, e^{\int_{x_0}^{x} \frac{q}{p}\,dt}$$

$$= \frac{1}{Kp}(py''_1 + qy'_1)\, e^{\int_{x_0}^{x} \frac{q}{p}\,dt}$$

$$= \frac{-r\,y_1}{K\,p}\, e^{\int_{x_0}^{x} \frac{q}{p}\,dt} = \frac{-ry_1}{pW} = -r\bar{y}_1,$$

i.e., $\quad (p\bar{y}_1)'' - (q\bar{y}_1)' + r\,\bar{y}_1 = 0,$

i.e., $\quad \overline{L}\,(\bar{y}_1) = 0.$

Using condition $\overline{B}_1$, we have

$$[a_1\, p(a) + a_2\, p'(a)]\,\bar{y}_1(a) + a_2[p(a)\,\bar{y}_1'(a) - q(a)\bar{y}_1(a)]$$

$$= a_1 p\,(a)\,\bar{y}_1(a) + a_2\,[p'\,(a)\bar{y}_1\,(a) + p(a)\,\bar{y}_1'\,(a) - q(a)\,\bar{y}_1\,(a)]$$

$$= a_1\frac{y_1\,(a)}{W(a)} + a_2\frac{y_1'\,(a)}{W\,(a)} \qquad \text{(by } \overline{C}\text{)}$$

$$= \frac{1}{W(a)}\{a_1\,y_1\,(a) + a_2\,y_1'\,(a)\} = 0 \qquad \text{(by } B_1\text{)}.$$

Thus $\bar{y}_1\,(x)$ satisfies condition $\overline{B}_1$, *i.e.*, $\bar{y}_1\,(x)$ is the left-hand solution of the adjoint problem.

Similarly we can prove that $\bar{y}_2\,(x)$ is the right-hand solution of the adjoint problem.

(*ii*) Following the procedure of the previous part, we can show that

$$y_1\,(x) = \frac{\bar{y}_1\,(x)}{p(x)\,\overline{W}\,(x)}\,,\quad y_2\,(x) = \frac{\bar{y}_1\,(x)}{p(x)\,\overline{W}\,(x)}$$

Hence by substituting $\bar{y}_1 = \dfrac{y_1}{p\,W}$ we get the result $p\,W = \dfrac{1}{p\,\overline{W}}$

(*iii*) The definition of $\overline{G}\,(x, t)$, namely

$$\overline{G}\,(x, t) = \begin{cases} \dfrac{\bar{y}_1\,(x)\,\bar{y}_2\,(t)}{p\,(t)\,\overline{W}\,(t)} & x \leqslant t \\[2ex] \dfrac{\bar{y}_1\,(t)\,\bar{y}_2\,(x)}{p\,(t)\,\overline{W}\,(t)} & x \geqslant t \end{cases}$$

gives the result

(*iv*) The result is proved by combining (*i*), (*ii*), (*iii*).

Example 1. Solve the boundary value problem

$$L(y) \equiv y'' + k^2 y = -x$$
$$B_1 : y(0) = 0$$
$$B_2 : y(1) = 0,$$

by using the Green's function of the corresponding homogeneous problem. What are the eigenvalues ?

Solution. Consider the problem

$$L(y) = y'' + k^2 y = 0 \quad \ldots(i)$$
$$y(0) = y(1) = 0 \quad \ldots(ii)$$

Since $y(x) \equiv 0$ is the only solution of the homogeneous problem, a unique Green's function exists.

$y_1(x) = \sin kx$ and $y_2(x) = \cos kx$ are two linearly independent solutions of (*i*).

Let $$G(x, t) = \begin{cases} C_1 \sin kx + C_2 \cos kx & x \leqslant t \\ C_3 \sin kx + C_4 \cos kx & x \geqslant t \end{cases}$$

where C_1, C_2, C_3, C_4 are functions of t to be determined. Continuity of G at $x=t$ implies

$$C_1 \sin kt + C_2 \cos kt = C_3 \sin kt + C_4 \cos kt \quad \ldots(a)$$

Discontinuity of $\frac{\partial G}{\partial x}$ at $x=t$ gives

$$(C_1 \cos kt - C_2 \sin kt) - (C_3 \cos kt - C_4 \sin kt) = \frac{-1}{k} \quad \ldots(b)$$

Boundary conditions lead to

$$C_2 = 0, \quad C_3 \sin k + C_4 \cos k = 0. \quad \ldots(c)$$

Solving the system (*a*), (*b*), (*c*) we get

$$C_1 = -\frac{\sin k(1-t)}{k \sin k}, \qquad C_2 = 0$$

$$C_3 = \frac{\sin kt \cos k}{K \sin K}, \qquad C_4 = -\frac{\sin kt}{k}$$

Hence $$G(x, t) = \begin{cases} -\dfrac{\sin kx \sin k(1-t)}{k \sin k} & x < t \\ -\dfrac{\sin kt \sin k(1-x)}{k \sin k} & x > t \end{cases} \quad \ldots(d)$$

Applying theorem 1 we shall have

$$y(x) = \int_0^1 G(x, t)\, t\, dt$$

$$= \frac{\sin k(1-x)}{k \sin k} \int_0^x t \sin kt\, dt + \frac{\sin kx}{k \sin k} \int_x^1 t \sin k(1-t)\, dt$$

$$=\frac{\sin kx}{k^2 \sin k}-\frac{x}{k^2}.$$

Note. Since $y_1=\sin kx$ and $y_2=\sin k(1-x)$ are the left-hand and right-hand solutions of the problem and $W(x)=-k \sin k$, $p(x)=1$, we obtain (*d*) by using the formula (*vi*) of section 8·2 also.

Example 2. Find the Green's function of the system

$$-iy'+\alpha y=0, \quad y(0)=y(1)$$

where $i=\sqrt{-1}$ and α is not an integer.

Solution. The equation in the standard form is

$$y'+i\alpha y=0. \quad \text{Here } p(x)=1.$$

A solution of the equation is $y=e^{-i\alpha x}$.

$$\text{Let} \qquad G(x, t)=\begin{cases} A e^{-i\alpha x} & x \leqslant t \\ B e^{-i\alpha x} & x > t \end{cases}$$

$G(x, t)$ discontinuous at $x=t$ with jump

$$A e^{-i\alpha t}-B e^{-i\alpha t}=-1,$$

i.e.,
$$A-B=-e^{i\alpha t}. \qquad \ldots(i)$$

The boundary condition gives

$$A=B e^{-i\alpha} \qquad \ldots(ii)$$

Solving (*i*) and (*ii*) we have

$$A=\frac{e^{i\alpha t}}{e^{i\alpha}-1}, \quad B=\frac{e^{i\alpha(t+1)}}{e^{i\alpha}-1}$$

Hence finally.

$$G(x, t)=\begin{cases} \dfrac{e^{i\alpha(t-x)}}{e^{i\alpha}-1} & x \leqslant t \\[2ex] \dfrac{e^{i\alpha(1+t-x)}}{e^{i\alpha}-1} & x > t \end{cases}$$

Example 3. Find the Green's function for the operator $\dfrac{d^2}{dx^2}+k^2$ with periodic boundary conditions $y(0)=y(1)$, $y'(0)=y'(1)$.

Does the operator have a Green's function with boundary conditions,

$$y(0)=y(1), \quad y'(0)=-y'(1).$$

Solution. Consider the equation,

$$y''+k^2 y=0.$$

$y_1=\sin kx$ and $y_2=\cos kx$ are two linearly independent solutions. Let

$$G(x, t)=\begin{cases} C_1 \sin kx+C_2 \cos kx & x \leqslant t \\ C_3 \sin kx+C_4 \cos kx & x > t \end{cases}$$

Proceeding as in example 1 and applying the boundary conditions we shall find

$$G(x, t)=\begin{cases} \dfrac{\cos\left(kx-kt+\dfrac{k}{2}\right)}{2k\sin\dfrac{k}{2}} & x \leqslant t \\[2ex] \dfrac{\cos\left(kt-kx+\dfrac{k}{2}\right)}{2k\sin\dfrac{k}{2}} & x \geqslant t \end{cases}$$

If we take the boundary conditions $y(0)=y(1)$, $y'(0)=-y'(1)$, then to determine C_3 and C_4 we shall have only one equation, *viz.*,

$$C_3 \sin\frac{k}{2}+C_4\cos\frac{k}{2}=\frac{\sin\left(kt+\dfrac{k}{2}\right)}{2k}.$$

Hence with these boundary conditions, the problem does not have a unique Green's function.

Note. In the case of periodic conditions, as in this example, it is not possible to find the left-hand and right-hand solutions and hence the formula (*vi*) of section 8·2 is not applicable.

Problem 1. Solve the BVP $y''=f(x)$; $y(0)=b, a, y(1)=b$

[**Hint** : Put $z=y+(a-b)x-a$]

Problem 2. Prove that the Green's function of

$$y''-\frac{1}{x}y'=h(x)$$

$$y(0)=y(1)=0$$

is
$$G(x, t)=\begin{cases} \dfrac{1}{2t}x^2(1-t^2) & x \leqslant t \\[1ex] \dfrac{1}{2}t(1-x^2) & x \geqslant t \end{cases}$$

Exercise 28

1. Find the Green's function of the following problems :

(*a*) $y''=0$; $y(0)=y(1)=0$.

(*b*) $y''=0$; $y(0)=y'(0)=0$.

(*c*) $y''=0$; $y(0)=y'(1)=0$.

(*d*) $y''=0$; $y(-1)=y(1)=0$.

(*e*) $y''-y=0$; $y(0)=y(1)=0$.

(*f*) $y''-K^2y=0$; $y(\pm\infty)$ is finite.

(*g*) $y''=0$, $y(0)=-y(1)$, $y'(0)=-y'(1)$.

2. Solve the non-homogeneous equation $y''=-f(x)$ with boundary conditions (*i*) $y(0)=y'(0)=0$.

(*ii*) $y(-1)=y(1)=0$.

3. Solve the system $y''-y=x$; $y(0)=y(1)=0$.

4. Solve the problem $y''+K^2 y=f(x)$; $y(0)=0$, $y(1)=0$.

5. Find the Green's function of the Bessel's operator $x\dfrac{d^2}{dx^2}+\dfrac{d}{dx}$ with boundary conditions $y(1)=0$, $y(0) < \infty$.

6. Reduce the following non-homogeneous problems to integral equations :

(*a*) $y''+\lambda y=x$; $y(0)=y\left(\frac{\pi}{2}\right)=0$.

(*b*) $y''=\lambda y+e^x$; $y(0)=y(1)=0$.

What are the eigenvalues of the corresponding homogeneous problems ?

7. Convert the BVP to an integral equation :

$$y''+\lambda y=0\ ;\quad y(0)=y'(1),\quad y'(0)=y(1).$$

What are the eigenvalues ?

8. Let $p(x)$ be the linear polynomial interpolating to $f(x)$ at $x=a$ and $x=b$ and $E(x)=f(x)-p(x)$.

Prove that

$$E(x)=\int_a^b G(x, t)\, f''(t)\, dt$$

where $G(x, t)$ is the Green's function of the system

$$y''=0\ ;\quad y(a)=y(b)=0.$$

9. $R_n(x)$ is the remainder after n terms in the Taylor's expansion of $f(x)$ in powers of $x-a$. If $a \leqslant x \leqslant b$, prove that

$$R_n(x)=\int_a^b G_n(x, t)\, f^{(n)}(t)\, dt$$

where $G_n(x, t)$ is the Green's function of the problem $\dfrac{d^n y}{dx^n}=0$.

$$\text{BC : } \left[\frac{d^r y}{dx^r}\right]_{x=a}=0 \qquad (r=0, 1, \ldots n-1).$$

10. Reduce the following BVP's to intergral equation and solve for $\lambda=0$:

(*a*) $y''=\lambda y+x^2$; $y(0)=y\left(\frac{\pi}{2}\right)=0$.

(*b*) $y''+\lambda y=2x+1$; $y(0)=y'(1)$, $y'(0)=y(1)$.

(*c*) $y''+\lambda y=e^x$; $y(0)=y'(0)$, $y(1)=y'(1)$.

(*d*) $y''+\lambda y=2x$; $y(0)=y(1)=0$, $y'(0)=y'(1)$

(*e*) $y^{(iv)}=\lambda y+1$; $y(0)=y'(0)=0$, $y''(1)=y'''(1)=0$.

11. Show that the solution of the two-point boundary value problem

$$\frac{d^2 y}{dx^2} - \omega^2 y = f(x) \qquad (x > 0),$$

$$y(0)=0, \quad y(x) \to 0 \text{ as } x \to \infty$$

is
$$y(x) = -\int_0^\infty G(x, t) f(t)\, dt$$

where
$$G(x, t) = \begin{cases} \dfrac{1}{\omega} \sin h\, \omega t\, e^{-\omega x} & 0 < t < x \\ \dfrac{1}{\omega} \sin h\, \omega x\, e^{-\omega t} & t > x \end{cases}$$

§8·4. Green's Function in the Generalised Sense

Consider the self-adjoint problem

$$L(y) = (py')' + qy = f(x) \qquad \ldots(i)$$

$$B_1 : a_1\, y(a) + a_2\, y'(a) = 0 \qquad \ldots(ii)$$

$$B_2 : b_1\, y(b) + b_2\, y'(b) = 0$$

$$I : a \leqslant x \leqslant b, \quad p(x) > 0 \text{ on I.}$$

In view of theorem 1 a unique solution to the non-homogeneous problem (*i*), (*ii*) is equivalent to the existence of a unique Green's function for the corresponding homogeneous problem. A unique solution to the non homogeneous problem exists if $y \equiv 0$ is the only solution to the homogeneous problem. If a non-trivial solution $y_0(x)$ satisfying (*i*) and both the boundary conditions exists the existence of the Green's function may not be unique or it may not exist. Then theorem (1) will not hold good. We sum up these observations in the following :

Theorem 3. Either the non-homogeneous problem (*i*), (*ii*) has a unique solution for all $f(x)$ or the corresponding homogeneous problem has a non trivial solution.

In the later case the non-homogeneous problem has a solution if and only if, the non-trivial solutions $y_0(x)$ of the homogeneous problem are orthogonal to $f(x)$ in the sense

$$(f, y_0) = \int_a^b f(x)\, y_0(x)\, dx = 0.$$

Proof. $$\int_a^b f(x)\, y_0(x)\, dx = \int_a^b L(y)\, y_0(x)\, dx$$

$$= \int_a^b \{L(y)\, y_0(x) - y(x)\, L(y_0)\}\, dx$$

$$= \Big[p(x)\, (y_0\, y' - y\, y_0') \Big]_a^b \quad \text{(Green's formula)}$$

$= 0$ (because $y(x)$ and $y_0(x)$ satisfy the same boundary conditions).

This proves the necessity. It may be observed that if $y(x)$ is a solution of the non-homogeneous problem and $y_0(x)$ a non-trivial solution of the homogeneous problem, then for any arbitrary constant μ, $z(x)=y(x)+\mu y_0(x)$ is also a solution of the non-homogeneous problem. The sufficiency of the theorem can be shown if in place of the Green's function, a 'generalised Green's' function' is introduced. In this case if $G(x, t)$ is the generalised Green's function, $\omega(x)$ is a function orthogonal to $y_0(x)$

$$\left(i.e., \qquad \int_a^b \omega(x)\, y_0(x)\, dx=0\right),$$

satisfies boundary condition B_1, B_2

and $$L(\omega)=f(x)$$

then $$\omega(x)=\int_a^b G(x, t) f(t)\, dt) \qquad ...(iii)$$

The fact that $y_0(x)$ is a non-trivial solution of the homogeneous problem also means that $y_0(x)$ is the eigenfunction corresponding to the eigenvalue $\lambda=0$ of the problem with $L(y)+\lambda y=0$.

We now define the generalised Green's function as follows :

Given the problem $L(y)=0$ with boundary conditions B_1, B_2. Let $y_0(x)$ be the non-trivial solution satisfying both the boundary conditions.

A Green's function in the generalised sense is a bivariate function $G(x, t)$ with the following properties :

(*a*) $G(x, t)$ is continuons in the domain $a \leqslant x, t \leqslant b$.

(*b*) $\dfrac{\partial G(x, t)}{\partial x}$ is discontinuous at $x=t$ with jump

$$\frac{\partial G(t-0, t)}{\partial x}-\frac{\partial G(t+0, t)}{\partial x}=\frac{-1}{p(t)}.$$

At other points of the domain, $\dfrac{\partial G}{\partial x}$ is continuous and hence bounded.

(*c*) $G(x, t)$ satisfies the boundary conditions B_1, B_2.

(*d*) For $x \neq t$ $G(x, t)$ satisfies the equation

$$L\, G(x, t)=y_0(x)\, y_0(t).$$

(*e*) $\displaystyle\int_a^b G(x, t)\, y_0(x)\, dx=0.$

Green's function in the generalised sense can be constructed on the same lines as in the case of ordinary Green's function. Since the homogeneous problem has a non-trivial solution $y_0(x)$, the problem with $L(y)=y_0(x)\, y_0(t)$ will have no such solutions.

If there are two non-trivial solutions to the homogeneous problem which are linearly independent on $a \leqslant x \leqslant b$, then all solutions to the equation $L(y)=0$ satisfy the boundary conditions, because of the linear homogeneity of the conditions. The boundary conditions become redundant. This trivial case is excluded.

Example 4. Find the Green's function of the problem

$$y''=0\,;\quad y'(0)=y'(1)=0,\ 0 \leqslant x \leqslant 1.$$

Solution. $y_0(x) \equiv 1$ is a non-trivial solution satisfying both boundary conditions. There is no other such non-trivial solution.

Green's function in the generalised sense, therefore, will exist. Since

$$L(G) \equiv \frac{\partial^2 G(x, t)}{\partial x^2} = y_0(x)\, y_0(t) = 1,$$

$$G(x, t) = \begin{cases} C_1 + C_2 x + \dfrac{x^2}{2} & x \leqslant t \\[2ex] C_3 + C_4 x + \dfrac{x^2}{2} & x > t \end{cases}$$

The boundary conditions and the continuity property of $G(x, t)$ at $x = t$ lead to $C_2 = 0$, $C_4 = -1$ and $C_3 - C_1 = t$. Also the orthogonality relation

$$\int_0^1 G(x, t)\, y_0(x)\, dx = 0$$

gives $\quad C_1 = 0$. Hence,

$$G(x, t) = \begin{cases} \dfrac{x^2}{2} & x \leqslant t \\[2ex] t - x + \dfrac{x^2}{2} & x > t \end{cases}$$

The jump relation is

$$\frac{\partial G(t-0, t)}{\partial x} - \frac{\partial G(t+0, t)}{\partial x} = 1.$$

For this to be equal to $\dfrac{-1}{p(t)}$ we should take $p(x) = -1$, *i.e.*, we should consider the operator $-\dfrac{d^2}{dx^2}$ instead of $\dfrac{d^2}{dx^2}$.

The result together with (*iii*) will enable us to solve the non-homogeneous problem

$$y'' = 1 - 2x, \quad y'(0) = y'(1) = 0.$$

Here $\quad f(x) = 2x - 1$, $p(x) = -1$. Hence

$$y(x) = \int_0^1 G(x, t) f(t)\, dt$$

$$= \int_0^x \left(t - x + \frac{x^2}{2}\right)(2t-1)\, dt + \int_x^1 \frac{x^2}{2}(2t-1)\, dt = \frac{x^2}{2} - \frac{x^3}{3}.$$

$\dfrac{x^2}{2} - \dfrac{x^3}{3}$ is not orthogonal to 1. Let

$\displaystyle\int_0^1 \left(\frac{x^2}{2} - \frac{x^3}{3} + A\right).\, 1\, dx = 0.$ Then $A = -\dfrac{1}{12}$.

Hence the solution of the problem which is orthogonal to $y_0 \equiv 1$ is $\dfrac{x^2}{2} - \dfrac{x^3}{3} - \dfrac{1}{12}$.

Problem 1. Find the Green's function of the problem

$$y''=0\ ;\ \ y(-1)=y(1),\ \ y'(-1)=y'(1).$$

Problem 2. Find the Green's function of the BVP

$$y''=0\ ;\ y'(0)-\alpha\, y(0)=0,\ y'(1)-\beta\, y(1)=0.$$

§8·5. One-sided Green's Function : Initial-Value Problems

In the earlier sections we have concentrated on the construction and application of a two-sided Green's function $G(x, t)$ of one dimension. Such functions arise in two-point boundary-value problems of linear differential equations. They are called one-dimensional, because one of the two arguments x is regarded as the functional variable and the other t as a parameter. In the case of a self-adjoint BVP the Green's function is symmetric *i.e.*, $G(x, t)=G(t, x)$. Hence in such a case it makes no difference which one of x or t is regarded as the functional variable. $G(x, t)$ is called two-sided as it had support on both sides of the parametric point t, the two branches of the function tied at t.

A Green's function satisfies homogeneous boundary conditions of the form $y(a)=y(b)=0$. In the limiting case when $b\to a$, the boundary conditions reduce to initial conditions $y(a)=y'(a)=0$. It is thus plausible that Green's function for an IVP should not only exist, the procedure for its construction should be on the same lines as that for BVP. Symmetry, however, is lost.

An IVP being a special case of a BVP is just a mathematical way of idealising an IVP. In the physical world, IVP's arise independently. Engineers have been using Green's function in IVP's and call it an *inpulsive response function.*

It is a characteristic property of the Green's function that the solution of a non-homogeneous differential system (BVP or IVP) can be explicitly obtained in the form of an integral the kernel of which is a Green's function.

It is also well-known that initial and boundary value problems can be solved by applying the so-called integral transforms, specially the Laplace and Fourier transforms. The solution involves the inversion of the transform in the form of a convolution integral. We thus expect that a convolution kernel may be a Green's function. In the Laplace and Fourier transforms, the kernel will be a Green's function $G(x-t)$ either explicitly (as in an IVP) or in a disguised form in conjunction with other terms (as in a BVP). In the sequel we shall look into these aspects in depth.

Green's Function Existence : A Heuristic Approach

As in the case of a BVP, it is possible to give a direct demonstration of the existence of a Green's function for an IVP, with the help of the method of variation of parameters. Given a system.

$$\text{DE : } L(D)\, y(x)=p_0\, D^n y+p_1\, D^{n-1} y+\ldots+p_n\, y=f(x) \qquad \ldots(1)$$

$$\text{ICs}: y(0)=y'(0)=\ldots=y^{(n-1)}(0)=0 \qquad \ldots(2)$$

where the coefficients p_i are continuous functions of x on an interval I containing the origin, $p_0(x)\neq 0$ on I and $D\equiv\frac{d}{dx}$. The interval I could be finite $[a, b]$, $[0, \infty)$, $(-\infty, 0]$ or the whole line $(-\infty, \infty)$.

Let $\{y_1, y_2, \ldots, y_n\}$ be a fundamental set of solutions of the corresponding homogeneous equation,

$$L(D)y=p_0 D^n y+p_1 D^{n-1} y+\ldots+p_n y=0 \qquad \text{(H)}$$

Assume the solution of the non-homogeneous system (1)—(2) to be

$$y(x)=A_1 y_1+A_2 y_2+\ldots+A_n y_n \qquad \ldots(3)$$

where the A_i are functions of x to be determined. Indeed proceeding as in §4·5 we shall have

$$A_i'(x)=\frac{W_i(x)}{W(x)} \quad (i=1, 2, \ldots, n) \qquad \ldots(4)$$

where the Wronskian $W(x)$ is give by

$$W(x)=\begin{vmatrix} y_1(x) & y_2(x) & \ldots & y_n(x) \\ y_1' & y_2' & \ldots & y_n' \\ \ldots & \ldots & \ldots & \ldots \\ y_1^{(n-1)} & y_2^{(n-1)} & \ldots & y_n^{(n-1)} \end{vmatrix} \neq 0 \; x\in I$$

and $W_i(x)$ is obtained from $W(x)$ by replacing its ith column by

$$\left(0, 0, \ldots, \frac{f(x)}{p(x)}\right)^T.$$

It can be shown that

$$W_1(x)=(-1)^{n-1}\frac{f(x)}{p_0(x)}\begin{vmatrix} y_2(x) & y_3(x) & \ldots & y_n(x) \\ y_2' & y_3' & \ldots & y_n' \\ \ldots & \ldots & \ldots & \ldots \\ y_2^{(n-2)} & y_3^{(n-2)} & \ldots & y_n^{(n-2)} \end{vmatrix}$$

$$W_2(x)=(-1)^{n-2}\frac{f(x)}{p_0(x)}\begin{vmatrix} y_1(x) & y_3(x) & \ldots & y_n(x) \\ y_1' & y_3' & \ldots & y_n' \\ \ldots & \ldots & \ldots & \ldots \\ y_1^{(n-2)} & y_3^{(n-2)} & \ldots & y_n^{(n-2)} \end{vmatrix} \quad \text{etc.}$$

In the above computations we have put

$$
\left.\begin{aligned}
A_1' y_1 + A_2' y_2 + \ldots + A_n' y_n &= 0\\
A_1' y_1' + A_2' y_2' + \ldots + A_n' y_n' &= 0\\
\ldots \quad \ldots \quad \ldots\\
A_1' y_1^{(n-2)} + A_2' y_2^{(n-2)} + \ldots + A_n' y_n^{(n-2)} &= 0\\
A_1' y_1^{(n-1)} + A_2' y_2^{(n-1)} + \ldots + A_n' y_n^{(n-1)} &= \frac{f(x)}{p_0(x)}
\end{aligned}\right\} \quad \ldots(5)
$$

We now show that

$$A_1(0) = A_2(0) = \ldots = A_n(0) = 0 \qquad \ldots(6)$$

Differentiating both sides of (3) $(n-1)$ times and using initial conditions (2) with relations (5) we get

$$
\left.\begin{aligned}
y(0) &= A_1(0)\, y_1(0) + \ldots + A_n(0)\, y_n(0) = 0\\
y'(0) &= A_1(0)\, y'_1(0) + \ldots + A_n(0)\, y'_n(0) = 0\\
\ldots \quad &\ldots \quad \ldots\\
y^{(n-1)}(0) &= A_1(0)\, y_1^{(n-1)}(0) + \ldots + A_n(0)\, y_n^{(n-1)}(0) = 0.
\end{aligned}\right\} \quad \ldots(7)
$$

The homogeneous system (7) in n quantities $A_1(0), A_2(0), \ldots, A_n(0)$ has a non-vanishing determinant $W(0)$. Hence the system (7) can have only the trivial solution (6). Hence

$$A_i(x) = \int_0^x A_i'(t)\, dt = \int_0^x \frac{W_i(t)}{W(t)}\, dt,$$

$$y(x) = \sum_{i=1}^{n} A_i(x)\, y_i(x)$$

$$= \int_0^x \frac{1}{W(t)} \left[\sum_{i=1}^{n} W_i(t)\, y_i(x) \right] dt.$$

The integrand in the above integral is

$$
= (-1)^{n-1} \frac{f(t)\, y_1(x)}{p_0(t)\, W(t)}
\begin{vmatrix}
y_2(t) & y_3(t) & \ldots & y_n(t)\\
y_2'(t) & y_3(t) & \ldots & y_n'(t)\\
\ldots & \ldots & \ldots & \ldots\\
y_2^{(n-2)}(t) & y_3^{(n-2)}(t) & \ldots & y_n^{(n-2)}(t)
\end{vmatrix}
$$

$$
+ (-1)^{n-2} \frac{f(t)\, y_2(x)}{p_0(t)\, W(t)}
\begin{vmatrix}
y_1(t) & y_3(t) & \ldots & y_n(t)\\
y_1'(t) & y_3'(t) & \ldots & y_n'(t)\\
\ldots & \ldots & \ldots & \ldots\\
y_1^{(n-2)}(t) & y_3^{(n-2)}(t) & \ldots & y_n^{(n-2)}(t)
\end{vmatrix}
+ \ldots
$$

$$= G(x, t)\, f(t)$$

$$\text{where } G(x, t) = \frac{(-1)^{n-1}}{p_0(t)W(t)} \begin{vmatrix} y_1(x) & y_2(x) & \cdots & y_n(x) \\ y_1(t) & y_2(t) & \cdots & y_n(t) \\ y_1'(t) & y_2'(t) & \cdots & y_n'(t) \\ \cdots & \cdots & \cdots & \cdots \\ y_1^{(n-2)}(t) & y_2^{(n-2)}(t) & \cdots & y_n^{(n-2)}(t) \end{vmatrix} \quad \ldots(8)$$

Thus finally

$$y(x) = \int_0^x G(x, t) f(t)\, dt \quad \ldots(9)$$

$G(x, t)$ is the Green's function of the IVP (1)—(2).

Assuming $0 \leqslant t \leqslant x < \infty$, $G(x, t)$ is characterised by the following properties

(*i*) $L(D)\, G(x, t) = 0 \qquad x \neq t$

(*ii*) $G(x, t)$ satisfies the homogeneous initial conditions (2)

i.e., $$G(0, t) = \frac{\partial G(0, t)}{\partial x} = \ldots = \frac{\partial^{n-1} G(0, t)}{\partial x^{n-1}} = 0$$

(*iii*) $G(x, t)$ has support only on one side of the parametric point t.

(*iv*) $G(x, t) \neq G(t, x)$ (Asymmetry).

(*v*) $$G(t, t) = \frac{\partial G(t, t)}{\partial x} = \ldots = \frac{\partial^{n-2} G(t, t)}{\partial x^{n-2}} = 0 \quad \ldots(10)$$

Also Differentiating the relation (8) $n-1$ times with respect to x and letting $x \to t+0$ we have

$$\frac{\partial^{n-1} G(t+0, t)}{\partial x^{n-1}} = \frac{(-1)^{n-1}}{p_0(t)\, W(t)} \begin{vmatrix} y_1^{(n-1)}(t) & y_2^{(n-1)}(t) & \cdots & y_n^{(n-1)}(t) \\ y_1(t) & y_2(t) & \cdots & y_n(t) \\ y_1'(t) & y'_2(t) & \cdots & y'_n(t) \\ \cdots & \cdots & \cdots & \cdots \\ y_1^{(n-2)}(t) & y_2^{(n-2)}(t) & \cdots & y_n^{(n-2)}(t) \end{vmatrix}$$

$$= \frac{1}{p_0(t)} \quad \ldots(11)$$

If we take $-\infty < x \leqslant t \leqslant 0$, relation (11) should be replaced by

$$\frac{\partial^{n-1} G(t-0, t)}{\partial x} = \frac{1}{p_0(t)}.$$

Procedure to Construct the Green's Function of an IVP

Exactly the same procedure to find the Green's function of an IVP is to be followed as that of a BVP.

Let $\{y_1(x), y_2(x),..., y_n(x)\}$ be a fundamental set of equation (1).

$$\text{Let } G(x, t) = \begin{matrix} a_1 y_1(x)+a_2y_2(x)+...+a_ny_n(x) & x \leqslant t \\ b_1 y_1(x)+b_2y_2(x)+...+b_ny_n(x) & x \geqslant t \end{matrix}$$

where the a_i, b_i are functions of t to be determined. $G(x, t)$ should satisfy the following relations:

(*i*) $L(D)\ G(x, t)=0,\ x \neq t$

(*ii*) $G(x, t)$ should satisfy the continuity relations

$$G(t-0, t)=G(t+0, t),\ \frac{\partial G(t-0, t)}{\partial x}=\frac{\partial G(t+0, t)}{\partial x}...,$$

$$\frac{\partial^{n-2} G(t-0, t)}{\partial x^{n-2}}=\frac{\partial^{n-2} G(t+0, t)}{\partial x^{n-2}} \qquad ...(12)$$

(*iii*) $G(x, t)$ should satisfy the jump relation,

$$\frac{\partial^{n-1} G(t-0, t)}{\partial x^{n-1}}-\frac{\partial^{n-1} G(t+0, t)}{\partial x^{n-1}}=\frac{-1}{p_0(t)} \qquad ...(13)$$

Setting $c_i=b_i-a_i$, we have from (12) and (13) the following system for c_i :

$$\begin{aligned} c_1y_1(t)+c_2\, y_2(t)+...+c_n\, y_n(t)&=0 \\ c_1y_1'(t)+c_2\, y_2'(t)+...+\, c_n\, y_n'(t)&=0 \\ ... \qquad ... \qquad ... \\ c_1y_1^{(n-2)}(t)+c_2\, y_2^{(n-2)}(t)+...+c_n\, y_n^{(n-2)}(t)&=0 \\ c_1y_1^{(n-1)}(t)+c_2\ \ y_2^{(n-1)}(t)+...+c_n\, y^{(n-1)}(t)&=\frac{1}{p_0(t)}\ . \end{aligned} \qquad ...(14)$$

The system (14) determines $c_1, c_2,..., c_n$ uniquely, as its determinant $W(t)$ never vanishes. One set of coefficients say a_i, can be determined in terms of b_i and c_i. At this stage the coefficients b_i remain undermined and the computed function $G(x, t)$ is called a *fundamental solution* of the differential equation (1) or the *fundamental function* of the operator $L(D)$.

The unknowns b_i and hence the Green's function $G(x, t)$ can be uniquely determined if we apply either n independent inital conditions or n independent boundary conditions. Boundary conditions were used in sections (8·2) to (8·4). We turn our attention to initial conditions now. Let $x=0$ to be the initial point and $0 \leqslant x,\ t<\infty$.

For $x \leqslant t$, $G(0, t)=G'(0,t)=...=G^{(n-1)}(0, t)=0$ lead to

$$\begin{aligned} a_1\, y_1(0)+a_2\, y_2(0)+...+a_n\, y_n(0)&=0 \\ a_1\, y'_1(0)+a_2\, y'_2(0)+...+a_n\, y'_n(0)&=0 \\ ... \qquad ... \qquad ... \\ a_1\, y_1^{(n-1)}(0)+a_2\, y_2^{(n-1)}(0)+...+a_n\, y_n^{(n-1)}(0)&=0 \end{aligned} \qquad ...(15)$$

Since the determinant of the system (15) $W(0) \neq 0$, it has only the trivial solution $a_1 = a_2 = \ldots = a_n = 0$. Also $b_i = c_i$. Hence the function $G_+(x, t)$ is uniquely determined and is called the Green's function of the IVP. It is right-sided as it has support only for $x \geqslant t$ and vanishes for $x < t$. For $t=0$, $G_+(x)$ has support in $[0, \infty)$.

If we assume $-\infty < x,\ t \leqslant 0$, then for $x \geqslant t$, $G(0, t) = G'(0, t) = \ldots = G^{(n-1)}(0, t) = 0$ will lead to the relations $b_1 = b_2 = \ldots = b_n = 0$, and $c_i = -a_i$. $G_-(x, t)$ will have support only for $x \leqslant t$. For $t=0$, $G_-(x)$ will have support in $(-\infty, 0]$.

The one-sided Green's function is a truncated function as defined below.

Truncated Function Notation

For $-\infty < x < \infty$ we define

$$x_+ = \begin{cases} x & x > 0 \\ 0 & x \leqslant 0 \end{cases}$$

For a function $\phi(x)$, $-\infty < x < \infty$, we define

$$\phi_+(x) = \begin{cases} \phi(x) & x > 0 \\ 0 & x \leqslant 0 \end{cases}$$

This notation can also be exhibited by the use of Heaviside's unit step function

$$H(x) = \begin{cases} 1 & x > 0 \\ 0 & x < 0 \end{cases}$$

The value of $H(x)$ at $x=0$ can be taken to be 1 or can remain undetermined. $H(x)$ is said to be *normalised* if $H(0) = \frac{1}{2}$. Thus

$$x_+ = H(x) \cdot x$$

$$\phi_+(x) = H(x) \cdot \phi(x)$$

For $-\infty < x,\ t < \infty$

$$(x-t)_+ = H(x-t)\,(x-t)$$

$$= \begin{cases} x-t & x > t \\ 0 & x < t \end{cases}$$

We can also define

$$x_- = \begin{cases} |x| & x < 0 \\ 0 & x \geqslant 0 \end{cases}$$

Hence $|x| = x_+ + x_-$.

The normalised unit function

$$H(x) = \begin{cases} 1 & x > 0 \\ \frac{1}{2} & x = 0 \\ 0 & x < 0 \end{cases}$$

is called "*quenching function*"

Problem Prove (*i*) $H'(x)=\delta(x)$

(*ii*) $H(a-|x|)$ has support $[-a, a]$.

Definition 1. Given a differential operator

$$L(D)=D^n+p_1D^{n-1}+\ldots+p_n$$

where the coefficients p_i are continuous functions and $D\equiv\frac{d}{dx}$. Let $G(x)$ satisfy the equation

$$L(D)G(x)=0 \qquad x\neq 0$$

and the conditions

$$G(0)=G'(0)=\ldots=G^{(n-2)}(0)=0 \quad G^{(n-1)}(0)=1.$$

Then $G(x)$ is called a *fundamental function* of the operator $L(D)$ and

$$G_+(x)=\begin{cases} G(x) & x>0 \\ 0 & x\leqslant 0\end{cases}$$

is called the right-sided Green's function for the IVP

$$L(D)y=0$$

$$y(0)=y'(0)=\ldots=y^{(n-1)}(0)=0.$$

§8·6. Miscellaneous Examples

Example 5. Find the Green's function and solve the IVP,

$$\text{DE} : D^n y=f(x)$$

$$\text{ICs} : y(0)=y'(0)=\ldots=y^{(n-1)}(0)=0.$$

Solution. Take $\{1, x, x^2,\ldots, x^{n-1}\}$ as a fundamental set of the homongeneous equation

$$D^n y=0.$$

$$W(t)=\prod_{k=0}^{n-1} k\,!=(n-1)!!$$

By formula (8),

$$G(x, t)=\frac{(-1)^{n-1}}{(n-1)!!}\begin{vmatrix} 1 & x & x^2 & \ldots & x^{n-1} \\ 1 & t & t^2 & \ldots & t^{n-1} \\ 0 & 1 & 2t & \ldots & (n-1)t^{n-2} \\ 0 & 0 & 2 & \ldots & \ldots \\ \ldots & \ldots & \ldots & \ldots & \ldots \\ 0 & 0 & 0 & \ldots & (n-1)\,!\,t \end{vmatrix} \quad x\geqslant t\geqslant 0.$$

The determinant is computed as follows Owing to conditions (10) $G(x, t)$ has a zero of order $n-1$ at $x=t$. As a polynomial in x, the determinant has degree $n-1$. Hence

$$G(x, t)=A(x-t)_+^{n-1}$$

Now the coefficient of x^{n-1} in the determinant is

$$\frac{(-1)^{(n-1)}}{(n-1)\,!!}(-1)^{n-1}\begin{vmatrix} 1 & t & t^2 & \cdots & t^{n-2} \\ 0 & 1 & 2t & \cdots & (n-2)t^{n-3} \\ 0 & 0 & 2\,! & \cdots & \cdots \\ \cdots & \cdots & \cdots & \cdots & \cdots \\ 0 & 0 & 0 & \cdots & (n-2)\,! \end{vmatrix}$$

$$=\frac{(n-2)\,!!}{(n-1)\,!!}=\frac{1}{(n-1)\,!}$$

Hence $\qquad A=\dfrac{1}{(n-1)\,!}$

and $\qquad G(x, t)=\dfrac{1}{(n-1)\,!}(x-t)_+^{n-1}$

The solution to the given system is

$$y(x)=\int_0^x G(x, t)\, f(t)dt$$

$$=\frac{1}{(n-1)\,!}\int_0^x (x-t)^{n-1} f(t)\, dt.$$

The reader can easily recognise this integral as Cauchy's repeated integral formula.

Example 6. Solve the IVP by using the Green's function

DE : $y''+y=-x \qquad (i) \qquad 0\leqslant x<\infty$

ICs : $y(0)=y'(0)=0 \qquad (ii)$

Solution. $\{\sin x, \cos x\}$ is a fundamental set of homogenous equation

$$y''+y=0.$$

Let $G(x, t)=\begin{cases} a\sin x+b\cos x & x\leqslant t \\ c\sin x+d\cos x & x\geqslant t \end{cases}$

$$0\leqslant x, t \leqslant\infty.$$

The initial conditions $G(0, t)=G'(0, t=0)$ lead to $a=b=0$.

Continuity property of $G(x, t)$ at $x=t$

i.e., $\qquad G(t+0, t)=0$

leads to $\qquad c \sin t + d \cos t = 1 \qquad \ldots(iii)$

The jump relation

$$\frac{\partial G(t+0, t)}{\partial x} = 1$$

gives $\qquad c \cos t - d \sin t = 0 \qquad \ldots(iv)$

From (*iii*) and (*iv*)

$$c = \cos t,\ d = -\sin t.$$

Hence $G(x, t) = \begin{cases} \sin(x-t) & x \geq t \\ 0 & x \leq t \end{cases}$

$$= H(x-t) \sin(x-t).$$

Solution of the IVP is

$$y(x) = -\int_0^x t \sin(x-t)\, dt = \sin x - x.$$

Remark 7. For $y_1 = \sin x$, $y_2 = \cos x$, we have

$$W(x) = -1$$

By formula (8)

$$G(x, t) = \frac{-1}{W(t)} \begin{vmatrix} y_1(x) & y_2(x) \\ y_1(t) & y_2(t) \end{vmatrix} = \sin(x-t) \qquad x \geq t.$$

Example 7. Find the Green's function and hence solve the IVP.

DE ; $y'' + 6y' + 9y = \sin x \qquad \ldots(i)$

ICs : $y(0) = y'(0) = 0 \qquad \ldots(ii)$

Solution. e^{-3x}, xe^{-3x} are two linearly independent solutions of the homogeneous equation.

Let $\quad G(x, t) = \begin{cases} ae^{-3x} + bx\, e^{-3x} & x \leq t \\ ce^{-3x} + dx\, e^{-3x} & x \geq t \end{cases}$

$$0 \leq x,\ t < \infty.$$

By the initial conditions $G(0, t) = G'(0, t) = 0$, we get

$$a = b = 0.$$

By the continuity property of $G(x, t)$ at $x = t$

$$G(t+0, t) = 0.$$

i.e., $\qquad c\, e^{-3t} + dt\, e^{3-t} = 0 \qquad \ldots(iii)$

The jump relation for G' is

$$G'(t+0, t) = 1$$

i.e., $$-3c\,e^{-3t}-3dt\,e^{-3t}+d\,.\,e^{-3t}=1$$

or $$-3c+(1-3t)d=e^{3t} \qquad \ldots(iv)$$

From (*iii*) and (*iv*)

$$c=-t\,e^{3t}\,,\ d=e^{3t}$$

Hence $$G(x,t)=\begin{cases}(x-t)\,e^{-3(x-t)} & x\geqslant t\\ 0 & x\leqslant t\end{cases}$$

The solution of the IVP is

$$y(x)=\int_0^x (x-t)\,e^{-3(x-t)}\sin t\,dt$$

$$=x\,e^{-3x}\int_0^x e^{3t}\sin dt-e^{-3x}\int_0^x t\,e^{3t}\sin t\,dt.$$

To evaluate the integrals we can use the formulae

$$\int e^{at}\sin bt\,dt=\frac{e^{at}(a\sin bt-b\cos bt)}{a^2+b^2}$$

$$\int e^{at}\cos bt\,dt=\frac{e^{at}(a\cos bt+b\sin bt)}{a^2+b^2}$$

Finally, we have the solution

$$y(x)=\frac{}{50}\left\{e^{-3x}(5x+3)+4\sin x-3\cos x\right\}.$$

Remark 8. Since in this example

$$W(x)=\begin{vmatrix}e^{-3x} & xe^{-3x}\\ -3e^{-2x} & -3xe^{-3x}+e^{-3x}\end{vmatrix}=e^{-6x},$$

by using formula (8), we have

$$G(x,t)=-e^{+6t}\begin{vmatrix}e^{-3x} & xe^{-3x}\\ e^{3t} & te^{-3t}\end{vmatrix}=(x-t)e^{-(3x-t)}\quad x\geqslant t$$

Example 8. Solve the Cauchy problem by using its Green's function,

DE : $-y'+y=f(x)$

ICs : $y(x)=o(e^x)$ as $x\to\infty$.

Solution. In this problem since the initial point $x=\infty$ lies on the right, the Green's function will be left-sided *i.e.*, will have support on the left of the parametric point.

$\{e^x\}$ is a fundamental set of the homogeneous equations

$$-y'+y=0$$

$$\text{Let } G(x, t)=\begin{cases} ae^x & x<t \\ be^x & x>t \end{cases} \qquad -\infty<x, t<\infty$$

By the initial condition

$$\underset{x\to\infty}{\text{Lt}} \frac{G(x, t)}{e^x}=0$$

We have $\quad b=0$

By the jump relation

$$G(t-0, t)-G(t+0, t)=1$$

We have $\quad ae^t=1$

Hence the Green's function is

$$G(x, t)=\begin{cases} e^{x-t} & x>t \\ 0 & x>t \end{cases}$$

or

$$G_+(x)=\begin{cases} e^x & x<0 \\ 0 & x>0 \end{cases}$$

Hence $\quad y(x)=\int_{-\infty}^{\infty} G(x-t) f(t)dt=e^x\int_x^{\infty} e^{-t} f(t)\, dt.$

Example 9. Using the Green's function solve the BVP.

DE : $(1-D^2)\, y=f(x)$

BCs : $y(x)=o\,(e^{\pm x})$ as $x\to\pm\infty$.

Solution. $\{e^x, e^{-x}\}$ is a fundamental set of the homogenous equation $-y''+y=0$.

$$\text{Let } G(x, t)=\begin{cases} ae^x+be^{-x} & x\leqslant t \\ ce^x+de^{-x} & x\geqslant t \end{cases}$$

$$-\infty<x, t<\infty$$

Here $-\infty$ and $+\infty$ are respectively the left and right boundaries. The boundary condition at $x=+\infty$ on $G(x, t)$ gives $c=0$ and the condition at $x=-\infty$ gives $b=0$.

By the continuity relation

$$G(t-0, t)=G(t+0, t)$$

we get $\qquad ae^t=de^{-t} \qquad \ldots(i)$

The jump relation

$$\frac{\partial G(t-0, t)}{\partial x}-\frac{\partial G(t+0, t)}{\partial x}=1$$

gives $\qquad ae^t+de^{-t}=1 \qquad \ldots(ii$

From (*i*) and (*ii*)

$$a=\tfrac{1}{2}e^{-t},\ d=\tfrac{1}{2}e^{t}$$

Hence $G(x,t)=\begin{cases} \tfrac{1}{2}e^{x-t} & x\leqslant t \\ \tfrac{1}{2}e^{t-x} & x\geqslant t \end{cases}$

$$=\tfrac{1}{2}e^{-|x-t|}\quad -\infty<x,\ t<\infty.$$

Also $\quad y(x)=\int_{-\infty}^{\infty} G(x,t)f(t)\,dt$

$$=\tfrac{1}{2}\int_{-\infty}^{\infty} e^{-|x-t|}\ f(t)\,dt.$$

Example 10. Find the fundamental solution and Green's function of the BVP.

DE : $y''=0$

BCs : $y(0)=y(1)=0.$

Solution. The fundamental solution satisfies the equation

$$y''=\delta(x-t)\qquad 0\leqslant x,\ t\leqslant 1$$

Successive integration leads to

$$y'(x)=H(x-t)+A$$
$$y(x)=H(x-t)\,.\,(x-t)+Ax+B$$

Hence the fundamental solution is

$$y(x)=(x-t)_{+}+Ax+B$$

A unique Green's function will result if A and B are determined by the boundary conditions.

$$G(0,t)=0 \Rightarrow B=0$$
$$G(1,t)=0 \Rightarrow A+(1-t)_{+}=0$$
$$\Rightarrow A=-(1-t)_{+}=-(1-t)=t-1.$$

Hence the required Green's function is

$$G(x,t)=(x-t)_{+}+x(t-1)$$

$$=\begin{cases} x(t-1) & x\leqslant t \\ t(x-1) & x\geqslant t \end{cases}\qquad 0\leqslant x,\ t\leqslant 1.$$

8·7. Green's Function as a Convolution Kernel

In the technique of operational calculus, we generally treat an operational symbol such as 'D' for differentiation as if it were a number. It is only in the last step that the original interpretation of the operational symbol is given. The justification for this procedure lies in the one-one correspondence between the laws of operation of the symbol on the one hand and laws of ordinary arithmetic on the other. This one-one correspondence can be established by means of suitable integral transforms. Laplace, Fourier, Mellin and Hankel transforms are examples of such transforms. For this reason the technique of integral transforms is also known as *operational calculus*. The following examples illustrate this connection.

A. **Laplace Transform (L.T.).** The unilateral transform is defined by

$$L(f)=\int_0^{\infty} e^{-sx} f(x)\, dx = F(s)$$

If $\quad f(0)=0$, then

$$L(f')=L(Df)=sF(s)$$

In general if $f(0)=f'(0)=\ldots\ldots=f^{(n-1)}(0)=0$

then $\quad L(f^{(n)})=s^n F(s)$

Thus in a sense the operator $D\equiv\frac{d}{dx}$ is mapped in a one-to-one manner into a number s (real or complex).

In the unilateral transform, it is assumed that the original function $f(x)$ has support in $[0, \infty)$.

For functions of unlimited support on both sides a *bilateral transform* is defined by

$$L(f)=\int_{-\infty}^{\infty} e^{-sx} f(x)\, dx = F(s).$$

B. **Fourier Transform (F.T.).** For functions f with bounded support, the Fourier Transform (F.T.) is defined by

$$\text{F.T. } (f) = F(\xi) = \int_{-\infty}^{\infty} e^{i\xi x} f(x)\, dx$$

For a differentiable function f

$$\text{F.T. } (D^n f) = (-i\xi)^n F(\xi)$$

The Fourier transform thus maps the operator D into the number $(-i\xi)$.

Convolution Algebra (Mikusinki's Operational Calculus)[1]

For the Laplace transform, a convolution operation between two functions is defined as follows .

(*a*) For the unilateral Laplace transformation for functions f, g, their convolution is

$$fog(x)=\int_0^x f(x-t)\, g(t) dt = gof(x),$$

satisfying the following properties :

(*i*) $(fog)oh=fo(goh)$ (Associativity)

(*ii*) For functions $f, g, h, \ldots\ldots$ vanishing at $x=0$

$$D(fog)=(Dfog)=(foDg)$$

1. A. Erdélyi (1962) : Operational Calculus and Generalised Functions' Holt, Rinehart & Winston, N.Y.

(*iii*) $L(fog)=F(s) \cdot G(s)$ (Product Theorem)

(*b*) For the bilateral Laplace transformation, convolution of f, g is

$$fog(x)=\int_{-\infty}^{\infty} f(x-t)\, g(t)\, dt=gof(x),$$

satisfying the following properties :

(*i*) $(fog)oh=fo(goh)$ (Associativity)

(*ii*) $D^n(fog)=(D^n fog)=(foD^n g)$

(*iii*) $L(fog)=F(s) \cdot G(s)$ (Product Theorem)

(*iv*) $f(-x)og(-x)=fog(-x)$ (Reflection in convolution)

(*v*) $f(x+h)og(x)=fog(x+h)$ (Shift of a convolution)

(*c*) For the Fourier transform, convolution is the same as that in a bilateral L.T. *viz.*,

$$fog(x)=\int_{-\infty}^{\infty} f(x-t)\, g(t)\, dt=gof(x).$$

The L.T. and the F.T. are linear mappings on suitable function spaces, members of which are called *test functions*. Owing to linearity the sum of original functions is mapped into sum of their images. Product of images, however, does not correspond to ordinary pointwise product of originals, but to another operation called *convolution poduct* defined above.

Now with pointwise addition of functions and convolution as product, a test function space becomes an algebra, which we call *convolution algebra*. No ordinary function is a unity in this algebra. However, if for the present we admit Dirac's delta function $\delta(x)$ into the fraternity of test functions, then since

$$fo\delta(x)=\int_{-\infty}^{\infty} f(t)\, \delta(x-t)\, dt=f(x)$$

$\delta(x)$ can be regarded as the unity in this algebra. In view of relation (*ii*), *the operator D assumes the role of a scalar multiplier in this algebra*.

The following results will be required in the sequel :

(*i*) L.T. $H(x)=\frac{1}{s}$

(*ii*) L.T. $H(x-a)=\frac{e^{-as}}{s}$

(*iii*) L.T. $\delta(x)=1$

(*iv*) L.T. $\delta(x-a)=e^{-as}$

(*v*) F.T. $\delta(x-a)=e^{i\xi a}$

(*vi*) F.T. $\delta(x)=1$.

In the Heaviside's operational calculus D is denoted by the letter p or s. Heaviside's calculus is thus just the analysis of convolution algebra of test function spaces. As pointed out earlier, the route from D to s and vice-versa is through the unilateral Laplace transform. Since $\delta(x)$ acts as the unity in the convolution algebra in the sense that

$$f o \delta = \delta o f = f,$$

a solution of a differential equation

$$L(D)y = D^n y + p_0 D^{n-1} y + \ldots\ldots + p_n y = \delta(x)$$

is a sort of a unit solution G(x) which we have called a fundamental solution. The exact solution to the equation

$$L(D)y = f(x)$$

is obtained by taking the convolution product of this function G(x), which with appropriate boundary or initial conditions is called Green's function of the problem. Therefore, the solution of the problem is expressed in the form

$$\int_0^x G(x-t)\, f(t)\, dt$$

or

$$\int_{-\infty}^{\infty} G(x-t)\, f(t)\, dt$$

depending upon whether the *unilateral* or *bilateral convolution* has been taken as the product operation.

Snmmary. To work in the convolution algebra we replace the operator D by a scalar s (or p) and functions by their Laplace transforms with a reverse interpretation. This explains the fact that *the unilateral Laplace transform is the operational calculus of the operator D.*

Note on Bilateral Convolution. If $f, g \in L^1(-\infty, \infty)$, then the bilateral or two-sided convolution $f*g$ of f and g is defined by the formula

$$(f*g)(x) = \int_{-\infty}^{\infty} f(x-t)\, g(t)\, dt$$

For functions with support on the right of $x=0$, it reduces to the ordinary Laplace convolution

$$\int_0^x f(x-t)\, g(t)\, dt.$$

The following properties are easily verified :

(*i*) $f, g \in L^1 \Rightarrow f*g \in L^1$ and $\| f*g \|_1 \leqslant \| f \|_1 . \| g \|_1$

(*ii*) $f*g = g*f$

(*iii*) $(f*g)*h = f*(g*h)$ for $f, g, h \in L^1$.

Besides these convolution has some marvellous properties which make it a versatile tool of analysis, some of which are :

(A) Convolution is a smoothness increasing operation.

For example if $g \in L^1$, then $f*g$ has at least those smoothness properties (*e.g.* continuity, Lipschitz continuity, modulus of continuity, moduli of smoothness, number of derivatives, total variation etc.) which are possesed by f.

(B) Various structural properties like period of a function f are inherited by $f*g$. To quote Shapiro[1] : "Convolution is like a marriage in which (unlike real life), the best properties of each parent is inherited by the offspring".

Example 11. What is the fundamental solution of the DE, $y''+a^2y=0$?

Soultion. If $G(x)$ is the fundamental solution. Then

$$D^2G+a^2G=\delta(x)$$

In the convolution algebra

$$s^2g(s)+a^2g(s)=1$$

i.e.,
$$g(s)=\frac{1}{s^2+a^2}$$

where
$$L(G)=g(s),\ L(\delta)=1$$

Hence $G(x)=\dfrac{\sin ax}{a}\quad x \geqslant 0.$

The above technique has been used to solve a differential system by the Laplace or Fourier transform. The choice of the appropriate transform depends upon the nature of conditions. For example, unilateral Laplace transform suits finite boundaries, but the bilateral Laplace transform and Fourier transform suit infinite boundaries.

The transform technique of solving a differential system consists in transforming the differential equation using the given conditions solve the image problem for $Y(s)$ algebraically and then invert to recover the original $y(x)$. For example, let the given differential system be

$$L(D)y=a_0D^ny+a_1D^{n-1}y+\ldots\ldots+a_ny=f(x) \qquad \ldots(1)$$

$$\text{ICs}: y(0)=y'(0)=\ldots\ldots=y^{(n-1)}(0)=0 \qquad \ldots(2)$$

where the coefficients a_i are constant.

1. Harold S. Shapiro, Smoothing and Approximation of Functions, Van Nostrand Reinhold Co. N.Y. 1969.

Taking the unilateral Laplace transform of both sides of (1) and using (2), we have

$$\Lambda(s) \,.\, Y(s) = F(s)$$

where
$$L(D^r y) = s^r Y(s)$$

and
$$\Lambda(s) = a_0 s^n + a_1 s^{n-1} + \ldots\ldots + a_n$$

Hence
$$Y(s) = \frac{F(s)}{\Lambda(s)} \quad \ldots(3)$$

If the inverse Laplace transform of $\frac{1}{\Lambda(s)}$ be $G(x)$, then inverting (3).

$$y(x) = G o f(x)$$

$$= \int_0^x G(x-t)\, f(t)\, dt \quad \ldots(4)$$

The convolution kernel G(x−t) is thus the Green's function of the IVP (1)−(2).

For example by taking L.T. the IVP

$$D^n y = f(x)$$

$$y(0) = y'(0) = \ldots\ldots = y^{(n-1)}(0) = 0$$

gives
$$Y(s) = \frac{F(s)}{s^n}$$

Inversion leads to the solution

$$y(x) = \frac{x_+^{n-1}}{\Gamma(n)}\, o f(x)$$

$$= \frac{1}{\Gamma(n)} \int_0^x \left(x - t \right)_+^{n-1} f(t)\, dt$$

The convolution kernel

$$G(x, t) = \frac{(x-t)_+^{n-1}}{\Gamma(n)}$$

$$= \begin{cases} \dfrac{(x-t)^{n-1}}{\Gamma(n)} & x \geqslant t \\ 0 & x \leqslant t \end{cases}$$

is thus the Green's function of the IVP.

A similar procedure applies in the case of Fourier transform also.

Summary. In the solution of a differential system in the form of a convolution integral by means of Laplace transform, the convolution kernel is the Green's function of the system. In the case of an IVP, the Green's function is the convolution kernel exactly. But in the case of a BVP the Green's function is a combination of the convolution kernel with an additional term arising out of the condition at another boundary point.

The use of the bilateral Laplace transform in solving a differential system can be illustrated as follows :

Let $y(x)$ be a function of bounded support and

$$L(D)y=a_0{}^n y+a_1D^{n-1}y+......+a_n y=f(x) \qquad ...(5)$$

Let $$\int_{-\infty}^{\infty} e^{-sx}\,G(x)\,dx=\frac{1}{L(s)} \qquad ...(6)$$

or $$G(x)=\frac{1}{2\pi}\int_{-i\infty}^{i\infty}\frac{e^{sx}}{L(s)}\,ds \qquad ...(7)$$

If we use the formal relation

$$e^{aD}f(x)=\sum_{k=0}^{\infty}\frac{a^k}{k\,!}\,D^k f=f(x+a)$$

then formally the solution of (5) is

$$\begin{aligned} y(x) &= \frac{1}{L(D)}\,f(x) \\ &= \left[\int_{-\infty}^{\infty} e^{-tD}\,G(t)\,dt\right]f(x) \\ &= \int_{-\infty}^{\infty} G(t)\,[e^{-tD}f(x)]\,dt \\ &= \int_{-\infty}^{\infty} G(t)\,f(x-t)\,dt \\ &= \int_{-\infty}^{\infty} G(x-t)\,f(t)\,dt. \end{aligned}$$

The following examples (12) and (13) illustrate the use of bilateral transform to find the Green's function.

Exampla 12. DE : $-y'+y=f(x)$

IC : $y(x)=0(e^x)$ as $x\to\infty$.

The DE can be written as

$$L(D)y=f(x)$$

where $$L(D)=1-D$$

Now it can be verified that

$$\int_{-\infty}^{\infty} e^{-sx}\, G(x)\, dx = \frac{1}{1-s} \qquad -\infty < \text{Re } s < 1$$

where
$$G(x) = \begin{cases} e^{x} & -\infty < x < 0 \\ 0 & 0 \leqslant x < \infty \end{cases}$$
$$= H(-x)\, e^{x}.$$

Hence $G(x-t)$ is the Green's function of the IVP and the solution is

$$y(x) = \int_{-\infty}^{\infty} G(x-t)\, f(t)\, dt \qquad x < t$$
$$= \int_{x}^{\infty} e^{x-t} f(t)\, dt$$
$$= e^{x} \int_{x}^{\infty} e^{-t} f(t)\, dt$$

Now $G(x) = 0\, (e^{x})$ as $x \to \infty$. Hence this property of $G(x)$ is inherited by $y(x)$ through the convolution operation.

(c.f. Ex. 8).

Example 13. DE : $(1-D^2)y = f(x)$
BCs : $y(x) = 0(e^{x})$ as $x \to +\infty$
$y(x) = 0(e^{-x})$ as $x \to -\infty$

It can be easily verified that

$$\frac{1}{2}\int_{-\infty}^{\infty} e^{-sx}\, e^{-|x|}\, dx = \frac{1}{1-s^2} \qquad -1 < \text{Re } s < 1.$$

Since $e^{-|x|}$ satisfies both the boundary conditions

$$G(x-t) = \tfrac{1}{2} e^{-|x-t|} = \begin{cases} \tfrac{1}{2} e^{x-t} & x \leqslant t \\ \tfrac{1}{2} e^{t-x} & x \geqslant t \end{cases}$$

is the Green's function of the BVP.

The solution of the system is

$$y(x) = \int_{-\infty}^{\infty} G(x-t)\, f(t)\, dt$$
$$= \tfrac{1}{2}\int_{-\infty}^{x} e^{t-x} f(t)\, dt + \tfrac{1}{2}\int_{x}^{\infty} e^{x-t} f(t)$$

The desired properties of $y(x)$ at $x = \pm\infty$ are inherited from the convolution kernel $G(x-t)$.

(c.f. Ex. 9).

Example 14. DE : $L(D)\, y = y'' + 6y' + 9y = f(x)$ ...(i)
ICs : $y(0) = 1,\ y'(0) = 0$ ...(ii)

We first make the initial conditions homogenous by making substitution

$$y(x)=z(x)+1$$

The given problem is transformed to

$$\text{DE}: z''+6z'+9z=f(x)-9 \qquad \ldots(iii)$$

$$\text{ICs}: z(0)=z'(0)=0 \qquad \ldots(iv)$$

The initial conditions (*iv*) suggest the use of unilateral Laplace transform. Thus form (*iii*)

$$(s^2+6s+9)\ X(s)=F(s)-\frac{9}{s}$$

i.e.

$$X(s)=\frac{F(s)}{(s+3)^2}-\frac{9}{s(s+3)^2} \qquad \ldots(v)$$

It can be easily verified that if

$$G(x)=xe^{-3x},$$

then

$$\int_0^\infty e^{-sx}\, G(x)\, dx=\frac{1}{(s+3)^2}$$

and

$$\int_0^\infty e^{-sx}\left[\int_0^x te^{-3t}\, dt\right] dx=\frac{1}{s(s+3)^2}$$

Hence inverting (*v*), we have the solution.

$$y(x)-1=z(x)=\int_0^x G(x-t)\, f(t)dt-9\int_0^x te^{-3t}\, dt$$

The Green's function of the IVP is, therefore

$$\begin{cases}(x-t)\, e^{-3(x-t)} & x\geqslant t\\ 0 & x\leqslant t\end{cases}$$

If $f(x)=\sin x$, then computing as in example 7, the solution is

$$y(x)=\frac{1}{50}\left\{(155x+53)\, e^{-3x}+4\sin x-\cos x\right\}.$$

Example 15. $\text{DE}: y''+y=f(x)$

$\text{ICs}: y(0)=y'(0)=0$

By taking L.T., $Y(s)=\dfrac{F(s)}{1+s^2}$

Inverting we get

$$y(x)=\int_0^x \sin(x-t)\, f(t)\, dt.$$

Hence the Green's function of the IVP is

$$G(x-t)=\begin{cases}\sin(x-t) & x\geqslant t\\ 0 & x\leqslant t\end{cases}$$

Example 16. DE : $y''+y=f(x)$ ...(i)

$$\text{BCs}: y(0)=y\left(\frac{\pi}{2}\right)=0 \quad ...(ii)$$

Taking the L.T., we have

$$(s^2+1)\,Y(s)-A=F(s)$$

where $$A=y'(0)$$

Hence $$Y(s)=\frac{F(s)}{1+s^2}+\frac{A}{1+s^2}$$

Inverting, we get

$$y(x)=\int_0^x \sin(x-t)\,f(t)\,dt+A\sin x$$

Before proceeding the discerning reader should notice that the convolution kernel $\sin(x-t)$ was found to be the Green's function of the IVP

$$y''+y=0$$
$$y(0)=y'(0)=0$$

To determine A, we use $y\left(\frac{\pi}{2}\right)=0$,

i.e. $$A=-\int_0^{\pi]2} \sin\left(\frac{\pi}{2}-t\right) f(t)\,dt$$

$$=-\int_0^{\pi/2} \cos t\, f(t)\,dt$$

Hence $$y(x)=\int_0^x \sin(x-t)\,f(t)-\int_0^{\pi/2}\sin x\cos t\, f(t)\,dt$$

$$=\int_0^{\pi/2} G(x,t)\,f(t)\,dt$$

where $$G(x,t)=\begin{cases}-\sin x\cos t & x\leqslant t\\ -\sin t\cos x & x\geqslant t\end{cases}$$

Now the fundamental solution of the equation $y''+y=0$ is obtained from solving

$$y''+y=\delta(x-t)$$

Taking LT., this gives

$$(s^2+1)Y(s)=e^{-st}$$

i.e., $$Y(s)=\frac{e^{-st}}{1+s^2}$$

Inverting we get the fundamental solution

$$y=H(x-t)\sin(x-t)=\begin{cases}\sin(x-t) & x>t\\ 0 & x\leqslant t\end{cases}$$

Hence the Green's function of the BVP is not a convolution kernel but a combination of a convolution kernel which is a fundamental solution and an additional term contributed by a boundary condition. The occurrence of the common part $\sin(x-t)$ is explained by the occurrence of the common end condition $y(0)=0$ in both IVP and BVP.

Example 17. D.E. $(1-D^2)\,y=f(x)$ ...(*i*)

BCs : $y(x)=0(e^{x})$ as $x\to+\infty$

$y(x)=0(e^{-x})$ as $x\to-\infty$

The boundary conditions suggest the suitability of Fourier transform in this case. Hence when we apply F.T. to solve (*i*) the stated BCs will have been incorporated in the solution. The Green's function of the given problem is thus the solution of

$$(1-D^2)\,G(x, t)=\delta(x-t) \qquad \text{...}(ii)$$

where $G(x, t)$ has Fourier transform $\overline{G}(p, t)$.

Taking F.T. of both sides of (*ii*) we get

$$(1+p^2)\,\overline{G}(p, t)=e^{ipt}$$

i.e.,
$$\overline{G}(p, t)=\frac{e^{ipt}}{1+p^2}$$

It can be verified that

F.T. of $\frac{1}{2}e^{-|x-t|}$ is

$$=\frac{1}{2}\int_{-\infty}^{\infty} e^{ipx}\,e^{-|x-t|}\,dx$$

$$=\frac{e^{ipt}}{1+p^2}$$

Hence $\quad G(x, t)=\frac{1}{2}e^{-|x-t|}$.

(c.f. Ex. 9, 13)

Variation Diminishing Convolution Kernel

The technique of using the bilateral transform in constructing the Green's function has implications for another important concept defined below.

Definition 2. A function $G(x)\in L(-\infty, \infty)$ is called a *frequency function* if it is non-negative and

$$\int_{-\infty}^{\infty} G(x)dx=1.$$

$$G_1(x)=\begin{cases} e^x & x<0 \\ 0 & x>0 \end{cases},$$

$$G_2(x)=\tfrac{1}{2}\, e^{-|x|}$$

$$G_3(x)=\frac{1}{\pi}\,\frac{1}{1+x^2},$$

$$G_4(x)=\frac{1}{\sqrt{\pi}}\, e^{-x^2}$$

are examples of frequency functions.

For any $a>0$, $G_a(x)=\begin{cases} 1/2a & |x|\leqslant a \\ 0 & |x|>a \end{cases}$ is a frequency function.

Definition 3. Let $G(x)$ be a frequency fuuction and let

$$f(x)=\int_{-\infty}^{\infty} G(x-t)\,\phi(t)\,dt$$

Then the kernel $G(x)$ is said to be variation diminishing if the number of sign changes in $f(x)$ never exceeds the number of sign changes in $\phi(x)$ on $-\infty<x<\infty$.

We stop this topic by taking a simple example which shows that such a variation diminishing convolution kernel is indeed the Green's function of a differential system.

Example 18. The Green's function of the IVP

$$(1-D)Y=f(x)$$

$$Y(x)=o(e^x) \text{ as } x\to\infty$$

was found to be

$$G(x)=\begin{cases} e^x & x<0 \\ 0 & x>0 \end{cases}$$

and the solution is

$$y(x)=\int_{-\infty}^{\infty} G(x-t)\, f(t)dt.$$

Also $\qquad f(x)=(1-D)\,y(x)$

$$=-e^x\, De^{-x}\, y(x).$$

Let $y(x)$ have n changes of sign at the zeros $x_1, x_2,\ldots, x_n$. Assume $-\infty<x_1<x_2\ldots x_n<\infty$. The given initial condition implies $y(\infty)=0$.

Hence $e^{-x}\, y(x)$ has zeros at $x_1, x_2,\ldots, x_n, \infty$.

By Rolle's theorem $D(e^{-x}\, y(x))$ has one zero in each of the intervals $(x_1, x_2), (x_2, x_3), \ldots, (x_{n-1}, x_n), (x_n, \infty)$.

Hence $-e^x D\, e^{-x}\, y = f(x)$ **has at least n changes of sign.**

Example 19. $L(D)\, y = f(x)$

$$\text{BC}: \quad y(x) = o(e^x) \quad \text{as } x \to +\infty$$

$$y(x) \Rightarrow 0 \qquad \text{as } x \to -\infty$$

where
$$L(D) = \prod_{k=1}^{n} \left(1 - \frac{D}{k}\right)$$

Here
$$\frac{L}{L(s)} = \prod_{k=1}^{n} \left(1 - \frac{s}{k}\right)^{-1}$$

$$= \sum_{k=1}^{n} (-1)^{k-1} \binom{n}{k} \frac{1}{1 - \frac{s}{k}} \qquad \textbf{(Prove !)}$$

$$= \sum_{k=1}^{n} (-1)^{k-1} \binom{n}{k} . k \int_{-\infty}^{0} e^{-(s-k)x}\, dx$$

(Re $s < k$)

$$= \int_{-\infty}^{\infty} e^{-sx}\, G(x)\, dx$$

where
$$G(x) = \sum_{k=1}^{n} (-1)^{k-1} \binom{n}{k} k\, e^{kx} \; (x < 0)$$

$$= -\frac{d}{dx} (1 - e^x)^n \qquad x < 0$$

$$= n e^x (1 - e^x)^{n-1} \quad \text{for } x < 0$$

$$= 0 \text{ for } x > 0.$$

$G(x)$ is the Green's function of the system. The case $n = 1$ was considered in example 18.

By a repetition of the argument of example 18, and by writing

$$\prod_{k=1}^{n} \left(1 - \frac{D}{k}\right) . y(x)$$

$$= \prod_{k=1}^{n} e^{kx} \left(-\frac{D}{k}\right) e^{-kx} . y(x)$$

it can be shown that the Green' function $G(x)$ is a variation diminishing convolution kernel.

Now for $x<0$

$$ne^{x}(1-e^{x})^{n-1}\geqslant 0$$

and
$$\int_{-\infty}^{\infty} G(x)dx=\int_{-\infty}^{0} ne^{x}(1-e^{x})^{n-1}\,dx$$
$$=1.$$

Hence $G(x)$ is a frequency function.

Problem 1. A real differentiable function $f(x)$ is said to be convex if $f''\geqslant 0$. Prove that if $G(x)$ is the Green's function of example 19, then $-\log G(x)$ is convex

Problem 2. Prove that the Green's function of the BVP

$$y''=0$$
$$y(-1)=y(1)=0,\quad y'(-1)=y'(1) \text{ is}$$

$$G(x,t)=\begin{cases}\dfrac{1-t}{4}(t-x-x^{2}) & x\leqslant t\\[2mm] \dfrac{1+t}{4}(t-x+x^{2}) & x\geqslant t\end{cases}$$

§8·8. Difference Equations : Discrete Green's Function

Analogous to differential equations, Green's function exists for difference equations too. We call the later a *discrete Green's function.* Although the like of such a function has occurred in the solution of difference problems in a disguised form, a systematic theory of a discrete Green's function has not yet been developed. For the first time, albiet to a limited extent, we here present a theory of such a function which parallels that of Green's function for differential systems. To get a clue to its existence we take resort to the method of variation of parameters.

A Difference Equation in Cubic Splines

To serve as a guide to the basic ideas of the theory, we discuss the solution of a difference equation occuring in the construction of cubic splines.

It is well-known that in cubic spline interpolation with equally spaced *knots* $x_k=kh$ $(k=0, 1, 2,\ldots, n)$, the following relations hold :

$$s''(x_{k+2})+4s''(x_{k+1})+s''(x_k)$$
$$=\frac{6}{h^{2}}\left\{s(x_{k+2})-2s(x_{k+1})+s(x_k)\right\} \qquad \ldots(i)$$

$$s'(x_{k+2})+4s'(x_{k+1})+s'(x_k)$$
$$=\frac{3}{h}\left\{s(x_{k+2})-s(x_k)\right\} \qquad \ldots(iii)$$

Equation (i) is for the curvature and equation (ii) is for the slope of the spline function $s(x)$. The spline function $s(x)$ is a piecewise polynomial of degree at most three and its derivatives are evaluated at points of interpolation (called *nodes*) which coincide with the knots x_k. The point of concern to our present context is that equations (*i*) and (*ii*) are examples of a difference equation of the second order of the form

$$u_{k+2}+4u_{k+1}+u_k=p_{k+1} \quad \text{...(1)}$$

For an exact solution of (1) we further require either the boundary values u_0, u_n or initial values u_0, Δu_0 where

$$\Delta u_k=u_{k+1}-u_k.$$

The homogeneous equation

$$u_{k+2}+4u_{k+1}+u_k=0 \quad \text{...(2)}$$

has a solution a^k if

$$a^2+4a+1=0$$

i.e.,

$$a=-2\pm\sqrt{3}$$

Two independent solutions of (2) are

$$u_k^1=(-2+\sqrt{3})^k \text{ and } u_k^2=(-2-\sqrt{3})^k$$

However, the general solution can be expressed in a more refined form if we put

$$e^{\alpha}=-2+\sqrt{3},\quad e^{-\alpha}=-2-\sqrt{3}$$

or

$$\sinh\alpha=\sqrt{3},\qquad \cosh\alpha=-2.$$

The complementary function of (1) is given by

$$u_k=\text{A}\cosh k\alpha+\text{B}\sinh k\alpha \quad \text{...(3)}$$

where A and B are arbitrary constants.

To find the complete solution of (1), we use the method of variation of parameters. Replace A and B by a_k and b_k respectively.

Let

$$u_k=a_k\cosh k\alpha+b_k\sinh k\alpha \quad \text{...(4)}$$

be a solution of (1). Then

$$u_{k+1}=a_k\cosh(k+1)\alpha+b_k\sinh(k+1)\alpha \quad \text{...(5)}$$

$$u_{k+2}=a_k\cosh(k+2)\alpha+b_k\sinh(k+2)\alpha \quad \text{...(6)}$$

where we have put

$$\Delta a_k\cosh(k+1)\alpha+\Delta b_k\sinh(k+1)\alpha=0 \quad \text{...(7)}$$

$$\Delta a_k\cosh(k+2)\alpha+\Delta b_k\sinh(k+2)\alpha=p_{k+1} \quad \text{...(8)}$$

Solving (7) and (8) for Δa_k, Δb_k

$$\Delta a_k=a_{k+1}-a_k=-\frac{\sinh(b+1)}{\sinh\alpha}p_{k+1} \quad \text{...(9)}$$

$$\Delta b_k=b_{k+1}-b_k=\frac{\cosh(k+1)\alpha}{\sinh\alpha}p_{k+1} \quad \text{...(10)}$$

The constants a_k, b_k can now be determined either by boundary conditions or by initial conditions

A. Boundary conditions. Let u_0 and u_n be given.

From (4) $\quad u_0=a_0$

$$u_n=a_n \cosh n\alpha+b_n \sinh n\alpha \qquad \text{...(11)}$$

From (9) and (10) we have

$$a_k=a_0-\sum_{r=1}^{k} \frac{\sinh r\alpha}{\sinh \alpha} p_r \qquad \text{...(12)}$$

$$a_n=a_0-\sum_{r=1}^{n} \frac{\sinh r\alpha}{\sinh \alpha} p_r \qquad \text{...(13)}$$

$$b_k=b_n-\sum_{r=k+1}^{n} \frac{\cosh r\alpha}{\sinh \alpha} p_r \qquad \text{...(14)}$$

From (11), (13) and (14) and we get

$$b_k \,.\, \sinh n\alpha=u_n-u_0 \cosh n\alpha+\cosh n\alpha \sum_{r=1}^{n} \frac{\sinh r\alpha}{\sinh \alpha} p_r - \sum_{r=k+1}^{n} \frac{\cosh r\alpha}{\sinh \alpha} p_r \qquad \text{...(15)}$$

Using values of a_k and b_k from (12) and (15) in (4) we get

$$u_k=u_0 \frac{\sinh (n-k)}{\sinh n\alpha}+u_n \frac{\sinh k\alpha}{\sinh n\alpha}+\sum^{n-1} G_{kr}\, p_r$$

$$G_{kr}=\begin{cases} -\dfrac{\sinh k\alpha \sinh (n-r)\alpha}{\sinh \alpha \sinh n\alpha}=A_{kr} & k\leqslant r \\[2ex] -\dfrac{\sinh r\alpha \sinh (n-k)\alpha}{\sinh \alpha \sinh n\alpha}=B_{kr} & k\geqslant r \end{cases} \qquad \text{...(16)}$$

The two-sided expression G_{kr} in (16) is called the *discrete Green's function* of the BVP.

With homogeneous boundary conditions $u_0=u_n=0$, we shall have as the solution of the BVP

$$u_k=\sum_{r=1}^{n-1} G_{kr}\, p_r$$

Properties of the Green's function G_{kr} for BVP

The following properties of G_{kr} can be verified :

(*i*) G_{kr} satisfies the homogeneous difference equation (2) except at $k=r$.

(*ii*) The two branches of G_{kr} join at $k=r$ *i.e.*, $A_{rr}=B_{rr}$.

(*iii*) G_{kr} satisfies the homogeneous boundary conditions *i.e.*, $G_{0r}=G_{nr}=0$.

(*iv*) $\triangle A_{rr}$ and $\triangle B_{rr}$ are not joined at $k=r$ and there is jump $\triangle A_{rr}-\triangle B_{rr}=-1$.

(*v*) G_{kr} is symmetric *i.e.*, $G_{kr}=G_{rk}$. Hence computation of G_{kr} is required only either for $k\leqslant r$ or for $r\leqslant k$. Also $G_{n-r,\, n-k}=G_{kr}$. Hence the $(n-1)\times(n-1)$ matrix $[G_{kr}]$ is symmetric about both diagonals.

Problem 1. Prove the following :

(*i*) For $n=2$, $[G_{kr}]=[G_{11}]=G_{11}=1$.

(*ii*) For $n=3$, $[G_{kr}]=\begin{bmatrix} G_{11} & G_{12} \\ G_{21} & G_{22} \end{bmatrix}=\frac{1}{15}\begin{bmatrix} 4 & -1 \\ -1 & 4 \end{bmatrix}$

(*iii*) For $n=4$ $[G_{kr}]=\begin{bmatrix} G_{11} & G_{12} & G_{13} \\ G_{21} & G_{22} & G_{23} \\ G_{31} & G_{32} & G_{33} \end{bmatrix}$

$$=\frac{1}{56}\begin{bmatrix} 15 & 4 & -1 \\ 4 & 16 & 4 \\ -1 & 4 & 15 \end{bmatrix}$$

Problem 2. Compute $[G_{kr}]_{4\times4}$.

B. Initial conditions. Let u_0 and $\triangle u_0$ be given.

From (9) and (10) we have

$$a_k=a_0-\sum_{r=1}^{k}\frac{\sinh r\alpha}{\sinh \alpha}p_r \qquad \text{...(17)}$$

$$b_k=b_0+\sum_{r=1}^{k}\frac{\cosh r\alpha}{\sinh \alpha}p_r \qquad \text{...(18)}$$

Making these substitutions in (4), the general solution of (1) is given by

$$u_k = a_0 \cosh k\alpha + b_0 \sinh k\alpha + P_k \qquad \ldots(19)$$

$$P_k = \sum_{r=1}^{k} \frac{\sinh (k-r)\alpha}{\sinh \alpha} p_r$$

P_k is a particular solution of (1).

We now determine a_0 and b_0 in terms of u_0 and Δu_0. From (4) and (5) we have

$$\Delta u_k = a_k \, \Delta \cosh k\alpha + b_k \, \Delta \sinh k\alpha$$

Hence

$$u_0 = a_0$$

$$\Delta u_0 = a_0 (\cosh \alpha - 1) + b_0 \sinh \alpha$$
$$= u_0 + \sqrt{3}\, b_0$$

i.e.,

$$b_0 = \frac{1}{\sqrt{3}} (\Delta u_0 - u_0).$$

In the case of homogeneous conditions $u_0 = \Delta u_0 = 0$ and the solution is

$$u_k = P_k = \sum_{r=1}^{k} \frac{\sinh (k-r)\alpha}{\sinh \alpha} p_r$$

$$= \sum_{r=1}^{k} G_{kr} \, p_r$$

where

$$G_{kr} = \begin{cases} \dfrac{\sinh (k-r)\alpha}{\sinh k} & 1 \leqslant r \leqslant k \\ 0 & k < r \end{cases}$$

G_{kr} is the discrete Green's function of the IVP.

Properties of the Green's Function of the IVP

(*i*) G_{kr} satisfies the homogeneous equation except at $k=r$.

(*ii*) G_{kr} satisfies the initial conditions $G_{0r} = \Delta G_{0r} = 0$.

(*iii*) $G_{rr} = 0$.

(*iv*) $\Delta G_{rr} = 1$.

(*v*) G_{kr} is one-sided and $G_{kr} \neq G_{rk}$ (asymmetry).

The $(n-1) \times (n-1)$ matrix G_{kr} is triangular.

C. Standard Procedure to Construct a Green's Function

The heuristic study of properties of the Green's function in (A) and (B) suggests the following common procedure for both BVPs and IVPs. We explain the case of a BVP.

Given the difference system

$$\text{DE}: u_{k+2}+p\,u_{k+1}+q\,u_k=0, \qquad 0\leqslant k\leqslant n \qquad \ldots(i)$$

$$\text{BCs}: u_0=u_n=0 \qquad \ldots(ii)$$

Let u_k and v_k be any two linearly independent solutions of (*i*).

$$\text{Let}\quad G_{kr}=\begin{cases} a\,u_k+bv_k=A_{kr}, & k\leqslant r \\ c\,u_k+dv_k=B_{kr}, & k\geqslant r \end{cases}$$

where a, b, c, d are functions of r to be determined.

I. By the equality of branches $A_{rr}=B_{rr}$

$$au_r+bv_r=cu_r+dv_r \qquad \ldots(iii)$$

II. Boundary conditions $A_{0r}=B_{nr}=0$ give

$$au_0+bv_0=0 \qquad \ldots(iv)$$

$$cu_n+dv_n=0 \qquad \ldots(v)$$

III. The jump relation $\triangle A_{rr}-\triangle B_{rr}=-1$ gives

$$a\,\triangle u_r+b\,\triangle v_r)-(c\,\triangle u_r+d\,\triangle v_r)=-1 \qquad \ldots(v)$$

The four independent relations (*iii*), (*iv*), and (*v*) and (*vi*) determine a, b, c, d uniquely.

In the case of an IVP instead of the boundary conditions $A_{0r}=B_{nr}=0$ we use the initial conditions $A_{0r}=0$, $\triangle A_{0r}=0$.

Remark 9. Combinations of the type $c_1\,\beta_1^{\,k}+c_2\,\beta^k_{\,2}$ can be converted into either $a\,\cos k\,\alpha+b\,\sin k\,\alpha$ or $a\,\cosh k\alpha+b\sinh k\alpha$ which are more suitable to work with.

Example 20. Find the Green's function of the BVP

$$\text{DE}: u_{k+2}+4\,u_{k+1}+u_k=0 \qquad \ldots(i)$$

$$\text{BCs}: u_0=0,\; u_n=0. \qquad \ldots(ii)$$

Solution. Two linearly independent solutions of the equation are

$$u_k=\cosh k\alpha,\; v_k=\sinh k\alpha$$

where

$$\cosh\alpha=-2,\; \sinh\alpha=\sqrt{3}.$$

$$\text{Let } G_{kr}=\begin{cases} a\cosh k\alpha+b\sinh k\alpha=A_{kr} & k\leqslant r \\ c\cosh k\alpha+d\sinh k\alpha=B_{kr} & k\geqslant r \end{cases}$$

By the relation $A_{rr}=B_{rr}$ we get

$$a\cosh r\alpha+b\sinh r\alpha=c\cosh r\alpha+d\sinh r\alpha \qquad \ldots(iii)$$

Boundary conditions give $A_{0r}=0$, $B_{nr}=0$,

i.e.,

$$a=0 \qquad \ldots(iv)$$

$$c\cosh n\alpha+d\sinh n\alpha=0 \qquad \ldots(v)$$

By the jump relation $\triangle A_{rr}-\triangle B_{rr}=-1$

$$b\,[\sinh_{(r+1)}\alpha-\sinh r\alpha]-[c\{\cosh_{(r+1)}\alpha-\cosh r\alpha\} +d\,\{\sinh_{(r+1)}\alpha-\sinh r\alpha\}]=-1 \qquad \ldots(iv)$$

From Eq. (*iii*), (*iv*), (*v*) and (*vi*) we ultimately get

$$a=0$$

$$b=\frac{\sinh(n-r)\alpha}{\sinh\alpha\sinh n\alpha}$$

$$c=-\frac{\sinh r\alpha}{\sinh\alpha}$$

$$d=\frac{\sinh r\alpha\cosh n\alpha}{\sinh\alpha\sinh n\alpha}$$

Hence
$$G_{kr}=\begin{cases}-\dfrac{\sinh k\alpha\,.\,\sinh(n-r)\alpha}{\sinh\alpha\,.\,\sinh n\alpha} & k\leqslant r\\[2ex] -\dfrac{\sinh r\alpha\,\sinh(n-k)\alpha}{\sinh\alpha\,\sinh n\alpha} & k\geqslant r\end{cases}$$

Example 21. Find the Green's function of the IVP

DE : $u_{k+2}+4\,u_{k+1}+u_k=0$

ICs : $u_0=0,\ \Delta u_0=0$

Solution. Proceeding as in example 20, we can find that in this case

$$a=b=0,$$

$$c=-\frac{\sinh r\alpha}{\sinh\alpha},\ d=\frac{\cosh r\alpha}{\sinh\alpha}.$$

Hence
$$G_{kr}=\begin{cases}\sinh(k-r)\alpha & k\geqslant r\\ 0 & k\leqslant r.\end{cases}$$

Example 22. Find the Green's function and hence solve the BVP

DE : $u_{k+2}+u_{k+1}+u_k=k\,.\,2^k$

BC : $u_0=A,\ u_n=B$

Solution. Consider the homogeneous problem

$$u_{k+2}+u_{k+1}+u_k=0 \quad ...(i)$$

$$u_0=0,\ u_n=0 \quad ...(ii)$$

Equation (*i*) is satisfied by $u_k=a^k$

if $$a^2+a+1=0$$

i.e., $$a=\frac{-1\pm i\sqrt{3}}{2}.$$

Put $$\frac{1+i\sqrt{3}}{2}=r(\cos\theta+\sin\theta).$$

Then $r=1,\ \cos\theta=-\frac{1}{2},\ \sin\theta=\frac{\sqrt{3}}{2}$

i.e., $$\theta=\frac{2\pi}{3}.$$

Hence $\cos k\theta$, $\sin k\theta$ are two linearly independent solutions of (*i*).

Let $$G_{k,r}=\begin{cases} a\cos k\theta+b\sin k\theta=A_{kr}, & k\leqslant r\\ c\cos k\theta+d\sin k\theta=B_{kr}, & k\geqslant r\end{cases}$$

Owing to boundary conditions $A_{or}=0$, $B_{nr}=0$

i.e. $$a=0 \qquad \ldots(iii)$$

$$c\cos n\theta+d\sin n\theta=0 \qquad \ldots(iv)$$

The relation $A_{rr}=B_{rr}$ gives

$$b\sin r\theta=c\cos r\theta+d\sin r\theta \qquad \ldots(v)$$

By the jump relation $\triangle A_{rr}-\triangle B_{rr}=-1$

$$b\,[\sin(r+1)]\,\theta-\sin r\theta[-c[\cos(r+1)\,\theta-\cos r\,\theta]$$
$$-d[\sin(r+1)\,\theta-\sin r\;\theta]=-1 \qquad \ldots(vi)$$

From (*iii*), (*iv*), (*v*) and (*iv*)

$$a=0$$

$$b=\frac{\sin(n-r)\,\theta}{\sin\theta\,.\,\sin n\theta}$$

$$c=\frac{\sin r\theta}{\sin\theta}$$

$$d=\frac{-\sin r\theta\cos n\theta}{\sin\theta\sin n\theta}$$

Hence $$G_{kr}=\begin{cases}\dfrac{\sin k\theta\sin(n-r)\,\theta}{\sin\theta\,.\,\sin n\theta} & k\leqslant r\\[2ex] \dfrac{\sin r\theta\sin(n-k)\,\theta}{\sin\theta\sin n\theta} & k\geqslant r\end{cases}$$

The general solution of the non-homogeneous problem is

$$u_k=A\;\sin(n-k)\;\theta+B\sin k\;\theta+\sum_{r=1}^{n}(r-1)2^{r-1}G_{kr}$$

From the relations $G_{o,r}=G_{n,r}=0$ by putting $k=0$, n, we get

$$A\sin n\theta=u_0,\; B\sin n\theta=u_n.$$

Hence finally the solution of the given BVP is

$$u_n=u_o\,\frac{\sin(n-k)\,\theta}{\sin n\theta}+u_n\,\frac{\sin k\theta}{\sin n\theta}+\sum_{r=2}^{n-1}(r-1)\,2^{r-1}G_{kr}$$

Example 23. Find the Green's function and solve the IVP

DE : $u_{k+2}+u_{k+1}+u_k=k\,.\,2^k$ ...(*i*)

ICs : $u_0=A,\qquad \Delta u_0=B$ ...(*ii*)

Solution. The Green's function is constructed for the homogeneous problem

$$u_{k+2}+u_{k+1}+u_k=0 \quad ...(iii)$$

$$u_0=0,\qquad \Delta u_0=0 \quad ...(iv)$$

Let
$$G_{kr}=\begin{cases} A_{k,r}=a\cos k\theta+b\sin k\theta & k\leqslant r\\ B_{kr}=c\cos k\theta+d\sin k\theta & k\geqslant r\end{cases}$$

Conditions $G_{o,r}=0,\qquad \Delta G_{o,r}=0$ give

$$a=0,\qquad b=0 \quad ...(v)$$

By the equality $A_{rr}=B_{rr}$, we get

$$c\cos r\,\theta+d\sin r\,\theta=0 \quad ...(vi)$$

By the jump relation $\Delta B_{rr}=0$

$$c\,\{\cos(r+1)\,\theta-\cos r\theta\}+d\,\{\sin(r+1)\,\theta-\sin r\theta\}=1$$

From (*v*) and (*vi*)

$$c=\frac{\sin r\theta}{\sin\theta},\qquad d=\frac{-\cos r\theta}{\sin\theta}$$

Hence
$$G_{kr}=\begin{cases} -\dfrac{\sin(k-r)\,\theta}{\sin\theta} & k\geqslant r\\[2ex] 0 & k\leqslant r\end{cases}$$

The general solutlon of (*i*) is

$$u_k=C\cos k\theta+D\sin k\theta-\sum_{r=1}^{k}(r-1)\,2^{r-1}\,\frac{\sin(k-r)\,\theta}{\sin\theta} \quad ...(vii)$$

For $k=0$, $\qquad u_0=A=C$

$$\Delta u_0=B$$

$$=A\,(\cos\theta-1)+D\sin\theta$$

$$=-\frac{3}{2}A+\frac{\sqrt{3}}{2}D$$

Hence
$$D=\frac{2}{\sqrt{3}}\left(B+\frac{3}{2}A\right).$$

Substituting for C and D in (*vii*) we get u_k.

Problem. Discuss the Green's functions for the following equations :

(*a*) $u_{k+2}-2\cos\alpha \,.\, u_{k+1}+u_k=0$

(*b*) $u_{k+2}-2\cos\alpha \,.\, u_{k+1}+u_k=0.$

§8·9. Notes

We have have explained in detail Green's function's role as the convolution kernel in Laplace and Fourier transforms. Some other seemingly different but essentially related roles are described in a nutshell in this section.

A. Green's Function as a Reproducing Kernel

Definition 4. Let X be a real or complex inner product space of functions on R. A function $K(x, t)$ of two variables $x, t \in R$ is called a *reproducing kernel* for the space X if

(*a*) for each fixed t $K(x, t) \in X$

(*b*) for every $f(x) \in X$, the reproducing property

$f(x)=(f(t), K(x, t))_t$ holds.

The reproducing property of the Green's function actually emanates from the reproducing (*i.e.*, sifting) property of the Dirac's delta function $\delta(x-t)$. For if we define an inner product in L^2 [0, 1] by

$$(f, g)_t=\int_0^1 f(t) \,.\, g(t)\, dt$$

the suffix in the bracket indicating variable of integration, then

$$f(x)=\int_{-\infty}^{\infty} f(t)\, \delta(x-t)\, dt$$

$$=(f(x), \delta(x-t))_t$$

The Green's function of an operator L(D) satisfies

$$L(D)\, G(x, t)=\delta(x-t).$$

Suppose we define an inner product of f, g by

$$(f, g)=\int_{-\infty}^{\infty} f(x)\, L(D) \,.\, g(t)\, dt$$

then
$$(f(t), G(x, t))_t=\int_{-\infty}^{\infty} [f(t) \,.\, L(D)\, G(x, t)]\, dt$$

$$=\int_{-\infty}^{\infty} f(t)\, \delta(x-t)\, dt$$

$$=f(x).$$

Thus with appropriate inner product, a Green's funct ion can be made to act as a reproducing kernel.

Example 24. Let X consist of all functions $f(x)$ of the form

$$f(x)=\int_0^x f(t)\,dt \qquad 0\leqslant x\leqslant 1 \qquad \ldots(i)$$

where $f \in L^2\,[0, 1[$.

If
$$G(x)=\int_0^x g(t)\,dt$$

then an inner product (f, g) in X is defined by

$$(F, G)_t=\int_0^1 f(t)\,g(t)\,dt$$

$$=\int_0^1 F'(t)\,.\,G'(t)\,dt \qquad \ldots(ii)$$

With this inner product

$$F(x)=\int_0^x F'(t)\,dt$$

$$=\int_0^1 F'(t)\,.\,\frac{\partial K(t, x)}{\partial t}\,dt$$

$$=(F(t), K(t, x))_t \qquad \ldots(iii)$$

where
$$\frac{\partial K(t, x)}{\partial t}=H(x-t)=\begin{cases} 1 & x<t \\ 0 & x<t \end{cases}$$

Hence
$$K(t, x)=\begin{cases} t & x\geqslant t \\ x & x\leqslant t \end{cases}$$

The reader may recall that $K(t, x)$ is the Green's function of the system $y''=0$, $y(0)=y(1)=0$.

Also $\frac{\partial K(t, x)}{\partial t}$ is the Green's function of the problem $y'=0$, $y(0)=0$.

Energy Inner Products. Some of the specific type of inner products are particularly suited to the variational methods in finite elements. Energy inner products is an example.

Definition 5. A linear operator A in an inner product space X is said to be *positive definite* if $(Ax, x) >$unless $x=0$.

The operator is said to be *symmetric* if $(Ax, y)=(x, Ay)$ for all $x, y, \in X$.

Example 25. An inner product in $L^2\,[0, 1]$ can be defined by

$$(f, g)_t=\int_0^1 (f(t)\,g(t)\,dt$$

Let $X=\{f . f \in c^2 [0, 1], f(0)=f(1)=0\}$

Then X is a manifold in L^2

It can be proved that operator A is a positive definite operator on X where

$$Ay=-\frac{d^2y}{dx^2}.$$

For
$$(Af, g)_t=-\int_0^1 f''(t)\, g(t)\, dt$$
$$=\int_0^1 f'(t)\, g'(t)\, dt$$
$$=(f, Ag)_t$$

Hence A is symmetric

Next if $(Af, f)=0$ then

$$\int_0^1 [f'(t)]^2\, dt=0$$

i.e., $f'(t)\equiv 0$ *i.e.,* $f(t)=c$

Since $f(0)=0$, $c=0$

Hence $f(t)\equiv 0$

Thus A is positive definite.

Definition 5. Let $X=L^2 [0, 1)$

where
$$(f, g)_t=\int_0^1 f(t)\, g(t)\, dt$$

Let A be a positive definite and symmetric operator on X. Then a new inner product on X can be induced by A by the relation

$$(f, g)_A=(Af, g)=\int_0^1 Af(t) . g(t)\, dt$$

clearly $(f, g)_A=(g, f)_A$.

Such inner products are called *energy inner products.* For $f=g$ we have the *energy norm* (or energy integral)

$$\| f \|_A=\left[\int_0^1 Af . f\, dt\right]^{1/2}$$

They are so called, because they are used to minimise the protential energy of physical systems.

Example 26. Let X denote the linear space of functions $F(x)$ in $L^2 [0, 1]$ which are *n*-fold integrals *i.e.,*

$$F(x)=\int_0^x \frac{(x-t)^{n-1}}{\Gamma(n)} f(t)\, dt.$$

Introduce in X an inner product by means a bilinear differential expression

$$(F, G)_A=\int_0^1 \left[\sum_{i=0}^{n} a_i(t)\, F^{(i)}(t)\, G^{(i)}(t)\right] dt \quad ...(i)$$

where $a_0>0$, $a_n>0$, $a_i \geqslant 0$ $i=1, 2, \ldots, n-1$ and A is an operater

$$AF=\sum_{i=1}^{n} (-1)^i [a_i(t)\, F^{(i)}(t)]^{(i)} \quad ...(ii)$$

Integration by parts in (*i*) leads to

$$(F, G)_A=P+\int_0^1 \left[\sum_{i=1}^{n} (-1)^i \{a^i(t)\, F^{(i)}(t)\}^{(i)}\right] G(t)\, dt$$

$$=Q+\int_0^1 \left[\sum_{i=1}^{n} (-1)^i \{a_i(t)\, G^{(i)}(t)\}^{(i)}\right] F(t)\, dt$$

where P, Q are constants. If conditions ars imposed such that

$P=Q=0$, then

$(F, G)_A=(A F, G)=(F, AG)=(G, F)$.

A is thus symmetric

Also $(A F, F)=0 \Rightarrow \sum_{i=1}^{n} a_i(t)\, [F^{(i)}(t)]^2=0$

$\Rightarrow F^{(i)}(t)=0 \qquad i=1, 2, \ldots, n$

Hence $\Rightarrow f=0$

$\Rightarrow F\equiv 0.$

Hence A is positive definite. Thus the inner product $(F, G)_A$ is an energy inner product.

Now let, $G(x, t)$ be the Green's function of the BVP,

DE : $Ay=0$

BCs: $\sum_{i=r}^{n} (-1)^i [a_i(x)\, .\, F^{(i)}(x)]^{i-r}=0, \quad r=1, 2, \ldots, n$

at $x=0$ and $x=1$.

It can be verified that G (x, t) is the reproducing kernel for the inner product space X.

Problem. Let $X=\{f \in C^2 [0, 1] : f(0)=f(1)=0\}$

$$Y=C [0, 1]$$

$$A : X \to Y$$

where $$Ay=y''+y.$$

Let $$(f, g)=\int_0^1 f(t)\, g(t)\, dt \text{ on X.}$$

Does the operator define an energy inner product on X ?

B. Green's Function as a Peano Kernel

Let L_x be an operator which annihilates polynomials of degree $\leqslant n-1$. D^n, Δ^n are examples of such operators. If $f \in C^n [a, b]$ and $x \in [a, b]$, then by Taylor's theorem with remainder,

$$f(x)=f(a)+f'(a)\,(x-a)+\ldots+\frac{f^{(n-1)}(a)\,(x-a)^{n-1}}{(n-1)\,!}$$

$$+\frac{1}{(n-1)\,!}\int_a^x f^{(n)}(t)\,(x-t)^{n-1}\, dt$$

Applying L_x on both sides we have

$$L_x(f)=\frac{L_x}{(n-1)\,!}\int_a^x f^{(n)}(t)\,(x-t)^{n-1}\, dt$$

$$=\frac{L_x}{(n-1)\,!}\int_a^b f^{(n)}(t)\,(x-t)_+^{n-1}\, dt$$

$$=\frac{1}{(n-1)\,!}\int_a^b f^{(n)}(t)\, L_x\,(x-t)_+^{n-1}\, dt$$

$$=\int_a^b f^{(n)}(t)\, K(t)\, dt$$

where $$\frac{1}{(n-1)\,!}\, L_x\,(x-t)_+^{n-1}=K(t)=L_x\, G\,(x, t).$$

The function $$G\,(x, t)=\frac{(x-t)_+^{n-1}}{(n-1)\,!}$$

$$=\begin{cases} \dfrac{(x-t)^{n-1}}{(n-1)\,!} & x \geqslant t \\ 0 & x \leqslant t \end{cases}$$

is the Green's function of the IVP

$$D^n y=0,$$

$$y(0)=y'(0)=\ldots=y^{(n-1)}=0.$$

The function $K(t)=L_x\, G(x, t)$ is called the *Peano Kernel* of the linear operator L_x. Also the remainder

$$R_n(x)=\int_a^b f^{(n)}(t)\, G(x, t)\, dt$$

Example 27. (Error in Lagrange Interpolation)

Let $x_0, x_1, x_2, \ldots, x_n$ be fixed nodes in $[a, b]$ and x any other fixed point. The Lagrange polynomial of degree n interpolating $f(x)$ at $x_0, x_1, \ldots, x_n$ is given by

$$p_n(x)=\sum_{k=0}^{n} f(x_k)\, l_k(x)$$

where
$$l_k(x)=\frac{W(x)}{(x-x_k)\, W'(x_k)}$$

$$W(x)=\prod_{k=0}^{n}(x-x_k)$$

Thus
$$l_k(x_k)=1.$$

More generally

$$l_k(x_i)=\delta_k^i=\begin{cases}1 & i=k\\ 0 & i\neq k\end{cases}$$

Let $E(f)=f(x)-p_n(x)$, then for given $x\in[a, b]$ E is the error functional on the function space.

E annihilates polynomials of degree $\leqslant n$.

Now consider the case of linear interpolation

i.e.,
$$n=1.$$

If
$$x_0=a, \quad x_1=b, \text{ then}$$

$$l_0(x)=\frac{x-b}{a-b}, \quad l_1(x)=\frac{x-a}{b-a}$$

$$p_1(x)=\frac{x-b}{a-b}f(a)+\frac{x-a}{b-a}f(b).$$

Hence $E_x(f)=f(x)-p_1(x)$ is an error functional which annihilates $f(x)$ if $f(x)$ has degree $\leqslant 1$. The peano kernel for E_x is $E_x\, H(x-t)$ where $H(x)$ is the unit step function.

It is possible to express the error function $E(f)$ also as an integral with a Green's function kernel.

By direct verification it can be seen that

$$E_x(f)=f(x)-\frac{x-b}{a-b}f(a)-\frac{x-a}{b-a}f(b)$$

$$=\frac{x-b}{b-a}\int_a^x (t-a)f''(t)\,dt+\frac{x-a}{b-a}\int_x^b (t-b)f''(t)\,dt$$

$$=\int_a^b G(x,t)f''(t)\,dt \qquad ...(i)$$

where
$$G(x,t)=\begin{cases}\dfrac{(x-a)(t-b)}{b-a} & x\leqslant t\\[2mm] \dfrac{(t-a)(x-b)}{b-a} & x\geqslant t\end{cases}$$

It is known that $G(x, t)$ is the Green's function of the BVP

$$y''=0, \qquad y(a)=y(b)=0.$$

Now put $y(x)=E_x(f)$, then we have

$$\text{DE}: y''(x)=f''(x) \qquad ...(ii)$$

$$\text{BCs}: y(a)=y(b)=0 \qquad ...(iii)$$

An interpretation of result (i) is that

$$\int_a^b G(x,t)f''(t)dt$$

is the solution of the BVP $(ii)-(iii)$.

The error function in Lagrange's linear interpolation of a function $f(x)$ is thus the solution of the BVP $(ii)-(iii)$ and is expressed as an integral with the Green's function of the BVP as kernel.

Problem Find the Peano kernel for $\Delta^4 f(x_0)$ where

$$\Delta f(x)=f(x+1)-f(x).$$

Remark 10. The above illustration indicates the symbiotic relationship of interpolation theory with linear differential equatious via the Green's function.

C. Green's Function as an L-Spline

Polynomial splines are functions with a prescribed degree of smoothness consisting of piecewise polynomials tied together at points called knots (or nodes). These splines belong to the kernel of a differential operator such as D^n. Generalised splines are obtained when they belong to kernels of more general differential operators. For example an operator of the type

$$p_n(D)=\prod_{k=1}^{n}(D^2+k^2)$$

generates trigonometric splines.

In practice to allow generalised integration by parts, the operator $p_n(D)$ is taken to be of the form

$$p_n(D)=L(D)\,.\,\bar{L}(D)=\bar{L}(D)\,L(D)$$

where L(D) and $\bar{L}(D)$ are adjoint operators.

For example, if

$$L(D)y=(D+2i)\,(D+3i)\,y=(D^2+5i\,D-6)y$$

then the adjoint operators $\bar{L}$ is given by

$$\bar{L}(D)z=(D-2i)\,(D-3i)\,z=(D^2-5i\,D-6)z.$$

Now $L\,\bar{L}(D)\,y=(D^2+4)\,(D^2+9)y$.

Hence the general solution of $L\bar{L}(D)\,y=0$ is given by

$$y(x)=A\sin 2x+B\cos 2x+C\sin 3x+D\cos 3x.$$

Before we relate the Green's function to such operators, we make the following

Definition 6. (Kernel of an operator)

Let $p_n(D)=D^n+a_1D^{n-1}+a_2D^{n-2}+\ldots+a_n$ be a polynomial operator. Then

$$\text{Ker}\,p_n(D)=\{f\in C^n : p_n(D)\,f=0\ a.e.\}$$

is called the *Kernel* of the operator $p_n(D)$.

Definition 7[1]. (L-Spline) Let a function $s(x)$ be defined on $[a, b]$. We say that $s(x)$ is an L-Spline of order n, if there exists an nth degree polynomial $p_n(x)$ and a sequence of knots

$$x_0=a\leqslant x_1\leqslant x_2\leqslant\ldots\leqslant x_k\leqslant x_{k+1}=b \text{ such that}$$

(*i*) on every non-empty interval $[x_i, x_{i+1}]$ $s(x)\in\text{Ker}\,p_n(D)$.

(*ii*) if every x_i has multiplicity j, then s has a continuous $(n-1-j)$th derivative in a neighbourhood of x_i.

If $j=n$, then $s(x_i)=s(x_i+)$.

Our concern in this note is to define an L-Spline as a Green's function.

Let $\phi\in\text{Ker}\,p_n(D)$ and

$$\phi(0)=\phi'(0)=\ldots=\phi^{(n-2)}(0)=0,\ \phi^{(n-1)}(0)=1.$$

If
$$\phi_+(t)=\begin{cases}\phi(t) & t\leqslant 0\\ 0 & t<0,\end{cases}$$

1. L. Schumaker : Spline functions : Basic Theory, 1981, John Wiley and Sons, NY.

then $\phi_+(t)$ is called the Green's function of the IVP

$$p_n(D)y=0$$

$$y(0)=y'(0)=\ldots=y^{(n-1)}(0)=0.$$

If the knots x_i have multiplicity r_i, then an L-spline $s(x)$ can be represented as

$$s(x)=g(x)+\sum_{i=1}^{k}\sum_{j=0}^{r_i-1} a_{ij}\,\phi_+^{(j)}(x-x_i)$$

where a_{ij} are real number and $g\in$ Ker p_n.

Problem. Let $L\bar{L}(D)=(D^2+1)(D^2+4)$. Then prove that the unique trignometric spline with only knots 0, π and satisfying

$$L\bar{L}(D)\,s(x)=0$$

$$s(0)=2,\ s'(0)=3,\ s(\pi)=0,\ s'(\pi)=1$$

is given by $\quad s(x)=\sin x+\cos x+\sin 2x+\cos 2x.$

For a detailed account of L-splines defined as Green's function the reader may consult Schumaker (1981), section 10.8.

D. Green's Function and Fractional Integrals

Definition 8. A function $f(x)$ is said to be *locally integrable* on I if

(*i*) $f(x)$ is defined on I with the possible exception of a set S_f of isolated points of I called the singularities of f

(*ii*) $f(x)$ is continuous in I except the singularities

(*iii*) the integral of f over any closed sub-interval of I containing no singularities of f in its interior exists either as a proper Riemann integral or as an absolutely convergent improper integral.

A locally integrable function f will be written as

$$f\in \text{Loc. (I)}$$

A piecewise continuous function is locally integrable.

Definition 9. (Riemann-Liouville Fractional Integral)

If $\alpha>0$ is any real number, the integral operator I^α called the Riemann-Liouville fractional integral of order α (RL-integral) is defined by

$$I^\alpha f(x(=\frac{1}{\Gamma(\alpha)}\int_0^x (x-t)^{\alpha-1} f(t)\,dt \qquad \ldots(1)$$

for $f\in C$ or $f\in$ Loc. $[0, \infty)$.

Definition 10. (Weyl Fractional Integral)

If α is any positive real number, the integral operator of order α W^α, called the *Weyl fractional integral* (W-integral) is defined by

$$W^\alpha f(x)=\frac{1}{\Gamma(\alpha)}\int_x^\infty (t-x)^{\alpha-1} f(t)\,dt \qquad \ldots(2)$$

where $f\in C$ or $f\in$ Loc. $[0, \infty)$.

In the theory of linear ordinary differential equation, the RL-integral has been known as the *Euler transform* of the first kind and the W-integral as the Euler transform of the second kind.

There is a natural motivation to these definitions. The RL-integral is a straightforward generalisation of Cauchy's *n* times repeated integration formula generated as follows :

$$I f(x)=\int_0^x f(t)\,dt,$$

$$I^2 f(x)=\int_0^x du \int_0^u f(t)\,dt = \int_0^x (x-t) f(t)\,dt$$

$$=\frac{1}{\Gamma(2)}\int_0^x (x-t) f(t)\,dt$$

$$I^3 f(x)=\frac{1}{\Gamma(3)}\int_0^x (x-t)^2 f(t)\,dt$$

$$\ldots \qquad \ldots \qquad \ldots$$

$$I^n f(x)=\frac{1}{\Gamma(n)}\int_0^x (x-t)^{n-1} f(t)\,dt \qquad \ldots(3)$$

Following Mikusinski we put for $\alpha>0$

$$h^\alpha(x)=\frac{x_+^{\alpha-1}}{\Gamma(\alpha)}=\begin{cases}\dfrac{x^{\alpha-1}}{\Gamma(\alpha)} & x>0\\ 0 & x\leqslant 0\end{cases} \qquad \ldots(4)$$

so that $h(x)$ represents the indefinite integration operator.

Unilateral Laplace convolution has been defined earlier as

$$f(x)og(x)=fog(x)=\int_0^x f(x-t)\,g(t)=gof(x) \qquad \ldots(5)$$

Using the notations (4) and (5) integrals (1) and (2) can be expressed as

$$I^\alpha f(x)=h^\alpha(x)0\,f(x) \qquad \ldots(5)$$

and

$$W^\alpha f(x)=h^\alpha(-x)0f(x) \qquad \ldots(7)$$

Both I^α and W^α obey the index law in the following sense :

For Re $\alpha>0$, Re $\beta>0$

$$I^\alpha I^\beta=I^{\alpha+\beta} \quad \text{and} \quad W^\alpha W^\beta=W^{\alpha+\beta} \qquad \ldots(8)$$

These laws are direct consequence of the well-known Euler's integral of the first kind

$$h^\alpha o h^\beta(x)=\int_0^x \frac{(x-t)^{\alpha-1}}{\Gamma(\alpha)}\frac{x^{\beta-1}}{\Gamma(\beta)}\,dt \qquad x\geqslant t\geqslant 0$$

$$=\frac{x_+^{\alpha+\beta-1}}{\Gamma(\alpha+\beta)}=h^{\alpha+\beta} \qquad \ldots(9)$$

An obvious distinction in RL and W integrals is in the support of their integrands. The *support* of a real function on R is defined as the smallest closed subset of R outside of which the function vanishes. While the support of $h^{\alpha}(x)$ is $[0, \infty)$, that of $h^{\alpha}(-x)$ is $(-\infty, 0]$. It is for this reason that $h^{\alpha}(x)$ and $h^{\alpha}(-x)$ are not combined.

In a certain sense I^{α} and W^{α} are adjoint operators. Let $L^2[0, \infty)$ be an inner product space with inner product

$$(f, g)=\int_0^{\infty} f(x)\, g(x)\, dx$$

for $f, g \in L^2$.

If $x^{\alpha}f$ and $x^{\alpha}g \in L^2$, then by changing the order of integration (justified by Fubini's theorem)

$$\begin{aligned}(I^{\alpha}f, g) &= \int_0^{\infty} (I^{\alpha}f(x))\, .\, g(x)\, dx \\ &= \int_0^{\infty}\left[\int_0^{\infty} h^{\alpha}(x-t)\, f(t)\, dt\right] g(x)\, dx \\ &= \int_0^{\infty}\left[\int_t^{\infty} h^{\alpha}(x-t)\, g(x)\, dx\right] f(t)\, dt \\ &= (f, W^{\alpha} g) \qquad \ldots(10)\end{aligned}$$

We can call the adjoint relationship (10) of I^{α} and W^{α} as a *generalised law of integration by parts.*

The main point of our concern in this note is that both the convolution kernels $h^{\alpha}(x)$ and $h^{\alpha}(-x)$ are generalisations of $h^{n}(x)$ and $h^{n}(-x)$ which are the Green's functions of the self-adjoint operator D^n.

Recall that the solution to the IVP.

$$\text{DE}: \; L(D)y=f(x) \qquad \ldots(11)$$

$$\text{ICs}: \; y(0)=y'(0)=\ldots\ldots=y^{(n-1)}(0)=0 \qquad \ldots(12)$$

where $L(D)=p_0D^n+p_1D^{n-1}+\ldots\ldots+p_n$.

$p_0(x)>0$, $p_i(x)$, f continuous on $[0, \infty)$ is given by

$$y(x)=\int_0^{x} G(x, t)\, f(t)\, dt$$

where $G(x, t)$ is the one-sided Green's function of the IVP (11)—(12).

The adjoint to the operator L(D) is given by

$$L(D)y=(-D)^n(p_0y)+(-D)^{n-1}(p_1y)+\ldots\ldots+p_ny$$

The one-sided Green's function $\bar{G}(x, t)$ to the adjoint IVP.

$$\text{DE}: \bar{L}(D)y=f(x)$$

$$\text{ICs}: \quad y(b)=y'(b)=\ldots\ldots=y^{(n-1)}(b)=0$$

is given by the relation

$$\bar{G}(x, t)=-G(t, x) \qquad \text{(Prove !)}$$

Solution is given by

$$y(x)=\int_b^x \bar{G}(x, t)\, f(t)\, dt=\int_x^b G(t, x)\, f(t)\, dt$$

Since b is arbitrary, it can be ∞.

In the case $L(D)=D^n$, $\bar{L}(D)=(-D)^n$, $L(D)$ is thus formally self adjoint as the factor $(-1)^n$ can be removed, We have seen that for the IVP $D^n y=f(x)$, $y(0)=y'(0)=\ldots\ldots=y^{(n-1)}=0$.

$$G(x, t)=\frac{(x-t)_+^{n-1}}{\Gamma(n)}$$

Hence Green's function of the adjoint problem

$$(-1)^n D^n y=f(x),\ y(b)=y'(b)=\ldots\ldots=y^{(n-1)}(b)=0$$

where we take $b=\infty$ is given by

$$\bar{G}(x, t)=-\frac{(t-x)_+^{n-1}}{\Gamma(n)}$$

and the solution is

$$y(x)=\frac{1}{\Gamma(n)}\int_x^\infty (t-x)^{n-1} f(t)\, dt \qquad t\geqslant x$$

The replacement of n by a real number α in these Green's functions kernels leads us to the stated definitions of RL and W integrals.

A question now forces itself : The function $h^n(x)$ was the Green's function of the operator D^n which is of a positive integral order. Shall we still call $h^\alpha(x)$ a Green's function for arbitrary α. We can do so if it is the solution to a differential equation of arbitrary order α, $D^\alpha y=f(x)$. There is no difficulty in defining a function,

$$h^\alpha(x)=x_+^\alpha \qquad \alpha>0.$$

But no satisfactory theory of a fractional order differential equation has yet been developed. Whatever, efforts have been made in this direction, they are at most partly successful. Among the most recent work in this field we can cite Prof. Al Bassam[1] (1987) of Kuwait University. He discusses solutions of differential equation of the type.

$$I^{-\alpha}y-\lambda y=f(x), \qquad \alpha>0$$

where λ is a parameter and $f\in \text{Loc}\,[0, \infty)$.

A theory of I^α and W^α as the special cases of the limit of a generalised difference operator has been given by Parashar[2] (1985).

Fractional Derivative. For $\alpha>0$, there is no hurdle in defining the integral I^α as

$$I^\alpha f(x)=\int_0^x \frac{(x-t)^{\alpha-1}}{\Gamma(\alpha)} f(t)\, dt$$

is assured for $\alpha>0$. The generalisation from I^n to I^α is thus natural. There is no such natural route to define D^α where $\alpha>0$ and $D\equiv\frac{d}{dx}$. The generalisation to D^α for $\alpha>0$ is achieved via I^α in the following alternate ways :

Definition 11.

(Weyl's definition) : $D^\alpha_R f=D^{n+1}\, I^{a+1-\alpha} f$

(Liouville's definition): $D^\alpha_R f=I^{n+1-\alpha}\, D^{n+1} f$

provided $D^{n+1}f$ exists and $[\alpha]=n$.

In the case $f(0)=f'(0)=\ldots\ldots=f^{(n)}(0)=0$ the two definitions are the same. The following examples have existed in classical analysis

$$I^{\frac{1}{2}} f(x)=\frac{1}{\sqrt{\pi}}\int_0^x \frac{f(t)\, dt}{\sqrt{x-t}}$$

$$D^{\frac{1}{2}} f(x)=DI^{\frac{1}{2}} f(x)=\frac{1}{\sqrt{\pi}}\, \frac{d}{dx}\int_0^x \frac{f)t)}{\sqrt{x-t}}\, dt$$

$$=\frac{1}{\sqrt{\pi}}\int_0^x \frac{f'(t)\, dt}{\sqrt{x-t}}$$

where $f(0)=0$

We have taken $n=[\alpha]=$ greatest integer $\leqslant\alpha$

so that $\quad n+1-\alpha>0$.

1, M.A. Al Bassam (1987) : "On Generalised Power Series and Generalised Operational Calculus and Its Application" in Nonlinear Analysis, Ed Rassias, World Scientific Pubilshing Company, Singapore, pp. 51—88.

2. B.P. Parashar (1985) : Generalised Differences and Their Applications, Indian J. of Pure and Applied Mathematics, 16/6), pp. 607—612.

In the case of the Weyl integral for $\alpha>0$ we define

$$(-D)^n W^{-\alpha}=W^{n-\alpha}$$

where $0<\alpha\leqslant 1,\ n-\alpha\geqslant 0$.

We conclude this note by citing an important smoothing property of I^α.

Definition 12. $f\in \text{Lip}\ \alpha$ on an interval I, $0<\alpha\leqslant 1$

$$\Rightarrow |f(x)-f(y)| \leqslant k\,|x-y|^\alpha$$

for all $x, y\in$ I and some constant $k>0$.

We say f has Lipschitz continuity of order α.

For $\alpha=1$, it is Lipschitz continuity and implies continuity of f. If $\alpha>1$, then f is a constant.

Lemma 1. $x>0,\ 1>\alpha>0$

$$\Rightarrow x^\alpha \in \text{Lip}_\alpha$$

Proof. Put $F(x)=(x+h)^\alpha-x^\alpha \qquad h>0$

Then $\qquad F(0)=h^\alpha$

$$F'(x)=\alpha[(x+h)^{\alpha-1}-x^{\alpha-1}]<0$$

Hence F(x) is a decreasing function and for $x>0$, we have

$$F(x)<F(0)=h^\alpha$$

The lemma is thus proved.

Theorem 3. (Weyl's Theorem). If $0<\alpha<1$ and f is a function of bounded variation on an interval I, then

$$I^\alpha f\in \text{Lip}\ \alpha \text{ on the interval I}$$

Proof. $I^\alpha f=\dfrac{1}{\Gamma(\alpha)}\displaystyle\int_0^x (x-t)^{\alpha-1} f(t)\, dt$

For $x>0$, $(x-t)^{\alpha-1}$ and $f(x)$ are bounded and integrable and $(x-t)^{\alpha-1}$ remains positive for $x>t$. Hence by the first mean value theorem of the integral there exists y, $0<y<x$ such that

$$I^\alpha f=\frac{f(y)}{\Gamma(\gamma+1)}\, x^\alpha, \qquad x>o$$

By lemma 1, $I^\alpha f\in \text{Lip}\ \alpha$.

Corollary. $I^\alpha f\in([\ 0, \infty]$.

In chapter 11, we shall study fractional integrals in more detail.

E. Eigenfunction Expansions and Green's Function

In section 4·2 of chapter 2, we considered the Sturm-Liouville problem,

$$\text{DE}: L(y)+\lambda y=[p(x)y']'+q[(x)y]+\lambda y=f(x) \qquad ...(i)$$

$$\text{BCs}: y(a)=y(b)=0 \qquad ...(ii)$$

where $x \in I=[a, b]$, $p(x)>0$ on I and $p(x)$, $q(x)$ are continuous on I.

Eigenvalues and eigenfunctions arise in solving the corresponding homogeneous system,

$$\text{DE} : L(y)+\lambda y=0 \qquad \ldots(iii)$$

$$\text{BCs} : y(a)=y(b)=0 \qquad \ldots(ii)$$

A sequence of eigenfunctions $\{\phi_n(x)\}$ corresponding to the eigenvalues $\{\lambda_n\}$ can be found which forms an orthonormal sequence on $[a, b]$. *i.e.*,

$$\int_a^b \phi_m(x)\, \phi_n(x)\, dx=\begin{cases} 1 & m=n \\ 0 & m\neq n \end{cases}$$

Functions $f(x)$ such that $f(a)=f(b)=0$, can be expanded in series of $\phi_n(x)$. Let

$$f(x)=\sum_{n=1}^{\infty} A_n\phi_n(x) \qquad \ldots(iv)$$

and

$$y(x)=\sum_{n=1}^{\infty} C_n\phi_n(x) \qquad \ldots(v)$$

Since $f(x)$ is known, A_n are known. But $y(x)$ is unknown, so C_n are unknown. In fact

$$A_n=\int_a^b f(t)\, \phi_n(t)\, dt \qquad \ldots(vi)$$

From (*iii*) $L(\phi_n)=-\lambda_n\phi_n$

Now consider the BVP

$$\text{DE} : L(y)=f(x) \qquad \ldots(vii)$$

$$\text{BCs} : y(a)=y(b=)0 \qquad \ldots(ii)$$

If $G(x, t)$ is the Green's function of the homogeneous problem

$$L(y)=0\ ;\ y(a)=y(b)=0$$

then by theorem 1 the solution of the problem (*vii*), (*ii*) is

$$y(x)=\int_a^b G(x, t)\, f(t)\, dt \qquad \ldots(viii)$$

If a non-trivial solution $y_0(x)$ of the homogenous BVP exists satisfying both the boundary conditions, then $G(x, t)$ appearing in ths relation (*viii*) should be deemed to be Green's function in the generalised sense.

Alternatively, the solution of (*vii*), (*ii*) can also be obtained as follows :

Making the substitutions (*iv*) and (*v*) in (*vii*), we get

$$\sum_{n=1}^{\infty} C_n\, L\phi_n(x)=\sum_{n=1}^{\infty} A_n\phi_n(x)$$

i.e.,
$$-\sum_{n=1}^{\infty} C_n\lambda_n\phi_n(x)=\sum_{n=1}^{\infty} A_n\phi_n(x)$$

Hence $\quad C_n=-\dfrac{A_n}{\lambda_n}$ if $\lambda_n\neq 0$.

Also
$$y(x)=-\sum_{n=1}^{\infty}\frac{A_n}{\lambda_n}\phi_n(x)$$
$$=-\sum_{n=1}^{\infty}\left[\frac{1}{\lambda_n}\int_a^b f(t)\,\phi_n(t)\phi_n(x)\,dt\right]$$
$$=-\int_a^b f(t)\left[\sum_{n=1}^{\infty}\frac{\phi_n(x)\,\phi_n(t)}{\lambda_n}\right]dt \qquad ...(ix)$$

Comparing thg relations (*viii*) and (*ix*), we get the bilinear relation,

$$G(x,t)=-\sum_{n=1}^{\infty}\frac{\phi_n(x)\,\phi_n(t)}{\lambda_n} \qquad ...(x)$$

If the Green's function $G_1(x, t)$ is related to the BVP (*iii*), (*ii*), then we proceed as follows :

$$L\varphi_n(x)=-\lambda_n\varphi_n(x)$$
$$L[\Sigma C_n\varphi_n]+\lambda\Sigma C_n\varphi_n(x)=\Sigma A_n\varphi_n(x)$$

i.e.,
$$\Sigma C_n(\lambda-\lambda_n)\,\varphi_n(x)=\Sigma A_n\varphi_n(x)$$

i.e.,
$$C_n=\frac{A_n}{\lambda-\lambda_n}$$

and we shall get
$$y(x)=\sum_{n=1}^{\infty}\frac{A_n\varphi_n(x)}{\lambda-\lambda_n} \qquad ...(xi)$$

If $\lambda=\lambda_n$, there is no solution to BVP (*i*), (*ii*) unless $A_n=0$ *i.e.*, $f(x)=0$.

When $\lambda\neq\lambda_n$, then solution to problem (*i*), (*ii*) is given by
$$y(x)=\int_a^b G_1(x,t)\,f(t)\,dt \qquad ...(xii)$$

Comparing (*xi*) and (*xii*), we have
$$G_1(x,t)=\sum_{n=1}^{\infty}\frac{\phi_n(x)\,\phi_n(t)}{\lambda-\lambda_n} \qquad ...(xiii)$$

For $\lambda=0$, $G_1(x, t)$ reduces to $G(x, t)$ as given by formula (x).

Relation (x) is what is known in Integral Equations as Mercer's theorem. We shall consider this theorem again in a later chapter.

Example 28. Consider the BVP

$$\text{DE}: \quad y''+\lambda y=0 \qquad \ldots(1)$$

$$\text{BCs}: \quad y(0)=y(1)=0 \qquad \ldots(2)$$

Solution. In this problem the eigenfunctions

$\phi_n(x)=\sqrt{2} \sin n\pi x$ correspond to the eigenvalues

$$\lambda_n=n^2\pi^2 \; ; \; n=1, 2, \ldots\ldots$$

The functions $\phi_n(x)$ form an orthonormal sequence on $[0, 1]$.

We know that the Green's function of the homogenoous system $y''=0$; $y(0)=y(1)=0$ is

$$G(x, t)=\begin{cases} x(t-1) & x\leqslant t \\ t(x-1) & x\geqslant t \end{cases}$$

From relation (x), we have

$$G(x, t)=-2\sum_{n=1}^{\infty}\frac{\sin nx \sin nt}{n^2\pi^2}$$

Hence we get

$$K(x, t)=\frac{2}{\pi^2}\sum_{n=1}^{\infty}\frac{\sin nx \sin nt}{n^2} \qquad \ldots(3)$$

where

$$K(x, t)=\begin{cases} x(1-t) & 0\leqslant x\leqslant t\leqslant 1 \\ t(1-x) & 0\leqslant t\leqslant x\leqslant 1 \end{cases} \qquad \ldots(4)$$

Corollary. Put $x=t=\frac{1}{2}$ in (3), (4).

Then $\quad K(x, t)=\frac{1}{4}$

Hence
$$2\sum_{n=1}^{\infty}\frac{\sin^2\frac{n\pi}{2}}{n^2}=\frac{\pi^2}{4}$$

i.e.,
$$\sum_{n=1}^{\infty}\frac{1}{(2n-1)^2}=\frac{\pi^2}{8} \qquad \ldots(5)$$

Now put
$$\sum_{n=1}^{\infty}\frac{1}{n^2}=K$$

Then $$K=\sum_{n=1}^{\infty}\frac{1}{(2n-1)^2}+\sum_{n=1}^{\infty}\frac{1}{(2n)^2}$$

$$=\frac{\pi^2}{8}+\frac{K}{4}$$

Hence $$K=\sum_{n=1}^{\infty}\frac{1}{n^2}=\frac{\pi^2}{6} \quad \ldots(6)$$

Result (6) could also be derived by integrating the relation

$$K(x, x)=x(1-x)=\sum_{n=1}^{\infty}\frac{1-\cos 2n\pi x}{n^2\pi^2}$$

with respect to x from 0 to 1. Term-by-term integration of the series is justified as it is absolutely convergent.

Problem 1. Derive from (3), (4)

$$\sum_{n=1}^{\infty}\frac{1}{n^4}=\frac{\pi^4}{90}$$

Problem 2. Prove that

$$\frac{2}{\pi^2}\sum_{n=1}^{\infty}\frac{\sin(n+\frac{1}{2})\pi x \,.\, \sin(n+\frac{1}{2})\pi t}{(n+\frac{1}{2})^2}=G(x, t)$$

where $$G(x, t)=\begin{cases} x & x\leqslant t \\ t & x\geqslant t,\ 0\leqslant x,\ t\leqslant 1\end{cases}$$

Example 29. Prove that if $P_n(x)$ is the Legendre polynomial, then

$$G(x, t)=\sum_{n=1}^{\infty}\frac{(n+\frac{1}{2})\,P_n(x)\,P_n(t)}{n(n+1)},\quad -1\leqslant x,\ t\leqslant 1$$

where

$$G(x, t)=\begin{cases}-\frac{1}{2}\log[(1-x)(1+t)]+\log 2-\frac{1}{2} & x\leqslant t\\ -\frac{1}{2}\log[(1+x)(1-t)]+\log 2-\frac{1}{2} & x\geqslant t\end{cases}$$

Solution. Consider the BVP

DE : $L(y)+\lambda y\equiv[(1-x^2)y']'+\lambda y=0$ ...(1)

BCs : $y(x)<\infty$ for $x=\pm 1$ ...(2)

It is known that when $\lambda=n(n+1)$. the system is satisfied by the Legendre polynomial $P_n(x)$. Hence the eigenfunctions of the system are $P_n(x)$ related to eigen values $\lambda_n=n(n+1)$. The case $n=0$ is left out of consideration as $\lambda_0=0$.

In view of the orthogonal property.

$$\int_{-1}^{1} P_m(x)\, P_n(x)\, dx=\begin{cases}\dfrac{2}{2n+1} & m=n\\[2mm] 0 & m\neq n\end{cases}$$

the sequence of polynomials $\phi_n(x)=\sqrt{n+\frac{1}{2}}\, P_n(x)$ is orthonormal over $[-1, 1]$.

We now construct the Green's function of the BVP,

$$\text{DE} : L(y)=[(1-x^2)y']'=0 \qquad \text{...(3)}$$

$$\text{BCs} : y(x)<\infty \text{ for } x=\pm 1 \qquad \text{...(2)}$$

Now $y_0(x)=\dfrac{1}{\sqrt{2}}$ is a solution of (3) which satisfies both the boundary conditions. There will, therefore, be no unique Green's function of the system and only a Green's function in the generalised sense will exist. The function $G(x, t)$ will satisfy the following relations :

(*a*) $G(x, t)$ is continous in the domain $-1\leqslant x, t\leqslant 1$.

(*b*) $G(x, t)$ satisfies the boundary conditions

i.e., $\qquad G(\pm 1, t)<\infty$

(*c*) $\dfrac{\partial G(x, t)}{dx}$ is discontinuous at $x=t$ with jump

$$\frac{\partial G(t-0, t)}{\partial x}-\frac{\partial G(t+0, t)}{\partial x}=\frac{-1}{p_0(t)} \qquad \text{...(4)}$$

(*d*) For $x\neq t$, $G(x, t)$ satisfies the non-homogeneous equation

$$LG(x, t)=y_0(x)\, y_0(t)=\tfrac{1}{2} \qquad \text{...(5)}$$

(*e*) $y_0(x)$ is orthogonal to $G(x, t)$, *i.e.,*

$$\int_{-1}^{1} G(x, t)\, y_0(x)\, dx=0$$

i.e.,

$$\int_{-1}^{1} G(x, t)\, dx=0 \qquad \text{...(6)}$$

Now $y_1=\log\dfrac{1+x}{1-x}$ and $y_2=1$ are two linearly independent solutions of the equation

$$L(y)=[(1-x^2)y']'=0$$

and

$$y_p=-\tfrac{1}{4}\log(1-x^2)$$

is a solution of $L(y)=\frac{1}{2}$.

Assume

$$G(x, t)=\begin{cases} A \log \dfrac{1+x}{1-x}+B-\dfrac{1}{4} \log (1-x^2) & x \leqslant t \\ C \log \dfrac{1+x}{1-x}+D-\dfrac{1}{4} \log (1-x^2) & x \geqslant t. \end{cases}$$

The boundary conditions imply that $A=\frac{1}{4}$, $C=-\frac{1}{4}$.

Hence we shall have

$$G(x, t)=\begin{cases} B-\frac{1}{2} \log (1-x) & x \leqslant t \\ D-\frac{1}{2} \log (1+x) & x \geqslant t \end{cases} \qquad ...(7)$$

By the continuity of $G(x, t)$ at $x=t$, we have

$$B-\tfrac{1}{2} \log (1-t)=D-\tfrac{1}{2} \log (1+t) \qquad ...(8)$$

Equation (1) can be rewritten as

$$[(x^2-1)y']'+\lambda y=0$$

Then $\lambda_n=-n(n+1)$, $p_0(x)=x^2-1$.

Hence the jump relation (4) will give

$$\frac{\partial G(t-0, t)}{\partial x}-\frac{\partial(t+0, t)}{\partial x}=\frac{1}{1-t^2} \qquad ...(9)$$

which is already true with (7).

By the orthogonality relation (6), we have

$$\int_{-1}^{t} [B-\tfrac{1}{2} \log (1-x)]\, dx+\int_{t}^{1} [D-\tfrac{1}{2} \log (1+x)]\, dx=0 \qquad ...(10)$$

By (8) and (10), we have

$$B=-\tfrac{1}{2} \log (1+t)+\log 2-\tfrac{1}{2}$$

$$D=-\tfrac{1}{2} \log (1-t)+\log 2-\tfrac{1}{2}$$

Hence by writing (7)

$$G(x, t)=\begin{cases} -\frac{1}{2} \log[(1-x)(1+t)]+\log 2-\frac{1}{2} & x \leqslant t \\ -\frac{1}{2} \log[(1+x)(1-t)]+\log 2-\frac{1}{2} & x \geqslant t \end{cases}$$

Using formula (X) and putting $\phi_n=\sqrt{n+\frac{1}{2}}\, P_n(x)$,

$\lambda_n=-n(n+1)$ we have finally,

$$G(x, t)=\sum_{n=1}^{\infty} \frac{(n+\frac{1}{2})\, P_n(x)\, P_n(t)}{n(n+1)}$$

APPENDIX

The Dirac Delta Function

In mathematical physics one often encounters impulsive forces acting for a very short time. A unit impulsive force can be idealised by a function $p(t)$ such that $\int_{-\infty}^{\infty} p(t)\, dt=1$. Paul Dirac in his book "The Principles of Quantum Mechanics", Oxford, (1947) has denoted such a function by $\delta(x)$ characterised by the properties

$$\delta(x)=0 \text{ if } x\neq 0 \qquad \dots(1)$$

$$\int_{-\infty}^{\infty} \delta(x)\, dx=1 \qquad \dots(2)$$

It has the important "sifting" property

$$\int_{-\infty}^{\infty} f(x)\, \delta(x-a)\, dx=f(a) \qquad \dots(3)$$

No ordinary function satisfies this property.

Mathematically, the delta function can be idealised as the limit of certain function sequetces. Let $\phi(x)$ be a frequency function, *i.e.*, $\phi(x)\geqslant 0$ and $\int_{-\infty}^{\infty} \phi(x)\, dx=1$. Then the function $n\phi(nx)\to\delta(x)$ as $n\to\infty$. Examples of such delta sequence are :

$$\frac{n}{\sqrt{\pi}}\, e^{-n^3x^3}, \quad \frac{\sin nx}{\pi x}, \quad \frac{1}{\pi}\,\frac{n}{1+n^2x^3}, \quad \frac{n}{2}\, e^{-n|x|}$$

The following lemma assumes importance whenever a change of variable is made in $\delta(x)$.

Lemma 2. If an ordinary function $f(x)$ has a simple zero at $x=a$, then

$$\delta\{f(x)\}=\frac{\delta(x-a)}{|f(a)|}$$

Proof. By the sifting property

$$I=\int_{-\infty}^{\infty} \phi(t)\, \delta(t)\, dt=\phi(0)$$

Put $\quad t=f(x)$, so that $f(a)=0$ and

$$I=\int_{-\infty}^{\infty} f'(x)\, \phi\{f(x)\}\, \delta\{f(x)\}\, dx=\phi(0)$$

Put $f'(x)\, \phi\{f(x)\}=\psi(x)$, then since $f(a)\neq 0$

$$f'(a)\, \phi(0)=\psi(a)$$

i.e.,

$$\phi(0)=\frac{\psi(a)}{\psi'(a)}$$

If $f'(a)>0$, then $f(x)$ is increasing at $x=a$.

Hence $$\int_{-\infty}^{\infty} \psi(x)\, \delta\{ f(x)\, dx=\frac{\psi(a)}{f'(a)}$$

If $f'(a)<0$, then $f(x)$ is decreasing at $x=a$. The negative sign of $f'(a)$ can be absorbed with direction of the integral to keep it from $-\infty$ to ∞. Thus

$$\int_{-\infty}^{\infty} \psi(x)\, \delta\{ f(x)\}\, dx=\frac{\psi(a)}{| f'(a) |}$$

Hence $$\delta\{ f(x)\}=\frac{\delta(x-a)}{| f'(a) |}$$

Properties :

(1) $\delta(ax)=\dfrac{1}{| a |}\,\delta(x),\ \delta(-x)=\delta(x).$

(2) $\delta(ax+b)=\dfrac{1}{| a |}\,\delta\left(x+\dfrac{b}{a}\right).$

(3) If $f(x)$ has simple zeros at $x_1, x_2, \ldots\ldots, x_n$, then

$$\delta\{ f(x)\}=\sum_{i=1}^{n} \frac{\delta(x-x_i)}{| f'(x_i) |}.$$

(4) $\delta(x^2-a^2)=\dfrac{1}{2a}\,\{\delta(x-a)+\delta(x+a)\}.$

(5) $\delta(\sin x)=\displaystyle\sum_{n=-\infty}^{\infty} \delta(x-n\pi).$

(6) $\delta(e^{ax}-b)=\dfrac{1}{ab}\,\delta\left(x-\dfrac{\log b}{a}\right).$

(7) $x^n\delta^{(n)}(x)=(-1)^n\,!\,\delta(x).$

(8) $x^{n+r}\delta^{(n)}(x)=0,\ r>0.$

(9) $x^n\delta^{(m)}(x)=\begin{cases}\dfrac{(-1)^n\, m\,!}{(m-n)\,!}\,\delta(x)^{(m-n)} & m\geqslant n.\\ 0 & m<u\end{cases}$

(10) $e^{-x}\delta^{(n)}(x)=\displaystyle\sum_{k=0}^{n}\binom{n}{k}\delta^{(k)}(x).$

(11) $\dfrac{\partial\delta\,(ax)}{\partial a}=-\dfrac{1}{a^2}\,\delta(x).$

(12) $\dfrac{\partial\delta'(ax)}{\partial a}=\dfrac{-2}{a^3}\,\delta'(x).$

(13) $f(x)\,\delta(x)=f(0)\,\delta(x)$

$$\begin{aligned} f(x)\delta'(x) &= (f\delta)'-f'\delta \\ &= (f(0)\,\delta(x))'-f'\delta \\ &= f(0)\,\delta'(x)-f'(0)\,\delta(x) \\ &\cdots \qquad \cdots \qquad \vdots \end{aligned}$$

$$f(x)\,\delta^{(n)}(x)=f(0)\,\delta^{(n)}(x)-nf'(0)\,\delta^{(n-1)}(x)+\binom{n}{2}f''(0)\delta^{(n-2)}+ \\ \ldots\ldots+(-1)^n f^{(n)}(0)\,\delta(x).$$

(14) For $f(x)=x$ we have

$$x\delta'(x)=-\delta(x)$$

i.e.,

$$x\delta'(x)+\delta(x)=0$$

Also

$$x\delta^{(n)}(x)=-n\delta^{(n-1)}(x)$$

$$x^n\delta^{(n)}(x)=(-1)^n\, n\,!\,\delta(x).$$

(15) The relations $H'(x)=\delta(x)$ and $\delta(x)=\delta(-x)$, show that $\delta(x)$ behaves like an even function of degree -1.

(16) By (15) and the relation

$$\int_{-\infty}^{\infty} f(x)\,\delta(x)\,dx=f(0)$$

we get for all functions $f(x)$

$$\int_{0}^{\infty} f(x)\,\delta(x)\,dx=\tfrac{1}{2}\,f(0).$$

9

Higher Dimensional Green's Function: Partial Differential Equations of Mathematical Physics

§9·1. Introduction

The aim of this chapter is to extend the study of one-dimensional Green's functions of linear ODEs to higher dimensional Green's functions of linear PDEs of mathematical physics.

Let D be a domain in an n-dimensional Euclidean space E_n of points $\bar{x}=(x_1, x_2 \ldots, x_n)$. A PDE of order m in D has its highest derivative of order m. A PDE of order m of the form

$$\sum_{i_1, i_2, \ldots, i_n} A_{i_1 i_2 \ldots i_n} \frac{\partial^m u(\bar{x})}{\partial x_1^{i_1} \partial x_2^{i_2} \ldots \partial x_n^{i_n}} + L\, u(\bar{x}) = f(\bar{x}) \qquad \ldots(1)$$

where $i_1+i_2+\ldots+i_n=m$ and L is a partial differential operator of order less than m is called

•**Quasilinear** if it is linear only in its highest derivatives.

•**Semilinear** if it is linear in all its derivatives.

•**Linear** if u and all its derivatives are linear and the coefficients are all functions of $\bar{x}$ only,

•**Non-linear** otherwise.

A general linear second order PDE in n independent variables can be expresses as

$$L(u) = \sum_{i,j=1}^{n} a_{ij}(\bar{x}) \frac{\partial^2 u(\bar{x})}{\partial x_i \partial x_j} + \sum_{i=1}^{n} b_i(\bar{x}) \frac{\partial u(\bar{x})}{\partial x_i} + c(\bar{x})\, u(\bar{x}) = f(\bar{x}) \qquad \ldots(2)$$

where $\bar{x}=(x_1, x_2, \ldots, x_n) \in D$, $u(\bar{x})=u(x_1, x_2, \ldots, x_n)$.

A real function $u(\bar{x})$ defined in the domain D of the given PDE is called a regular *solution* if u and all its derivatives contained in the equation are continuous in D and turn the equation into an identity.

Not all solusions of a PDE are regular. Solutions which are not regular at isolated point or on some manifolds of special type, are called *fundamental* (or *elementary*) *solutions*. It is also called *free space Green's function* or *influence function*.

Theoretically the inverse of a linear partial differential operator L is an integral operator, the kernel of which is the Green's function. It satifies the equation.

$$L(u)=\delta(\bar{x})=\delta(x_1, x_2,\ldots, x_n) \qquad \ldots(3)$$

where $\delta(\bar{x})$ is the n-dimensional Dirac delta function. Salient properties of $\delta(\bar{x})$ will be described in the sequel.

Linear second order PDEs can be classified for n independent variables as we did in the case $n=2$. The *principal part* of (2) is

$$G\left(\frac{\partial}{\partial x_i}, \frac{\partial}{\partial x_j}\right)u=\sum_{i,j=1}^{n} a_{ij}(\bar{x})\frac{\partial^2 u}{\partial x_i\,\partial x_j} \qquad \ldots(4)$$

and

$$Q(\lambda_i, \lambda_j)=\sum_{i,j=1}^{n} a_{ij}(\bar{x})\,\lambda_i\,\lambda_j \qquad \ldots(5)$$

is defined to be the characteristic form of (4).

For the classification of (2) we transform the principal part (5) in the canonical from

$$Q=\sum_{i=1}^{n} \alpha_i\,\xi_i^2$$

by transformations $\lambda_i=\lambda_i(\xi_1, \xi_2\ldots, \xi_n)$, $i=1, 2,\ldots n$. Since the $n/2$ $(n-1)$ coefficients a_{ij} cannot be uniquely transformed by these n transformations for $n>3$, we consider (4) when a_{ij} are constants. The problem is thus reduced to transforming the n^2-matrix $[a_{ij}]$ into the diagonal form. This is a standard problem in linear algebra. In the canonical form α_i have values 1, -1 or 0.

The total number of nonzero coefficients α_i is called *rank*, the number of positive α_i is called *index* and the number of positive α_i minus the number of negative α_i is called the *signature* of quadratic form Q.

When all α_i are positive or all are negative, the quadratic form Q is said to *positive definite* or *negative definite* respectively and PDE (2) is said to be *elliptic* at the given point $\bar{x}$ at which the coefficients $\alpha_i(\bar{x})$ were considered.

If all α_i except one, say α_k, are of one sign and α_k of the opposite sign, then (2) is said to be *hyperbolic*.

In case l $(1>l<n-1)$ coefficients are of one sign and the rest of opposite sign, then (2) is *ultrahyperbolic*.

(2) is said to be parabolic if at least one of the coefficients α_i is zero.

We say that (2) is elliptic hyperbolic or parabolic in a domain D if it is so at every point of D.

(2) is called *uniformly elliptic* if there exist non-zero numbers k_0 and k_1 of one sign such that

$$k_0 \sum_1^n \lambda_i^2 \leqslant Q(\lambda_1, \lambda_2, \ldots, \lambda_n) \leqslant k_1 \sum_1^n \lambda_i^2$$

In sections 9·2 through 9·10 we present an inventory of those mathematical prerequisites, which are required for a proper comprehension of the conceptual and constructive aspects of higher dimensional Green's function. This makes the exposition self-contained besides putting into clearer focus the context in which the role of Green's function is significant. However, the knowledgeable reader may start from section 9·12 and refer to the earlier sections when required.

§9·2 Partial Differential Equations of Mathematical Physics

In applications to physical problems, we often encounter linear PDEs of the form

$$\nabla^2 u = \lambda \frac{\partial^2 u}{\partial t^2} + \mu \frac{\partial u}{\partial t} + \nu u + h \qquad \ldots(1)$$

where h is a given function of position ; λ, μ, ν are certain physical constants and ∇^2 is the Laplace operator in coordinates of the relevant space. For example in Cartesian coordinates.

- in one dimension : $\nabla^2 = \dfrac{\partial^2}{\partial x^2}$
- in two dimensions : $\nabla^2 = \dfrac{\partial^2}{\partial x^2} + \dfrac{\partial^2}{\partial y^2}$
- in three dimensions : $\nabla^2 = \dfrac{\partial^2}{\partial x^2} + \dfrac{\partial^2}{\partial y^2} + \dfrac{\partial^2}{\partial z^2}$

in n dimensions : $\nabla^2 = \dfrac{\partial^2}{\partial x_1^2} + \dfrac{\partial^2}{\partial x_2^2} + \ldots + \dfrac{\partial^2}{\partial x_n^2}$

The operator ∇^2 can be transformed from Cartesian to polar coordinates.

The unknown function u besides being a function of position variables (*i.e.*, space coordinates), also depends on the time variable t.

The operator ∇^2, pronounced "nabla squared" is called the *Laplace operator* or the *Laplacian*. It is also sometimes denoted by Δ, pronounced 'delta'. The differential expression ∇^2 is called the Laplacian of u and solutions of Laplace's equation $\nabla^2 u = 0$ are called *harmonic* or *potential functions*. A potential function is called *regular* if it has second order continuous partial derivatives. Throughout this chapter we shall be concerned with the following equations involving the operator ∇^2. Out of these the Laplace, wave and diffusion are three basic equations.

(*i*) *Laplace's* (or *Potential*) *Equation*

$$\nabla^2 u(\bar{x})=0$$

(*ii*) *Poisson's Equation*

$$\nabla^2 u(\bar{x})=h(\bar{x})$$

(*iii*) *Wave Equation*

$$\nabla^2 u(\bar{x})=\frac{1}{c^2}\ \frac{\partial u^2(\bar{x}, t)}{\partial t^2}$$

(*iv*) *Heat or Diffusion Equation*

$$\nabla^2 u(\bar{x}, t)=\frac{1}{\alpha^2}\ \frac{\partial u(\bar{x}, t)}{\partial t}$$

(*v*) *Helmoltz Equation*

$$(\nabla^2+k^2)\, u(\bar{x})=0$$

(*vi*) *Telegraph Equation*

$$\frac{\partial^2 u}{\partial x^2}-\text{LC}\ \frac{\partial^2 u}{\partial t^2}-(\text{RC}+\text{LG})\ \frac{\partial u}{\partial t}-\text{RG}\ u=0.$$

(*vii*) *Bi-Laplacian* or *Bi-harmonic Equation*

$$\nabla^4 u(x, y)=\frac{\partial^4 u}{\partial x^4}+2\ \frac{\partial^4 u}{\partial x^2\, \partial y^2}+\frac{\partial^4 u}{\partial y^4}=0$$

The reader might be familiar with the following examples of harmonic functions :—

(*a*) If $f(\bar{z})=u(x, y)+iv(x, y)$ is an analytic function in a domain D, then owing to Cauchy-Riemann equations, $u(x, y)$ and $v(x, y)$ are both real harmonic functions, called conjugate harmonic functions. $W=f(\bar{z})$ is then called the *complex potential.*

(*b*) If $f(z+ix\ \cos t+iy\ \sin t, t)$ is differentiable in x, y, z in a region D, and

$$u(x, y, z)=\int_a^b f(z+ix\cos t+i\,x\cos t+iy\,\text{sin}t, t)\, dt$$

then f and $u(x, y, z)$ are harmonic in D.

Summation Convention. On certain occasions, we shall find it convenient to dispense with the sigma sign and write expressions

$$\sum_{i,j=1}^{n} A_{ij} \quad \text{and} \quad \sum_{i=1}^{n} B_i$$

simply as A_{ij} and B_i respectively.

The equation (2) can be written as

$$a_{ij}\frac{\partial^2 u}{\partial x_i\,\partial x_j}+b_i\frac{\partial u}{\partial x_i}+cu=f$$

and
$$\nabla^2 u=\frac{\partial^2 u}{\partial x_i\,\partial x_i}.$$

This convention has to be used with caution as $\dfrac{\partial^2 u}{\partial x_i\,\partial x_j}$ also stands for $u_{x_i x_j}$ and $\dfrac{\partial u}{\partial x_i}$ for u_{x_i}. In the later x_i, x_j are fixed.

§9.3. Manifolds : Curves, Surfaces and Hypersurfaces

Any consideration of *a* PDE draws heavily upon geometrical concepts. Hence we explain certain terms related to geometrical structures in a Euclidean n-space.

In the Cartesian xy-plane, an equation of the form $y=f(x)$ denotes a curve. Such a form is not convenient when ambiguity may arise owing to the many-valuedness of f. For example both functions $y=\pm(1-x^2)^{1/2}$ represent the same circle $x^2+y^2=1$. To avoid such ambiguities recourse is taken to parametric representation such as

$$x=x(t), \qquad y=y(t) \qquad a\leqslant t\leqslant b$$

where $x(t)$ and $y(t)$ are continuous functions of a real parameter t. In a special case the parameter may be arc length s.

The parameteric representation of a given curve may not be apparently unique. For example,

$$r_1(t)=(\cos t, \sin t),\ r_2(t)=(\sin t, \cos t) \qquad -\infty<t<\infty$$
$$r_3(t)=(\cos 2\pi t, \sin 2\pi t),\ r_4(t)=(\cos t^2, \sin t^2)$$

all represent the same circle $x^2+y^2=1$.

In R^3, equations $x=x(t)$, $y=y(t)$, $z=z(t)$ denote a space curve. The basic idea of a space curve is that of a twisted wire which can be unwinded into the shape of a linear interval.

A curve $x_i=x_i(t)$ $(i=1, 2,\ldots, n)$, $a\leqslant t\leqslant b$ in n-space is called a C^n curve if each $x_i(t)\in C^n$. Since the mapping $u=a+(b-a)\,t$ maps the unit interval $I=[0, 1]$ onto $[a, b]$ in a one-one continuous manner, there is no loss of generality in assuming the paramter interval to be I *i.e.*, $t\in[0, 1]$.

A curve in n-space is defined to be a continuous map $\bar{x} : I\to R^n$. It is a *closed* curve if $\bar{x}(0)=\bar{x}(1)$. A curve $\bar{x} : I\to R^2$ is a *plane curve*.

In the three-dimensional space R^3, a surface may be represented by two parameters u, v as

$$x=x(u, v),\ y=y(u, v),\ z=z(u, v)$$

provided
$$\frac{\partial(x, y)}{\partial(x, v)}\neq 0,\quad \frac{\partial(y, z)}{\partial(u, v)}\neq 0.$$

In general if $m<n$, then for real independent parameters $s_1, s_2, \ldots, s_m$, the equations

$$\bar{x}=\bar{x}\,(s_1, s_2, \ldots, s_m)$$

i.e., $$x_i=x_i\,(s_1, s_2, \ldots, s_m) \qquad (i=1, 2, \ldots n) \qquad \ldots(2)$$

represent a geometrical structure called an m-dimensional *manifold* or *surface* in the n-space. If $m=n-1$, the manifold is called a *hypersurface*. A *curve* is a one-dimensional manifold. An equntion of the form $f(\bar{x})=f(x_1, x_2, \ldots, x_n)=c$ (real) represents a hypersurface.

If the parameters $s_1, s_2, \ldots, s_n$ appear in the first degree, the manifolds are called planes. In particular :

- a zero dimensional plane is a *point*
- a one-dimensional plane is a *straight line*
- an $(n-1)$-dimensional plane is a *hyperplane*
- when $m=n$ the plane is the whole space.

If L_m is an m-dimensional subspace of R^n and $\bar{x}_0 \in R^n$ is a fixed vector, then $\bar{x}_0+L_m$ is an m-dimensional plane parallel to L_m.

If $\bar{x} \in R^n$, then the equation

$$a(\bar{x}, \bar{x})+2b(\bar{x})+c=0 \qquad \ldots(3)$$

is called a *quadratic hypersurface*, where $a(\bar{x}, \bar{x})$ is a quadratic form, $b(\bar{x})$ is a linear form and c is a constant. If the running coordinates of the point $\bar{x}$ are $x_1, x_2, \ldots, x_n$, then (3) can be written in the form

$$\sum_{i,j=1}^{n} a_{ij}\, x_i x_j+2 \sum_{i=1}^{n} b_i\, x_i+c=0 \qquad \ldots(4)$$

In our work we shall designate a quadric hypersurface simply as a surface. The following canonical forms of (6) are noteworthy :

(*a*) $x_1^2+x_2^2+\ldots+x_n^2=r^2$ (Sphere)

(*b*) $\dfrac{x_1^2}{a_1^2}+\dfrac{x_2^2}{a_2^2}+\ldots+\dfrac{x_n^2}{a_n^2}=1$ (Ellipsoid)

(*c*) $\dfrac{x_1^2}{a_1^2}+\ldots+\dfrac{x_k^2}{a_k^2}-\dfrac{x^2_{k+1}}{a^2_{k+1}}-\ldots-\dfrac{x_n^2}{a_n^2}$ (Hyperboloid)

(*d*) $\lambda_1 x_1^2+\lambda_2 x_2^2+\ldots+\lambda_{n-1}\, x^2_{n-1}-2\mu x_n=0.$ (Paraboloid)

The equation

$$|\,\bar{x}-\bar{a}\,|^2=(x_1-a_1)^2+\ldots+(x_n-a_n)^2=r^2$$

is the sphere with radius r and centre $\bar{a}=(a_1, a_2, \ldots, a_n)$.

We can now introduce the notation of a family of curves and surfaces. If t is a real parameter, the equation

$$f\,(x, y\,;\,t)=0 \qquad \ldots(5)$$

represents a single infinity (or a one-parameter family) of curves on plane. Fixing t results in fixing a particular curve of the family. This curve in turn may be represented by another parameter s. The curve will then be represented by

$$x=x(s\,;\,t_0),\ y=y(s\,;\,t_0)$$

In general the system

$$x_i=x_i(s\,;\,t_1, t_2,\dots, t_k)\ (i=1, 2,\dots, n)$$

defines a k-parameter or k-fold infinity of curves in n-space. More generally, for $m<n$, the system

$$x_i=x_i\,(s_1, s_2,\dots, s_m\,;\,t_1, t_2,\dots, t_k)\ (i=1, 2,\dots, n)$$

defines a k-fold family of manifolds.

If $m=2$ the system

$$x_i=x_i\,(s_1, s_2\,;\,t_1, t_2,\dots, t_k)\ (i=1, 2,\dots, n)$$

defines a k-fold infinity of surfaces.

Envelopes of families of curves and surfaces are *singular manifolds.*

An ordinary differential equation is supposed to be definedon an interval I. On the same lines solutions of a PDE will be supposed to be defined in a region T. For example, T is an n-dimensional region in the n-space R^n for the Laplace's equation and $\bar{x}=(x_1, x_2,\dots \dots x_n)$. For the wave and heat equation T is an $(n+1)$-dimensional region of n space coordinates $x_1, x_2,\dots, x_n$ and one time coordinate t.

A point $P(\bar{a})$ is said to be *interior* to a region T if it is possibe to draw a sphere $\Sigma(x_i-a_i)^2=r^2$ such that all points $\bar{x}=(x_1, x_2,\dots, x_n)$ for which $\Sigma(x_i-a_i)^2<r^2$ belong to T. P is called an *exterior* point of T if there exists a sphere centred at P none of whose interior points belong to T. If every sphere about P contains both interior and exterior points, then P is called a *boundary point* of T. The set of all boundary points of T is called the *boundary* of T and is denoted by ∂T (sometimes by bdry T). The set of all points in T and ∂T is denoted by $\bar{T}=T\cup\partial T$ and is called the *closure* of T. A region T is called *closed* if it contains all its boundary points *i.e.*, $\bar{T}=T$. A region T is called *open* if all its points are interior points. As an example,

T is a *closed ball* : $x_1^2+x_2^2+\dots+x_n^2\leqslant R^2$

T is an *open ball* : $x_1^2+x_2^2+\dots+x_n^2<R^2$

∂T is the *sphere* : $x_1^2+x_2^2+\dots+x_n^2=R^2$

While T is an n-dimensional region, ∂T is $(n-1)$-dimensional.

A region T is called *bounded* if it can be contained in the interior of a ball of finite radius. T is said to be *convex* (or *connected* or *arcwise connected*) if any two points of T can be joined by an arc of a curve all of whose points belong to T. The portion of the curve $\bar{x}=\bar{x}(t)$, $a\leqslant t\leqslant b$ is called an *arc* or *segment* of the curve. An open connected region is called a *domain.*

A region is called *simply connected* if every closed curve lying in the region can be shrunk continuously to a point without leaving the region. As an example, a ball of finite radius is a bounded simply connected region.

§9·4. Domain and Range of Operators

Let dV denote the Euclidean volume element of an n-dimensional region T. For example, in n-space with Cartesian coordinates

$$dV = dx_1\, dx_2 \ldots\ldots dx_n.$$

For $n=3$ in polar coordinates

$$dV = r^2 \sin\theta\, dr\, d\theta\, d\phi$$

$f \in C^n$ if f and all its derivativas with respect to $x_1, x_2, \ldots, x_n$ of order n are continuous. It is piecewise continuous if there are only a finite number of finite discontinuities in every bounded region.

A function $u(\bar{x})$ defined on a domain T is said to be square-integrable in the Lebesgue sense if

$$\int_T u^2\, dV < \infty.$$

The space of all such square integrable functions is denoted by $L^2(T)$.

An inner product in $L^2(T)$ is defined by

$$(u, v) = \int_T uv\, dV$$

which induces the norm

$$\| u \| = (u, u)^{1/2} = \left(\int_T u^2\, dV \right)^{1/2}$$

The operators we are going to discuss are $\nabla^2, \frac{\partial}{\partial t}$ and $\frac{\partial^2}{\partial t^2}$ and their domains will be subsets of $L^2(T)$ satisfying additional conditions.

The domain of ∇^2 is a subset of $L^2(T)$ with its members having piecewise continuous, square integrable second derivatives and satisfy certain boundary data on the boundary of T.

The domain of the operator $\frac{\partial}{\partial t}$ is the subset of $L^2(T)$ whose members u are such that $\frac{\partial u}{\partial t}$ is piecewise continuous and square integrable over T and satisfies initial condition such as $u(\bar{x}, 0)=0$.

The domain of the operator $\frac{\partial^2}{\partial t^2}$ is a subset of $L^2(T)$ such that its members u have piecewise continuous second derivative $\frac{\partial^2 u}{\partial t^2}$.

Also $\frac{\partial^2 u}{\partial t^2}$ are square integrable and initial conditions such as

$$u(\bar{x}, 0)=0, \quad \frac{\partial u}{\partial t}(\bar{x}, 0)=0$$

are satisfied.

The domain of a combination of these operators is the intersection of the domains of the operators involved.

§9·5. Boundary and Initial Value Problems

The general solution *i.e.*, the totality of all solutions of a PDE is not much relevent in mathematical physics. The interest is directed to picking out that solution which satisfies certain restrictions conforming to the physical solution. If the number of independent variables is n, these restrictions relate to $(n-1)$ dimensional manifolds E_{n-1} of the n-dimensional region T. These manifolds may be the whole or part of the boundary ∂T.

The problem of finding a solution with conditions related to the boundary is called a *boundary value problem* (BVP).

Let G be a domain in the $(n-1)$-dimensional subspace E_{n-1} of the variables, say $x_1, x_2, \ldots, x_{n-1}$. Then the following is a *Cauchy problem*.

Find the regular solution $u(x)$ of the equation

$$\sum_{i=1}^{n-1} \frac{\partial^2 u}{\partial x_i^2} - \frac{\partial^2 u}{\partial x_n^2} = 0 \qquad \ldots(1)$$

satisfying the conditions

$$u(x_1, x_2 \ldots, x_{n-1}, 0)=f(\bar{x}) \qquad \ldots(i)$$

$$\left[\frac{\partial u(x_1, x_2, \ldots, x_{n-1}, x_n)}{\partial x_n}\right]_{x_n=0} = g(\bar{x}) \qquad \ldots(ii)$$

for $\bar{x}=(x_1, x_2, \ldots, x_{n-1}) \in$ G and f, g are sufficiently smooth functions defined in G.

Conditions (*i*), (*ii*) are called *Cauchy conditions* or *initial conditions*, f, g are called *Cauchy data* and the system (1), (*i*)—(*ii*) is called a *Cauchy problem* or *initial value probldm* (IVP). G is called the *initial manifold* or *initial domain.* In the IVP G is the hypersurface obtained by the intersection of the n-dimensional region T and the hyperplane $x_n=0$. An initial domain may not be a proper subset of the boundary. For example in E_2 consisting of points (x, t), the initial domain may be $t=0$ or a subset of it.

In general elliptic equations are associated with boundary conditions and hyperbolic and parabolic equations with initial conditions.

The following types of BVPs are met with in potential theory :

I. **The first boundary value problem or the Dirichlet problem.** The solution u is a regular hormonic function in a regular harmonic in a region T and u assumes assigned values on the boundary ∂T.

II. **The second boundary value problem or the Neumann problem.** The solution is harmonic in T and the normal derivative $\frac{\partial u}{\partial n}$ takes assigned values on ∂T.

III. **The third boundary value problem or the Churchill problem (or the Robbin problem).** A linear combination of the harmonic function and its normal derivative takes assigned values on the boundary.

The solution of a BVP or IVP to be meaningful, must conform to the physical situation. The following requirements ensure the relevance of the solution of a problem :

(*i*) *Existence.* The problem must have a solution.

(*ii*) *Uniqueness.* The solution must be unique.

(*iii*) *Stability.* The solution must depend on the data continuously. That is to say, a small change in the data should cause a small change in the solution.

A problem which satisfies all these three requirements is called *well-posed* or *properly posed* or *correctly set*, a notion first introduced by Hadamard.

A problem which is not well-posed is called *ill-posed or improperly posed.*

Boundary value problem for the elliptic equation and Cauchy problem for a hyperbolic equation are well-posed.

A Cauchy problem for the elliptic equation and boundary value problem for the hyperbolic equation are not well-posed.

Example 1. PDE : $\frac{\partial^2 u}{\partial x^2}-\frac{\partial^2 u}{\partial t^2}=0$

ICs : $u(x, 0)=f(x),\ \frac{\partial u}{\partial t}=g(x).$

The D' Alembert's solution to the Cauchy problem is

$$u(x, t)=\tfrac{1}{2} f(x+t)+\tfrac{1}{2} f(x-t)+\tfrac{1}{2}\int_{x-t}^{x+t} g(s)\, ds,\quad t>0$$

The solution exists, is unique and depends continuously on the data $f(x)$ and $g(x)$. Hence the Cauchy problem for the wave equation is well-posed.

Example 2. For the hyperbolic equation $\frac{\partial^2 y}{\partial x \partial y}=0$, a boundary value problem for a rectangle does not exist.

Example 3. (Hadamard[1])

$$\text{PDE}: \quad \frac{\partial^2 u}{\partial x^2}+\frac{\partial^2 u}{\partial y^2}=0 \quad -\infty<x<\infty, -\infty<y<\infty$$

$$\text{ICs} - u(x, 0)=0, \frac{\partial u(x, 0)}{\partial y}=\frac{\sin nx}{n}.$$

Here $$u(x, y)=\frac{\sin nx \sinh ny}{n^2}$$

The solution thus exists and is unique. However, it is uot stable. As $n\to\infty$ the Cauchy data become

$$u(x, 0)=0, \quad \frac{\partial u(x, 0)}{\partial y}=0$$

and $$u(x, y)=0$$

is the solution corresponding to these Cauchy data. But as $n\to\infty$,

$$u(x, y)=\frac{\sin nx \sinh ny}{n^2}$$

does not tend to zero. A small change in Cauchy data thus causes a large change in the solution. The Cauchy problem for the potential equation thus is not well-posed.

Problem. Argue as in Ex. 3 for the solution

$$u_n(x, y)=\frac{\cosh nx \cos ny}{n^2}$$

$$u_n(0, y) \Rightarrow 0 \text{ as } n\to\infty$$

$$\frac{\partial u_n(0, y)}{\partial x}=0.$$

§9·6. Results in Higher Dimensional Calculus

A. Vectorial Differentiation : Inner and Outer Derivatives.

Let $u(\bar{x})$ be a scalar-valued C^1 function of position $\bar{x}=(x_1, x_2, \ldots, x_n)$ *i.e.*, the first order derivatives $\frac{\partial u}{\partial x_i}$ are continuous. $P(\bar{a})$ is a given point in space with position vector $\bar{a}=(a_1, a_2, \ldots\ldots a_n)$ and $\bar{b}=(b_1, b_2, \ldots, b_n)$ a given unit vector. The vectorial equation of the ray from P parallel to $\bar{b}$ is given by

$$\bar{x}=\bar{a}+s\bar{b}, \quad s\geqslant 0$$

1. J.S. Hadamard (1952): Lectures on Cauchy's Problem in Linear Partial Differential Equations, Dover, N.Y.

If $Q(\bar{x})$ is any point on this ray, then the coordinates of Q are given by the equations

$$x_i = a_i + sb_i \quad (i=1, 2, \ldots, n) \qquad \ldots(2)$$

The limit

$$\frac{\partial u}{\partial s} = \lim_{s \to 0} \frac{u(\bar{x}) - u(\bar{a})}{s}$$

if exists, is called the *directional derivative* of u at P in the direction of the unit vector $\bar{b}$. For example, in 3-space $\frac{\partial u}{\partial x}$, $\frac{\partial u}{\partial y}$, $\frac{\partial u}{\partial z}$ represent the rates of change of u in the directions of the x- and y- and z-axes respectively. In an n-space

$$\frac{\partial u}{\partial s} = \sum_{i=1}^{n} \frac{\partial u}{\partial x_i} \frac{\partial x_i}{\partial s}$$

$$= \sum_{i=1}^{n} b_i \frac{\partial u}{\partial x_i} \qquad \ldots(3)$$

Thus
$$\frac{\partial}{\partial s} = \sum_{i=1}^{n} b_i \frac{\partial}{\partial x_i}$$

denotes differentiation in the direction of the vector b.

It is possible to express the directional derivative $\frac{\partial u}{\partial s}$ in (3) in terms of a vector differential operator Δ defined by

$$\text{grad } u = \nabla u = \left(\frac{\partial u}{\partial x_1}, \frac{\partial u}{\partial x_2}, \ldots, \frac{\partial u}{\partial x_n} \right)$$

∇u is called the gradient vector of the scalar function u.

Using the inner product notation

$$u \,.\, v = v_1 v_1 + u_2 v_2 + \ldots, u_n v_n$$

where
$$u = (u_1, u_2, \ldots, u_n), \; v = (v_1, v_2, \ldots, v_n)$$

we can express the relation (3) in the form

$$\frac{\partial u}{\partial s} = \bar{b} \,.\, \nabla u$$

$$= \frac{d\bar{x}}{ds} \cdot \nabla u \qquad \ldots(4)$$

If the angle between the vectors $\bar{b}$ and ∇u is γ, then,

$$\left| \frac{\partial u}{\partial s} \right| = | \nabla u | \cos \gamma \qquad \ldots(5)$$

$\left|\dfrac{\partial u}{\partial s}\right|$ is maximum when $\gamma=0$, *i.e.*, b is parallel to grad u. This means the vector grad u is in the direction of maximum increase of u.

Let $u(\bar{x})$ be a scalar C^1 function and $u(\bar{x})=0$ be a given surface in n-space. Then $\dfrac{d\bar{x}}{ds}$ is a unit tangential vector to S at the point $\bar{x}$.

Now
$$\frac{\partial u}{\partial s}=\frac{\partial u}{\partial x_1}\frac{\partial x_1}{\partial s}+\ldots+\frac{\partial u}{\partial x_n}\frac{\partial x_n}{\partial s}$$
$$=\nabla u\cdot\frac{d\bar{x}}{ds}=0.$$

Hence ∇u is a normal vector to the surface $u=0$,
$$\bar{n}=\frac{\nabla u}{|\nabla u|}$$
is a unit normal vector to S, where
$$|\nabla u|=\left(\sum\left(\frac{\partial u}{\partial x_i}\right)^2\right)^{1/2}$$

If $|\nabla u|\neq 0$, then the directional derivative of f along the unit normal $\bar{n}$ to $u=0$ is given by
$$\frac{\partial f}{\partial n}=\bar{n}\cdot\nabla f$$
and is called the *normal derivative* of the scalar function $f(\bar{x})$ at $\bar{x}$ of the surface $u(\bar{x})=0$.

Since $\dfrac{d\bar{x}}{ds}$ is a unit tangent vector to the surface S : $u(\bar{x})=0$, if $f(\bar{x})$ is a scalar function, then
$$\frac{\partial f}{\partial s}=\frac{d\bar{x}}{ds}\cdot\nabla f$$
is a *tangential derivative* or *inner derivative* of f on the surface S.

In case $\dfrac{d\bar{x}}{ds}\cdot\nabla u\neq 0$, then $\dfrac{d\bar{x}}{ds}$ is not a tangent vector to S and $\dfrac{\partial f}{\partial s}$ will not be a tangential derivative. Such a derivative is called an *outer derivative*.

The directional derivative of a vector function can also be defined on the same lines as that of a scalar function as given in (4) Let
$$\bar{w}(\bar{x})=(w_1, w_2,\ldots, w_n)$$
be a vector function with components $w_i\in C^1$ being functions of $(x_1, x_2,\ldots, x_n)$. Then the directional derivative of $\bar{w}$ in the direction of the vector $\bar{b}$ is given by

$$\frac{\partial \bar{w}}{\partial s} = \sum \frac{\partial \bar{w}}{\partial x_i} \frac{dx_i}{ds}$$

$$= \sum \left(\frac{dx_i}{ds} \cdot \frac{\partial}{\partial x_i} \right) \bar{w}$$

$$= \left(\frac{d\bar{x}}{ds} \cdot \nabla \right) \bar{w} \qquad \ldots(6)$$

where $\bar{x} = \bar{a} + s\bar{b}$. The difference in the formula (4) and (6) should be marked.

B. Vectorial Integration—Green's theorem

Gauss' Divergence Theorem Let $\bar{w}$ be a vector field in a region T bounded by a closed smooth surface ∂T such that

$$\bar{w} \in C^1 \text{ in T and } \bar{w} \in C \text{ in } T \cup \partial T (= \bar{T})$$

and $\bar{n}$ is the outward drawn unit normal to ∂T, then by the divergence theorem of Gauss

$$\int_T \nabla \bar{w} \, dV = \int_{\partial T} (\bar{n} \cdot \bar{w}) dS \qquad \ldots(7)$$

where dV is the volume element of T and dS is the surface element of ∂T. The integral on the left of (7) denotes multiple integral of the nth order in T and that on the right denotes multiple integral of the $(n-1)$th order over ∂T.

If v is a scalar function and $\bar{n}$ is a unit normal to the surface $v = c$, then the normal derivative of v (*i.e.*, the directional derivative of v along $\bar{n}$), which we denote by $\frac{\partial v}{\partial n}$ is given by

$$\frac{\partial v}{\partial n} = \bar{n} \cdot \nabla v \qquad \ldots(8)$$

If we put $\bar{w} = u \cdot \nabla v$, then

$$\nabla \bar{w} = \nabla u \cdot \nabla v + u \nabla^2 v$$

and

$$\bar{n} \cdot \bar{w} = \bar{n} \cdot (u \cdot \nabla v)$$

$$= u \cdot (\bar{n} \cdot \nabla v)$$

$$= u \frac{dv}{dn}.$$

Hencc (7) leads to the *first form of Green' theorem*

$$\int_T (\nabla u \cdot \nabla v + u \nabla^2 v) dV = \int_{\partial T} u \frac{\partial v}{\partial n} dS \qquad \ldots(9)$$

It is assumed that in (9) $u, v \in C^2$ in T and $u, v \in C$ in T,

$v \in C^1$ in $T \cup \partial T$, $v \in C^2$ in T and $u \in C^1$ in T.

Interchanging u and v in (9) and subtracting we get the second form of Green's theorem

$$\int_T (u\nabla^2 v - v\nabla^2 u)dV = \int_{\partial T}\left(u\frac{\partial v}{\partial n} - v\frac{\partial u}{\partial n}\right)dS \quad \ldots(10)$$

where $u, v \in C^1$ in $T \cup \partial T$ and $u, v \in C^2$ in T.

Let $u = v$ and be harmonic in T, then (9) gives

$$\int_T (\nabla u)^2\, dV = \int_{\partial T} u\frac{du}{dn}\, dS \quad \ldots(11)$$

If we take $v = 1$ in (10) then we get

$$\int_T \nabla^2 u\, dV = \int_{\partial T} \frac{\partial u}{\partial n}$$

Remark 1. (*a*) For a three-dimensional closed volume T, ∂T is a surface. In Cartesian coordinates $dV = dx\,dy\,dz$. dS is a surface element.

(*b*) For a two-dimensional closed plane region T, $dV = dx\,dy$ and ∂T is a curve and $dS = ds$.

Problem. Verify formula (9) for $u = v = x$ and T a circular disc of radius a and centre origin.

Example 4. The two-dimensional form of the Gauss' theorem (7) is

$$\int_T \int \nabla \bar{\omega}\, dx\, dy = \int_C \bar{n}\,\bar{\omega}\, ds.$$

Here $\nabla = i\frac{\partial}{\partial x} + j\frac{\partial}{\partial y}$, T is a plane region bounded by a closed curve C and ds the element of arc. The unit tangent vector $\bar{t}$ and the unit normal vector n at a ponit $\bar{x} = (x, y)$ of C are given respectively by

$$\bar{t} = i\frac{dx}{ds} + j\frac{dy}{ds} = \left(\frac{dx}{ds}, \frac{dy}{ds}\right)$$

$$\bar{n} = \left(\frac{dy}{ds}, -\frac{dx}{ds}\right) = (n_1, n_2).$$

If $\bar{\omega}(x, y) = (f(x, y), g(x, y))$

then $$\nabla\omega = \frac{\partial f}{\partial x} - g\frac{\partial g}{\partial y},$$

$$\bar{n}\,\bar{\omega} = f\frac{dy}{ds} - g\frac{dx}{ds}$$

Hence we get two-dimensional form of *Stoke's theorem*

$$\int_T \int \left(\frac{\partial f}{\partial x} + \frac{\partial g}{\partial y} \right) dx \ \ dy = \int_C (f\,dy - g\,dx) \qquad ...(12)$$

Example 5. In three-dimensional space

$$\bar{n} \cdot \bar{\omega} = (n_1, n_2, n_3) \cdot (\omega_1, \omega_2, \omega_3)$$
$$= n_1\omega_1 + n_2\omega_2 + n_3\omega_3$$

$$\nabla \bar{\omega} = \frac{\partial \omega_1}{\partial x} + \frac{\partial \omega_2}{\partial y} + \frac{\partial \omega_3}{\partial z}$$

Hence by (7), we have

$$\int_T \iint \left(\frac{\partial \omega_1}{\partial x} + \frac{\partial \omega_2}{\partial y} + \frac{\partial \omega_3}{\partial z} \right) dx\,dy\,dz$$

$$= \iint_{\partial T} (\omega_1\,dy\,dz + \omega_2\,dz\,dx + \omega_3\,dx\,dy) \qquad ...(13)$$

In the *n*-dimensional region in Cartesian coordinates we have

$$dV = x_1\,dx_2 ... dx_{i-1}\,dx_i\,dx_{i+1} ... dx_n$$
$$n_i dS = dS_i = dx_1 . dx_2 ... dx_{i-1}\,dx_{i+1} ... dx_n$$

and
$$\int_T \nabla \bar{\omega}\,dx_1\,dx_2 ...,\ dx_n = \int_{\partial T} \left(\sum \omega_i\,dS_i \right) \qquad ...(14)$$

Problem. Using (14) prove the "*gradient theorem*

$$\int_T \nabla \phi dV = \int_{\partial T} \phi ds$$

[**Hint** : Use (14) component-wise]

§9·7. Adjoint operator and Green's Theorems

In section 4·2 we considered the idea of adjoint to an ordinary differential oparator. The idea was developed through the process of integration by parts. In higher dimensional calculus the role of integration by parts is taken over by the Green's theorems and their special cases.

Let L be a linear partial differential operator with *n* independent variables. If *u, v* are continuously differentiable functions, then we can integrate *v* L(*u*) by parts till all derivatives of *u* are removed. The process at this stage gives rise to another linear partial differential operator $\bar{L}$ giving a relation of the form

$$vL(u) = u\bar{L}(u) + \nabla \bar{\omega} \qquad ...(15)$$

Relation (15) determines $\bar{L}$ uniquely upto an additive vector function $f(\bar{x})$ such that $\nabla f(\bar{x}) = 0$. (15) can be rewritten as

$$vL(u) - u\bar{L} = \nabla \bar{\omega} \qquad ...(16)$$

where $\bar{\omega}=P(u, v)$ is an n-dimensional bilinear vector function. L and $\bar{L}$ are called adjoint operators. The reader may recall that the analogue of relation (16) in the case of, ordinary differential operator was derived in the form of Lagrange's identity

$$v\,L(u)-u\,\bar{L}(v)=\frac{d}{dx}\{P(u, v)\}.$$

The role of the differential operator $\frac{d}{dx}$ is assumed by the operator ∇ in higher dimension.

Integrating (16) in the region T we get

$$\int_T \{v\,L(u)-u\,\bar{L}(v)\}\,dV=\int_T \nabla\,\bar{\omega}\,dV$$

$$=\int_{\partial T} \bar{n}\,\bar{\omega}\,dS \qquad \ldots(17)$$

(17) is the analogue of Green's formula.

Let $u, v \in L^2(T)$ and that $L(u)$ and $L(v)$ be continuous in T, then using the inner product

$$(f, g)=\int_T f g\,dV,$$

the result (17) can expressed as

$$(v, L(u))-(\bar{L}(v), u)=\int_{\partial T} \bar{n}\,\bar{\omega}\,dS.$$

In case the right hand side of (17) vanishes then

$$(v, L(u))=(\bar{L}(v), u) \qquad \ldots(18)$$

In view of relation (18), the use of the term adjoint operators for L and $\bar{L}$ is justified.

Multi-index. If $p_1, p_2, \cdots, p_n$ are non-negative integers, then

$$p=(p_1, p_2, \cdots, p_n)$$

is called a *multi-index*. The following conventions are adopted in its use :

(*a*) $|p|=p_1+p_2+\cdots+p_n$.

(*b*) $\bar{x}^p=x_1^{p_1}\,x_2^{p_2}\ldots x_n^{p_n}$ where $\bar{x}=(x_1, x_2, \cdots, x_n)$

(*c*) $p!=p_1\,!\,p_2\,!\,\ldots\,p_n\,!$

(*d*) $\binom{p}{q}=\binom{p_1}{q_1}\binom{p_2}{q_2}\cdots\binom{p_n}{q_n}$

$$=\frac{p\,!}{(p-q)\,!\,q\,!}$$

(*e*) If $D_i = \dfrac{\partial}{\partial x_i}$, then the *multi-differential operator* $D=(D_1, D_2, \ldots, D_n)$ operates as follows :

$$D^p = D_1^{p_1} D_2^{p_2} \ldots D_n^{p_n} = \frac{\partial^{|p|}}{\partial x_1^{p_1}\, \partial x_2^{p_2} \ldots \partial x_n^{p_n}}$$

(*f*) $a_p = a_{p_1 p_2 \ldots p_n}$.

If we denote unit vectors along the orthogonal coordinate vectors x_i by l_i then $\Sigma\, l_i^2 = 1$ and

$$\nabla = l_1 \frac{\partial}{\partial x_1} + l_2 \frac{\partial}{\partial x_2} + \ldots + l_n \frac{\partial}{\partial x_n}.$$

The multioperator $D=(D_1, D_2, \ldots, D_n)$ is not to be confused with the vectorial differential operator

$$\nabla = \left(\frac{\partial}{\partial x_1}, \frac{\partial}{\partial x_2}, \ldots, \frac{\partial}{\partial x_n}\right) = \sum l_i \frac{\partial}{\partial x_i}$$

While powers of ∇ appear in the sense of inner product, a of power D is product of its consituents. For example, for $n=3$

$$\nabla^2 u = \frac{\partial^2 u}{\partial x_1^2} + \frac{\partial^2 u}{\partial x_2^2} + \frac{\partial^2 u}{\partial x_3^2}.$$

But if $p=(1, 0, 1)$, then $|p| = 2$ and

$$D^p u = D_1 D_2^0 D_3 u = \frac{\partial^2 u}{\partial x_1 \partial x_3}.$$

The convention of multi-index enables us to make the following definition of adjoint :

Definition 1. A linear partial differential operation of order *m* can be represented as :

$$L(u) = \sum_{|p| \leq m} a_p(\bar{x})\, D^p u, \qquad \bar{x} \in T \qquad \ldots(i)$$

where $\quad p=(p_1, p_2, \ldots, p_n) \quad D=(D_1, D_2, \ldots, D_r)$

$\bar{x} = (x_1, x_2, \ldots, x_n)$.

The *adjoint operator* to L is given by

$$\bar{L}(v) = \sum_{|p| \leq m} (-1)^{|p|} D^p (a_p v) \qquad \ldots(ii)$$

In our work we shall be concerned with first and second order operators.

If $L=\bar{L}$, the operator is called self-adjoint.

Example 6. Let $L(u) = \displaystyle\sum_{i=1}^{n} a_i(\bar{x}) \frac{\partial u(\bar{x})}{\partial x_i} + b(\bar{x})\, u(\bar{x})$

where $a_i(\bar{x})$ and $b(\bar{x})$ are scalar functions of point $\bar{x}=(x_1, x_2,\ldots, x_n)$. The adjoint operator is given by

$$\bar{L}(v)=-\sum_{i=1}^{n} \frac{\partial(a_i v)}{\partial x_i}+b u$$

Here $$vL(u)-u\bar{L}(v)=\sum_{i=1}^{n} \frac{\partial(a_i uv)}{\partial x_i}$$

$$=\left(\frac{\partial}{\partial x_1}, \frac{\partial}{\partial x_2},\ldots, \frac{\partial}{\partial x_n}\right). \quad (a_1uv,\ldots, a_n uv)$$

$$=\nabla \cdot \bar{\omega}$$

where $\bar{\omega}=(a_1 u v,\ldots, a_n u v)=p(u, v)$ is the *bilinear concomitant vector.*

Example 7. In § 4·2 we saw that if $D\equiv\frac{d}{dx}$, then the operators given by

$$L(y)=p(x) D^2y+q(x) Dy+v(x)y \qquad \ldots(i)$$

$$\bar{L}(z)=D^2(pz)-D(qz)+r z \qquad \ldots(ii)$$

are adjoint and satisfy the Lagrange's identity

$$z L(y)-y\bar{L}(z)=\frac{d}{dx} P(y, z) \qquad \ldots(iii)$$

where $$P(y, z)=p(z Dy-y Dz)+(q-Dp) yz \qquad \ldots(iv)$$

is the bilinear concomitant. Now an n-dimensional generalisation of the ordinary differential operator D is the partial differential operator

$$\nabla=\left(\frac{\partial}{\partial x_1}, \frac{\partial}{\partial x_2},\ldots, \frac{\partial}{\partial x_n}\right).$$

Hence the natural generalisations of operators (*i*) and (*ii*) are respectively

$$L(u)=p(\bar{x}) \nabla^2u+\bar{q}(\bar{x}) \nabla u+r(\bar{x})u \qquad \ldots(v)$$

$$\bar{L}(v)=\nabla^2(p v)-\nabla(\bar{q} v)+rv \qquad \ldots(vi)$$

where the coefficients $p(\bar{x})$ and $r(\bar{x})$ are scalar-valued functions of $\bar{x}=(x_1, x_2,\ldots, x_n)$ and $\bar{q}(\bar{x})=(q_1, q_2,\ldots, q_n)$ is a vector-valued function. These assumptions ensure the dimensional equality of each term in (*v*) and (*vi*).

From (*v*) and (*vi*) we shall have

$$vL(u)-u\bar{L}(v)=[pv \nabla^2u-u \nabla^2(pv)]+[\bar{q}v \nabla u+u \nabla (\bar{q} v)]$$

$$=\nabla[pv\,\nabla u-u\nabla(pv)]+\nabla(quv)$$

$$=\nabla[p\,(v\nabla u-u\,\nabla v)-uv\,\nabla\,p]+\nabla(\bar{q}uv)$$

$$=\nabla\bar{\omega} \qquad ...(vii)$$

which is the analogue of (*vi*). Here

$$\omega=p\,(v\nabla u-u\nabla v)+(\bar{q}-\nabla p)\,uv.$$

Now $\bar{L}(u)=p\nabla^2 u+\bar{q}\nabla u+ru+2(\nabla p-\bar{q})\nabla u+(\nabla^2 p-\nabla\bar{q})u.$

Hence $\bar{L}(u)=L(u)$ *i.e.*, L is *self-adjoint* if $\bar{q}=\nabla p$ (*cf.* $q=p'$ in ODE §4·2).

In this case we have

$$vL(u)-uL(v)=\nabla[p(v\nabla u-u\nabla v)] \qquad ...(viii)$$

If further u and v are solutions of the equation $L(u)=0$, then

$$\nabla[p(v\,\Delta u-u\,\nabla v)]=0$$

i.e.,
$$p(v\nabla u-u\nabla v)=\text{constant} \qquad ...(ix)$$

Relation (*ix*) is the analogue of the *Abel's formula.*

As an example it is easy to see that the equation

$$L(u)=\nabla^2 u+r(\bar{x})u=0$$

is self-adjoint.

If we integrate both sides of (*vii*) in region T and use the divergence theorem, then

$$\int_T [v\,L(u)-u\,\bar{L}(v)]\,dV$$

$$=\int_T \nabla\,\bar{\omega}\,dV$$

$$=\int_{\partial T} \bar{n}\bar{\omega}\,dS$$

$$=\int_{\partial T} \bar{n}\,[p(v\,\nabla\,u-u\nabla v)+(\bar{q}-\nabla p)\,uv]\,dS$$

$$=\int_{\partial T}\left[p\left(v\,\frac{\partial u}{\partial n}-u\,\frac{\partial v}{\partial n}\right)+\left(q_n-\frac{\partial p}{\partial n}\right)u\,v\right]dS \qquad ...(x)$$

where $\bar{n}\nabla u=\dfrac{\partial u}{\partial n}$ and $q_n=\bar{n}\cdot\bar{q}.$

In the particular case when L is self-adjoint, $q=\nabla p$ and (*x*) reduces to

$$\int_T [v\,L(u)-u\,L(v)]dV$$

$$=\int_{\partial T} p\left(v\,\frac{\partial u}{\partial n}-u\,\frac{\partial v}{\partial n}\right)dS \qquad \ldots(xi)$$

(*xi*) is the *Green's formula* (or *Green's theorem*).

Example 8. (Generalised form of Green's theorem). Consider a linear second order operator

$$L(u)=\sum_{i=1}^{n}\sum_{j=1}^{n} A_{ij}(\bar{x})\,\frac{\partial^2 u}{\partial x_i\,\partial x_j}$$

$$+\sum_{i=1}^{n} B_i(\bar{x})\,\frac{\partial u}{\partial x_i}+C(\bar{x})\,u \qquad \ldots(i)$$

where $A_{ij}=A_{ji}$.

The adjoint operator $\bar{L}$ is given by

$$\bar{L}(v)=\sum_{i=1}^{n}\sum_{j=1}^{n}\frac{\partial^2(A_{ij}\,v)}{\partial x_i\,\partial x_j}-\sum_{i=1}^{n}\frac{\partial(B_i\,v)}{\partial x_i}+Cv \qquad \ldots(ii)$$

It is easily verified that

$$vL(u)-u\,\bar{L}(v)=\sum_{i=1}^{n}\frac{\partial}{\partial x_i}\left[\sum_{j=1}^{n}\left\{A_{ij}\,v\,\frac{\partial u}{\partial x_j}-u\,\frac{\partial(A_{ij}\,v)}{\partial x_j}\right\}+B_i\,uv\right]$$

$$=\nabla\bar{\omega} \qquad \text{(Divergence of } \bar{\omega}) \qquad \ldots(iii)$$

where $$\nabla=\left(\frac{\partial}{\partial x_1},\frac{\partial}{\partial x_2},\ldots,\frac{\partial}{\partial x_n}\right)$$

$$=l_1\frac{\partial}{\partial x_1}+l_2\frac{\partial}{\partial x_2}+\ldots+l_n\frac{\partial}{\partial x_n}$$

$$\bar{\omega}=P(u,v)=(P_1,P_2,\ldots,P_n) \qquad \ldots(iv)$$

$$P_i=\sum_{j=1}^{n}\left\{A_{ij}\,v\,\frac{\partial u}{\partial x_j}-u\,\frac{\partial(A_{ij}\,v)}{Ax_j}\right\}+B_i\,uv$$

The condition for self-adjointness of $\bar{L}$ is $L(u)=\bar{L}(u)$

i.e., $$\sum_{i=1}^{n}\sum_{j=1}^{n}A_{ij}\,\frac{\partial^2 u}{\partial x_i\,\partial x_j}+\sum_{i=1}^{n}B_i\,\frac{\partial u}{\partial x_i}$$

$$=\sum_{i=1}^{n}\sum_{j=1}^{n}\frac{\partial^2(A_{ij}u)}{\partial x_i\,\partial x_j}-\sum_{i=1}^{n}\frac{\partial(B_i u)}{\partial x_i}$$

$$=\sum_{i=1}^{n}\sum_{i=1}^{n}\left[A_{ij}\frac{\partial^2 u}{\partial x_i\,\partial x_j}+\left(\frac{\partial A_{ij}}{\partial x_i}\frac{\partial u}{\partial x_j}\right.\right.$$

$$\left.\left.+\frac{\partial A_{ij}}{\partial x_j}\frac{\partial u}{\partial x_i}\right)+\frac{\partial^2 A_{ij}}{\partial x_i\,\partial x_j}u\right]$$

$$-\sum_{i=1}^{n}\left[B_i\frac{\partial u}{\partial x_i}+\frac{\partial B_i}{\partial x_i}u\right] \qquad ...(v)$$

Since $A_{ij}=A_{ji}$,

$$\sum_{i=1}^{n}\sum_{j=1}^{n}\frac{\partial A_{ij}}{\partial x_i}\frac{\partial u}{\partial x_j}=\sum_{i=1}^{n}\sum_{j=1}^{n}\frac{\partial A_{ji}}{\partial x_j}\frac{\partial u}{\partial x_i}$$

$$=\sum_{i=1}^{n}\sum_{j=1}^{n}\frac{\partial A_{ij}}{\partial x_j}\frac{\partial u}{\partial x_i}.$$

If we equate the coefficients of $\frac{\partial^2 u}{\partial x_i\,\partial x_j}$, $\frac{\partial u}{\partial x_i}$, u on both sides of (v), we have

$$B_i=2\sum_{j=1}^{n}\frac{\partial A_{ij}}{\partial x_j}-B_i \qquad ...(vi)$$

$$0=\sum_{j=1}^{n}\frac{\partial^2 A_{ij}}{\partial x_i\,\partial x_j}-\frac{\partial B_i}{\partial x_i} \qquad ...(vii)$$

Both (vi) and (vii) are satisfied if

$$B_i=\sum_{j=1}^{n}\frac{\partial A_{ij}}{\partial x_j} \qquad i=1, 2,\ldots, n) \qquad ...(viii)$$

or

$$B_i=\left(\frac{\partial}{\partial x_1},\frac{\partial}{\partial x_2},\ldots,\frac{\partial}{\partial x_n}\right)(A_{i1}, A_{i2},\ldots, A_{in}) \qquad ...(ix)$$

$$=\text{div } A_i$$

where $A_i=(A_{i1}, A_{i2},\ldots, A_{in})$.

Relation (ix) for self-adjointness corresponds to the relation $q=p'$ in one-dimensional case.

Again the relation (*iii*) can be said to be *generalised Lagrange's identity*. Integrating in T,

$$\int_T [vL(u)-u\,\bar{L}(v)]\,dV=\int_T \nabla\,\bar{\omega}\,dV \qquad \ldots(x)$$

$$=\int_{\partial T}\bar{n}\bar{\omega}\,dS$$

$$=\int_{\partial T}\left(\sum_{1}^{n} n_i\,P_i\right)dS \qquad \ldots(xi)$$

$\bar{n}=(n_1, n_2,\ldots, n_n)$ is the unit normal at $\bar{x}\in\partial T$. (*xi*) can be called the *generalised Green's theorem.*

The cases $n=2, 3$ are considered in the next two examples.

Example 9. The operators L and $\bar{L}$ given by

$$L(u)=A\frac{\partial^2 u}{\partial x^2}+2B\frac{\partial^2 u}{\partial x\,\partial y}+C\frac{\partial^2 u}{\partial y^2}+D\frac{\partial u}{\partial x}+E\frac{\partial u}{\partial y}+Fu$$

and $\bar{L}(v)=\dfrac{\partial^2(Av)}{\partial x^2}+2\dfrac{\partial^2(Bu)}{\partial x\,\partial y}+\dfrac{\partial^2(Cv)}{\partial y^2}-\dfrac{\partial(Dv)}{\partial x}-\dfrac{\partial(Ev)}{\partial y}+Fv$

are adjoint and

$$vL(u)-u\bar{L}(v)=\nabla\bar{\omega} \qquad \ldots(i)$$

where
$$\nabla=\left(\frac{\partial}{\partial x}, \frac{\partial}{\partial y}\right)=i\frac{\partial}{\partial x}+j\frac{\partial}{\partial y}$$

$$\bar{\omega}=(P, Q) \qquad =i\,P+j\,Q$$

$$P=A\left(v\frac{\partial u}{\partial x}-u\frac{\partial v}{\partial u}\right)+B\left(v\frac{\partial u}{\partial y}-u\frac{\partial v}{\partial y}\right)+\left(D-\frac{\partial A}{\partial x}-\frac{\partial B}{\partial y}\right)uv$$

$$Q=C\left(v\frac{\partial u}{\partial y}-u\frac{\partial v}{\partial y}\right)+B\left(v\frac{\partial u}{\partial x}-u\frac{\partial v}{\partial x}\right)+\left(E-\frac{\partial B}{\partial x}-\frac{\partial C}{\partial y}\right)uv$$

Integrating both sides of (*i*) in T,

$$\int_T\int[vL(u)-u\,\bar{L}(v)]\,dx\,dy=\int_T\int\nabla\,\bar{\omega}\,dV$$

$$=\int_T\int\left(\frac{\partial P}{\partial x}+\frac{\partial Q}{\partial y}\right)dx\,dy$$

$$=\int_C (P\,dy-Q\,dx) \qquad \ldots(ii)$$

(by Stoke's theorem).

In the special case if L is self-adjoint, then

$$D=\frac{\partial A}{\partial x}+\frac{\partial B}{\partial y},\ E=\frac{\partial B}{\partial x}+\frac{\partial C}{\partial y}$$

and
$$L(u)=\frac{\partial}{\partial x}\left(A\frac{\partial u}{\partial x}+B\frac{\partial u}{\partial y}\right)+\frac{\partial}{\partial y}\left(B\frac{\partial u}{\partial x}+C\frac{\partial u}{\partial y}\right)$$
$$=\nabla\bar{\omega}.$$

The results are directly extended to three dimensions in the next example.

Exmaple 10. Given

$$L(u)=A\frac{\partial^2 u}{\partial x^2}+B\frac{\partial^2 u}{\partial y^2}+C\frac{\partial^2 u}{\partial z^2}+2F\frac{\partial^2 u}{\partial y\,\partial z}+2G\frac{\partial^2 u}{\partial z\,\partial x}$$
$$+2H\frac{\partial^2 u}{\partial z\,\partial x}+M\frac{\partial u}{\partial x}+N\frac{\partial u}{\partial y}+P\frac{\partial u}{\partial z}+Qu.$$

Then $vL(u)-u\bar{L}(v)=\nabla\bar{\omega}$ ...(*iv*)

where
$$\nabla\equiv\left(\frac{\partial}{\partial x},\frac{\partial}{\partial y},\frac{\partial}{\partial z}\right)=i\frac{\partial}{\partial x}+j\frac{\partial}{\partial z}+k\frac{\partial}{\partial z}$$

$$\bar{\omega}=(\omega_1,\omega_2,\omega_3)=i\omega_1+j\omega_2+k\omega_3$$

$$\omega_1=A\left(v\frac{\partial u}{\partial x}-u\frac{\partial v}{\partial x}\right)+F\left(v\frac{\partial u}{\partial y}-u\frac{\partial u}{\partial y}\right)$$
$$+H\left(v\frac{\partial u}{\partial z}-u\frac{\partial v}{\partial z}\right)+\left(M-\frac{\partial A}{\partial x}-\frac{\partial F}{\partial y}-\frac{\partial H}{\partial z}\right)uv$$

$$\omega_2=F\left(v\frac{\partial u}{\partial x}-u\frac{\partial v}{\partial x}\right)+B\left(v\frac{\partial v}{\partial y}-u\frac{\partial v}{\partial y}\right)$$
$$+G\left(v\frac{\partial u}{\partial z}-u\frac{\partial v}{\partial z}\right)+\left(N-\frac{\partial F}{\partial x}-\frac{\partial B}{\partial y}-\frac{\partial G}{\partial z}\right)uv$$

$$\omega_3=H\left(v\frac{\partial u}{\partial x}-u\frac{\partial v}{\partial x}\right)+G\left(v\frac{\partial u}{\partial y}-u\frac{\partial v}{\partial y}\right)$$
$$+C\left(v\frac{\partial u}{\partial z}-u\frac{\partial v}{\partial z}\right)+\left(P-\frac{\partial H}{\partial x}-\frac{\partial G}{\partial y}-\frac{\partial C}{\partial z}\right)uv$$

Integrating both sides of (*iv*)

$$\int_T [v\,L(u)-u\,\bar{L}(v)]\,dx\,dy\,dz$$
$$=\int_{\partial T}(\omega_1\,dy\,dz+\omega_2\,dz\,dx+\omega_3\,dx\,dy)\qquad ...(v)$$

Problem. Find the conditions under which the operator L is self-adjoint in the above example.

Exercise 31

1. If $L(u)=A\dfrac{\partial^2 u}{\partial x^2}+C\dfrac{\partial^2 u}{\partial y^2}+D\dfrac{\partial u}{\partial x}+E\dfrac{\partial u}{\partial y}+Fu,$

then find the adjoint operator $\bar{L}$. Prove that

$$\int_T\int [v\,L(u)-u\,\overline{L}(v)]\,dx\,dy=\int_C (P\cos\alpha+Q\cos\beta)ds$$

where $P=Av\dfrac{\partial u}{dx}-u\dfrac{\partial(Av)}{\partial x}+Duv$

$Q=Cv\dfrac{\partial u}{\partial y}-u\dfrac{\partial(Cv)}{\partial y}+Euv.$

2. Prove that the following operators are self-adjoint and compute $vL(u)-uL(v)$:

(*a*) $L(u)=x^3\dfrac{\partial^2 u}{\partial x^2}+(y^2+yz)\dfrac{\partial^2 u}{\partial y^2}+(x+y)^2\dfrac{\partial^2 u}{\partial z^2}$
$\qquad+3x^2\dfrac{\partial u}{\partial x}+(2y+z)\dfrac{\partial u}{\partial y}$

(*b*) $L(u)=(y^2-z^2)\dfrac{\partial^2 u}{\partial x^2}+(z^2-x^2)\dfrac{\partial^2 u}{\partial y^2}+(x^2-y^2)\dfrac{\partial^2 u}{\partial z^2}.$

3. Find the adjoints and compute $vL(u)-u\,\overline{L}\,(v)$:

(*a*) $x^2\dfrac{\partial^2 u}{\partial x^2}+y^2\dfrac{\partial^2 u}{\partial y^2}+z^2\dfrac{\partial^2 u}{\partial z^2}+2xy\dfrac{\partial^2 u}{\partial x\partial y}+2yz\dfrac{\partial^2 u}{\partial y\partial z}$
$\qquad+2zx\dfrac{\partial^2 y}{\partial z\partial x}+2x\dfrac{\partial u}{\partial x}+2y\dfrac{\partial u}{\partial y}+2z\dfrac{\partial u}{\partial z}$

(*b*) $(x^2+y^2)\dfrac{\partial^2 u}{\partial x^2}+(y^2+z^2)\dfrac{\partial^2 u}{\partial y^2}+(z^2+x^2)\dfrac{\partial^2 u}{\partial z^2}.$

4. Prove that if u, v are twice continuously differentiable functions of x, y in a plane region T bounded by a closed curve C, then

$$\int_T\left[v\frac{\partial^2 u}{\partial x\partial y}-u\frac{\partial^2 v}{\partial x\partial y}\right]dxdy=\int_C\left(v\frac{\partial u}{\partial y}dy+u\frac{\partial v}{\partial x}dx\right)$$
$$=-\int_C\left(u\frac{\partial u}{\partial x}dx+u\frac{\partial v}{\partial y}dy\right)$$

[**Hint.** Use Stoke's theorem $\displaystyle\int_T\int\frac{\partial F}{\partial x}dxdy=\int_C F\,dy,$

$\displaystyle\int_T\int\frac{\partial G}{\partial Y}dxdy=-\int_C G\,dx$ and put $F=G=\phi\psi$ etc.]

5. If $L(u)=\frac{\partial^2 u}{\partial x \partial y}+D\frac{\partial u}{\partial x}+E\frac{\partial u}{\partial y}+Fu$,

then prove that

$$\int_T \int [v\,L(u)-u\,L(v)]\,dxdy$$

$$=-\int_C \left[v\left(\frac{\partial u}{\partial x}+Eu\right)dx+u\left(\frac{\partial v}{\partial y}-Dv\right)dy\right].$$

6. In the space-time domain T : $a \leqslant x \leqslant b$, $0 \leqslant t \leqslant t_0$, the operator associated with the wave equation is given by

$$L=c^2\frac{\partial^2}{\partial x^2}-\frac{\partial^2}{\partial t^2}$$

Prove that $\overline{L}=L$ and $vL(u)-uL(v)=\nabla \overline{w}$

where $$\nabla=i\frac{\partial}{\partial x}+j\frac{\partial}{\partial t},$$

$$\overline{w}=c^2(vu_x-uv_x)i-(vu_t-uv_t)\,j.$$

Hence deduce

$$\int_0^{t_0}\int_a^b [vL(u)-uL(v)]\,dtdx=c^2\int_0^{t_0}(vu_x-uv_x)\,dt$$

$$-\int_a^b (vu_t-uv_t)\,dx.$$

§9·8. Coordinate Transformations in the Laplace Operator ∇^2

We shall consider transformation of coordinates from Cartesian to polar in ∇^2 in two, three and n dimensions. It should be kept in mind that a transformation from $x, y, \ldots\ldots$ coordinates to $u, v, \ldots\ldots$ coordinates is *non-singular* if the Jacobian $J=\frac{\partial(x, y, \ldots)}{\partial(u, v, \ldots)}\neq 0$. It is called singular if $J=0$. Points at which $J=0$ are called *singularities of the transformation.*

(A) Two-Dimensional Space

Put $x=r\cos\theta$, $y=r\sin\theta$

$0 \leqslant r < \infty$, $0 \leqslant \theta \leqslant 2\pi$

When $r\neq 0$, $J=\frac{\partial(x, y)}{\partial(r, \theta)}=r.$

(0, 0) is the only point of singularity of the transformation.

Take $u_1=x^2+y^2=r^2$

$$\frac{\partial V}{\partial x}=\frac{\partial V}{\partial r}\frac{\partial r}{\partial x}+\frac{\partial r}{\partial \theta}\frac{\partial \theta}{\partial x}=\left(\cos\theta\frac{\partial}{\partial r}-\frac{\sin\theta}{r}\frac{\partial}{\partial \theta}\right)V \qquad \ldots(i)$$

$$\frac{\partial V}{\partial y}=\left(\sin\theta\,\frac{\partial}{\partial r}+\frac{\cos\theta}{r}\,\frac{\partial}{\partial\theta}\right)V \qquad ...(ii)$$

$$\frac{\partial^2 V}{\partial x^2}=\left(\cos\theta\,\frac{\partial}{\partial r}-\frac{\sin\theta}{r}\,\frac{\partial}{\partial\theta}\right)\left(\cos\theta\,\frac{\partial V}{\partial r}-\frac{\sin\theta}{r}\,\frac{\partial V}{\partial\theta}\right)$$

$$=\cos^2\theta\,\frac{\partial^2 V}{\partial r^2}-\frac{2\sin\theta\cos\theta}{r}\,\frac{\partial^2 V}{\partial r\partial\theta}+\frac{\sin^2\theta}{r^2}\,\frac{\partial^2 V}{\partial\theta^2}$$

$$+\frac{\sin^2\theta}{r}\,\frac{\partial V}{\partial r}+\frac{2\sin\theta\cos\theta}{r^2}\,\frac{\partial V}{\partial\theta} \qquad ...(iii)$$

By changing θ to $\frac{\pi}{2}-\theta$, we get

$$\frac{\partial^2 V}{\partial y^2}=\sin^2\theta\,\frac{\partial^2 V}{\partial r^2}+\frac{2\sin\theta\cos\theta}{r}\,\frac{\partial^2 V}{\partial r\partial\theta}+\frac{\cos^2\theta}{r^2}\,\frac{\partial^2 V}{\partial\theta^2}$$

$$+\frac{\cos^2\theta}{r}\,\frac{\partial V}{\partial r}-\frac{2\sin\theta\cos\theta}{r^2}\,\frac{\partial V}{\partial\theta} \qquad ...(iv)$$

Adding (*iii*) and (*iv*), we get

$$\frac{\partial^2 V}{\partial x^2}+\frac{\partial^2 V}{\partial y^2}=\frac{\partial^2 V}{\partial r^2}+\frac{1}{r}\,\frac{\partial V}{\partial r}+\frac{1}{r^2}\,\frac{\partial^2 V}{\partial\theta^2} \qquad ...(v)$$

or

$$\nabla^2 V=\frac{1}{J}\left\{\frac{\partial\left(J\,\frac{\partial V}{\partial r}\right)}{\partial r}+\frac{\partial\left(\frac{J}{u_1}\,\frac{\partial V}{\partial\theta}\right)}{\partial\theta}\right\} \quad I$$

(B) Three-Dimensional Space

Put $x=r\sin\theta\cos\phi,\ y=r\sin\theta\sin\phi,\ z=r\cos\theta$

$$0\leqslant r<\infty,\ 0\leqslant\theta\leqslant\pi,\ 0\leqslant\phi<2\pi$$

$$J=\frac{\partial(x, y, z)}{\partial(r, \theta, \phi)}=r^2\sin\theta$$

$$u_1=x^2+y^2+z^2=r^2$$

$$u_2=x^2+y^2=r^2\sin^2\theta.$$

Points on the z-axis are singularities of the transformation. Taking $\rho=r\sin\theta,\ x=\rho\cos\phi,\ y=\rho\sin\phi$ in (*v*)

$$\frac{\partial^2 V}{\partial x^2}+\frac{\partial^2 V}{\partial y^2}=\frac{\partial^2 V}{\partial\rho^2}+\frac{1}{\rho}\,\frac{\partial V}{\partial\rho}+\frac{1}{\rho^2}\,\frac{\partial^2 V}{\partial\phi^2} \qquad ...(vi)$$

Using the pair $z=r\cos\theta,\ \rho=r\sin\theta$ in (*v*),

$$\frac{\partial^2 V}{\partial z^2}+\frac{\partial^2 V}{\partial\rho^2}=\frac{\partial^2 V}{\partial r^2}+\frac{1}{r}\,\frac{\partial V}{\partial r}+\frac{1}{r^2}\,\frac{\partial^2 V}{\partial\theta^2} \qquad ...(vii)$$

Also
$$\frac{1}{\rho}\,\frac{\partial V}{\partial\rho}=\frac{1}{r}\,\frac{\partial V}{\partial r}+\frac{\cos\theta}{r^2\sin\theta}\,\frac{\partial V}{\partial\theta}$$

Adding (*vi*) and (*vii*)

$$\frac{\partial^2 V}{\partial x^2}+\frac{\partial^2 V}{\partial y^2}+\frac{\partial^2 V}{\partial z^2}$$

$$=\frac{\partial^2 V}{\partial r^2}+\frac{2}{r}\frac{\partial V}{\partial r}+\frac{1}{r^2}\frac{\partial^2 V}{\partial \theta^2}+\frac{\cos\theta}{r^2\sin\theta}\frac{\partial V}{\partial\theta}$$

$$+\frac{1}{r^2\sin^2\theta}\frac{\partial^2 V}{\partial\phi^2} \qquad \ldots(viii)$$

or $$\nabla^2 V=\frac{1}{r^2\sin\theta}\left\{\frac{\partial\left(r^2\sin\theta\,\frac{\partial V}{\partial r}\right)}{\partial r}+\frac{\partial\left(\sin\theta\,\frac{\partial V}{\partial\theta}\right)}{\partial\theta}+\frac{\partial}{\partial\phi}\left(\frac{r^2\sin\theta}{r^2\sin^2\theta}\,\frac{\partial V}{\partial\phi}\right)\right\}$$

$$=\frac{1}{J}\left\{\frac{\partial\left(J\,\frac{\partial V}{\partial r}\right)}{\partial r}+\frac{\partial\left(\frac{J}{u_1}\frac{\partial V}{\partial\theta}\right)}{\partial\theta}+\frac{\partial\left(\frac{J}{u_2}\frac{\partial V}{\partial\phi}\right)}{\partial\phi}\right\} \qquad \text{II}$$

(C) n-Dimensional Space

$$x_1=r\sin\theta_1\sin\theta_2\ldots\ldots\sin\theta_{n-2}\sin\theta_{n-1}$$
$$x_2=r\sin\theta_1\sin\theta_2\ldots\ldots\sin\theta_{n-2}\cos\theta_{n-1}$$
$$x_3=r\sin\theta_1\sin\theta_2\ldots\ldots\sin\theta_{n-3}\cos\theta_{n-2}$$
$$\ldots \qquad \ldots \qquad \ldots$$
$$x_{n-1}=r\sin\theta_1\cos\theta_2$$
$$x_n=r\cos\theta_1$$
$$0\leqslant r<\infty,\ 0\leqslant\theta_i\leqslant\pi,\ i=1,2,\ldots\ldots,n-2$$
$$0\leqslant\theta_{n-1}<2\pi$$

Put $u_1=x_1^2+x_2^2+\ldots\ldots+x_n^2 \;=r^2$

$$u_2=x_1^2+x_2^2+\ldots\ldots+x^2_{n-1}=r^2\sin^2\theta_1$$
$$u_2=x_1^2+x_2^2+\ldots\ldots+x^2_{n-2}=r^2\sin^2\theta_1\sin^2\theta_2$$
$$\ldots \qquad \ldots \qquad \ldots$$
$$u_{n-1}=x_1^2+x_2^2=r^2\sin^2\theta_1\sin^2\theta_2\ldots\ldots\sin^2\theta_{n-2}$$

$$J=\frac{\partial(x_1,x_2,\ldots,x_n)}{\partial(r,\theta_1,\theta_2,\ldots,\theta_{n-1})}=r^{n-1}\sin^{n-2}\theta_1,\ \sin^{n-3}\theta_2\ldots$$
$$\ldots\sin^2\theta_{n-3}\sin\theta_{n-2}$$

The formula holds :

$$\nabla^2 V=\frac{\partial^2 V}{\partial x_1^2}+\frac{\partial^2 V}{\partial x_2^3}+\ldots\ldots+\frac{\partial^2 V}{\partial x_n^2}$$

$$=\frac{1}{J}\left\{\frac{\partial\left(J.\frac{\partial V}{\partial r}\right)}{\partial r}+\frac{\partial\left(\frac{J}{u_1}\frac{\partial V}{\partial\theta_1}\right)}{\partial\theta_1}+\ldots+\frac{\partial\left(\frac{J}{u_{n-1}}\frac{\partial V}{\partial\theta_{n-1}}\right)}{\partial\theta_{n-1}}\right\} \qquad \text{III}$$

Formula III is proved by induction. The result is true for $n=2, 3$. Assume it to be true for $n-1$.

Taking $r \sin \theta_1 = \rho$

$$x_1 = \rho \sin \theta_2 \sin \theta_3 \ldots\ldots \sin \theta_{n-2} \sin \theta_{n-1}$$

$$x_2 = \rho \sin \theta_2 \ldots\ldots \sin \theta_{n-2} \cos \theta_{n-1}$$

$$x_3 = \rho \sin \theta_2 \ldots\ldots \cos \theta_{n-2}$$

$$\ldots \quad \ldots \quad \ldots$$

$$x_{n-1} = \rho \cos \theta_2$$

Take $\quad J' = \dfrac{\partial(x_1, x_2, \ldots x_{n-1})}{\partial(\rho, \theta_2, \theta_3, \ldots, \theta_{n-1})}$

$$= \rho^{n-2} \sin^{n-3} \theta_2 \sin^{n-4} \theta_3 \ldots\ldots \sin^2 \theta_{n-3} \,.\, \sin \theta_{n-2}$$

$$u_1' = u_2 = r^2 \sin^2 \theta_1$$

$$u_2' = u_3 = r^2 \sin^2 \theta_1 \sin^2 \theta_2$$

$$u'_{n-2} = u_{n-1} = r^2 \sin^2 \theta_1 \sin^2 \theta_2 \ldots \sin^2 \theta_{n-2}$$

Then $\quad \dfrac{\partial^2 V}{\partial x_1^2} + \ldots + \dfrac{\partial^2 V}{\partial x^2_{n-1}}$

$$= \frac{1}{J'} \left\{ \frac{\partial}{\partial \rho}\left(J' \frac{\partial V}{\partial \rho} \right) + \frac{\partial\left(\dfrac{J'}{u_1'} \dfrac{\partial V}{\partial \theta_2} \right)}{\partial \theta_2} + \ldots + \frac{\partial \left(\dfrac{J'}{u'_{n-2}} \dfrac{\partial V}{\partial \theta_{n-1}} \right)}{\partial \theta_{n-1}} \right\}$$

$$= \frac{\partial^2 V}{\partial \rho^2} + \frac{n-2}{\rho} \frac{\partial V}{\partial \rho} + \frac{1}{J'} \frac{\partial \left(\dfrac{J'}{u_2} \dfrac{\partial V}{\partial \theta_2} \right)}{\partial \theta_2} + \ldots$$

$$+ \frac{1}{J'} \frac{\partial \left(\dfrac{J'}{u'_{n-2}} \dfrac{\partial V}{\partial \theta_{n-1}} \right)}{\partial \theta_{n-2}} \qquad \ldots\text{(IV)}$$

Using (*ii*) and (*v*) with $x_n = r \cos \theta_1$, $\rho = r \sin \theta_1$ we have

$$\frac{\partial^2 V}{\partial x_n^2} + \frac{\partial^2 V}{\partial \rho^2} = \frac{\partial^2 V}{\partial r^2} + \frac{1}{r} \frac{\partial V}{\partial r} + \frac{1}{r^2} \frac{\partial^2 V}{\partial \theta_1^2} \qquad \ldots(ix)$$

$$\frac{(n-2)}{\rho} \frac{\partial V}{\partial \rho} = \frac{n-2}{r} \frac{\partial V}{\partial r} + \frac{(n-2) \cos \theta_1}{r^2 \sin \theta_1} \frac{\partial V}{\partial \theta_1} \qquad \ldots(x)$$

Adding $\dfrac{\partial^2 V}{\partial x_n^2}$ to both sides of (IV) and using (*ix*) and (*x*) we get

$$\nabla^2 V = \frac{\partial^2 V}{\partial r^2} + \frac{n-1}{r} \frac{\partial V}{\partial r} + \frac{1}{r^2} \left\{ \frac{\partial^2 V}{\partial \theta_1^2} + \frac{(n-2) \cos \theta_1}{\sin \theta_1} \frac{\partial V}{\partial \theta_1} \right\}$$

$$+ \frac{1}{J'} \frac{\partial \left(\dfrac{J'}{u_2} \dfrac{\partial V}{\partial \theta_2} \right)}{\partial \theta_2} + \ldots + \frac{1}{J'} \frac{\partial \left(\dfrac{J'}{u_{n-1}} \dfrac{\partial V}{\partial \theta_{n-1}} \right)}{\partial \theta_{n-1}} \qquad \ldots(xi)$$

It can be verified that

$$\frac{\partial^2 V}{\partial r^2}+\frac{n-1}{r}\frac{\partial V}{\partial r}=\frac{1}{J}\frac{\partial\left(J\frac{\partial V}{\partial r}\right)}{\partial r}$$

$$\frac{1}{r^2}\left\{\frac{\partial^2 V}{\partial \theta_1^2}+\frac{(n-2)\cos\theta_1}{\sin\theta_1}\frac{\partial V}{\partial \theta_1}\right\}=\frac{1}{J}\frac{\partial\left(\frac{J}{u_1}\frac{\partial V}{\partial \theta_1}\right)}{\partial \theta_1}$$

$$\frac{1}{J'}\frac{\partial\left(\frac{J'}{u_i}\frac{\partial V}{\partial \theta_i}\right)}{\partial \theta_i}=\frac{1}{J}\frac{\partial\left(\frac{J}{u_i}\frac{\partial V}{\partial \theta_i}\right)}{\partial \theta_i}\qquad i=2, 3,\ldots, n-1.$$

Making these substitutions in (*xi*) we obtain result III for n.

Corollary 1. III can be written as

$$\nabla^2 V=\frac{\partial^2 V}{\partial r^2}+\frac{n-1}{r}\frac{\partial V}{\partial r}+\frac{1}{r^2}\text{ (angular terms)}. \qquad \ldots(xii)$$

If V is a function of radial symmetry, it is a function of r only and the angular terms in (*xii*) vanish. Thus

$$\nabla^2 V=\frac{\partial^2 V}{\partial r^2}+\frac{n-1}{r}\frac{\partial V}{\partial r}.$$

This result can be proved independently also.

9·9. Volume and Surface Area of n-Dimensional Spherical Region

Let T be the spherical region (or ball)

$$x_1^2+x_2^2+\ldots\ldots+x_n^2\leqslant 1 \qquad \ldots(i)$$

If u_n is the volume of this region, then

$$u_n=\int_T dV=\int_T dx_1\, dx_2\ldots dx_n \qquad \ldots(ii)$$

It can be easily seen that the volume v_n of the ball of radius r is given by

$$v_n=u_n\,.\,r^n \qquad \ldots(iii)$$

The hyperplane $x_n=t$ $(-1<t<1)$ will cut the unit ball in the $(n-1)$-dimensional spherical region $x_1^2+x_2^2\ldots+x^2_{n-1}\leqslant 1-t^2$.

By (*iii*) the volume of this $(n-1)$-dimensional ball is

$$u_{n-1}(1-t^2)^{\frac{n-1}{2}}$$

Hence the volume element of T is

$$dV=u_{n-1}(1-t)^{\frac{n-1}{2}}\,dt$$

Since the minimum and maximum values of t are -1 and 1,

$$u_n=u_{n-1}\int_{-1}^{1}(1-t^2)^{\frac{n-1}{2}}\,dt$$

$$=u_{n-1}\int_0^1 u^{-1/2}(1-u)^{\frac{n-1}{2}}\,du \qquad (t^2=u)$$

$$=u_{n-1}\frac{\Gamma\left(\frac{n+1}{2}\right)\Gamma\left(\frac{1}{2}\right)}{\Gamma\left(\frac{n+2}{2}\right)}$$

we get the reduction formula

$$\Gamma\left(\frac{n+2}{2}\right)u_n=\pi^{1/2}\Gamma\left(\frac{n+1}{2}\right)u_{n-1} \qquad \text{...(iv)}$$

Iteration of formula (*iv*) leads to

$$\Gamma\left(\frac{n+2}{2}\right)u_n=C\,\pi^{\frac{n}{2}}$$

For $n=1$, T is the interval $[-1, 1]$ hence $u_1=2$. This gives $C=1$ and

$$u_n=\frac{\pi^{\frac{n}{2}}}{\Gamma\left(\frac{n+2}{2}\right)} \qquad \text{...(vi)}$$

The volume of the sphere of radius r is

$$v_n=\frac{\pi^{\frac{n}{2}}}{\Gamma\left(\frac{n+2}{2}\right)}r^n \qquad \text{...(vi)}$$

If S is the surface area of sphere of radius r, then

$$S=\frac{dv_n}{dr}=\frac{2\pi^{\frac{n}{2}}}{\Gamma\left(\frac{n}{2}\right)}r^{n-1} \qquad \text{...(vii)}$$

Hence the area w_n of the surface of the unit sphere is given by

$$w_n=\frac{2\pi^{\frac{n}{2}}}{\Gamma\left(\frac{n}{2}\right)} \qquad \text{...(viii)}$$

w_n is also called the *solid angle* at a point in n-space.

Using (*viii*) v_n can be expressed as

$$v_n=\frac{w_n r^n}{n} \qquad \text{...(ix)}$$

Let dw denote the hypersurface element on the unit sphere. By Using polar coordinates

$$dw=(\sin^{n-2}\theta_1 \sin^{n-3}\theta_2 \ldots \sin\theta_{n-2})d\theta_1\, d\theta_2 \ldots d\theta_{n-1} \qquad \ldots(x)$$

For example, for $n=2$, T is a circular region and $dw=d\theta$.

For $n=3$, $dw=\sin\theta\, d\theta\, d\phi$.

The general formula (x) can be established as follows : The n-dimensional volume element is

$$dV=dx_1\, dx_2 \ldots dx_n.$$

If one of the coordinates is constant, we get a hypersurface and the volume element dV will reduce to hypersurface element dw. If $x_1, x_2, \ldots, x_n$ are changed to polars and $r=1$, then

$$\begin{aligned} dw &= J\, d\theta_1\, d\theta_2 \ldots d\theta_{n-1} \\ &= (\sin^{n-2}\theta_1 \sin^{n-3}\theta_2 \ldots \sin\theta_{n-2})d\theta_1 \ldots d\theta_{n-1} \end{aligned}$$

For a sphere of radius r the *volume element* is

$$\begin{aligned} dV &= (r^{n-1}\sin^{n-2}\theta_1 \sin^{n-3}\theta_2 \ldots \sin\theta_{n-2})dr\, d\theta_1 \ldots d\theta_{n-1} \\ &= r^{n-1}\, dw\, dr \end{aligned}$$

and $\quad d\Omega=r^{n-1}\, dw$ is the *surface element.*

§9·10. Change of Variables in the Higher Dimensional Delta Function

It is easy to see that the n-dimensional delta function is the product of n one-dimensional delta functions. We define the δ-function $\delta(x_1, x_2, \ldots, x_n)$ to be such that for a continuous function $\phi(x_1, x_2, \ldots, x_n)$

$$\int_T \phi(x_1, x_2, \ldots, x_n)\, \delta(x_1, x_2, \ldots, x_n) dV=\phi(0, 0, \ldots, 0) \qquad \ldots(i)$$

where T is the concerned n-dimensional domain in which φ has compact support. In Cartesian coodinates

$$dV=dx_1\, dx_2 \ldots, dx_n \text{ and}$$

$$\int_T \varphi(x_1, n_2, \ldots, x_n)\, \delta(x_1)\delta(x_2)\ldots\delta(x_n)\, dx_1\, dx_2 \ldots dx_n$$

$$=\phi(0, 0, \ldots, 0) \qquad \ldots(ii)$$

Comparing (i) and (ii) we get

$$\delta(x_1, x_2, \ldots, x_n)=\delta(x_1)\delta(x_2)\ldots\delta(x_n) \qquad \ldots(iii)$$

In coordinates other than Cartesian, the expressions are more complicated.

We now generalize the lemma 2 of the previous chapter to effect change of arguments in the n-dimensional δ-function.

Let $\phi\,(x_1, x_2 \ldots, x_n)$ be a continuous function of compact support in an n-dimensional region T. If $(a_1, a_2, \ldots, a_n)$ is a given point in T, then

$$\int_T \phi(x_1, x_2, \ldots, x_n)\, \delta(x_1-a_1)\, \delta(x_2-a_2) \ldots \delta(x_n-a_n)\, dx_1\, dx_2 \ldots dx_n$$
$$=\phi(a_1, a_2, \ldots, a_n) \qquad \ldots(iv)$$

Let us make the transformation

$$x_i = x_i(\xi_1, \xi_2, \ldots, \xi_n), \quad (i=1, 2, \ldots, n),$$

where $$J=\frac{\partial(x_1, x_2, \ldots, x_n)}{\partial(\xi_1, \xi_2, \ldots, \xi_n)} \neq 0.$$

Then $dV = dx_1\, dx_2 \ldots dx_n = |\,J\,|\, d\xi_1\, d\xi_2 \ldots d\xi_n$.

Let $x_i = a_i$ correspond to $\xi_i = b_i$ $(i=1, 2, \ldots, n)$, and let

$$\phi(x_1, x_2, \ldots, x_n) = F(\xi_1, \xi_2, \ldots, \xi_n)$$

Then $\quad \phi(a_1, a_2, \ldots, a_n) = F(b_1, b_2, \ldots, b_n)$.

Relation (iv) will be changed to

$$\int_{T'} F(\xi_1, \xi_2, \ldots, \xi_n)\, \delta(x_1-a_1) \ldots \delta(x_n-a_n)\, |\,J\,|\, d\xi_1 \ldots d\xi_n$$
$$=F(b_1, b_2, \ldots, b_n)$$
$$=\int_{T'} F(\xi_1, \xi_2 \ldots, \xi_n)\, \delta(\xi_1-b_1) \ldots \delta(\xi_n-b_n)\, d\xi_1 \ldots d\xi_n$$

Hence $$\delta(x_1-a_1) \ldots \delta(x_n-a_n) = \frac{1}{|\,J\,|}\, \delta(\xi_1-b_1) \ldots \delta(\xi_n-b_n) \qquad \ldots(v)$$

(v) is the desired formula.

The result fails at a *point of singularity* where $J=0$. In the neighbourhood of such a point, the transformation is not one-one and one or more of the transformation variables may be many-valued or indeterminate. Such a variable may be ignored in the transformation and will be called a *redundant coordinate* (also called *ignorable coordinate*).

For example, let

$$x_1 = x_1(\xi_1, \xi_2), \quad x_2 = x_2(\xi_1, \xi_2).$$

If for $\xi_1 = b_1$ J vanishes, then x_1, x_2 are not independent and ξ_2 will remain undetermined. In the neighbourhood of the points for which $\xi_1 = b_1$, ξ_2 is redundant and δ is function of ξ_1 only. The following examples illustrate the procedure.

Example 11. Take $x = r\cos\theta$, $y = r\sin\theta$

$$J=\frac{\partial(x, y)}{\partial(r, \theta)} = r.$$

In the neighbourhood of the point $x_0 = r_0\cos\theta_0$, $y_0 = r_0\sin\theta_0$, we have

$$\delta(x-x_0)\, \delta(y-y_0) = \frac{\delta(r-r_0)\, \delta(\theta-\theta_0)}{r}\,; \; r \neq 0.$$

At the origin $r=0$ for all values of θ.

Hence $\delta(x)\,\delta(y)$ is a function of r alone.

Let $\delta(x)\,\delta(y)=A(r)\,\delta(r)$.

If T is a small closed circle of radius $\in$ enclosing (0, 0), then since $dxdy=r\,dr\,d\theta$,

$$1=\int_T\int \delta(x)\,\delta(y)\,dx\,dy$$

$$=\int_{r=\in}\int_0^{2\pi} A(r)\,\delta(r)\,r\,dr\,d\theta$$

$$=\Big[\,2\pi\,r\,A(r)\,\Big]_{r=0}$$

$$\Rightarrow \qquad A(r)=\frac{1}{2\pi r}\,.$$

Finally $\qquad \delta(x)\,(\delta(y)=\dfrac{\delta(r)}{2\pi r}$ ...(*vi*)

Example 12. Take $x=r\sin\theta\cos\phi$, $y=r\sin\theta\sin\phi$, $z=r\cos\theta$.

Then $J=r^2\sin\theta$.

For $x_0=r_0\sin\theta_0\cos\phi_0$, $y_0=r_0\sin\theta_0\sin\phi_0$, $z_0=r_0\cos\theta_0$

$$\delta(x-x_0)\,\delta(y-y_0)\,\delta(z-z_0)=\frac{\delta(r-r_0)\,\delta(\theta-\theta_0)\,\delta(\phi-\phi_0)}{r^2\sin\theta}$$

$$r\neq 0,\quad \theta\neq 0.$$

If either $r=0$ or $\theta=0$, then $J=0$.

Case 1. $r=0$. The origin (0, 0, 0) is singular point. θ and ϕ are ignorable.

Let $\quad \delta(x)\,\delta(y)\,\delta(y)=B(r)\,\delta(r)$.

Then in the unit sphere T with centre origin,

$$1=\int\int_T\int \delta(x)\,\delta(y)\,\delta(z)\,dx\,dy\,dz$$

$$=\int_0^1\int_0^{\pi}\int_0^{2\pi}[B(r)\,\delta(r)\,.\,r^2\sin\theta]\,dr\,d\theta\,d\phi$$

$$=\Big[\,4\pi\,r^2\,B(r)\,\Big]_{r=0}$$

Hence $\qquad B(r)=\dfrac{1}{4\pi r^2}$

and $\qquad \delta(x)\,\delta(y)\,\delta(z)=\dfrac{\delta(r)}{4\pi r^2}$ ...(*vii*)

For the n-dimensional sphere T

$$r^2=x_1^{\,2}+x_2^{\,2}+\ldots+x_n^{\,2}.$$

Let $\delta(\bar{x}) = \delta(x_1)\, \delta(x_2) \ldots \delta(x_n)$

$$= C(r)\, \delta(r).$$

$$1 = \int_T \delta(x_1) \ldots \delta(x_n) dx_1 \ldots dx_n$$

$$= \int_0^1 r^{n-1} C(r)\, \delta(r) \left[\int_{\partial T} d\omega_n \right] dr$$

$$= \left[r^{n-1} C(r)\, \omega_n \right]_{r=0}$$

Hence $C(r) = \dfrac{1}{\omega_n\, r^{n-1}}$

and
$$\delta(\bar{x}) = \frac{\delta(r)}{\omega_n\, r^{n-1}} \qquad \ldots(viii)$$

Case 2. On the z-axis $x=0$, $y=0$, $z=r \neq 0$ and $\theta=0$. Hence ϕ is the only ignorable coordinate

Let $\delta(x)\, \delta(y)\, \delta(z-z_0) = A(r, \theta) \,.\, \delta(\theta) \,.\, \delta(r-r_0)$.

Hence $1 = \int_T \delta(x)\, \delta(y)\, \delta(z-z_0)\, dx\, dy\, dz$

$$= \int \left[\int_0^\pi \int_0^{2\pi} A(r, \theta)\, \delta(\theta)\, \delta(r-r_0)\, r^2 \sin\theta\, d\theta\, d\phi \right] dr$$

$$= \int \left[\int_0^\pi \int 2\pi r^2 \sin\theta\, A(r, \theta)\, d\theta \right] \delta(r-r_0)\, dr$$

$$= \left[2\pi r^2 \sin\theta \,.\, A(r, \theta) \right]_{\substack{r=r_0 \\ \theta=0}}$$

Hence $A(r, \theta) = \dfrac{1}{2\pi r^2 \sin\theta}$.

Thus finally

$$\delta(x)\, \delta(y)\, \delta(z-z_0) = \frac{\delta(r-r_0)\, \delta(\theta)}{2\pi r^2 \sin\theta} \qquad \ldots(ix)$$

The following theorem stated without proof generalises the above results.

Theorem 1. Let $x_1, x_2, \ldots x_n$ be the Cartesian coordinates of a point in the Euclidean space R^n and $\xi_1, \xi_2, \ldots, \xi_n$ any other coordinate system. Let the coordinates of a given point P be α_i in the x-coordinate system and β_i in the ξ-coordinate system. If J is the Jacobian of the transformation $\bar{x} \to \bar{\xi}$, and $J \neq 0$ at P, them

$$\delta(x_1-\alpha_1)\, \delta(x_2-\alpha_2) \ldots \delta(x_n-\alpha_n) = \frac{\delta(\xi_1-\beta_1) \ldots \delta(\xi_n-\beta_n)}{|J|}.$$

Let J=0 at P for $\xi_1=\beta_1,\ldots,\ \xi_k=\beta_k$ and arbitrary $\xi_{k+1},\ldots,\ \xi_n$ (redundant).

Then $\delta(x_1-\alpha_1)\ \delta(x_2-\alpha_2)\ldots\delta(x_n-\alpha_n)$

$$=\frac{\delta_1(\xi_1-\beta_1)\delta(\xi_2-\beta_2)\ldots\delta(\xi_k-\beta_k)}{|\ J_k\ |}$$

where
$$J_k=\int J\ d\xi_{k+1}\ d\xi_{k+2}\ldots d\xi_n.$$

9·11. Potential Theory

Consider the scalar function of point $u(\bar{x})=u(x_1\ x_2,\ldots,\ x_n)$, $\bar{x}=(x_1,\ x_2,\ldots,\ x_n)$ in a given domain T with boundary ∂T. Then the equation

$$\nabla^2 u=\frac{\partial^2 u}{\partial x_1^2}+\frac{\partial^2 u}{\partial x_2^2}+\ldots+\frac{\partial^2 u}{\partial x_n^2}=0$$

is called the *Laplace's equation* for T. It is also called the *potential* or *harmonic equation.* A solution $u(\bar{x})\in C$ in $T\cup\partial T$ of Laplace's equation is called a *potential* or *harmonic function.* If further $u(\bar{x})\in C^2$ in T, it is called a *regular harmonic functian.* Solutions which are not regular are called *fundamental* or *elementary solutions.* A *Green's function* for T is a fundamental solution satisfying appropriate boundary or initial conditions.

A non-homogeneous equation

$$\Delta^2 u=h(\bar{x})$$

is called Poisson's equation.

The following theorem shows that it is possible to extend the property of a regular harmonic function from the interior of the region T to its exterior if T does not contain the origin. Before taking the *n*-dimensional case, we prove the theorem in 3-space, which is known as *Kelvin's inversion theorem.*

Kelvin's Inversion Theorem

If $u\ (x,\ y,\ z)$ is a harmonic function in $(x,\ y,\ z)$ in a domain T, then

$$v(x,\ y,\ z)=\frac{a}{r}\ u\left(\frac{a^2x}{r},\ \frac{a^2y}{r},\ \frac{a^2z}{r}\right)\ r\neq 0$$

is a harmonic function of $x,\ y,\ z$ in the domain T′ obtained by the inversion of T into the sphere of radius a.

Proof. Let C be the centre of a sphere of radius a and $\bar{c}$ the position vector of C. Points P, Q with position vectors $\bar{r}$ and $\bar{\rho}$ are said to be *inverse* in (or with respect to) the sphere if points C, P, Q are collinear and $a^2=r\rho$ where $r=|\ \bar{r}\ |$, $\rho=|\ \bar{\rho}\ |$. Collinearity of C, P, Q implies existence of scalars $\lambda,\ \mu,\ \nu$ such that

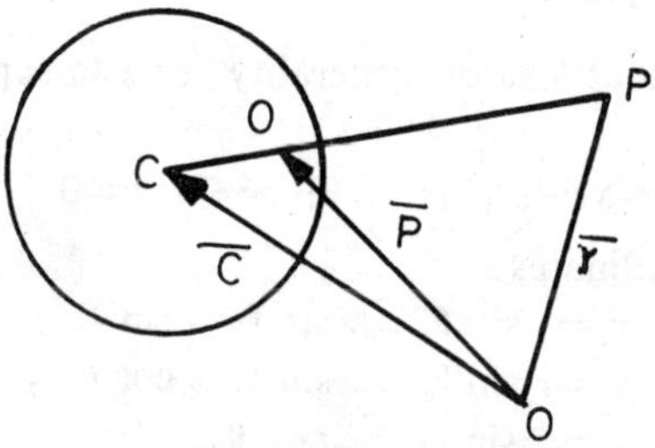

Fig. 9·1,

$\lambda+\mu+\nu=1$ and $\qquad \lambda\bar{c}+\mu\bar{r}+\nu\bar{p}=0.$

If $\bar{r}=(x, y, z)$, and $\rho=(\xi, \eta, z)$, then

$$r^2=x^2+y^2+z^2,\ \rho^2=\xi^2+\eta^2+\zeta^2$$

and
$$\xi=\frac{a^2x}{r^2},\ \eta=\frac{a^2y}{r^2},\ z=\frac{a^2z}{r^2} \qquad \ldots(i)$$

Using Lame's transformation of the Laplacian (which can be found in any standard book on advanced calculus) we have

$$\frac{\partial^2 V}{\partial\xi^2}+\frac{\partial^2 V}{\partial\eta^2}+\frac{\partial^2 V}{\partial\zeta^2}=\frac{r^6}{a^6}\left[\frac{\partial}{\partial x}\left(\frac{a^2}{r^2}\frac{\partial V}{\partial x}\right)\right.$$
$$\left.+\frac{\partial}{\partial y}\left(\frac{a^2}{r^2}\frac{\partial V}{\partial y}\right)+\frac{\partial}{\partial z}\left(\frac{a^2}{r^2}\frac{\partial V}{\partial z}\right)\right].$$

$$\frac{r^6}{a^6}\left[\frac{a}{r}\sum\frac{\partial^2\left(\frac{a}{r}\right)}{\partial x^2}-\frac{av}{r}\sum\frac{\partial^2\left(\frac{a}{r}\right)}{\partial x^2}\right]$$

$$=\frac{r^5}{a^5}\nabla^2\left(\frac{aV}{r}\right)$$

from which the result follows.

We now prove the more general theorem by changing to polar coordinates.

Theorem 2. $u(x_1, x_2, \ldots\ldots x_n)$ is a regular harmonic function in an n-dimensional bounded region T not containing the origin, then

$$v(x_1, x_2, \ldots\ldots, x_n)=\frac{1}{r^{n-2}}\,u\left(\frac{x_1}{r^2}, \frac{x_2}{r^2}, \ldots\ldots, \frac{x_n}{r^2}\right),$$
$$r^2=x_1^2+x_2^2+\ldots\ldots+x_n^2\ (r\neq 0) \qquad \ldots(i)$$

is also harmonic in the region T′ obtained from T by inversion with respect to the unit sphere $x_1^2+x_2^2+\ldots\ldots+x_n^2=1$.

Proof. Without loss of generality we can suppose T to be the spherical region

$$r^2=x_1^2+x_2^2+\ldots\ldots+x_n^2\leqslant 1,\ r\neq 0$$

In polar coordinates

$$\begin{aligned} x_1&=r\sin\theta_1\ldots\ldots\sin\theta_{n-2}\sin\theta_{n-1}\\ x_2&=r\sin\theta_1\ldots\ldots\sin\theta_{n-2}\cos\theta_{n-1}\\ x_3&=r\sin\theta_1\ldots\ldots\cos\theta_{n-2}\\ &\ldots\quad\ldots\quad\ldots\\ x_{n-1}&=r\sin\theta_1\cos\theta_2\\ x_n&=r\cos\theta_1\end{aligned}\qquad\ldots(ii)$$

If we put $y_i=\dfrac{x_i}{r^2}$ and $\rho=\dfrac{1}{r}$, then y_i can be expressed in polar coordinates with r replaced by ρ *i.e.*,

$$\begin{aligned} y_1&=\rho\sin\theta_1\ldots\ldots\sin\theta_{n-2}\sin\theta_{n-1}\\ y_2&=\rho\sin\theta_1\ldots\ldots\sin\theta_{n-2}\cos\theta_{n-1}\\ &\ldots\quad\ldots\quad\ldots\\ y_n&=\rho\cos\theta_1\\ \rho^2&=y_1^2+y_2^2+\ldots\ldots+y_n^2\end{aligned}\qquad\ldots(iii)$$

For $r\leqslant 1$, we have $\rho\geqslant 1$.

If we put $u(x_1, x_2, \ldots\ldots, x_n)=u(r, \theta_1, \ldots\ldots, \theta_n)$, then

$$v(x_1, \ldots\ldots, x_n)=v(r, \theta_1, \ldots\ldots, \theta_n)$$

and $$u\left(\frac{x_1}{r^2}, \ldots\ldots, \frac{x_n}{r^2}\right)=u(\rho, \theta_1, \ldots\ldots, \theta_n)$$

Thus $$v(r, \theta_1, \ldots\ldots, \theta_n)=\rho^{n-2}\,u(\rho, \theta_1, \ldots\ldots, \theta_n)$$

or $$v(\rho, \theta_1, \ldots\ldots, \theta_n)=r^{n-2}\,u(r, \theta_1, \ldots\ldots, \theta_n)$$

Put $u(r_1\,\theta_1, \ldots\ldots, \theta_n)=u(r)$,

Then $$v(\rho)=r^{n-2}\,u(r)\qquad\ldots(iv)$$

Now $$\nabla^2 u(r)=\frac{\partial^2 u}{\partial r^2}+\frac{n-1}{r}\frac{\partial u}{\partial r}+\frac{1}{r^2}$$

(angular derivatives of u)

and $$\nabla^2 v(\rho)=\frac{\partial^2 v}{\partial\rho^2}+\frac{n-1}{\rho}\frac{\partial v}{\partial\rho}+\frac{1}{\rho^2}$$

(angular derivatives of v)

We prove that $\nabla^2 u(r)=0\Leftrightarrow\nabla^2 v(\rho)=0$.

Differentiating both sides of (iv) with respect to ρ,

$$\frac{\partial v}{\partial\rho}=-r^n\,\frac{\partial u}{\partial r}-(n-2)\,r^{n-1}\,u$$

$$\frac{\partial^2 v}{\partial\rho^2}=r^{n+2}\,\frac{\partial^2 u}{\partial r^2}+2(n-1)\,r^{n+1}\,\frac{\partial u}{\partial r}+(n-2)(n-1)\,r^n\,u.$$

Using corollary 1 of section 9·8 and (*iv*), we have

$$\nabla^2 v(\rho)=\frac{\partial^2 v}{\partial \rho^2}+\frac{n-1}{\rho}\frac{\partial v}{\partial \rho}+\frac{1}{\rho^2}\text{ (angular terms)}$$

$$=r^{n+2}\frac{\partial^2 u}{\partial r^2}+(n-1)\, r^{n+1}\frac{\partial u}{\partial r}+r^n \text{ (angular terms in } u)$$

$$=r^{n+2}\,\nabla^2 u(r)$$

$$=0.$$

Q.E.D.

Problem. (*a*) Show that if $u(x, y)$ is regular harmonic in a plane bounded region not containing the origin, then

$$v(x, y)=u\left(\frac{x}{r^2}, \frac{y}{r^2}\right), \; (r^2=x^2+y^2)$$

is also harmonic at all points of the plane where $v(x, y)$ is defined.

(*b*) Show that if $u(x, y, z)$ is regular harmonic in a bounded region in space not containing the origin, then

$$v(x, y, z)=\frac{1}{r}\, u\left(\frac{x}{r^2}, \frac{y}{r^2}, \frac{z}{r^2}\right), \; (r^2=x^2+y^2+z^2)$$

is also harmonic in the region where it is defined.

Remarks 2. (*a*) The harmonic property is also invariant under similarity transformations, translations and simple reflections across planes.

Rotations and magnifications of the form $x'=\alpha x$, $y'=\beta y$ are examples of *similarity* (or *perspective*) transformation of the plane.

(*b*) In view of Theorem 2, u is called harmonic in T if v is harmonic in T′ obtained by inverting T in the unit sphere.

The function $u(x, y, z)=\dfrac{A}{r}$ is harmonic in any closed and bounded region not containing the origin. Hence

$$\frac{1}{r}\, u\left(\frac{x}{r^2}, \frac{y}{r^2}, \frac{z}{r^2}\right)$$

is also harmonic except at infinity.

(*c*) In vectors (*i*) can be written as $v(\bar{x})=\left|\bar{x}\right|^{2-n} u\left(\dfrac{\bar{x}}{|\bar{x}|^2}\right)$.

Properties of Harmonic Functions in Bounded Regions

The integral identities known by the names Green's theorems derived in section 9·7 are the most important tool in potential theory. We reproduce the identities as follows :

$$\text{I.}\quad \int_T (\nabla u \cdot \nabla v+u\nabla^2 v)\, dV=\int_{\partial T} u\frac{\partial v}{\partial n}\, dS$$

II. $\int_T (u\nabla^2 v - v\nabla^2 u)\, dV = \int_{\partial T} \left(u \frac{\partial v}{\partial n} - v \frac{\partial u}{dn} \right) dS$

III. If u is harmonic in T, then

$$\int_T (\nabla u)^2\, dV = \int_{\partial T} u \frac{\partial u}{\partial n}\, dS$$

IV. $\int_T \nabla^2 u\, dV = \int_{\partial T} \frac{\partial u}{\partial n}\, dS$

The integral $\int_T (\nabla u)^2\, dV$ in III is known as the Dirichlet's integral.

Lemma 1. Let u be a regular harmonic function in a bounded region T and $u \in C^1(\bar{T})$, $\bar{T} = T \cup dT$. Then if $u=0$ on ∂T, then $u \equiv 0$ in T.

Proof. By formula III $u=0$ on ∂T implies $\nabla u = 0$ in T *i.e.*, $\frac{\partial u}{\partial x} = \frac{\partial u}{\partial y} = \frac{\partial u}{\partial z} = 0$. Hence u=constant. Since $u=0$ on ∂T and u is continuous in $\bar{T}$, $u=0$ in T.

Corollary 2. The solution of a Dirichlet problem, if exists, is unique. For, if ϕ and ψ are two harmonic functions in T such that $\phi=\psi$ on ∂T. Then $u=\phi-\psi$ is harmonic in T and $u=0$ on ∂T. Hence $u \equiv 0$ in T *i.e.*, $\phi=\psi$ in T.

Lemma 2. Let u be a regular harmonic function in a bounded region T, $u \in C^1$ in $\bar{T}$ and let the normal derivative $\frac{\partial u}{\partial n} = 0$ on ∂T. Then u=constant in T.

Proof. By III, $\nabla u \equiv 0$ in T, *i.e.*, u=constant in T.

Corollary 3. The solution of a Neumann problem, if exists, is unique upto an additive constant.

Problem. The lemmas fail for the function $u = \log(x^2+y^2)$ in the disc. : $x^2+y^2 \leqslant 1$. Why ?

Lemma 3. The surface integral of the normal derivative of a harmonic function vanishes.

The lemma follows from formula IV, because $\nabla_u^2 = 0$. The converse is not true.

Corollary 4. The solution to the Neumann problem

$$\text{DE} : \nabla^2 u = 0 \text{ in T}$$

$$\text{BC} : \frac{\partial u}{\partial n} = f \text{ on } \partial T$$

will exist if $\int_{\partial T} f dS = 0$.

Mean-Value Property of Harmonic Functions

Theorem 3. The mean-value of a regular harmonic function over the surface of a sphere is equal to its value at the centre of the sphere.

Proof. Let T be the spherical region

$$|\bar{x}|^2=\sum_{i=1}^{n} x_i^2 \leqslant R^2,\ \bar{x}=(x_1, x_2, \ldots\ldots, x_n):$$

Then ∂T is the spherical surface (or sphere)

$$|\bar{x}|^2=R^2$$

The surface area of T is

$$\Omega_n=\omega_n R^{n-1}$$

where $\omega_n=\dfrac{2\pi^{\frac{n}{2}}}{\Gamma\left(\frac{n}{2}\right)}$ is the surface area of the n-dimensional unit sphere.

In a sphere the normal derivatives are along radius vectors, hence

$$\frac{\partial u}{\partial n}=\frac{\partial u}{\partial r}.$$

By lemma 3,

$$\int_{\partial T} \frac{\partial u(\bar{x})}{\partial r}\, dS=0.$$

On ∂T $r=R=$constant. Hence the surface element dS is indepedent of r. Thus

$$0=\left[\frac{\partial}{\partial r}\int_{\partial T} u(\bar{x})\, dS\right]_{r=R}$$

Hence $\displaystyle\int_{\partial T} u(\bar{x})\, dS$

is independent of r. Since $u(\bar{x})$ is a function of r,

$$\int_{\partial T} u(\bar{x})\, dS=\int_{\partial T} u(\bar{0})\, dS=\Omega_n\, u(\bar{0}),\ \bar{0}=(0, 0, \ldots\ldots, 0)$$

Hence $\displaystyle\frac{1}{\Omega_n}\int_{\partial T} u(\bar{x})\, dS=u(\bar{0})$

If the centre of the sphere is $\bar{a}=(a_1, a_2, \ldots\ldots, a_n)$, then

$$\frac{1}{\Omega n}\int_{\partial T} u(\bar{x})\, dS=u(\bar{a}) \qquad \ldots(1)$$

(1) is sometimes called the *first mean-value theorem*. Q.E.D.

Theorem 4. (Second Mean-Value Theorem)

The average value of a harmonic function in a ball equals its value at the centre.

Proof. Take the region T to be $|\bar{x}-\bar{a}| \leqslant \rho$. Then the first mean-value theorem (1) can be written in the form

$$u(\bar{a})\,\rho^{n-1}=\frac{1}{\omega_n}\int_{|\bar{x}-\bar{a}|=\rho} u(\bar{x})\,dS$$

Hence integrating both sides with respect to ρ,

$$u(\bar{a})\int_0^R \rho^{n-1}\,d\rho=\frac{1}{\omega_n}\int_0^R\left[\int_{|\bar{x}-\bar{a}|=\rho} u(\bar{x})\,dS\right]d\rho$$

$$=\frac{1}{\omega_n}\int_{|\bar{x}-\bar{a}|\leqslant R} u(\bar{x})\,dV$$

Hence $u(\bar{a})\,\dfrac{R^n}{n}=\dfrac{1}{\omega_n}\displaystyle\int_{|\bar{x}-\bar{a}|\leqslant R} u(\bar{x})\,dV,$

or $$u(\bar{a})=\frac{1}{v_n}\int_T u(\bar{x})\,dV \qquad \ldots(2)$$

where $v_n=\dfrac{\omega_n R^n}{n}$ is the volume of the sphere T.

Conversely, (1) can be deduced from (2).

This is done by writing (2) in the form

$$\frac{r^n\,u(\bar{a})}{n}=\frac{1}{\omega_n}\int_0^r d\rho\int_{|\bar{x}-\bar{a}|=\rho} u(\bar{x})\,dS$$

and differentiating both sides with respect to r. Q.E.D.

Formulas (1) and (2) are also known as mean-value formulas for harmonic functions, for a sphere and for a ball respectively.

(A). If T is the circular region

$$(x-a)^2+(y-b)^2\leqslant R^2$$

then ∂T is the circle

$$(x-a)^2+(y-b)^2=R^2.$$

In polar coordinates a point on ∂T is given by

$$x=a+R\cos\theta,\ y=b+R\sin\theta$$

$$dS=ds=R\,d\theta,\ \Omega_2=2\pi R$$

$$dV=dx\,dy=r\,dr\,d\theta,\ v_2=\pi R^2$$

By formula (1)

$$\frac{2}{2\pi R}\int_{\partial T} u(a+R\cos\theta,\ b+R\sin\theta)\,R\,d\theta$$

$$=\frac{1}{2\pi}\int_0^{2\pi} u(a+R\cos\theta,\ b+R\sin\theta)\ d\theta$$

$$=u(a, b)$$

By formula (2),

$$\frac{1}{\pi R^2}\int_0^R\int_0^{2\pi} u(a+r\cos\theta,\ b+r\sin\theta) r dr\ d\theta=u(a, b)$$

If T is the ball $(x-a)^2+(y-b)^2+(z-c)^2 \leqslant R^2$ then ∂T is the sphere $(x-a)^2+(y-b)^2+(z-c)^2=R^2$.

In polar coordinates a point on ∂T is given by

$$x=a+R\sin\theta\sin\phi,\ y=b+R\sin\theta\cos\phi,\ z=c+R\cos\theta$$

$$dS=R^2\sin\theta\ d\theta\ d\phi,\ \Omega_3=4\pi R^2$$

$$dV=dx\ dy\ dz=r^2\sin\theta\ dr\ d\theta\ d\phi$$

By formula (1),

$$\frac{1}{4\pi}\int_0^{\pi} d\theta\int_0^{2\pi}\left[u(a+R\sin\theta\sin\phi,\ b+R\sin\theta\cos\phi,\ c+R\cos\theta)\right]d\phi$$

$$=u(a, b, c)$$

Theorem 5. (The Maximum—Minimum Principle).

The maximum (or minimum) of a regular harmonic function occurs on the boundary.

Proof. If u is constant in $T\cup\partial T$, there is nothing to prove.

Let u be a non-constant function harmonic in $T\cup\partial T$ such that its maximum M occurs at a point of T. Let F the set of points of T at which $u=M$. We shall prove that F is both open and closed relative to T.

To prove that F is open take $\bar{x}_0\in F$, so that $u(\bar{x}_0)=M$. If γ is the closed sphere $|\bar{x}-\bar{x}_0|\leqslant\rho$ contained in T, then vol $\gamma=V\gamma$

$$=\frac{\omega_n\rho^n}{n}.$$

We can prove that $u=M$ in the whole of γ. If not let $\bar{x}\in\gamma$ such that $u(\bar{x})<M$. But by the second mean-value theorem

$$M=u(\bar{x}_0)=\frac{1}{V\gamma}\int_\gamma u(\bar{x})\ dV<\frac{M}{V\gamma}\int_\gamma dV=M$$

This is a contradiction. Hence $u(\bar{x})=M$ in the whole of γ, *i.e.*, x_0 is an interior point of F. Hence F is open.

That F is closed follows from the continuity of u in T. Let P_i be a sequence of points in F with limit point P. Since $u=M$ at

P_i and u is continuous, u=M at P also. Hence P∈F *i.e.*, F is closed.

Thus F is both open and closed in T. Hence either F=ϕ or F=T. If F=T, then u=M in T and owing to the continuity of u in TU∂T u=M on ∂T also. Thus u=constant M throughout TU∂T. This is a contradiction. Hence F≠T and F=ϕ. Hence u attains its maximum only on ∂T.

The result for the minimum is proved on similar lines if in place of u we take $-u$. Q.E.D.

Corollary 5. If a regular harmonic function in T is constant on ∂T, then it is constant in T.

Problem. (1) Using the maximum principle, prove that the

BVP
$$\nabla^2 u = h(\bar{x}) \text{ in T}$$
$$u(\bar{x}) = f(x) \text{ on } \partial\text{T}$$

has a unique solution where $h(\bar{x})$ ∈C in TU∂T and f ∈C on ∂T.

(2) Prove that $u(\bar{x})$ is harmonic in T, if and only if it has the mean-value property.

Potential Equation for a Circle : Poisson's Integral

In example 22 of chapter 1, we solved the potential equation in a rectangular strip parallel to the y-axis. The mean-value theorem for a circular region determines the value of the harmonic function at the centre as the mean of its values at the circumference. We now solve the potential equation for any point inside the circle other than the centre. To be exact we solve the Dirichlet problem for the circle.

$$\text{DE : } \nabla u^2 = \frac{\partial u}{\partial x^2} + \frac{\partial^2 u}{\partial y^2} = 0 \text{ in } 0 \leqslant r \leqslant \text{R}, \qquad \ldots(i)$$
$$0 \leqslant \theta \leqslant 2\pi$$

$$\text{BC : } u(r, \theta) = f(\theta) \text{ on } \partial\text{T } r=\text{R}, 0 \leqslant \theta \leqslant 2\pi \qquad \ldots(ii)$$

f a given function.

In polar coordinates $\nabla^2 u = 0$ becomes

$$u_{rr} + \frac{1}{r} u_r + \frac{1}{r^2} u_{\theta\theta} = 0 \qquad \ldots(iii)$$

Put $u(r, \theta) = \text{U}(r) \,.\, \text{W}(\theta)$

Then (*iii*) gives

$$\text{U}''\text{W} + \frac{1}{r} \text{U}'\text{W} + \frac{1}{r^2} \text{UW}'' = 0, \qquad \text{U}' = \frac{d\text{U}}{dr}, \text{W}' = \frac{d\text{W}}{d\theta}.$$

or
$$\frac{1}{\text{U}}(r^2\text{U}''r + \text{U}') = -\frac{\text{W}''}{\text{W}} = \lambda \quad \text{(say)}.$$

Hence $r^2\text{U}''r + \text{U}'' - \lambda\text{U} = 0, \quad \text{W}'' + \lambda\text{W} = 0.$

Since we want the solutions to be periodic of period 2π, $\lambda=n^2$ (n an integer).

Hence for U we have the Cauchy-Euler homogeneous equation

$$r^2 U''+rU'-n^2U=0$$

Hence $$U(r)=Ar^n+Br^{-n}$$

We want the solution to exis for $r=0$. Hence $B=0$.

The equation for W will be

$$W''+n^2W=0.$$

This gives $$W=a_n \cos n\theta+b_n \sin n\theta \quad n=0, 1, 2\ldots\ldots$$

Hence $$u(r, \theta)=\sum_{n=0}^{\infty} r^n (a_n \cos n\theta+b_n \sin n\theta) \qquad \ldots(iv)$$

To determine a_n and b_n we use the BC

$$u(R, \theta)=f(\theta).$$

Hence $$f(\theta)=\sum_{n=0}^{\infty} R^n (a_n \cos n\theta+b_n \sin n\theta) \qquad \ldots(v)$$

From (v), we have

$$a_0=\frac{1}{2\pi}\int_0^{2\pi} f(\phi)\, d\phi$$

$$a_n=\frac{1}{\pi R^n}\int_0^{2\pi} f(\phi) \cos n\phi\, d\phi$$

$$b_n=\frac{1}{\pi R^n}\int_0^{2\pi} f(\phi) \sin n\phi\, d\phi$$

Making these substitutions in (iv) we have

$$u(r, \theta)=\frac{1}{2\pi}\int_0^{2\pi} f(\phi)\, d\phi+\frac{1}{\pi}\sum_{n=0}^{\infty}\left(\frac{r}{R}\right)^n$$

$$\left[\cos n\theta \int_0^{2\pi} f(\phi) \cos n\phi\, d\varphi+\sin n\theta \int_0^{\pi 2} f(\varphi) \sin n\phi\, d\phi\right]$$

$$=\frac{1}{\pi}\int_0^{2\pi}\left[\frac{1}{2}+\sum_{n=1}^{\infty}\left(\frac{r}{R}\right)^n \cos n(\theta-\phi)\right] d\phi$$

$$=\frac{R^2-r^2}{2\pi}\int_0^{\pi}\frac{f(\phi)\, d\phi}{R^2-2Rr\cos(\theta-\phi)+r^2} \qquad \ldots(vi)$$

The integral in (*vi*) called *Posson's integral.*

The coefficient of $f(\varphi)$ *viz* $\dfrac{R^2-r^2}{2\pi[R^2-2Rr\cos(\theta-\phi)+r^2]}$

is called the *Poisson Kernel.*

Now let R=1. Then $u(r, \theta)\to f(\theta)$ as $r\to 1$.

Hence

$$f(\theta)=\frac{1}{2\pi}\lim_{r\to 1}\int_0^{2\pi} f(\phi)\,.\,\frac{1-r^2}{1-2r\cos(\theta-\phi)+r^2}\,d\phi \quad ...(vii)$$

Since in this case

$$f(\theta)=\sum_{n=0}^{\infty}(a_n\cos n\theta+b_n\sin n\theta)$$

the integral on the right of (*vii*) is said to be the *Poisson sum* of the Fourier series for $f(\theta)$.

From the relation (*vii*) and the sifting property of the δ-function

$$f(\theta)=\int f(\phi)\,\delta(\phi-\theta)d\varphi,$$

it follows that the Poisson kernel behaves like the Dirac delta function.

§9·12. Characteristic Singularity and Fundamental Solutions of the Potential Equation

Recall the second form of Green's theorem considered in section 9·6 :

$$\int_T (u\nabla^2 v-v\nabla^2 u)\,dV=\int_{\partial T}\left(u\frac{\partial v}{\partial n}-v\frac{\partial u}{\partial n}\right)dS \quad ...(i)$$

The minimum requirements for (*i*) to hold are that $u, v\in C^2$ in T. If either u or v has a singularity inside T, the theorem cannot be applied as given in (*i*). If u or v has an integrable singularity say $\bar{a}$, to apply (*i*), we must delete $\bar{a}$, from T and consider the new domain $T'=T-\{\bar{a}\}$. Since the deletion of one point cannot affect the value of a Riemann integral, we shall have

$$\int_{T'}(u\nabla^2 v-v\nabla^2 u)\,dV=\int_T (u\nabla^2 v-v\nabla^2 u)\,dV$$

The removal of a single point $\bar{a}$ has not altered the left hand integral of (*i*). But an important modification will have to be made in the integral on the right hand side.

Delete from the region T the interior of a closed ball $\sigma : |\bar{x}-\bar{a}| \leqslant \epsilon$ of small radius $\epsilon > 0$. We get a new region T′ with boundary $\partial T' = \partial T \cup \partial\sigma$. (see Fig. 9·2)

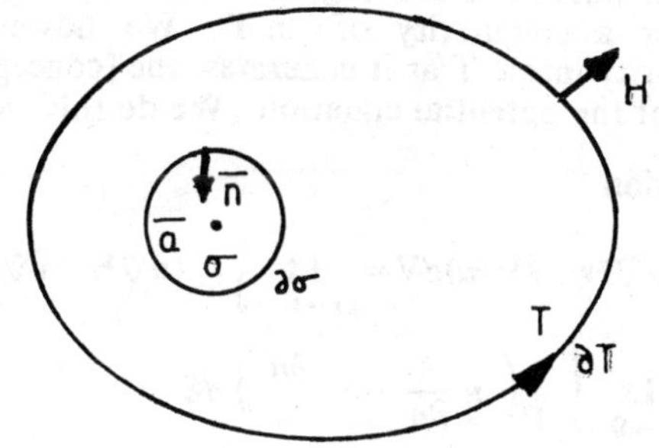

Fig. 9·2,

Now take u to be a regular harmonic function in T and v a potential function with singularity at $\bar{a}$ inside T *i.e.*,

$$\nabla^2 u = 0, \qquad u \in C^2 \text{ in T}$$

and $\nabla^2 v = 0$ in T except at $\bar{x} = \bar{a}$.

In particular, choose v to be a potential function of spherical symmetry *i.e.*, v is a function of r only.

We now actually compute such a function. The function will be called a *singularity function* and point $\bar{a}$ its *characterstic singularity.*

The technique is used to determine a harmonic function u at a point inside the domain of its harmonicity. The singularity function helps to 'pick up' that value of u.

As before let $\bar{a}$ be a fixed point inside T and $\bar{x}$ any point in T. In n-space

$$r^2 = |\bar{x}-\bar{a}|^2 = \sum_{i=1}^{n} (x_i - a_i)^2.$$

Then as shown in corollary 1 of section 9·8 the potential function $v = v(r)$ will satisfy the equation in T,

$$v'' + \frac{n-1}{r} v' = 0, \quad (r \neq 0) \qquad \ldots(ii)$$

The general solution of (ii) is given by

$$v(r) = \begin{cases} c_1 + \dfrac{c_2}{r^{n-2}} & n > 2 \\ c \log r + d & n = 2 \end{cases} \quad (r \neq 0) \qquad \ldots(iii)$$

The fixed point $\bar{a}$ could have been taken to be the origin. From (*iii*) it is obvious that all potential functions of radial symmetry have the origin as a singularity if it is contained in T. Now, the potential equation is invariant under translations. Hence we could transfer the origin $\bar{a}$ to the outside of the region T and there will then be no need to consider a singularity of v in T. We, however, choose to retain the singularity inside T as it generates the concept of a fundamental solution of the potential equation. We do this as follows :

In the relation

$$0=\int_T (u\nabla^2 v-v\nabla^2 u)dV=\underset{\varepsilon\to 0}{\mathrm{Lt}}\int_{T'}(u\nabla^2 v-v\nabla^2 u)\,dV$$

$$=\underset{\varepsilon\to 0}{\mathrm{Lt}}\int_{T'}\left(u\frac{\partial v}{\partial n}-v\frac{\partial u}{\partial n}\right)dS$$

put $v=\frac{1}{r^{n-2}}$. Then referring to Fig. 9·1

$$0=\underset{\varepsilon\to 0}{\mathrm{Lt}}\int_{\partial T'}\left[u\frac{\partial(1/r^{n-2})}{\partial n}-\frac{1}{r^{n-1}}\frac{\partial u}{\partial r}\right]dS$$

$$=\int_{\partial T}\left[u\frac{\partial(1/r^{n-2})}{\partial r}-\frac{1}{r^{n-2}}\frac{\partial u}{\partial r}\right]dS$$

$$=\underset{\varepsilon\leftarrow 0}{\mathrm{Lt}}\int_{\partial\sigma}\left[u\frac{\partial(1/r^{n-2})}{\partial r}-\frac{1}{r^{u-2}}\frac{\partial u}{\partial r}\right]dS$$

since on ∂T $\frac{\partial}{\partial n}=\frac{\partial}{\partial r}$ and on $\partial\sigma$ $\frac{\partial}{\partial n}=-\frac{\partial}{\partial r}$.

On the sphere $\partial\sigma$ $r=\epsilon$, $\frac{1}{r^{n-2}}=\frac{1}{\epsilon^{n-2}}$

and $\left[\frac{\partial(1/r^{n-2})}{\partial r}\right]_{r=\epsilon}=\frac{2-n}{\varepsilon^{n-1}}$.

Hence $\int_{\partial T}\left(u\frac{\partial(1/r-^{n-2})}{\partial r}-\frac{1}{r^{n-2}}\frac{\partial u}{\partial r}\right)dS$

$$=-\underset{\varepsilon\to 0}{\mathrm{Lt}}\int_{\partial\sigma}\left[\frac{(n-2)u}{\varepsilon^{n-1}}+\frac{1}{\varepsilon^{n-2}}\frac{\partial u}{\partial r}\right]dS$$

$$=-\underset{\varepsilon\to 0}{\mathrm{Lt}}\int_{\partial\sigma}\left[(n-2)\,u+\varepsilon\frac{\partial u}{\partial r}\right]\frac{dS}{\epsilon^{n-1}}$$

$$=-\underset{\varepsilon\to 0}{\mathrm{Lt}}\,\omega_n\int_{\partial\sigma}\left[(n-2)\,u+\epsilon\frac{\partial u}{\partial r}\right]\frac{dS}{\omega_n\,\varepsilon^{n-1}}$$

$$=-\underset{\varepsilon\to 0}{\mathrm{Lt}}\,\omega_n\left[(n-2)\,u(\bar{a})+\epsilon\frac{\partial u(\bar{a})}{\partial r}\right]\qquad\ldots(iv)$$

Here ω_n is the surface area of the unit n-sphere. Now $T' \to T$ and $\partial T' \to \partial T$ when $\in \to 0$. Hence

$$\int_{\partial T}\left[u \frac{\partial(1/r^{n-2})}{\partial r} - \frac{1}{r^{n-2}} \frac{\partial u}{\partial r}\right] dS = -(n-2)\,\omega_n\, u(\bar{a}) \qquad \ldots(v)$$

Since the integral on the left hand side of (v) is

$$= \int_T \left(u\nabla^2(1/r^{n-2}) - \frac{1}{r^{n-2}} \nabla^2 u\right) dV$$

and $\nabla^2 u = 0$, we have

$$\int_T u\nabla^2(1/r^{n-2})\, dV = -(n-2)\,\omega_n\, u(\bar{a}) \qquad \ldots(vi)$$

Also as per property of the n-dimensional δ function, we have

$$\int_T u\,.\,\delta(\bar{x} - \bar{a})\, dV = u(\bar{a}) \qquad \ldots(vii)$$

Comparing (vi) and (vii).

$$\nabla^2\left(\frac{1}{r^{n-2}}\right) = -(n-2)\,\omega_n\,\delta(\bar{x} - \bar{a})\cdot \qquad \ldots(viii)$$

$$n \geqslant 3$$

In the special case $n = 3$, $\omega_3 = 4\pi$ and

$$\nabla^2\left(\frac{1}{r}\right) = -4\pi\,\delta(x - a_1, y - a_2, z - a_3)$$

$$= -4\pi\,(x - a_1)\,.\,\delta(y - a_2)\,.\,\delta(z - a_3) \qquad \ldots(ix)$$

In the case $n = 2$, putting $c = -1$, $d = 0$ in (iii)

$$v = -\log r$$

Hence

$$0 = \int_T \left[-u\nabla^2(\log r) + \log r \nabla^2 u\right] dV$$

$$= \int_{\partial T}\left[-u \frac{\partial(\log r)}{\partial r} + \log r \frac{\partial u}{\partial r}\right] dS$$

$$= \int_{\partial\sigma}\left[-u \frac{\partial(\log r)}{\partial r} + \log r \frac{\partial u}{\partial r}\right] dS$$

$$= \int_{\partial\sigma}\left(-u \frac{1}{\epsilon} + \log \varepsilon \frac{\partial u}{\partial r}\right) dS$$

$$= -2\pi \int_{\partial\sigma}\left(u - \epsilon \log \varepsilon \frac{\partial u}{\partial r}\right) \frac{dS}{2\pi\epsilon}$$

$$= -2\pi\left[u(\bar{a}) - \varepsilon \log \epsilon\,.\,\frac{\partial u(\bar{a})}{\partial r}\right]$$

$$= -2\pi\, u(\bar{a}) \text{ as } \varepsilon \to 0.$$

Since $\nabla^2 u=0$ in T, we have

$$\nabla^2 \log \frac{1}{r}=-2\pi\delta(\bar{x}-\bar{a}) \qquad ...(x)$$

If we put

$$\gamma(\bar{x}, \bar{a})=\gamma(r)=\begin{cases} \dfrac{1}{(n-2)\,\omega_n\,.\,r^{n-2}} & n\geqslant 3 \\ \dfrac{1}{2\pi}\log\dfrac{1}{r} & n=2 \end{cases} \qquad ...(xi)$$

Then $\gamma(r)$ is called the *singularity function* for $\nabla^2 u=0$ with singularity at $r=0$ *i.e.*, $\bar{x}=\bar{a}$.

It has the property

$$\nabla^2\gamma(\bar{x}, \bar{a})=-\delta(\bar{x}-\bar{a}) \qquad ...(xii)$$

If $h(\bar{x})$ is any regular harmonic function in T, then for a function

$$E(\bar{x}, \bar{a})=\gamma(\bar{x}, \bar{a})+h(\bar{x})$$
$$\nabla^2 E(\bar{x}, \bar{a})=-\delta(\bar{x}-\bar{a}) \qquad ...(xiii)$$

Such a function $E(\bar{x}, \bar{a})$ is called a *fundamental* (or *elementary*) *solution* of the Laplace's equation corresponding to the characteristic singularity $\bar{x}=\bar{a}$ and $\gamma(\bar{x}, \bar{a})$ is called its *principal part*.

Since $h(\bar{x})$ is a bounded function, from (*xi*) it is clear that

$$E(\bar{x}, \bar{a})\to+\infty \text{ as } r=|\bar{x}-\bar{a}|\to 0$$

Remark 3. It is interesting to see that the familiar case of one dimension ($n=1$) is also included in formula (*xi*). Choose $a=0$.

Then the 'unit ball' will be $|x|\leqslant 1$ *i.e.*, the closed interval $[-1, 1]$ and $r=|x|$.

Also $\omega_1=2$ and $\nabla^2=\dfrac{d^2}{dx^2}$.

Hence by (*xi*)

$$\gamma(r)=-|x|/2$$

$$\frac{d^2}{dx^2}\gamma(r)=-\frac{1}{2}\frac{d^2}{dx^2}|x|=\frac{1}{2}\frac{d}{dx}[H(-x)-H(x)]$$
$$=-\delta(x)$$

where $H(x)=\begin{cases}1 & x>0\\ 0 & x<0\end{cases}$, $\dfrac{dH(x)}{dx}=\delta(x)$ and $\delta(-x)=\delta(x)$.

Formula (*xi*) for the fundamental solution of $\nabla^2 u=0$ is thus valid for $n=1, 2, 3, \ldots\ldots$

Green's Function for the Dirichlet Problem

Consider the relation (v). It affords a unique solution $u(\bar{a})$ of the Laplace's equation at an interior point $\bar{a}$ of T on the basis of a singularity function with a characteristic singularity at $\bar{a}$. The formula involves both the values of u and $\frac{\partial u}{\partial n}$ on the boundary ∂T. This makes the system overdetermined as either the values of u or those of $\frac{\partial u}{\partial n}$ only on ∂T are required to determine u. For the Dirichlet problem we have to get rid of the occurrence of $\frac{\partial u}{\partial n}$ by choosing $E(\bar{x}, \bar{a})$ in such a way that it vanishes on ∂T. This is accomplished by taking $h(\bar{x})=-\gamma(\bar{x}, \bar{a})$ on ∂T. Such a function E is called *Green's function* of the Dirichlet problem and is denoted by $G(\bar{x}, \bar{a})$. We have

$$-\int_{\partial T} u \frac{\partial G}{\partial n} dS=u(\bar{a}) \qquad \ldots(xiv)$$

(*xiv*) provides an explicit solution of the Dirichlet problem for u. But this presupposes solution of a Dirichlet problem for h. Once this h and consequently the Green's function G has been found for a given domain, then other Dirichlet problems for the same domain can be found by a formula. There are cases of domains when the determination of the Green's function is easy. In the sequel we shall give examples of several domains for which Green's function will be determined and the results standardised.

Definition 2. Given the Dirichlet problem

DE : $\nabla^2 u(\bar{x})=0$ in T

BC : $u(\bar{x})$ is given on ∂T

By the *Green's function* of the potential equation for the domain T we mean a two-point function $G(\bar{x}, \bar{a})$ satisfying the following properties :

(*i*) $G(\bar{x}, \bar{a})$ is harmonic in T except at $\bar{x}=\bar{a}$

(*ii*) $G(\bar{x}, \bar{a})-\gamma(\bar{x}, \bar{a})$ is regular harmonic in T where $\gamma(\bar{x}, \bar{a})$ is the fundamental solution of the Laplace equation

(*iii*) $G(\bar{x}, \bar{a})=0$ on ∂T

$G(,\bar{x}, \bar{a})\to 0$ as $|\bar{x}-\bar{a}|\to\infty$ if T is unbounded

(*iv*) In the neighbourhood of the point $\bar{x}=\bar{a}$

$|\bar{x}-\bar{a}|^{n-2} G(\bar{x}, \bar{a})$ for $n\geqslant 3$ and

$$\frac{G(\bar{x}, \bar{a})}{\log|\bar{x}-\bar{a}|} \text{ for } n=2 \text{ are bounded.}$$

It is indicated from the above definition that the Green's function of a potential problem depends on three factors :

(*a*) the PDE

(*b*) the domain

(*c*) the nature of BC *i.e.*, whether it involves Dirichlet or Neumann conditions.

Thus to obtain the Green's function of a Neumann problem, we replace condition (*iii*) by the conditions,

(*iv*) $\dfrac{\partial G}{\partial n}=0$ on ∂T.

The difficulties created by condition (*iv*) will be discussed in section 9·14.

Besides the defining properties stated above the Green's function also satisfies the following properties :

I. Positivity. $G(\bar{x}, \bar{a}) \geqslant 0$ throughout T.

Proof. Consider the figure 9·2. It is already given that $G(\bar{x}, \bar{a})=0$ on ∂T. Since $G(\bar{x}, \bar{a}) \to \infty$ as $\bar{x} \to \bar{a}$, ϵ can be so chosen that $G(\bar{x}, \bar{a})>0$ when $|\bar{x}-\bar{a}|<\epsilon$. Hence $G(\bar{x}, \bar{a}) \geqslant 0$ on the boundary $\partial T \cup \partial \sigma$ of T′. Now in T′ $G(\bar{x}, \bar{a})$ is regular harmonic and by the extremum principle, $G(\bar{x}, \bar{a})$ cannot assume a negative value in T′. Since ϵ is arbitrary, the property holds in T.

II. Symmetry : $G(\bar{a}, \bar{b})=G(\bar{b}, \bar{a})$.

Proof. Let $\bar{a}, \bar{b}$ be two points in the domain T. Remove from T the interior of the balls $\sigma : |\bar{x}-\bar{a}| \leqslant \epsilon$, $\sigma' : |\bar{x}-\bar{b}| \leqslant \epsilon$ of a small radius $\epsilon>0$. Denote the remaining part by T_1. Then $\partial T_1=\partial T \cup \partial\sigma \cup \partial\sigma'$. (See Fig 9·3). In T_1 both $G(\bar{x}, \bar{a})$ and $G(\bar{x}, \bar{b})$ are are regular hormonic. Hence $\nabla^2 G(\bar{x}, \bar{a})=0$ and $\nabla^2 G(\bar{x}, \bar{b})=0$ in T_1. Thus by the Green's theorem

$$0=\int_{T_1}\left[G(\bar{x}, \bar{a})\, \nabla^2 G(\bar{x}, \bar{b})-G(\bar{x}, \bar{b})\nabla^2 G(x, \bar{a})\right]dV$$

$$=\int_{\partial T_1}\left[G(\bar{x}, \bar{a})\, \frac{\partial G(\bar{x}, \bar{b})}{\partial n}-G(\bar{x}, \bar{b})\, \frac{\partial G(\bar{x}, \bar{a})}{\partial n}\right]dS.$$

This leads to

$$\int_{\partial T}\left[G(\bar{x}, \bar{a})\, \frac{\partial G(\bar{x}, \bar{b})}{\partial n}-G(\bar{x}, \bar{b})\, \frac{\partial G(\bar{x}, \bar{a})}{\partial n}\right]dS$$

$$=\int_{\partial\sigma}\left[G(\bar{x}, \bar{a})\, \frac{\partial G(\bar{x}, \bar{b})}{\partial r}-G(\bar{x}, \bar{b})\, \frac{\partial G(\bar{x}, \bar{a})}{\partial c}\right]dS$$

$$+\int_{\partial\sigma'}\left[G(\bar{x}, \bar{a})\, \frac{\partial G(\bar{x}, \bar{b})}{\partial r}-(G, \bar{b})\, \frac{\partial G(x, \bar{a})}{\partial r}\right]dS$$

By the defining properties of the Green's function both

$G(\bar{x}, \bar{a})=0$, $G(\bar{x}, \bar{b})=0$ on ∂T. Hence

$$\int_{\partial\sigma}\left[G(\bar{x}, \bar{a})\, \frac{\partial G(\bar{x}, \bar{b})}{\partial r}-G(\bar{x}, \bar{b})\, \frac{\partial G(\bar{x}, \bar{a})}{\partial r}\right]dS$$

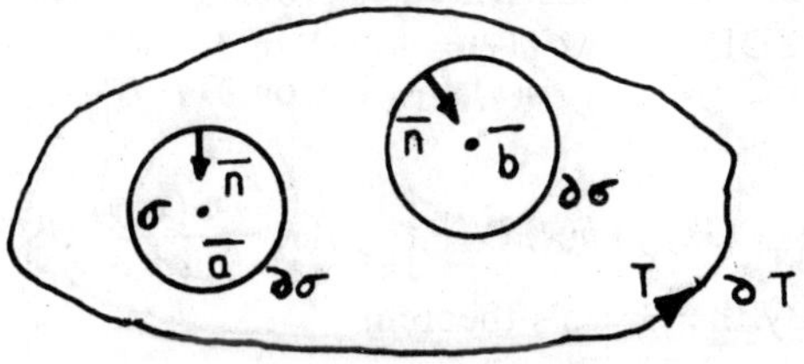

Fig. 9·3.

$$=\int_{\partial\sigma'}\left[G(\bar{x}, \bar{b})\frac{\partial G(\bar{x}, \bar{a})}{\partial r}-G(\bar{x}, \bar{a})\frac{\partial G(\bar{x}, \bar{b})}{\partial r}\right]dS$$

Let $$G(\bar{x}, \bar{a})=\gamma(\bar{x}, \bar{a})+h(\bar{x}, \bar{a})$$

and $$G(\bar{x}, \bar{b})=\gamma(\bar{x}, \bar{b})+h(\bar{x}, \bar{b}).$$

Then if we use the integral formula (v) and let $\in\to 0$, then we have

$$G(\bar{a}, \bar{b})=G(\bar{b}, \bar{a}).$$

The essence of the above proof is contained in the defining property of the Green's function viz,

$$\nabla^2 G(\bar{x}, \bar{a})=-\delta(\bar{x}-\bar{a})$$

and $$\nabla^2 G(\bar{x}, \bar{b})=-\delta(\bar{x}-\bar{b}).$$

If we use these relations in

$$0=\int_{T_1}\left[G(\bar{x}, \bar{a})\nabla^2 G(\bar{x}, \bar{b})-G(\bar{x}, \bar{b})\nabla^2 G(\bar{x}, \bar{a})\right]dV$$

$$=\int_{T}\left[G(\bar{x}, \bar{a})\nabla^2(\bar{x}, \bar{b})-G(\bar{x}, \bar{b})\nabla^2 G(\bar{x}, a)\right]dV$$

$$=\int_{T} G(\bar{x}, \bar{a})\,\delta(\bar{x}-\bar{b})dV-\int_{T} G(\bar{x}, \bar{b})\,\delta(\bar{x}-\bar{a})\,dV$$

$$G(\bar{b}, \bar{a})-G(\bar{a}, \bar{b})$$

i.e., $$G(\bar{a}, \bar{b})=G(\bar{b}, \bar{a})$$

where we have used the fact that the Riemann integrals over T and T_1 are equal and also the sifting property of the delta function.

One immediate consequence of the symmetry property of $G(\bar{x}, \bar{y})$ is that either of the two $\bar{x}, \bar{y}$ can be regarded as the functional variable and the other as parametric point. If $\bar{x}\in R^n$, then $G(\bar{x}, \bar{y})$ is called an n-dimensional Green's function.

We now state and prove the theorem which can be regarded as the main reason for creating the Green's function theory *i.e.*, converting a BVP to an integral.

Theorem 6. Given the Dirichlet problem

$$\text{PDE}: \quad \nabla^2 u(\bar{x}) = -h(\bar{x}) \text{ in T}$$
$$\text{BC}: \quad u(\bar{x}) = -f(\bar{x}) \text{ on } \partial\text{T}.$$

Then

$$u(\bar{y}) = \int_T G(\bar{x}, \bar{y})\, h(\bar{x}) dV + \int_{\partial T} f(\bar{x}) \frac{\partial G(\bar{x}, \bar{y})}{\partial n} dS \qquad ...(1)$$

Proof. By the Green's theorem

$$\int_T u(\bar{x})\, \nabla^2 G(\bar{x}, \bar{y}) dV$$
$$= \int_T G(\bar{x}, \bar{y})\, \nabla^2 u(\bar{x})\, dV + \int_T \left(u \frac{\partial G}{\partial n} - G \frac{\partial u}{\partial n} \right) dS$$

Now it is given that

$$\nabla^2 u(\bar{x}) = -h(\bar{x}) \text{ and } \nabla^2 G(\bar{x}, \bar{y}) = -\delta(\bar{x} - \bar{y}) \text{ in T}$$

and $\qquad u(\bar{x}) = f(\bar{x}),\ G(\bar{x}, \bar{y}) = 0$ on ∂T.

Using the sifting property

$$\int_T u(\bar{x})\, \delta(\bar{x} - \bar{y})\, dV = u(\bar{y})$$

we shall have (1). **Q.E.D.**

The above theorem can be used to convert an eigenvalue problem of PDE to an integral equation.

For example, let

$$\nabla^2 u(\bar{x}) + \lambda r(\bar{x})\, u(\bar{x}) = -h(\bar{x}) \text{ in T}$$
$$u(\bar{x}) = -f(\bar{x}) \text{ on } \partial\text{T}.$$

The problem can be reduced to

$$u(\bar{y}) = F(\bar{y}) + \lambda \int_T r(\bar{x})\, u(\bar{x})\, G(\bar{x}, \bar{y})\, dV \qquad ...(2)$$

where $\qquad F(\bar{y}) = \int_T h(\bar{x})\, G(\bar{x}, \bar{y}) dV + \int_{\partial T} f(\bar{x}) \frac{\partial G}{\partial n} dS$

In practice we shall take the sign of G to be such that

$$\nabla^2 G(\bar{x}, \bar{a}) = \delta(\bar{x} - \bar{a}).$$

Consequently in theorem (6) we shall take

$$\nabla^2 u = h(\bar{x}) \text{ in T and } u = f \text{ on } \partial\text{T}.$$

§9·13. Construction of Green's Function for Dirichlet Problems

The most commonly used methods of constructing a Green's function for the Laplace operator are :

(*i*) the method of reflection (or images)

(*ii*) the method of Laplace and Fourier transforms.

The method of reflection is suitable for potential problems while the transform method is suitable for wave and diffusion problems or for those potential problems in which the function vanishes at infinity. We describe and illustrate these two methods in this section.

Method of Reflection. The method is an adaptation of the relation (*xiv*) viz, $G(\bar{x}, \bar{a}) = \gamma(r) + h(\bar{x})$ where $h(\bar{x})$ has a singularity $\bar{b}$

at the optical image of $\bar{a}$ in the boundary ∂T of the given domain T. Also the form of $h(\bar{x})$ is so chosen that $h(\bar{x})=-\gamma(r)$ on ∂T and $G(\bar{x}, \bar{a})\to 0$ as $r\to\infty$. In the case of a straight line boundary $\bar{a}$, $\bar{b}$ are points of reflection or images in the line. In the case of a sphere $|\bar{x}-\bar{a}|=R$, $\bar{a}$ and $\bar{b}$ are inverse points *i.e.*,

$$\bar{b}=\frac{R^2}{|\bar{a}|^2}\bar{a}$$

In general for an n-dimensional region T, the boundary ∂T will be a hypersurface For example, in n-space if T is the half-space consisting of points $\bar{x}=(x_1, x_2, \ldots\ldots x_n)$ for which $x_i\geqslant 0$, then ∂T is the hyperplane $x_i=0$. Usual examples of half-spaces in two and three dimensions are respectively :

upper half-plane $y\geqslant 0$ with boundary $y=0$
upper half-space $z\geqslant 0$ with boundary $z=0$.

In the case of plane regions, advantage is taken of the technique of conformal transformations. We first derive the Green's function for the upper half-plane, say $G(\bar{x}, \bar{a})$. If a conformal transformation maps the upper half-plane onto a region T with boundary ∂T, then since a conformal transformation does not alter harmonicity the transformation G′ of G is the Green's function for the domain T. Thus in the two dimentional space, the problem of finding a Green's function reduces to that of determining a suitable conformal mapping. In view of the usefulness of this technique we shall later give a list of conformal mappings which map the upper half-plane onto other regions say disks, strips, wedges etc. and vice-versa.

Example 13. Find the Green's function of the equation

$$\nabla^2 u=\frac{\partial^2 u}{\partial x^2}+\frac{\partial^2 u}{\partial y^2}=0$$

for the upper half-plane.

Solution. Here T is given by $y\geqslant 0$ and ∂T is $y=0$. Let (ξ, η) be a fixed point and (x, y) any point in T. Then

$$r=\sqrt{(x-\xi)^2+(y-\eta)^2},\ \eta\geqslant 0$$

and
$$G(x, y\,;\,\xi,\eta)=\frac{1}{2\pi}\log r+g\,(x, y\,;\,\xi, \eta)$$

where g is to be determined so that $G(x, y\,;\,\xi, \eta)$ satisfies the requirements,

(*i*) $\nabla^2 G=\delta(x-\xi)\,\delta(x-\eta)$.
(*ii*) $G=0$ on ∂T. *i.e.* when $y=0$
(*iii*) $G\to 0$ when $\sqrt{x^2+y^2}\to\infty$.

The image of $P(\xi, \eta)$ in the y-axis is $Q(\xi, -\eta)$. If we take

$$g(x, y)=-\frac{1}{2\pi}\log\sqrt{(x-\xi)^2+(y+\eta)^2},$$

then g is a regular harmonic function in T, G=0 where $y=0$. Since

$$G(x, y\,;\,\xi, \eta)=\frac{1}{4\pi}\log\frac{(x-\xi)^2+(y-\eta)^2}{(x-\xi)^2+(y+\eta)^2} \qquad \ldots(1)$$

$G\to 0$ as $\sqrt{x^2+y^2}\to\infty$, the expression in (1) is the required Green's function.

Using theorem 6, we can now solve the problem

PDE : $\nabla^2 u=h(x, y)$ in $-\infty<x<\infty,\ y\geqslant 0$

BC : $u(x, 0)=f(x),\ u\to 0$ as $x^2+y^2\to\infty$.

$$u(\xi, \eta)=\int_{-\infty}^{\infty}\int_0^{\infty}(G(x, y\,;\,\xi, \eta)\,h(x, y)\,dx\,dy$$

$$+\int_{-\infty}^{\infty} f(x)\,\frac{\partial G}{\partial n}\,dx \qquad \ldots(2)$$

Along the boundary $y=0$,

$$\frac{\partial G}{\partial n}=\left[-\frac{\partial G}{\partial y}\right]_{y=0}=\frac{1}{\pi}\,\frac{\eta}{(x-\xi)^2+\eta^2}$$

Hence from (2)

$$u(\xi, \eta)=\int_{-\infty}^{\infty}\int_0^{\infty}G(x, y\,;\,\xi, \eta)\,h(x, y)\,dx,\,dy$$

$$+\frac{1}{\pi}\int_{-\infty}^{\infty}\frac{\eta\, f(x)\,dx}{(x-\xi)^2+\eta^2}$$

We took $\dfrac{\partial G}{\partial n}=-\dfrac{\partial G}{\partial y}$ as $\bar{n}$ is inward drawn.

Remark 4. Adopting the above procedure, the Green's function the problem.

PDE : $\dfrac{\partial^2 u}{\partial x^2}+\dfrac{\partial^2 u}{\partial y^2}=0$ in $-\infty<y<\infty,\ x\geqslant 0$

BC : $u(0, y)=0$ is given by

$$G(x, y\,;\,\xi, \eta)=\frac{1}{4\pi}\log\frac{(x-\xi)^2+(y-\eta)^2}{(x+\xi)^2+(y-\eta)^2}$$

Generalising to n-space, $(n\geqslant 3)$, if

$$\bar{x}=(x_1, x_2, \ldots, x_n),\ \bar{y}=(y_1, y_2, \ldots, y_n)$$

$$r^2=|\bar{x}-\bar{y}|^2=\sum_{i=1}^{n}(x_i-y_i)^2$$

then the n-dimensional Green's function for the problem,

PDE : $\nabla^2 u=0$ in $x_1\geqslant 0,\ -\infty<x_i<\infty\ (i=2, 3, \ldots, n)$

BC : $u(0, x_2, x_3, \ldots, x_n)=0$ is given by

$$G(\bar{x}, \bar{y}) = \frac{1}{(2-n)\omega_n} \frac{1}{r^{n-2}} + g(\bar{x}, \bar{y})$$

$$\text{where } g = \frac{-1}{(2-n)\,\omega_n} \left\{ (x_1 + y_1)^2 + \sum_{i=2}^{n} (x_i - y_i)^2 \right\}^{-\frac{(n-2)}{2}} \quad n \geqslant 3.$$

The example illustrates the method for a domain whose boundary is a hyperplane. In 3-space this hyperplane was a coordinate plane and in 2-space it was a coordinate axis. In the next example, we illustrate the method for a domain which is an n-dimensional region. In particular the region is a ball T of given radius R and the boundnry ∂T is a sphere of radius R *i.e.*,

$$T: \ x_1^2 + x_2^2 + \ldots + x_n \leqslant R^2$$
$$\partial T: \ x_1^2 + x_2^2 + \ldots + x_n^2 = R^2.$$

In this case for the parametric point $\bar{y}$ of G, we shall take the inverse point $\frac{R^2}{|\bar{y}|^2} \cdot \bar{y}$ for g. That G vanishes on the sphere ∂T, is obvious from the fact that on the boundary ∂T, the point $\bar{y}$ and its inverse point $\frac{R^2}{|\bar{y}|^2} \bar{y}$ coincide. However, to be able to apply the formula

$$G(\bar{x}, \bar{y}) = \{\gamma(r) + g(\bar{x})\}$$

we should find a relation between the distances r of $\bar{x}$, $\bar{y}$ and r_1 of $\bar{x}$ and $\frac{R^2\bar{y}}{|\bar{y}|^2}$. We do this in the following lemma.

Lemma 4 $x = (x_1, x_2, \ldots, x_n)$ is any point and $\bar{y} = (y_1, y_2, \ldots, y_n)$ a given point in the ball

$$T: \quad |\bar{x}|^2 = x_1^2 + x_2^2 + \ldots + x_n^2 \leqslant R^2$$

$$\text{Let} \quad r^2 = |\bar{x} - \bar{y}|^2 = \sum_{i=1}^{n} (x_i - y_i)^2$$

$$r_1^2 = \left| \bar{x} - \frac{R^2}{K^2} \bar{y} \right|^2 = \sum_{i=1}^{n} \left(x_i - \frac{R^2}{K^2} y_i \right)^2$$

$$K^2 = |\bar{y}|^2 = y_1^2 + y_2^2 + \ldots + y_n^2.$$

Then on the sphere $\partial T: \ |\bar{x}|^2 = R^2$, r and r_1 are related by

$$r = \frac{K}{R} r_1.$$

Proof. Now $r^2 = \sum x_i^2 + \sum y_i^2 - 2 \sum x_i y_i$.

If the point $\bar{x}$ lies on ∂T then $|\bar{x}|^2 = \sum x_i^2 = R^2$ (see Fig. 9·4).

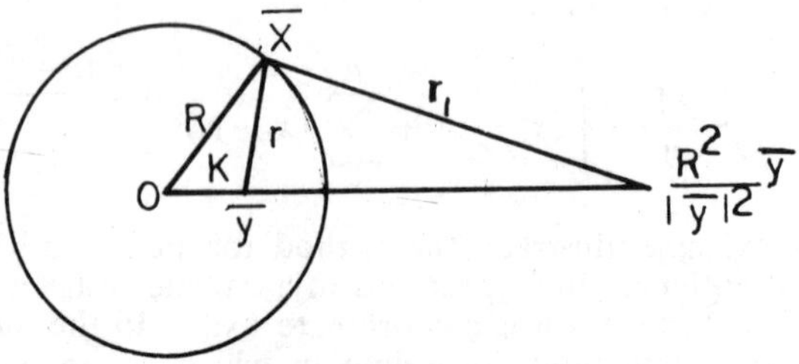

Fig. 9·4.

Hence $r^2 = R^2 + K^2 - 2\sum_{1}^{n} x_i y_i$

Also $r_1^2 = \sum x_i^2 + \frac{R^4}{K^4}\sum y_i^2 - 2\frac{R^2}{K^2}\sum_{1}^{n} x_i y_i$

$$= R^2 + \frac{R^4}{K^2} - 2\frac{R}{K^2}\sum x_i y_i$$

$$= \frac{R^2}{K^2}\left(R^2 + K^2 - 2\sum x_i y_i\right)$$

$$= \frac{R^2}{K^2} r^2$$

i.e., on ∂T $r = \frac{K}{R} r_1$.

Example 14. Solve the Dirichlet problem for the disc.

PDE : $\nabla^2 u = h(x, y)$ in T $x^2 + y^2 \leqslant R^2$

BC : $u = f(\theta)$ on ∂T $x^2 + y^2 = R$.

Solution. We first construct the Green's function of the corresponding homogeneous problem. The Green's function is given by

$$G(x, y\,;\,\xi, \eta) = \gamma(r) - \gamma\left(\frac{K}{R} r_1\right)$$

where $\gamma(r) = \frac{1}{2\pi}\log r$

$$r^2 = (x-\xi)^2 + (y-\eta)^2$$

$$K^2 = \xi^2 + \eta^2$$

$$r_1^2 = \left(x - \frac{R^2}{K^2}\xi\right)^2 + \left(y - \frac{R^2}{K^2}\eta\right)^2$$

Owing to lemma 4, on ∂T $r=\frac{K}{R}r_1$. Hence the requirement that $G=0$ on ∂T, is fulfilled.

Hence $G(x, y\,;\, \xi, \eta)$

$$=\frac{1}{2\pi}\log\left(\frac{r}{r_1}\cdot\frac{R}{K}\right)$$

$$=\frac{1}{4\pi}\log\frac{R}{K}\sqrt{\frac{(x-\xi)^2+(y-\eta)^2}{\left(x-\frac{R^2}{K^2}\xi\right)^2+\left(y-\frac{R^2}{K^2}\eta\right)^2}}$$

$$=\frac{1}{4\pi}\log R^2 K^2\frac{(x-\xi)^2+(y-\eta)^2}{(K^2x-R^2\xi)^2+(K^2y-R^2\eta)^2} \quad \ldots(i)$$

The expression can also be put in the polar form, if we make the substitutions.

$x=\rho\cos\theta,\ y=\rho\sin\theta\,;\ \xi=K\cos\phi,\ \eta=K\sin\phi.$

Thus

$$G(x, y\,;\, \xi, \eta)=\frac{1}{4\pi}\log R^2\frac{\rho^2-2\rho K\cos(\theta-\phi)+K^2}{R^4-2\rho K^2R^2\cos(\theta-\phi)+K^2\rho^2} \quad \ldots(ii)$$

On ∂T, $\frac{\partial G}{\partial n}=\left[\frac{\partial G}{\partial \rho}\right]_{\rho=R}$

$$=\frac{(R^2-K^2)}{2\pi R\,[R^2-2RK\cos(\theta-\phi)-K^2]}$$

and $ds=R\,d\theta.$

Hence by formula (1) of the theorem 6, we get the solution

$$u(\xi, \eta)=\frac{1}{2\pi}\int_0^{2\pi} f(\theta)\,\frac{R^2-K^2}{R^2-2RK\cos(\theta-\phi)+K^2}\,d\theta$$

$$+\int_0^{2\pi}\int_0^{R} G(\rho, \theta\,;\, \xi, \eta)\,h(\rho, \theta)\,d\theta\,d\rho$$

If $h=0$, then we have just the Possion's integral formula

$$u(\xi, \eta)=u(K, \phi)=\frac{R^2-K^2}{2\pi}\int_0^{2\pi}\frac{f(\theta)\,d\theta}{R^2-2RK\cos(\theta-\phi)+K^2} \quad \ldots(iii)$$

found in section 9·11.

In view of the symmetry $G(x, y, ; \xi, \eta)=G(\xi, \eta\,;\, x, y)$ we can interchange ξ, y with x, y in the above derivation without altering the result.

The Use of Conformal Mapping in Two-Dimensional Problems

If T is a simply connected plane region, and $G(x, y\,;\, \xi\ \eta)$ its Green's function, then if the conformal mapping of T onto the half-plane $y \geqslant 0$ is known, G can be determined from the Green's function of the half-plane.

Let $z=x+iy$, $\zeta=\xi+i\eta$, then the Green's function of the upper half-plane is

$$G(x, y ; \xi\ \eta)=\frac{1}{4\pi} \log \frac{(x-\xi)^2+(y-\eta)^2}{(x-\xi)^2+(y+\eta)^2}$$

$$=\frac{1}{2\pi} \log \left| \frac{z-\zeta}{z-\bar{\zeta}} \right|$$

$$=\frac{1}{2\pi} \operatorname{Re}\left[\log \frac{z-\zeta}{z-\bar{\zeta}} \right]$$

We now give two lemmas which exhibit the existence and uniqueness of the Green's function via conformal mapping.

Lemma 5. Let $\omega=F(z)$ be an analytic function which maps the region T of the z-plane onto the upper half of the ω-plane, with $F'(z) \neq 0$ in T so that F is one-one, then the Green's function which satisfies the properties.

(*i*) $\nabla^2 G(x, y ; \xi, \eta)=0$ except at $(x, y)=(\xi, \eta)$ in T.

(*ii*) $G(x, y ; \xi, \eta)=0$ on ∂T.

(*iii*) $G(x, y ; \xi, \eta)=G(\xi, \eta ; x, y)$ (Symmetry)

(*iv*) $G \sim \frac{1}{2\pi} \log r$ as $r \to 0$ where $r^2=(x-\xi)^2+(y-\eta)^2$ is given by

$$G(x, y ; \xi, \eta)=\frac{1}{2\pi} \log \left| \frac{F(z)-F(\zeta)}{F(z)-\bar{F}(\zeta)} \right| \quad \ldots(1)$$

$$=\frac{1}{2\pi} \operatorname{Re}\left\{ \log \frac{F(z)-F(\zeta)}{F(z)-\bar{F}(\zeta)} \right\}$$

where $z=x+iy$, $\zeta=\xi+i\eta$.

Proof. We can also write, since $\log z=\frac{1}{2} \log (z\bar{z})$

$$G(x, y ; \xi, \eta)=\frac{1}{4\pi} \log \left\{ \frac{\bar{F}(z)-F(\zeta)}{\bar{F}(z)-\bar{F}(\zeta)} \times \frac{F(z)-\bar{F}(\zeta)}{\bar{F}(z)-F(\zeta)} \right\} \quad \ldots(2)$$

Symmetry in z and ζ follows from (2). Hence
$G(x, y ; \xi, \eta)=G(\xi, \eta ; x, y)$.

In T, $F(z)=F(\zeta)$ only if $z=\zeta$, in which case $F(z) \neq \bar{F}(\zeta)$. Since $F(z)$ is analytic in T, $\log \frac{F(z)-F(\zeta)}{F(z)-\bar{F}(\zeta)}$ is analytic in T except at $z=\zeta$. Since by the Cauchy-Riemann equations the real part of an analytic function is harmonic except at its singular points, the function

$$\operatorname{Re}\left\{ \log \frac{F(z)-F(\zeta)}{F(z)-\bar{F}(\zeta)} \right\}$$

is harmonic in T except when $F(z)=F(\zeta)$ *i.e.*, when $z=\zeta$.

Since F(z) maps the boundary ∂T onto the x—axis, F(z) is real when z lies on ∂T. Thus on ∂T, since $F(\zeta)=\bar{F}(\zeta)$

$$G(x, y ; \xi, \eta)=\frac{1}{2\pi} \text{ Re } \log \left\{\frac{F(z)-F(\zeta)}{F(z)-\bar{F}(\zeta)}\right\}$$

$$=\frac{1}{2\pi} \text{ Re } \log 1=0.$$

Finally, we write (1) as

$$G(x, y ; \xi, \eta)=\frac{1}{2\pi} \log \left| F(z)-F(\zeta) \right|-\frac{1}{2\pi} \log \left| F(z)-\bar{F}(\zeta) \right|$$

By Taylor's theorem

$$\left| F(z)-F(\zeta) \right|=\left| z-\zeta \right| \left| F'(\zeta)+\frac{z-\zeta}{2\,!} F''(\zeta)+\ldots \right|$$

and $$F(z)\neq\bar{F}(\zeta),$$

Hence $$G\sim\frac{1}{2\pi} \log r \text{ as } z\to\zeta$$

The lemma is thus proved and G is indeed the Green's function for T

Lemma 6. If $w=F_1(z)$ and $w=F_2(z)$ are analytic functions in the region T both mapping T onto the upper half of the w-plane, then F_1 and F_2 yield the same Green's function for T.

Proof. For real a, b, c, d and $ad-bc<0$, the bilinear transformation

$$w=\frac{az+b}{cz+d}$$

is a conformal mapping which maps the upper half of the z-plane. onto the upper half of the w-plane. Hence if F(z) maps the region T of z-plane onto the upper half-plane, then

$$w=\frac{aF(z)+b}{cF(z)+d}$$

will map the upper half of the t-plane, where $t=F(z)$, onto the upper half of the w-plane. Hence F_1, F_2 will be related by

$$F_2=\frac{aF_1+b}{cF_1+d}\,.$$

Hence

$$\frac{F_2(z)-F_2(\xi)}{F_2(z)-\bar{F}_2(\zeta)}=\frac{F_1(z)-F_1(\zeta)}{F_1(z)-\bar{F}_1(\zeta)}$$

and $$\frac{1}{2\pi} \log \left|\frac{F_2(z)-F_2(\zeta)}{F_2(z)-\bar{F}_2(\zeta)}\right|=\frac{1}{2\pi} \log \left|\frac{F_1(z)-F_1(\zeta)}{F_1(z)-\bar{F}_1(\zeta)}\right|.$$

As an application of the conformal technique, we derive the Green's function of the disk from that of the upper half-plane.

If $$F(z)=i\,\frac{R-z}{R+z}$$

then $w=F(z)$ maps the disk $|z| \leqslant R$ onso the upper half w-plane.

Hence by lemma 5, the Green's function for the disc is

$$G(x, y; \xi, \eta)=\frac{1}{2\pi} \log \left| \frac{\dfrac{R-z}{R+z}-\dfrac{R-\zeta}{R+\zeta}}{\dfrac{R-z}{R+z}+\dfrac{R-\bar{\zeta}}{R+\bar{\zeta}}} \right|$$

$$=\frac{1}{4\pi} \log \left| \frac{R(z-\zeta)}{R_2-z\bar{\zeta}} \right|^2$$

which after simplification leads to the formula (*i*) derived in example 14.

We now give a list of some conformal mappings which will help us to construct the Green's function of one plane region from that of another simply connected plane region.

(A) Mappings onto the Upper Half-Plane

In the mappings listed below a specified z-region is mapped into the upper half ($v \geqslant 0$) of the w-plane. The chart also mentions the Green's function of the z-region. Again $z=x+iy$ and $\zeta=\xi+i\eta$.

(B) Mappings onto the Unit Disk

S. No.	*Region*	*Mapping*
1.	Half-plane $y \geqslant 0$	$w=\dfrac{i-z}{i+z}$
2.	Infinite strip $-\infty<x<\infty$, $0 \leqslant y \leqslant b$	$w=\dfrac{i-e^{\pi z/b}}{i+e^{\pi z/b}}$
3.	Infinite sector $0 \leqslant \theta \leqslant \alpha$, $0 \leqslant r<\infty$	$w=\dfrac{i-z^{\pi/\alpha}}{i+z^{\pi/\alpha}}$

The following lemmas will be useful in constructing Green's function of one region from that of another.

Lemma 7. If the analytic function $G(z)$ is such that mapping $w=G(z)$ maps a region T onto the upper half of w-plane, then the mapping

$$w=\frac{i-G(z)}{i+G(z)}$$

maps T onto the unit disk $|w| \leqslant 1$.

Proof. The mapping $t=G(z)$ maps the z-region T onto the upper t-plane.

Also the mapping

$$w=\frac{i-t}{i+t}$$

maps the upper t-plane onto the unit disk $|w| \leqslant 1$. Combining the two mappings, the desired transformation is obtained.

Lemma 8. $w=F(z)$ maps the interior of a z-region T conformally onto the upper half of the w-plane. with $F'(z) \neq 0$ in T. Then the mapping which maps T conformally onto the unit disk

$$|w| \leqslant 1$$

and also the point $z=\zeta$ onto $w=0$, is given by

$$w=\frac{F(z)-F(\zeta)}{F(z)-\bar{F}(\zeta)}$$

Proof. The mapping $t=F(z)$ maps the z-region T onto the upper half of the t-plane. It is easy to see that the mapping

$$w=\frac{t-\tau}{t-\bar{\tau}}$$

maps the upper half of the t-plane onto the unit disk $|w| \leqslant 1$ and also maps the point τ in the upper t-plane onto the point $w=0$. Putting $t=F(z)$, $\tau=F(\zeta)$, we find that

$$w=\frac{F(z)-F(\zeta)}{F(z)-\bar{F}(\zeta)}$$

maps the region T of z-plane onto $|w| \leqslant 1$ and the point $z=\zeta$ onto $w=0$.

Lamma 9. If $w=F(z)$ maps the z-region T conformally onto the unit disk $|w| \leqslant 1$. Then the Green's function for T is

$$G(x, y, \xi, \eta)=\frac{1}{2\pi} \log \left| \frac{F(z)-F(\zeta)}{1-F(z) \cdot \bar{F}(\zeta)} \right|$$

Proof. The mapping $t=F(z)$ maps the region T onto the unit disc $|t| \leqslant 1$. Also the mapping

$$w=i \cdot \frac{1-t}{1+t}$$

maps the unit t-disk onto the upper half of w-plane, Hence

$$w=i\,\frac{1-F(z)}{1+F(z)}$$

maps the region T onto upper half of w-plane. The Green's function for T is

$$G(x, y, \xi, \eta)=\frac{1}{2\pi} \log \left| \frac{F(z)-F(\zeta)}{1-F(z) \cdot \bar{F}(\xi)} \right|.$$

S. No	Region	Mapping	$G(x, y\,;\,\xi, \eta)$
1.	Disk $\lvert z \rvert \leqslant R$ Bdry : $x^2+y^2=R^2$ (see ex. 14)	$w=i\,\dfrac{R-z}{R+z}$	$\dfrac{1}{2\pi}\log\left\lvert\dfrac{R(z-\zeta)}{R^2-z\bar{\zeta}}\right\rvert$
2.	Semidisk $\lvert z \rvert \leqslant, y \geqslant 0$ Bdry : $x^2+y^2=R^2$ $y\geqslant 0$	$w=-\dfrac{1}{2}\left(\dfrac{z}{R}+\dfrac{R}{z}\right)$	$\dfrac{1}{2\pi}\log\left\lvert\dfrac{z\zeta-R^2}{z\bar{\zeta}-R^2}\right\rvert$ For $R=1$ $\dfrac{1}{2\pi}\log\left\lvert\dfrac{z\zeta-1}{z\bar{\zeta}-1}\right\rvert$ $=\dfrac{1}{4\pi}\log\dfrac{\rho^2k^2-2\rho k\cos(\theta-\varphi)+1}{\rho^2k^2-2\rho k\cos(\theta+\varphi)+1}$ where $x=\rho\cos\theta,\quad y=\rho\sin\theta$ $\xi=k\cos\phi,\quad \eta=k\sin\phi$
3.	Half plane indented by semicircle : $\lvert z_1 \rvert \geqslant R,\ y\geqslant 0$	$w=\dfrac{1}{2}\left(\dfrac{z}{R}+\dfrac{R}{z}\right)$	As in (2)
4.	Infinite strip $-\infty<x<\infty$, $0\leqslant y\leqslant b$	$w=e^{\frac{\pi z}{b}}$ $=e^{\frac{\pi x}{b}}\left(\cos\dfrac{\pi y}{b}+i\sin\dfrac{\pi y}{b}\right)$	$\dfrac{1}{4\pi}\log\left[\dfrac{\cosh\frac{\pi}{b}(\xi-x)-\cos\frac{\pi}{b}(\eta-y)}{\cosh\frac{\pi}{b}(\xi-x)-\cos\frac{\pi}{b}(\eta+y)}\right]$ $\dfrac{\partial G}{\partial n}=\left[-\dfrac{\partial G}{\partial y}\right]_{y=0}$ $=\dfrac{1}{2b}\,\dfrac{\sin\frac{\pi}{b}\eta}{\cosh\frac{\pi}{b}(\xi-x)-\cos\frac{\pi\eta}{b}}$ on the x-axis

5.	Infinite strip $0 \leqslant x \leqslant b$ $-\infty < y < \infty$	$w = e^{i\frac{\pi z}{b}}$	$\frac{1}{4\pi}\log\left[\frac{\cosh\frac{\pi}{b}(y-\eta)-\cos\frac{\pi}{b}(x-\xi)}{\cosh\frac{\pi}{b}(y-\eta)-\cos\frac{\pi}{b}(x+\xi)}\right]$ $\frac{\partial G}{\partial n}=\left[-\frac{\partial G}{\partial x}\right]_{x=0}$
6.	Infinite sector $0 \leqslant \theta \leqslant \alpha$ $0 \leqslant r < \infty$	$w = z^{\frac{\pi}{\alpha}}$ $0 \leqslant \arg z \leqslant \alpha$ $0 \leqslant \mathrm{rg}^2\ w \leqslant \pi$	$\frac{1}{2\pi}\log\left\lvert\frac{z^{\pi/\alpha}-z^{\pi/\alpha}}{z^{\pi/\alpha}-\overline{\zeta}^{\pi/\alpha}}\right\rvert$
7.	Circular sector $r = \lvert z \rvert \leqslant R$ $0 \leqslant \theta \leqslant \alpha$	$w = -\frac{1}{2}\left[\left(\frac{z}{R}\right)^{\frac{\pi}{\alpha}}+\left(\frac{R}{z}\right)^{\frac{\pi}{\alpha}}\right]$	
8.	Exterior of disk $\lvert z \rvert \geqslant R$	$w = -i\,\frac{R-z}{R+z}$	Same as 1 (interior of disk)

Example 15. Obtain the Green's function for the first quadrant $0\leqslant x<\infty$, $0\leqslant y<\infty$. Hence solve the BVP,

$$\text{PDE}: \nabla^2 u=\frac{\partial^2 u}{\partial x^2}+\frac{\partial^2 u}{\partial y^2}=0$$

$$\text{BCs}: u(x, 0)=f(x), u(0, y)=0$$

$$u\to 0 \text{ as } x^2+y^2\to\infty$$

Solution. It is easy to see that $F(z)=z^2$ maps the first quadrant onto the upper half plane, Hence by lemma 5, the Green's function of the first quadrant is given by

$$G(x, y; \xi, \eta)=\frac{1}{2\pi}\log\left|\frac{z^2-\zeta^2}{z^2-\bar{\zeta}^2}\right|; z=y+iy, \zeta=\xi+i\eta$$

$$=\frac{1}{2\pi}\log\left|\frac{z-\zeta}{z-\bar{\zeta}}\cdot\frac{z+\zeta}{z+\bar{\zeta}}\right|$$

$$=\frac{1}{4\pi}\log\left[\frac{(x-\xi)^2+(y-\eta)^2}{(x-\xi)^2+(y+\eta)^2}\right]\left[\frac{(x+\xi)^2+(y+\eta)^2}{(x+\xi)^2+(y-\eta)^2}\right]$$

The boundary of T consists of the positive half of x-and y-axis. On the x-axis, we have

$$\frac{\partial G}{\partial n}=\left[-\frac{\partial G}{\partial y}\right]_{y=0}=\frac{4}{\pi}\frac{x\xi\eta}{[(x-\xi)^2+\eta^2][(x+\xi)^2+\eta^2]}$$

and on the y-axis

$$\frac{\partial G}{\partial n}=\left[-\frac{\partial G}{\partial x}\right]_{x=0}=\frac{4}{\pi}\frac{y\xi\eta}{[(y-\eta)^2+\xi^2][(y+\eta)^2+\xi^2]}$$

By theorem 6, the solution of the given BVP is

$$u(\xi, \eta)=\int_{\partial T} f(z)\frac{\partial G}{\partial n}dz$$

Since ∂T consists of x and y-axes from 0 to ∞ and $u=0$ on the y-axis.

$$u(\xi, \eta)=\frac{4}{\pi}\int_0^\infty f(x)\frac{x\xi\eta dx}{[x-\xi)^2+\eta^2][(x+\xi)^2+\eta^2]}$$

Interchanging (x, y) and (ξ, η), we have

$$u(x, y)=\frac{4}{\pi}\int_0^\infty f(\xi)\frac{\xi xy d\xi}{[(x-\xi)^2+y^2][x+\xi)^2+y^2]}$$

$$=\frac{1}{\pi}\int_0^\infty f(\xi)\left[\frac{y}{(x-\xi)^2+y^2}-\frac{y}{[(x+\xi)^2+y^2}\right]d\xi.$$

Example 16, Obtain the Green's function for the infinite sector $0\leqslant\theta\leqslant\alpha$, $0\leqslant r<\infty$ in the polar form. Hence solve the BVP,

$$\text{PDE}: \nabla^2 u=0$$

$$\text{BCs}: u(r, 0)=f(r), u(r, \alpha)=g(r).$$

Solution. It was found that the Green's function of the infinite sector is

$$G(x, y\,;\,\xi, \eta)=\frac{1}{2\pi}\log\left|\frac{z^{\pi/\alpha}-\zeta^{\pi/\alpha}}{z^{\pi/\alpha}-\bar{\zeta}^{\,-\pi/\alpha}}\right| \qquad \ldots(i)$$

Put $z=re^{i\theta}$, $\zeta=\rho e^{i\phi}$, then (i) reduces to

$$G(r, \theta\,;\,\rho, \phi)=\frac{1}{4\pi}\log\frac{r^{2\pi/\alpha}-2r^{\pi/\alpha}\,.\,\rho^{\pi/\alpha}\cos(\theta-\phi)\frac{\pi}{\alpha}+\rho^{2\pi/\alpha}}{r^{2\pi/\alpha}-2r^{\alpha}\,\rho^{\pi/\alpha}\cos(\theta+\phi)\frac{\pi}{\alpha}+\rho^{2\pi/\alpha}} \qquad \ldots(ii)$$

To solve the given BVP we have to find $\frac{\partial G}{\partial n}$ at $\theta=0$ and $\theta=\alpha$ which form the boundary of T (see Fig. 9·5).

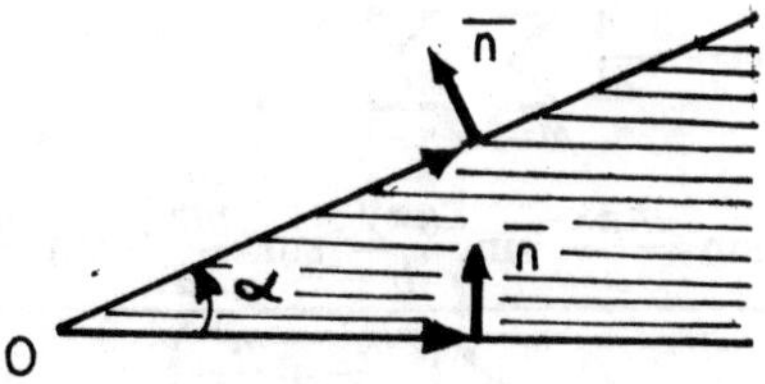

Fig. 9·5.

Since perpendicular to the radius vector the variation is in θ, $dn=rd\theta$.

$$\frac{\partial G}{\partial n}=\frac{1}{r}\frac{\partial G}{\partial \theta}.$$

Hence for the boundaries $\theta=0$, $\theta=\alpha$, we have respectively

$$\frac{\partial G}{\partial n}=\left[-\frac{\partial G}{r\partial\theta}\right]_{\theta=0}$$

(minus sign is used as with increasing θ, n is inward drawn at $\theta=0$)

$$=\frac{1}{r\alpha}\cdot\frac{r^{\pi/\alpha}\,\rho^{\pi/\alpha}\sin\frac{\pi\phi}{\alpha}}{r^{2\pi/\alpha}-2r^{\pi/\alpha}\,\rho^{\pi/\alpha}\cos\frac{\pi}{\alpha}\varphi+\rho^{2\pi/\alpha}} \qquad \ldots(iii)$$

and

$$\frac{\partial G}{\partial n}=\left[+\frac{1}{r}\frac{\partial G}{\partial\theta}\right]_{\theta=\alpha}$$

(plus sign is used because with increasing θ, $\bar{n}$ is outward drawn at $\theta=\alpha$)

$$=\frac{1}{r\alpha}\cdot\frac{r^{\pi/\alpha}\cdot\rho^{\pi/\alpha}\sin\frac{\pi\phi}{\alpha}}{r^{2\pi/\alpha}+2r^{\pi/\alpha}\rho^{\pi/\alpha}\cos\frac{\pi}{\alpha}\phi+\rho^{2\pi/\alpha}}\qquad\ldots(iv)$$

Using theorem 6, the solution to the given BVP is given by

$$u(\rho,\phi)=\frac{1}{\pi}\left[\int_0^\infty\frac{f(u^{\alpha/\pi})\,v\sin\phi_1\,du}{u^2\sin^2\phi_1+(u-u\cos\varphi_1)^2}\right.$$

$$\left.+\int_0^\infty\frac{g(u^{\alpha/\pi})\,v\sin\phi_1\,du}{u^2\sin^2\varphi_1+(u+v\cos\varphi_1)^2}\right]$$

where $\quad r^{\pi/\alpha}=u,\ e^{\tau/\alpha}=v,\ \frac{\pi}{\alpha}\phi=\phi_1$

Example 17. Obtain the Green's function for the rectangle R : $0\leqslant x\leqslant a$; $0\leqslant y\leqslant b$ in the form

$$G(x,y;\xi,\eta)=-\frac{4}{ab}\sum_{m=1}^{\infty}\sum_{n=1}^{\infty}$$

$$\frac{\sin\frac{m\pi x}{a}\sin\frac{n\pi y}{b}\sin\frac{m\pi\xi}{a}\sin\frac{n\pi\eta}{b}}{\pi^2\left(\frac{m^2}{a^2}+\frac{n^2}{b^2}\right)}\qquad\ldots(i)$$

Solution. The use of conformal transformation to find G requires the use of elliptic functions. An alternative procedure is to generalise the technique of example (27) of chapter 8.

Consider the eigenvalue problem

$$\nabla^2\phi+\lambda\phi=0\quad\text{in R}$$
$$\phi=0\quad\text{on }\partial\text{R}.$$

It is easily verified that

$$\phi=\phi_{mn}(x,y)$$
$$=2\sin\frac{m\pi x}{a}\sin\frac{n\pi y}{b},\ m,n=1,2,3,\ldots$$

are eigenfunctions of the problem corresponding to the eigenvalues

$$\lambda_{mn}=\pi^2\left(\frac{m^2}{a^2}+\frac{n^2}{b^2}\right),\ m,n=1,2,3,\ldots.$$

The functions $\phi_{mn}(x,y)$ form an orthonormal system on R and vanish on ∂R.

The Green's function of the rectangle R is given by

$$\nabla^2 G = \delta(x-\xi)\,\delta(y-\eta) \text{ in R}$$
$$G=0 \quad \text{on } \partial R.$$

Let $\delta(x-\xi)\,\delta(y-\eta)$ be expanded in an orthonormal series

$$\delta(x-\xi)\,\delta(y-\eta)=2\sum_{m=1}^{\infty}\sum_{n=1}^{\infty} A_{mn} \sin\frac{m\pi x}{a} \sin\frac{n\pi y}{b}$$

Then $A_{mn}=2\int_R\int \sin\frac{m\pi x}{a} \sin\frac{n\pi y}{b}\,\delta(x-\xi)\,\delta(y-\eta)\,dx\,dy$

$$=2\sin\frac{m\pi\xi}{a}\sin\frac{n\pi\eta}{b}.$$

Hence $\frac{\partial^2 G}{\partial x^2}+\frac{\partial^2 G}{\partial y^2}=4\sum_{m=1}^{\infty}\sum_{n=1}^{\infty}$

$$\sin\frac{m\pi x}{a}\sin\frac{n\pi y}{b}\sin\frac{m\pi\xi}{a}\sin\frac{n\pi\eta}{b} \quad \ldots(ii)$$

If we put $\frac{\partial}{\partial x}=d_1$, $\frac{\partial}{\partial y}=d_2$ and use the formula

$$\frac{1}{d_1^2+d_2^2}\sin\frac{m\pi x}{a}\sin\frac{n\pi y}{b}=\frac{-\sin\frac{m\pi x}{a}\sin\frac{n\pi y}{b}}{\frac{\pi^2 m^2}{a^2}+\frac{\pi^2 n^2}{b^2}}$$

(see 1·3)

then integrating (*ii*) and using the condition $G=0$ an ∂R we obtain (*i*).

Problem. Show that the Green's function for the equation $(\nabla^2+k)u=0$ on the rectangle $0\leqslant x\leqslant a$, $0\leqslant y\leqslant b$ is :

$$G(x,y\,;\,\xi,\eta\,;\,k)=\frac{-4}{ab}\sum_{m,\,n=1}^{\infty}$$

$$\frac{\sin\frac{m\pi x}{a}\sin\frac{n\pi y}{b}\sin\frac{m\pi\xi}{a}\sin\frac{n\pi\eta}{b}}{\pi^2\left(\frac{m^2}{a^2}+\frac{n^2}{b^2}\right)-k}$$

Other Two-Dimensional Operators

The Green's function of a two dimensional second order

operator L is related to its adjoint L derived in example 9. Indeed, if

$$L(u)=A\frac{\partial^2 u}{\partial x^2}+2B\frac{\partial^2 u}{\partial x\partial y}+C\frac{\partial^2 u}{\partial y^2}+D\frac{\partial u}{\partial x}+E\frac{\partial u}{\partial y}+Fu, \quad ...(i)$$

then the adjoint operator $\bar{L}$ is given by

$$\bar{L}(v)=\frac{\partial^2(Av)}{\partial x^2}+2\frac{\partial^2(Bv)}{\partial x\partial y}+\frac{\partial^2(Cv)}{\partial y^2}-\frac{\partial(Dv)}{\partial x}-\frac{\partial(Ev)}{\partial y}+Fv \quad ...(ii)$$

and $$\int_R\int\left\{vL(u)-u\bar{L}(v)\right\}dx\,dy=\int_{\partial R}(P\,dy-Q\,dx)$$

If the right hand integral vanishes, then

$$\int_R\int vL(u)\,dx\,dy=\int_R\int u\,\bar{L}(v)\,dx\,dy \quad ...(iii)$$

The Green's function of the operator L for the region R if it exists, is defined to be that function $G=G(x, y\,;\,\xi, \eta)$ which satisfies the equation

$$\bar{L}G=\delta(x-\xi)\,\delta(y-\eta) \quad ...(iv)$$

and $$G=0 \text{ on } \partial R.$$

If we are given the problem

$$L(u)=f(x, y) \quad \text{in } R$$

$$u=0 \quad \text{on } \partial R$$

then putting $v=G$ in (*iii*) and using the relation

$$\int_R\int u\,\bar{L}(G)\,dx\,dy$$

$$=\int_R\int u\,.\,\delta(x-\xi)\,\delta(y-\eta)\,dx\,dy$$

$$=u(\xi, \eta)$$

we have the solution

$$u(\xi, \eta)=\int_R\int G(x, y\,;\,\xi, \eta)\,f(x, y)\,dx, dy$$

In a special case $L=\bar{L}$ (*i.e.*, L is self-adjoint). Then G is symmetric in (x, y) and (ξ, η) if the adjoint boundary condition is also the same. As before while searching for the adjoint operator we may use the following two-dimensional analogies of integration by parts,

$$\int_R\int u\frac{\partial v}{\partial x}\,dx\,dy=\int_R\int\frac{\partial(uv)}{\partial x}\,dx\,dy-\int_R\int v\frac{\partial u}{\partial x}\,dx\,dy$$

$$=\int_{\partial R} u\, v\cos\alpha\, ds-\int_R\int v\frac{\partial u}{\partial x}\,dx\,dy$$

and $$\int_R\int u\frac{\partial v}{\partial y}\,dx\,dy=\int_{\partial} u\,v\cos\beta\,dx-\int_R\int v\frac{\partial u}{\partial y}\,dx\,dy$$

where $\bar{n}=i\cos\alpha+j\cos\beta$ is the outword drawn unit normal to ∂R. One of the variables say y, may denote time and in that case R is a *space-time domain.*

Example 18. Find the Green's function of the modified Helmoltz operator ∇^2-k^2 for R the entire plane.

Solution. The Green's function $G(x, y\,;\,\xi, \eta)$ satisfies

$$\nabla^2 G-k^2 G=0 \text{ in } R,\ (x, y)\neq(\xi, \eta) \qquad \ldots(i)$$

$$G\to 0 \text{ as } (x-\xi)^2+(y-\eta)^2\to\infty.$$

Put $\quad x-\xi=\rho\cos\phi,\ y-\eta=\rho\sin\phi \qquad \ldots(ii)$

Hence $\quad \rho^2=(x-\xi)^2+(x-\eta)^2$

In a circular disk $(x-\xi)^2+(y-\eta)^2\leqslant\rho^2$

$$G\sim\frac{1}{2\pi}\log\rho \text{ as } \rho\to 0,$$

i.e., G is independent of ϕ.

Hence in polar coordinates (*ii*), equation (*i*) is

$$\frac{d^2G}{d\rho^2}+\frac{1}{\rho}\frac{dG}{d\rho}-k^2 G=0 \text{ except when } \rho=0 \qquad \ldots(iii)$$

The general solution of (*iii*) is given by

$$G=AI_0(k\rho)+BK_0(k\rho)$$

where I_0 and K_0 are modified Bessel functions of first and second kind respectively.

As $x\to\infty$ $\quad I_0(x)\sim\frac{e^x}{\sqrt{2\pi x}}$ and $K_0(x)\sim\frac{e^{-x}}{\sqrt{2\pi x}}$,

we must have A=0.

Also since $K_0(x)\sim-\log x$ as $x\to 0$

and $\quad G\sim\frac{1}{2\pi}\log\rho$ as $\rho\to 0$, we have

$$B=-\frac{1}{2\pi}$$

Hence finally $\quad G=-\frac{1}{2\pi}K_0(k\rho).$

As an application, the solution to the BVP

$$\nabla^2 u - k^2 u = h(x, y), \qquad -\infty < x < \infty, -\infty < y < \infty$$

$$u \to 0 \text{ as } x^2 + y^2 \to \infty$$

is given by

$$u(\xi, \eta) = -\frac{1}{2\pi} \int_{\infty}^{\infty} \int_{\infty}^{\infty} K_o \left(k\sqrt{(x-\xi)^2 + (y-\eta)^2}\right) h(x, y)\, dx\, dy.$$

Using the method of images it can be seen that the Green's function for the upper half-plane is

$$G(x, y\,;\, \xi, \eta) = \frac{-1}{2\pi}\Big[K_o\left(k\sqrt{(x-\xi)^2 + (y-\eta)^2}\right)$$

$$- K_o \left(k\sqrt{(x-\xi)^2 + (y+\eta)^2}\right)\Big]$$

It may be noticed that for this operator advantage of a conformal transformation $u + i\, v = F(x + i\, y)$ cannot be taken as the equation

$$\frac{\partial^2 V}{\partial x^2} + \frac{\partial^2 V}{\partial^2 y} - k^2 V = 0$$

does not change to

$$\frac{\partial^2 V}{\partial u^2} + \frac{\partial^2 V}{\partial v^2} - k^2 V = 0$$

unless $k = 0$.

§9·14. The Neumann Problem

PDE : $\nabla^2 u = 0$ in T

BC : $\dfrac{\partial u}{\partial n} = f$ on ∂T.

If a Green's function $G(\bar{x}, \bar{y})$ for this problem is to exist, it will satisfy the following properties (see Def. 2).

(*a*) As a function of $\bar{x}$, in T

$$\nabla^2 G(\bar{x}, \bar{y}) = 0 \text{ except at } \bar{x} = \bar{y}$$

(*b*) $|\,\bar{x} - \bar{y}\,|^{n-2}\, G(\bar{x}, \bar{y})$ is bounded and finite near $\bar{x} = \bar{y}$.

(*c*) $\dfrac{\partial G}{\partial n} = 0$ on ∂T.

However, such a function cannot exist. By Gauss theorem

$$\int_T \nabla^2 G\, dV = \int_{\partial T} \frac{\partial G}{\partial n}\, dS \qquad \text{...}(i)$$

Since $\nabla^2 G = \delta(\bar{x} - \bar{y})$ and $\dfrac{\partial G}{\partial n} = 0$ on ∂T, the left hand side of (*i*) is 1 and the right hand side is 0. This is a contradiction. Hence the Green's function in the usual sense for the Neumann problem does not exist. In fact, the contradiction has its origin to a more general problem:

PDE : $\Delta^2 u = h$ in T ...(*ii*)

BC : $\dfrac{\partial u}{\partial n} = g$ on ∂T

which because of the relation (*i*) requires

$$\int_T h\,dV = \int_{\partial T} g\,dS \qquad \ldots(iii)$$

Thus unless the *compatibility* condition (*iii*) holds, the BVP can have no solution. Even in the case (*iii*) holds, though solution for (*ii*) exists, it is unique only upto an additive constant. Hence Green's function is modified to remove this contradiction. This is done in two ways defining two modifications of the Green's function, *Neumann function* and *Hilbert's function.*

(*a*) *Neumann Function.* $N(\bar{x}, \bar{y})$. It is required to satisfy

$$\nabla^2 N(\bar{x}, \bar{y}) = \delta(\bar{x} - \bar{y}) \text{ in T} \qquad \ldots(iv)$$

and

$$\frac{\partial N}{\partial n} = \alpha(S) \text{ on } \partial T. \qquad \ldots(v)$$

where α(S) is any function for which

$$\int_{\partial T} \alpha(S)\,dS = 1 \qquad \ldots(v)$$

Using the Green's second theorem, we have

$$\int_T (u\nabla^2 N - N\nabla^2 u)\,dV = \int_{\partial T}\left(u\,\frac{\partial N}{\partial n} - N\,\frac{\partial u}{\partial n}\right) dS$$

Hence $u(\bar{y}) = A - \int_{\partial T} N\,.\,g\;dS + \int_{\partial T} N\,.\,h\;dV$

where $A = \int_{\partial T} \alpha u\,dS$ is an arbitrary constant.

(*b*) *Hilbert's Function* $H(\bar{x}, \bar{y})$. This satisfies

$$\nabla^2 H(\bar{x}, \bar{y}) = \delta(\bar{x} - \bar{y}) - \beta(\bar{x}, \bar{y}) \text{ in T} \qquad \ldots(vii)$$

$$\frac{\partial H}{\partial n} = 0 \text{ on } \partial T.$$

where β is any function for which

$$\int_{\partial T} \beta(\bar{x}, \bar{y})\,dV = 1 \qquad \ldots(viii)$$

Remark 5. The term Neumann function also applies to Bessel function of the second kind.

§9·15. Eigenfunction Expansions and Green's Function

The device explained in section 8·9 of the preceeding chapter for expanding the one-dimensional Green's function has a direct generalisation to higher dimensional Green's function. The result is summed up in the following :

Theorem 7. The Green's function $G(\bar{x}, \bar{y})$ of the Dirichlet problem

$$\text{PDE}: Lu(\bar{x})=0 \text{ in T} \qquad ...(i)$$
$$\text{BC}: u(\bar{x})=0 \text{ on } \partial\text{T} \qquad ...(ii)$$

has the bilinear expansion

$$G(\bar{x}, \bar{y})=\sum_{n=1}^{\infty} \frac{u_n(\bar{x})\, u_n(\bar{y})}{\lambda_n} \qquad ...(iii)$$

where $u_n(\bar{x})$ are orthonormal eigenfunctions corresponding to the eigenvalues λ_n of the problem.

$$\text{PDE}: L\, u(\bar{x})=\lambda u(\bar{x}) \text{ in T} \qquad ...(iv)$$
$$\text{BC}: u(\bar{x})=0 \text{ on } \partial\text{T} \qquad ...(ii)$$

Proof. If we are given the BVP

$$\text{PDE}: L\, u(\bar{x})=f(\bar{x}) \text{ in T} \qquad ...(v)$$
$$\text{BC}: u(\bar{x})=0 \text{ on } \partial\text{T} \qquad ...(ii)$$

then formally

$$u(\bar{x})=\int_T G(\bar{x}, \bar{y})\, f(\bar{y})\, dV \qquad ...(vi)$$

$$\text{Let } u(\bar{x})=\sum_{n=1}^{\infty} C_n u_n(\bar{x}) \qquad ...(vii)$$

$$f(\bar{x})=\sum_{n=1}^{\infty} A_n u_n(\bar{x}) \qquad ...(viii)$$

Then C_n are unknown and A_n are known and given by the formula

$$A_n=\int_T f(\bar{y})\, u_n(\bar{y})\, dV$$

From (*vii*), we have

$$L\, u(\bar{x})=\sum_{n=1}^{\infty} C_n\, L\, u_n(\bar{x})$$

$$=\sum_{n=1}^{\infty} C_n \lambda_n u_n(\bar{x})=f(\bar{x})$$

$$=\sum_{n=1}^{\infty} A_n u_n \,($$

Hence $\quad C_n = \dfrac{A_n}{\lambda_n}$.

Also $\quad u(\bar{x}) = \sum_{n=1}^{\infty} \dfrac{A_n}{\lambda_n} u_n(\bar{x})$

$$= \sum_{n=1}^{\infty} \left[\frac{1}{\lambda_n} \int_T u_n(\bar{x})\, u_n(\bar{y})\, f(\bar{y})\, dV \right] \qquad ...(ix)$$

Comparing relations (*vi*) and (*ix*), we have

$$G(\bar{x}, \bar{y}) = \sum_{n=1}^{\infty} \frac{u_n(\bar{x})\, u_n(\bar{y})}{\lambda_n}$$

Corollary 6. The Green's function $G(\bar{x}, \bar{y}\,;\, \lambda)$ defined by the equation

$$LG - \lambda G = \delta(\bar{x} - \bar{y}) \text{ in } T$$

and $\quad G = 0$ on ∂T

has the bilinear form

$$G(\bar{x}, \bar{y}\,;\, \lambda) = \sum_{n=1}^{\infty} \frac{u_n(\bar{x})\, u_n(\bar{y})}{\lambda_n - \lambda}$$

provided λ is not an eigenvalue of the system (*iv*), (*ii*).

§9·16. The Method of Integral Transforms to Construct Green's Function

Our aim in this section is to illustrate rather than to prove the main ingredients of the theory of integral transforms in the construction of a Green's function. It is, however, imperative to keep in mind that each transform and its inversion holds under specific analytic conditions. On the other hand, a Green's function depends on the differential operator and the nature of the boundary. It does not depend on specific boundary values. The choice of the transform is, therefore, dictated by the choice of the boundary condition in the problem. For example, if Fourier transform is to be used, the boundary variable must be allowed to vary from $-\infty$ to ∞ and the solution of the problem must be bounded at $\pm\infty$. With this broad guideline we consider the following examples. The contour of the inversion integral depends on the BCs.

Example 19. Find the Green's function of the BVP,

DE : $(a^2 - D^2)y = 0 \qquad a > 0$

BCs : $\quad y(x) = 0\ (e^{x}) \qquad$ as $x \to +\infty$

$\qquad\quad y(x) = 0\ (e^{-x}) \qquad$ as $x \to -\infty$

Solution. The Green's function of the BVP is given by

$$(a^2-D^2)\, G(x, \xi)=\delta(x-\xi) \qquad ...(i)$$

$$G(x, \xi)\to 0 \text{ as } |x| \to\infty \qquad ...(ii)$$

The Fourier transformation (F.T.) is given by

$$\text{F.T. } f=\bar{f}(p)=\int_{-\infty}^{\infty} e^{ipt}\, \bar{f}(t)\, dt \qquad \alpha<\text{Im}p<\beta$$

The inversion formula is

$$f(t)=\frac{1}{2\pi}\int_{i\gamma-\infty}^{i\gamma+\infty} e^{ipt}\, \bar{f}(p)dp \qquad \alpha<\gamma<\beta$$

Taking the F.T. of both sides of (*i*), we have

$$(p^2+a^2)\bar{G}(p, \xi)=e^{ip\xi}$$

i.e.,
$$\bar{G}(p, \xi)=\frac{e^{ip\xi}}{p^2+a^2}$$

Hence
$$G(x, \xi)=\frac{1}{2\pi}\int_{-\infty}^{\infty} \frac{e^{ip(\xi-x)}}{p^2+a^2}\, dp \qquad ...(iii)$$

The contour of the inversion integral (*iii*) has to conform to the BCs given. Such a contour in the present case is a large upper semicircle of radius R bounded by the x-axis when ultimately $R\to\infty$. By the standard procedure

$$G(x, \xi)=\frac{1}{2a}\, e^{-a|x-\xi|} \qquad a>0 \qquad ...(iv)$$

If only $G\to 0$ as $x\to-\infty$ is the boundary condition, then the suitable inversion integral is given by

$$G(x, \xi)=\frac{1}{2\pi}\int_{i\gamma-\infty}^{i\gamma+\infty} \frac{e^{ip(\xi-x)}}{p^2+a^2} \qquad \gamma>a$$

$$=\begin{cases} 0 & x<\xi \\ \dfrac{1}{a}\sinh\,[a(x-\xi)] & x>\xi \end{cases}$$

Such a Green's function which is unbounded at ∞ is, however, of little significance.

Example 20. Find out a fundamental solution of the two dimensional potential equation, using the Fourier transform.

Solution. The fundamental solution $e(x, y\,;\, \xi, \eta)$ is given by

$$\frac{\partial^2 e}{dx^2}+\frac{\partial^2 e}{\partial y^2}=\delta(x-\xi)\,\delta(y-\eta) \qquad ...(i)$$

Taking the Fourier transform of both sides of (*i*) with respect to x, we get

$$\frac{d^2}{dy^2}\bar{e}(p, y ; \xi, \eta)-p^2\bar{e}(p, y ; \xi, \eta) = e^{ip\xi}\,\delta(y-\eta)$$

or

$$\left(\frac{d^2}{dy^2}-p^2\right)\bar{e}=e^{ip\xi}\,\delta(y-\eta) \qquad \ldots(ii)$$

The part of the fundamental solution which corresponds to the complementary function of the ordinary differential equation (*ii*) satisfies $\nabla^2 u=0$ and hence is a harmonic function. We do not include this in the fundamental solution. The desired function will be obtained as the particular integral of (*ii*).

Again solving (*ii*) by the use of Fourier transform with respect to y, as in the previous example, we shall get

$$\bar{e}(p, y ; \xi, \eta)=\begin{cases}\dfrac{-e^{ip\xi}}{2p}\,e^{-p\,|\,y-\eta\,|} & \text{Rep.} >0\\[2ex] \dfrac{e^{ip\xi}}{2p}\,e^{p\,|\,y-\eta\,|} & \text{Rep.} <0\end{cases} \qquad \ldots(iii)$$

Taking the inverse Fourier transform of $\bar{e}$ we have

$$e(x, y ; \xi, \eta)=\frac{1}{2\pi}\int_{-\infty}^{\infty}\bar{e}\,(p, y ; \xi, \eta)\,e^{-ipx}\,dp \qquad \ldots(iv)$$

Owing to (*iii*), the integral in (*iv*) has to be split in two parts, one for Rep. <0 and the other for Rep. >0. Since $p=0$ is a singularity of $\bar{e}$, the contour should be such that it begins at $-\infty$, avoids $p=0$ by cutting the Imp axis in it where $t\neq 0$ and ends at $+\infty$. We can thus write (*iv*) as

$$e(x, y ; \xi, \eta)=\frac{1}{4\pi}\int_{-\infty}^{it}\frac{e^{-ip(x-\xi)+p\,|\,y-\eta\,|}}{p}\,dp - \frac{1}{4\pi}\int_{it}^{\infty}\frac{e^{-ip(x-\xi)-p\,|\,y-\eta\,|}}{p}\,dp \qquad \ldots(v)$$

Now the exponential integral is defined by

$$E_i(z)=\int_{z}^{\infty}\frac{e^{-p}}{p}\,dp \qquad \ldots(vi)$$

By an appropriate change of variable, the two integrals in (*v*) can be related to (*vi*) and we shall have

$$e(x, y ; \xi, \eta)=-\frac{1}{4\pi}\,[E_i\{t(x-\xi)+it\,|\,y-\eta\,|\} + E_i\{t(x-\xi)-it\,|\,y-\eta\,|\}] \qquad \ldots(vii)$$

Again, $E_i(z)$ has the asymptotic representation for large z given by

$$E_i(z) = -\log z - \gamma - \sum_{n=1}^{\infty} \frac{(-1)^n z^n}{n \cdot \lfloor n} \qquad \ldots(viii)$$

where γ is the Euler's constant ($\gamma = 0{\cdot}577215\ldots$) and the series is an analytic function of z.

If we put $Z = X + iY = t(x-\xi) + it \mid y-\eta \mid$, then (*vii*) can be written as

$$e(x, y\,;\,\xi, \eta) = -\frac{1}{4\pi}\,[E_i(Z) + E_i(\overline{Z})]$$

$$= -\frac{1}{2\pi}\,\text{Re}\,E_i(Z)$$

Now $\log Z = \log \mid Z \mid + i \tan^{-1} \dfrac{Y}{X}$

Hence using (*viii*), we have

$$e(x, y\,;\,\xi, \eta) = \frac{1}{2\pi} \log \{(x-\xi)^2 + (y-\eta)^2\}^{1/2}$$

$$+ \text{harmonic terms}$$

Retaining the non-harmonic part, we ultimately we have

$$e(x, y\,;\,\xi, \eta) = \frac{1}{2\pi} \log \mid r - r' \mid \qquad \ldots(ix)$$

where $\quad r = (x, y),\ r' = (\xi, \eta)$

Remark 6. The term fundamental (or elementary) solution is due to the property that solution of problems with boundaries are obtained on its basis. Another significance is that if

$$u(x, y) = \int_T e(x, y, \xi, \eta)\, f(\xi, \eta)\, d\xi dy$$

then $u(x, y)$ satisfies the Poisson's equation

$$\frac{\partial^2 u}{dx^2} + \frac{\partial u^2}{\partial y^2} = f(x, y) \qquad \text{(Prove)}$$

Example 21. Find the Green's function determined by the equations,

$$\nabla^2 G = \delta(x-\xi)\,\delta(y-\eta), \qquad T: \begin{array}{c} -\infty < x < \infty \\ 0 \leqslant y \leqslant b \end{array}$$
$$G = \text{on } \partial T$$

Solution. The problem has been solved earlier by the method of images. Here we use the Fourier transform.

Taking F.T in x as in the previous example, we have

$$L\overline{G}=\left(\frac{\partial^2}{\partial y^2}-p^2\right)\overline{G}(p, y ; \xi, \eta)=e^{i\xi p}\delta(y-\eta) \quad ...(i)$$

(*i*) is in fact an ordinary differential equation in y, $0\leqslant y\leqslant b$ such that

$$\bar{G}(p, 0 ; \xi, \eta)=0 \text{ and } G(p, b ; \xi, \eta)=0.$$

With these conditions $\overline{G}(p, y ; \xi, \eta)$ will be a Green's function and can be found as described in chapter 9. In fact let

$$\overline{G_1}(p, y ; \xi, \eta)=\begin{cases} A \sinh py+B \cosh py & y\leqslant\eta \\ C \sinh py+D \cosh py & y\geqslant\eta \end{cases}$$

be the Green's function of $L\overline{G}=0$.

(*a*) The boundary conditions give $B=0$,

$$C \sinh pb+D \cosh pb=0.$$

(*b*) By the continuity at $y=\eta$

$$A \sinh p\eta=C \sinh p\eta+D \cosh p\eta$$

The jump of $\dfrac{\partial\bar{G}}{\partial y}$ at $y=\eta$ gives

$$A \cosh p\eta-C \cosh p\eta-D \sinh p\eta=-\frac{1}{p}$$

Solving for A, C, D we have

$$A=\frac{\sinh p(\eta-b)}{p \sinh pb}, \qquad C=\frac{\sinh p\eta \cosh pb}{p \sinh pb}$$

$$D=-\frac{\sinh p\eta}{p}.$$

Thus $\overline{G}$ for (*i*) will be given by

$$\bar{G}(p, y ; \xi, \eta)=\begin{cases} e^{ip\xi}\dfrac{\sinh py \sinh p(\eta-b)}{p \sinh pb} & y\leqslant\eta \\ e^{ip\xi}\dfrac{\sinh p\eta \sinh p(y-b)}{p \sinh pb} & y\geqslant\eta \end{cases} \quad ...(ii)$$

The inverse F.T. of (*ii*) will give the required Green's function G. Formally

$$G(x, y ; \xi, \eta)=\frac{1}{2\pi}\int_{-\infty}^{\infty} e^{-ip(x-\xi)}\,\overline{G}(p, y ; \xi, \eta)dp \quad ...(iii)$$

Example 22. Solve the wave equation

$$L(u)=\frac{c^2\partial^2 u}{\partial x^2}-\frac{\partial^2 u}{\partial t^2}=h(x, t) \quad ...(i)$$

in the space-time domain $-\infty < x < \infty$, $0 \leqslant t \leqslant T$ subject to the initial conditions

$$u(x, 0) = f(x), \quad u_t(x, 0) = g(x). \qquad \text{...(ii)}$$

Solution. We solve the problem by using the Green's function of the corresponding homogeneous problem viz.

$$\text{PDE : } L(u) = \frac{c^2 \partial^2 u}{\partial x^2} - \frac{\partial^2 u}{\partial t^2} = 0 \qquad \text{...(iii)}$$

$$\text{ICs : } u(x, 0) = 0, \quad u_t(x, 0) = 0 \qquad \text{...(iv)}$$

In this case the operator L is self-adjoint and the Green's function $G(x, t ; \xi, \tau)$ is given by

$$\text{DE : } LG = \delta(x-\xi), \ \delta(t-\tau) \qquad \text{(v)}$$

$$\text{DCs : } G(x, T ; \xi, \tau) = 0 = G_t(x, T ; \xi, \tau) \qquad \text{(vi)}$$

Though not specified clearly, we would assume that the temperature vanishes at $x = \pm\infty$. Hence we also assume that $G = G_x = 0$ as $x \to \pm\infty$. This makes the Fourier transform a right choice to obtain the solution.

Taking the F.T. of both sides of (*v*) with respect to x, we have

$$c^2 p^2 \bar{G}(p, t ; \xi, \tau) + \frac{\partial^2 \bar{G}(p, t ; \xi, \tau)}{\partial t^2} = -e^{ip\xi} \, \delta(t-\tau) \qquad \text{...(vii)}$$

Transforms of condition (*vi*) are

$$\bar{G}(p, T ; \xi, \tau) = 0, \quad \frac{\partial \bar{G}}{\partial t}(p, T ; \xi, \tau) = 0. \qquad \text{...(viii)}$$

Applying the method of variation of parameters the solution of the system (*vii*), (*viii*) is given by

$$\bar{G}(p, t ; \xi, \tau) = \frac{e^{ip\xi}}{cp} \int_t^T \sin cp(t-u) \, \delta(u-\tau) du$$

$$= \begin{cases} \dfrac{e^{ip\xi}}{cp} \sin cp(t-\tau) & t < \tau \\ 0 & t > \tau \end{cases}$$

Inverse F.T. gives

$$G(x, t ; \xi, \tau) = \begin{cases} -\tfrac{1}{2}c & |x-\xi| < c(t-\tau) \\ -\tfrac{1}{4}c & |x-\xi| = c(t-\tau) \\ 0 & \text{otherwise} \end{cases}$$

Now $vL(u) - uL(v)$

$$= c^2 \frac{\partial}{\partial x}\left(v \frac{\partial u}{\partial x} - u \frac{\partial v}{\partial x} \right) - \frac{\partial}{\partial t}\left(v \frac{\partial u}{\partial t} - u \frac{\partial v}{\partial t} \right)$$

Hence integrating over the given domain

$$\int_0^{t_0}\int_{-\infty}^{\infty} [v\mathrm{L}(u)-u\mathrm{L}(v)]\,dx\,dt$$

$$=c^2\int_0^{t_0}\left(v\frac{\partial u}{\partial x}-u\frac{\partial v}{\partial x}\right)dt-\int_{-\infty}^{\infty}\left(v\frac{\partial u}{\partial t}-u\frac{\partial v}{\partial t}\right)dx$$

Put $v=G(x, t\,;\,\xi, \tau)$ to get the solution.

Exercise 32

1. "The product of two harmonic functions is harmonic."
Give a proof if the statement is true.
Give a counter-example if the statement is false.

2. Obtain the solution $u(x, y)$ of the Laplace's equation in the half-plane $y\geqslant 0$ with

$$u(x, 0)=f(x), \quad \lim_{x^2+y^2\to\infty} u(x, y)=0$$

by use of the Fourier transform as follows :

$$\bar{u}(p, y)=\int_{-\infty}^{\infty} e^{ipx}\, u(x, y)dx$$

$$\nabla^2 u=0 \Rightarrow \frac{\partial^2 \bar{u}}{\partial y^2}(p, y)-p^2\bar{u}(p, y)=0$$

$$\Rightarrow \bar{u}=A(p)\,e^{-py}+B(p)\,e^{py}$$

Since $\bar{u}(p, o)=\bar{f}(p)$ and u is bounded for $y\to\infty$

$$u(p, y)=f(p)\,.\,e^{-|p|y}$$

Now $\text{F.T.}\left\{\frac{y}{\pi(x^2+y^2)}\right\}=e^{-|p|y}$

By convolution theorem

$$u(x, y)=\frac{1}{\pi}\int_{-\infty}^{\infty}\frac{yf(\xi)\,d\xi}{(x-\xi)^2+y^2}.$$

3. Find the Green's function and hence solve the BVP

$$\nabla^2 u=0,\quad 0\leqslant x<\infty,\quad 0\leqslant y<\infty,$$
$$u(0, y)=g(y)\quad 0\leqslant y<\infty$$
$$u_y(x, 0)=f(x)\quad 0\leqslant x<\infty.$$

4. Explain the paradoxes :

(*a*) The Dirichlet problem has a unique solution. Yet $u=1$ and $u=\frac{1}{r}$ are both solutions of $\nabla^2 u=0$ in the unit ball and both take boundary values 1 on the boundary.

(*b*) The maximum of a non-constant harmonic function occurs on the boundary. $-\log(x^2+y^2)$ is harmonic on the unit circle. Yet it vanishes on the boundary and is positive inside.

5. Show that the transformation $z=w^{1/2}$ maps the half-plane Re $w \geqslant 0$ on the segment $|\arg z| \leqslant \frac{\pi}{4}$. Hence or otherwise show that the Green's function for Laplace equation, when the boundary of the region is $x \pm y=0$, $x>0$ and the region is $|y|<|x|$ is

$$G(x, y\,;\,\xi\,\eta)=\frac{1}{4}\log\left[\frac{(x^2-y^2-\xi^2+\eta^2)^2+4(xy-\xi\eta)^2}{(x^2-y^2+\xi^2-\eta^2)^2+4(xy-\zeta\eta)^2}\right]$$

6. Show that the transformation $z=\sinh^{-1} w$ maps the half-plane Re $w \geqslant 0$ onto the semi-inflnite strip $x \geqslant 0$, $-\frac{\pi}{2} \leqslant y \leqslant \frac{\pi}{2}$. Hence or otherwise show that the two-dimensional Green's function of the Laplace's equation in this region is

$$\begin{aligned} G(x, y\,;\,\xi, \eta)=\tfrac{1}{2}\log\{&(\sinh x \cos y-\sinh \xi \cos \eta)^2\\ &+(\cosh x \sin y-\cosh \xi \sin \eta)^2\}\\ -\tfrac{1}{2}\log\{&(\sinh x \cos y+\sinh \xi \cos \eta)^2\\ &+(\cosh x \sin y-\cosh \xi \sin \eta)^2\} \end{aligned}$$

7. Show that the three-dimensional Green's function of the Helmoltz equation

$$\nabla^2 u-k^2 u=0$$

in the region $z \geqslant 0$ is given by

$$G(x, y, z\,;\,\xi, \eta, \zeta)=\frac{e^{-k}}{r}-\frac{e^{-kr_1}}{r_1}$$

where

$$r^2=(x-\xi)^2+(y-\eta)^2+(z-\zeta)^2$$
$$z_1{}^2=(x-\xi)^2+(y-\eta)^2+(z+\zeta)^2$$

8. Show that the harmonic Green's function for the half-space $z \geqslant 0$ is

$$G(x, y, z\,;\,\xi, \eta, \zeta)=\frac{1}{4\pi r}-\frac{1}{4\pi r_1}$$

where r_1, r_1 are as in ex. 7.

9. Show that the harmonic Green's and Neumann's functions for the half-plane $y \geqslant 0$ are respectively

$$G(x, y\,;\,\xi, \eta)=-\frac{1}{2\pi}\log\frac{r}{r'}$$

$$N(x, y\,;\,\xi, \eta)=-\frac{1}{2\pi}\log rr'$$

where

$$r^2=(x-\xi)^2+(y-\eta)^2$$
$$r'^2=(x-\xi)^2+(y+\eta)^1$$

§9·17. The One-Dimensional Wave Equation

The equation is given by

$$\frac{\partial^2 u}{\partial x^2}-\frac{1}{c^2}\frac{\partial^2 u}{\partial t^2}=0 \qquad \ldots(i)$$

where c is a constant. In physical context it represents the velocity of wave propagation. Such an equation occurs for example, in the vibration of a string fixed at both ends. The equation can be solved by the method of separation of variables which in turn transforms itself into an eigenvalue problem (see ex. 3 in Exercise 9). In this section we are concerned with d' Alembert's solution. This can be done by the procedure explained in section 1·3 of chapter 1.

Factorising the operator we can write (1) as

$$\left(D_x-\frac{1}{c}D_t\right)\left(D_x+\frac{1}{c}D_t\right)u=0 \qquad \ldots(ii)$$

where $\quad D_x=\dfrac{\partial}{\partial x},\quad D_t=\dfrac{\partial}{\partial t}$

(*ii*) can be interpreted as the product of two directional derivatives. The first is

$$v=\left(D_x+\frac{1}{c}D_t\right)u$$

in the direction $\left(1, \dfrac{1}{c}\right)$ of the xt-plane and the second is

$$\left(D_x-\frac{1}{c}D_t\right)v$$

in the direction $\left(1, \dfrac{-1}{c}\right)$. These are called characteristic directions and suggest new coordinate directions. We put $x+ct=\xi$, $x-ct=\eta$ and call ξ, η characteristic variables.

In the new variables the equation (*i*) reduces to

$$\frac{\partial^2 u}{\partial\xi\partial\eta}=0$$

which leads to the general solution of (*i*) viz,

$$u(x, t)=\phi(x+ct)+\psi(x-ct) \qquad \ldots(iii)$$

where ϕ and ψ are arbitrary functions. For an infinite string the functions ϕ, ψ can be determined if the following initial conditions are given :

$$u(x, 0)=f(x),$$
$$u_t(x, 0)=g(x)$$

The ICs give

$$\phi(x)+\psi(x)=f(x) \qquad \ldots(iv)$$
$$c\varphi'(x)-c\psi'(x)=g(x) \qquad \ldots(v)$$

i.e., the solution u at (x_0, t_0) is the average of values at A and B plus the integral of $g/2c$ between A, B. Thus the value of u at P depends only on AB. AB is called the *domain of dependence* of P.

Now let a segment AB be given on the initial line $t=0$. Through each of A and B two characteristics will pass. One pair of these characteristics cuts at a point P and the other pair diverges. The region of the xt-plane between these diverging characteristics is called the domain or *range of influence of* AB (see Fig. 9·7) and the triangle ABP formed by the converging charactertics is called the *domain of determinancy* of AB or the *domain of propogation* of the wave $u(x, t)$ (see Fig. 9·8).

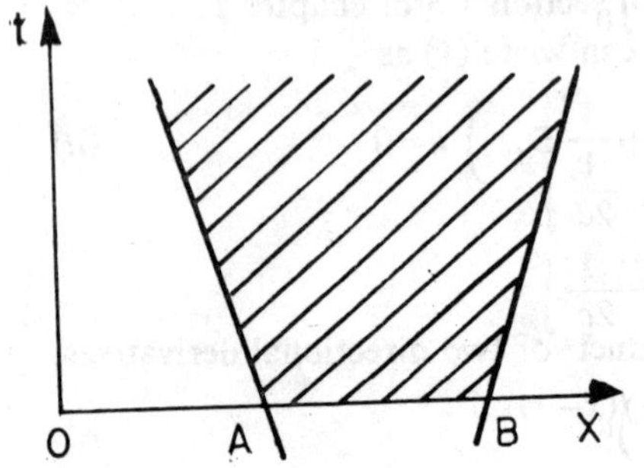

Fig. 9·7.

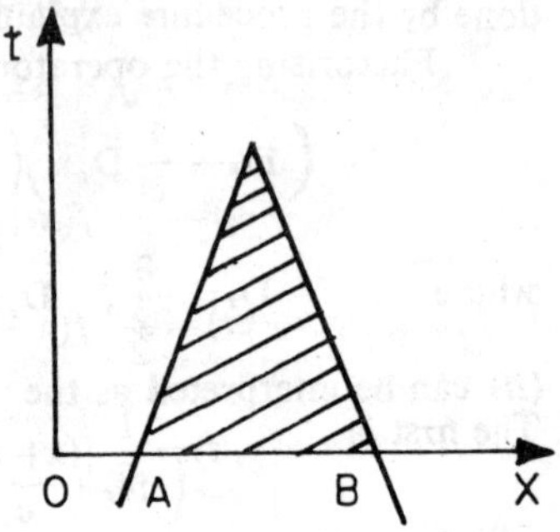

Fig. 9·8.

The above discussion has centred around infinite interval I$(-\infty<x<\infty)$ and no boundary conditious were involved. If I is a semi-infinite interval say $0\leqslant x<\infty$, then a single boundary condition is prescribed at $x=0$ in the form of either $u(0, t)$ or $\dfrac{\partial u(0, t)}{\partial x}$ given. If however, I is a finite closed interval, then two boundary conditions of the type mentioned above, one at each end are prescribed. In what follows we illustrate the use of the Laplace transform in t in solving the wave equation.

Use of Laplace Transform

(A) Infinite Range. No boundary conditions are given. We are to solve the IVP.

$$\text{PDE}: \frac{\partial^2 u}{\partial x^2}-\frac{1}{c^2}\frac{\partial^2 u}{\partial t^2}=0\ ;\ -\infty<x<\infty,\ 0\leqslant t<\infty \qquad \ldots(i)$$

$$\text{ICs}: \quad u(x, 0)=f(x),\ u_t(x, 0)=g(x) \qquad \ldots(ii)$$

$$\text{Let L.T. } \{u(x, t)\}=\text{U}(x, p)=\int_0^\infty e^{-pt}\, u(x, t)\, dt$$

Using the formula

$$\text{L. T.}\ \left\{\frac{\partial^2 u(x, t)}{\partial t^2}\right\}=p^2\, \text{U}(x, p)-pu(x, 0)-\frac{\partial u\,(x, 0)}{\partial t}$$

Solving (*iv*) and (*v*) for ϕ' and ψ'

$$\phi'(x)=\frac{1}{2}f'(x)+\frac{1}{2c}g(x)$$

$$\psi'(x)=\frac{1}{2}f'(x)-\frac{1}{2c}g(x)$$

Integrating we get

$$\phi(x)=\frac{1}{2}f(x)+\frac{1}{2c}\int_0^x g(z)dz+A$$

$$\psi(x)=\frac{1}{2}f(x)-\frac{1}{2c}\int_0^x g(z)dz+B$$

Since $\phi+\psi=f$, $A+B=0$.
Hence

$$\phi(x+ct)=\frac{1}{2}f(x+ct)+\frac{1}{2c}\int_0^{x+ct} g(z)dz+A$$

$$\psi(x-ct)=\frac{1}{2}f(x-ct)-\frac{1}{2c}\int_0^{x-ct} g(z)dz-A$$

Finally $u(x,t)=\frac{1}{2}[(x+ct)+f(x-ct)]$

$$+\frac{1}{2c}\int_{x-ct}^{x+ct} g(z)\,dz \qquad \ldots(vi)$$

(*vi*) is commonly known as *d'Alembert's solution* to the wave equation.

Interpretation. On the basis of the characteristic lines $x-ct=$ constant and $x+ct=$constant the d'Alembert's solution has an interesting geometrical intepretation. Choose a point $P(x_0, t_0)$ in the *xt*-plane and draw the two characteristics $x\pm ct=x_0\pm ct_0$ rhrough P to cut the initial line in A, B (see Fig. 9·6).

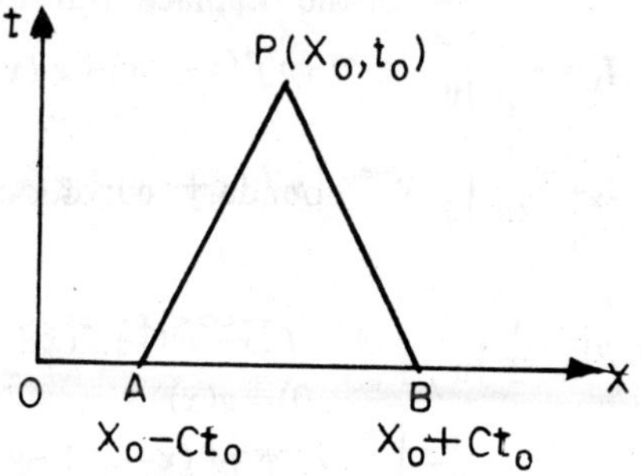

Fig. 9·6.

At P, we have

$$u(x_0,t_0)=\frac{f(x_0+ct_0)+f(x_0-ct_0)}{2}+\frac{1}{2c}\int_{x_0-ct_0}^{x_0+ct_0} g(z)\,dz$$

we have by taking L.T. of (*i*) with (*ii*)

$$\frac{\partial^2 U(x,p)}{\partial x^2}-\frac{1}{c^2}\left[p^2 U(x,p)-pu(x,0)-\frac{\partial u(x,0)}{\partial t}\right]=0$$

i.e.,

$$\frac{\partial^2 U(x,p)}{\partial x^2}-\frac{p^2}{c^2}U(x,y)=-\frac{1}{c^2}\left\{pf(x)+g(x)\right\} \quad (iii)$$

Equation (*iii*) can be solved by using the Green's function of of the homogeneous equation with the condition that $(U(\pm\infty, p)$ is finite. The required Green's function is

$$G(x,y)=\frac{-c}{2p}\; e^{-\frac{p}{c}|x-y|}$$

Using theorem 1 of chapter 8, the solution of (*iii*) can be expressed as

$$U(x,p)=\frac{1}{2cp}\int_{-\infty}^{\infty} e^{-\frac{p}{c}|x-y|}\{pf(y)+g(y)\}\,dy \quad \ldots(iv)$$

(*iv*) can be expressed as

$$U(x,p)=\frac{1}{2cp}\int_{-\infty}^{x} e^{-\frac{p}{c}(x-y)}\{pf(y)+g(y)\}\,dy$$

$$+\frac{1}{2cp}\int_{x}^{\infty} e^{-\frac{p}{c}(y-x)}\{p(y)+g(y)\}\,dy$$

$$=I_1+I_2 \text{ (say).}$$

If we put $\frac{x-y}{c}=w$ in the integral I_1 and $\frac{y-x}{c}=w$ in I_2, then we find that

$$I_1=\frac{1}{2p}\int_0^{\infty} e^{-pw}\{pf(x-cw)+g(x-cw)\}\,dw$$

$$I_2=\frac{1}{2p}\int_0^{\infty} e^{-pw}\{pf(x+cw)+g(x+cw)\}\,dw$$

Hence

$$U(x,p)=\frac{1}{2}\int_0^{\infty} e^{-pw}\{f(x-cw)+f(x+cw)\}\,dw$$

$$+\frac{1}{2p}\int_0^{\infty} e^{-pw}\{g(x-cw)+g(x+cw)\}\,dw \quad \ldots(v)$$

Taking the inverse L.T. of both sides of (*v*), we have

$$U(x,t)=\frac{1}{2}\left\{f(x-ct)+f(x+ct)\right\}$$

$$+\frac{1}{2}\int_0^t \{g(x-cw)+g(x+cw)\}\, dw$$

$$=\frac{1}{2}\{f(x-ct)+f(x+ct)\}+\frac{1}{2c}\int_{x-ct}^{x+ct} g(y)\, dy \quad \ldots(vi)$$

which we found earlier by separating variables.

Problem 1. Solve the IVP if $g=0$ and $f=\begin{cases} K & |x|<a \\ 0 & |x|>0 \end{cases}$

Problem 2. By using the method of variation of purameters or othewise, prove that the general solution of the equation

$$z''(x)-w^2 z(x)=f(x), \qquad f(x)\in C(I)$$

is given by

$$z(x)=a\, e^{wx}+b\, e^{-wx}-z_0(x)$$

where a, b are arbitrary constants,

$$z_0(x)=-\frac{1}{2w}\left[\int_\alpha^x e^{(y-x)w} f(y)\, dy+\int_x^\beta e^{(x-y)w} f(y)\, dy\right]$$

is a particular integral and α, β are any two given points of I. Use this result in problem A.

(B) Semi-infinite Range. One boundary condition is prescribed. We are to solve the problems,

$$\text{PDE}: \quad \frac{\partial^2 u}{\partial x^2}-\frac{1}{c^2}\frac{\partial^2 u}{\partial t^2}=0\ ;\ 0\leqslant x<\infty,\ 0\leqslant t<\infty \quad \ldots(i)$$

$$\text{ICs}: \quad u(x, 0)=f(x),\ u_t(x, 0)=g(x) \quad \ldots(ii)$$

$$\text{BC}: \quad u(0, t)=h(t) \quad \ldots(iii)$$

For compatibility of the ICs and the BC we assume $f(0)=h(0)$ and $h'(0)=g(0)$.

Proceeding as in case (A), the L.T. of (i) with conditions (ii) is

$$\frac{\partial^2 U(x, p)}{\partial x^2}-\frac{p^2 U(x, p)}{c^2}=-\frac{1}{c^2}\{p\, f(x)+g(x)\} \quad \ldots(iv)$$

If we choose $\alpha=0$, $\beta=\infty$ in the problem cited above, the general solution of (iv) is given by

$$U(x, p)=b\, e^{-\frac{p}{c}x}+\frac{1}{2cp}\left[\int_0^x e^{-\frac{p}{c}(x-y)}\{p\,(f(y)+g(y)\}\, dy\right.$$

$$\left.+\int_x^\infty e^{-\frac{p}{c}(y-x)}\{p\, f(y)+g(y)\}\, dy\right] \ldots(v)$$

In view of the finiteness of U(x, p) at $x=\infty$, the constant $a=0$. For b, using condition (*iii*)

$$\text{L.T. } \{h(t)\}=\bar{h}(p)=b+\frac{1}{2cp}\int_0^\infty e^{-\frac{p}{c}y}\{p\,f(y)+g(y)\}\,dy$$

Hence

$$U(x, p)=\bar{h}(p)\,.\,e^{-\frac{p}{c}x}-\frac{e^{-\frac{p}{c}x}}{2cp}\int_0^\infty e^{-\frac{p}{c}y}\{p\,f(y)+g(y)\}\,dy$$

$$+\frac{1}{2cp}\int_0^x e^{-\frac{p}{c}(x-y)}\{p\,f(y)+g(y)\}\,dy$$

$$+\frac{1}{2cp}\int_x^\infty e^{-\frac{p}{c}(y-x)}\{p\,f(y)+g(y)\}\,dy \quad ...(vi)$$

Changing the variable of integeration in (*vi*), we have

$$U(x, p)=\bar{h}(p)\,.\,e^{-\frac{p}{c}x}-\frac{e^{-\frac{p}{c}x}}{2p}\int_0^\infty e^{-py}\{p\,f(cy)+g(cy)\}\,dy$$

$$+\frac{1}{2p}\int_0^{x/c} e^{-py}\{p\,f(x-cy)+g(x-cy)\}\,dy$$

$$+\frac{1}{2p}\int_0^\infty e^{-py}\{p\,f(x+cy)+g(x+cy)\}\,dy$$

$$\text{Now } \bar{h}(p)\,e^{-\frac{p}{c}x}-\frac{e^{-\frac{p}{c}x}}{2}\int_0^\infty e^{-py}f(cy)\,dy-\frac{1}{2p}\,e^{-\frac{p}{c}x}\int_0^\infty e^{-py}g(cy)\,dy$$

$$=\text{L.T.}\left[\left\{h(t)-\frac{1}{2}f(ct)-\frac{1}{2}\int_0^t g(cy)\,dy\right\}H\left(t-\frac{x}{c}\right)\right]$$

where H (x) is the unitary function

$$H(x)=\begin{cases}1 & x>0\\ 0 & x<0\end{cases}$$

$$\frac{1}{2p}\int_0^{x/c} e^{-py}\{p\,f(x-cy)+g(x-cy)\}\,dy$$

$$=\frac{1}{2p}\int_0^\infty e^{-py}\{p\,f(x-cy)+g(x-cy)\}\,H\left(\frac{x}{c}-y\right)dy$$

$$=\frac{1}{2}\int_0^\infty e^{-py} f(x-cy)\, \mathrm{H}\left(\frac{x}{c}-y\right) dy+\frac{1}{2p}$$

$$\int_0^\infty e^{-py}\, g(x-cy)\, \mathrm{H}\left(\frac{x}{c}-y\right) dy$$

$$=\text{L.T.}\left\{\frac{1}{2} f(x-ct)\,.\,\mathrm{H}\left(\frac{x}{c}-t\right)\right.$$

$$\left.+\frac{1}{2}\int_0^t g(x-cy)\ \mathrm{H}\left(\frac{x}{c}-y\right) dy\right.$$

$$\frac{1}{2p}\int_0^\infty e^{-py}\,\{p\,f(x+cy)+g(x+cy)\}\,dy$$

$$=\text{L.T.}\,\frac{1}{2}\left\{f(x+ct)+\int_0^t g(x+cy)\,dy\right\}$$

Hence taking inverse L.T. we have

$$u(x,t)=\left[\ h(t)-\frac{1}{2}\ f(ct)-\frac{1}{2}\int_0^t g(cy)\,dy\ \right]\mathrm{H}\left(t-\frac{x}{c}\right)$$

$$+\frac{1}{2} f(x-ct)\,\mathrm{H}\left(\frac{x}{c}-t\right)+\frac{1}{2}\int_0^t g(x-ct)\ \mathrm{H}\left(\frac{x}{c}-y\right) dy$$

$$+\frac{1}{2}(x-ct)+\frac{1}{2}\int_0^t g(x+cy)\,dy,\ x\geqslant 0.$$

One special case of zero initial values

$f=g=0$ is worth mentioning :

$$u(x,t)=h(t)\,\mathrm{H}\left(t-\frac{x}{c}\right)$$

$$=h\left(t-\frac{x}{c}\right)\mathrm{H}\left(t-\frac{x}{c}\right),\ x\geqslant 0.$$

Problem 1. Try if you succeed in solv'ng this problem with zero ICs by the method of separation of variables.

Problem 2. The semi-infinite range is part of the infinite range the solution of which did not require a BC of the type (*iii*). Then why do we still require the BC in the semi-infinite case ?

(Hint : One of the two given functions ϕ, ψ had argument $x-ct$ which ranges from $-\infty$ to ∞ for $0\leqslant x<\infty$, $0\leqslant t<\infty$ and the values of that function for $x<0$ will remain undetermined if a BC is not prescribed.

Problem 3. Solve the problem of semi-infinite range without the use of L·T.

Problem 4. Solve the problem of infinite range by using the Fourier transform with respect to x.

Remark 7. The case of finite range $a \leqslant x \leqslant b$ is complicated and is not discussed here.

Remark 8. Slightly more general equations than the wave equation do occur in practice. They are :

(*a*) **The Telegraphist's Equation**

$$\frac{\partial^2 u}{\partial x^2} - \frac{1}{c^2}\frac{\partial^2 u}{\partial t^2} - 2a\frac{\partial u}{\partial t} - bu = 0 \qquad \ldots(i)$$

The term $2a \dfrac{\partial u}{\partial t}$ is a dissipative term and can be removed by a change of dependent variable $v(x, t) = e^{at}\, u(x, t)$. The resulting equation is

$$\frac{\partial^2 v}{\partial x^2} - \frac{1}{c^2}\frac{\partial^2 v}{\partial t^2} - (b - a^2)\, v = 0.$$

The term $2\,a \dfrac{\partial u}{\partial t}$ corresponds to a damping force and tne term bu produces dispersion in waves.

(*b*) **Spherical and Plane Waves.** The equation of *Spherically Symmetric Waves* in n dimenions is

$$\frac{\partial^2 u}{\partial r^2} - \frac{1}{c^2}\frac{\partial^2 u}{\partial t^2} + \frac{n-1}{r}\frac{\partial u}{\partial r} = 0 \qquad \ldots(ii)$$

where r is the distance from the centre of the waves. The solutions of (*ii*) are called *spherical waves*.

A plane wave is a solution of the wave equation which depends only on time t and one space coordinate.

§9·18. The One-Dimensional Diffusion Equation

We have seen that an IVP for the potential equation is not well-posed. Hence no ICs were involved in its solution. For the wave equation we always give two ICs. These equations are typical of two classes of second order linear PDEs : the elliptic and the hyperbolic respectively. In this section we study *the heat equation* or *the diffusion equation* which is typical of the third category, the parabolic equations. In one-dimension the equation is

$$\frac{\partial^2 u}{\partial x^2} = \frac{\partial u}{\partial t} \qquad \ldots(i)$$

In higher dimension it is

$$\nabla^2 u = \frac{\partial u}{\partial t} \qquad \ldots(ii)$$

The equation is governed by one BC and one IC. Factorisation of operators in equation (i) can be resorted to for its solution as we did in the case of the wave equation. But the process inyolves interpretation of the fractional derivative $\frac{\partial^{1/2}}{\partial t^{1/2}}$ as given in the previous chapter. We do not adopt this procedure here. Instead we adopt the method of separation of variables or L.T.

(A) Heat Conduction in an Infinite Bar

We consider this case first owing to its simplicity. Another advantage as in the case of the wave equation in an infinite range is the absence of any BC.

Given the *Cauchy-Dirichlet Problem*,

$$\text{PDE : } \frac{\partial^2 u}{\partial x^2}=\frac{\partial u}{\partial t} \qquad \ldots(i)$$

$$-\infty<x<\infty$$
$$0\leqslant t<\infty$$

$$\text{IC : } u(x, 0)=f(x) \qquad -\infty<x<\infty \qquad \ldots(ii)$$

We assume $f(x)\to 0$ and $|u(x, t)|<M$ as $|x|\to\infty$ as unbounded temperatures are unrealistic.

Let $u(x, t)=X(x)\,Y(t)$ be a solution. Then

$$\frac{X''(x)}{X(x)}=\frac{Y'(t)}{Y(t)}=\lambda \quad \text{(say)} \qquad \ldots(iii)$$

For $Y(t)$, we have $Y=ce^{\lambda t}$. If $\lambda>0$, then $Y(t)$ and hence $u(x, t)$ grows exponentially in time. Since we discard this, $\lambda<0$. Put $\lambda=-\mu^2$.

Then $\quad Y=ce^{-\mu^2 t} \qquad \ldots(iv)$

and $\quad X''+\mu^2X=0 \qquad \ldots(v)$

The general solution is

$$X(x)=A\,e^{i\mu x}+B\,e^{-i\mu x} \qquad \ldots(vi)$$

It may be recalled that in the case of the wave equation, the separation procedure entailed the constant, μ to be discrete *i.e.*, integers. Iu the present case μ can take all real values from $-\infty$ to ∞. Consequently it is unnecessary to include both $e^{i\mu x}$ and $e^{-i\mu x}$. We choose $e^{i\mu x}$ and write

$$X=A(\mu)\,e^{i\mu x} \qquad \ldots(vii)$$

where $A(\mu)$ is an arbitrary constant depending on μ. Combining (iv) and (vii) we find that the general solution of (i) is

$$u(x, t)=w(\mu)\,e^{i\mu x-\mu^2 t} \qquad \ldots(viii)$$

Since μ is a continuous variable $-\infty<\mu<\infty$, instead of a discrete sum, we can integrate over μ from $-\infty$ to ∞ and the resulting expression

$$u(x, t)=\int_{-\infty}^{\infty} w(\mu)\, e^{i\mu x-\mu^2 t}\, d\mu \qquad \ldots(ix)$$

is a solution. The given IC determines the arbitrary function $w(\mu)$. Thus

$$f(x)=\int_{-\infty}^{\infty} w(\xi)\,.\, e^{ix}\, d\xi \qquad \ldots(x)$$

$w(\mu)$ is obtained by Fourier inversion formula.

Thus $$w(\mu)=\frac{1}{2\pi}\int_{-\infty}^{\infty} f(\xi)\, e^{-i\mu\xi}\, d\xi \qquad \ldots(xi)$$

From (ix) and (xi), we get

$$u(x, t)=\frac{1}{2\pi}\int_{-\infty}^{\infty}\int_{-\infty}^{\infty} f(x)\, e^{-t\mu^2+i\mu(x-\xi)}\, d\xi\, d\mu$$

The integral in μ is written as follows :

$$I=\int_{-\infty}^{\infty} e^{-t\mu^2+i\mu(x-\xi)}\, d\mu$$

$$=\int_{-\infty}^{\infty}\left[e^{-t\left\{\mu^2-\frac{i\mu(x-\xi)}{t}\right\}}\right] d\mu$$

$$=e^{-\frac{(x-\xi)^2}{4t}}\int_{-\infty}^{\infty}\left[e^{-t\left\{\mu-\frac{i(x-\xi)}{2t}\right\}^2}\right] d\mu$$

If we put $z=\mu-\dfrac{i(x-\xi)}{2i}$ and $\gamma=-\dfrac{x-\xi}{2t}$, then

$$I=e^{-\frac{(x-\xi)^2}{4t}}\int_{i\gamma-\infty}^{i\gamma+\infty} e^{-tz^2}\, dz$$

$$=e^{-\frac{(x-\xi)^2}{4t}}\int_{-\infty}^{\infty} e^{-tz^2}\, dz$$

$$=\sqrt{\frac{\pi}{t}}\; e^{-\frac{(x-\xi)^2}{4t}}$$

Hence finally,

$$u(x, t)=\frac{1}{2\sqrt{\pi t}}\int_{-\infty}^{\infty} f(\xi)\,.\, e^{-\frac{(x-\xi)^2}{4t}}\, d\xi \qquad \ldots(xii)$$

Problem. Using (*xii*) prove that

$$| u(x, t)-f(x) | \to 0 \text{ as } t \to 0$$

Now consider the function

$$G(x, t ; \xi, \tau)=G(x, t-\tau ; \xi, 0)$$

where
$$G(x, t ; \xi, 0)=\frac{1}{2\sqrt{\pi t}} e^{-\frac{(x-\xi)^2}{4t}} \quad \ldots(xiiii)$$

Then
$$f(x)= \lim_{t\to 0} (u, t)$$

$$= \lim_{t\to 0} \int_{-\infty}^{\infty} f(\xi) \,.\, G(x, t ; \xi, 0)\, d\xi$$

$$=\int_{-\infty}^{\infty} f(\xi) \left\{ \lim_{t\to\tau} G(x, t ; \xi, \tau) \right\} d\xi$$

$$=\int_{-\infty}^{\infty} f(\xi)\, \delta(x-\xi)\, d\xi$$

Hence
$$\lim_{t\to\tau} G(x, t\, \xi; \tau)=\delta(x-\xi) \quad \ldots(xiv)$$

Owing to its relation (*xiv*) with the Dirac delta function, the kernel (*xiii*) is called a *fundamental solution* for the heat equation.

Note. (*i*)–(*ii*) is called a *Cauchy-Dirichlet problem* because the condition (*ii*) can be regarded both as an IC (time=0) and a BC (boundary $t=0$ for the plane $t>0$).

(B) Semi-infinite Bar

The Laplace transform is best suited to solve the heat equation in this case. We solve

$$\text{PDE} : \frac{\partial^2 u}{dx^2}-\frac{\partial u}{\partial t}=0 \qquad 0\leqslant x<\infty, 0\leqslant t<\infty$$

$$\text{IC} \;:\; u(x, 0)=0 \qquad 0\leqslant x<\infty$$

$$\text{BCs} :\; u(0, t)=f(t),\; u(\infty, t)<\infty$$

By taking the L.T. of the equation in t

$$\frac{\partial^2 \bar{u}\;(x, p)}{\partial x^2}-p\,\bar{u}\,(x, p)=0$$

Hence $\bar{u}\,(x, p)=A\, e^{x\sqrt{p}}+B\, e^{-x\sqrt{p}}$

where $\bar{u}$ =L.T.u

Since $u(\infty, t)<\infty$, $A=0$.

Also $\bar{u}(0, p)=\bar{f}(p)$. Hence $B=\bar{f}(p)$.

Thus $\bar{u}(x, p)=\bar{f}(p) \cdot e^{-x\sqrt{p}}$

Taking inverse L.T., we get

$$u(x, t)=f(t)O\left\{\frac{x\, e^{-\frac{x^2}{4t}}}{2\sqrt{\pi t^3}}\right\}$$

$$=\frac{x}{2\sqrt{\pi}}\int_0^t f(t-w) \cdot \frac{e^{-\frac{x^2}{4w}}}{\sqrt{w^3}}\, dw$$

If $f=u_0$ (constant)

then $u(x, t)=u_0 \cdot erfc\left(\frac{x}{2\sqrt{t}}\right)$.

Problem. Solve the Cauchy-Dirichlet problem for an infinite bar by using the Fourier transform with respect to the variable x.

(C) Heat Polynomials and Temperature Functions

It is easy to verify that the heat equation

$$\frac{\partial^2 u}{\partial x^2}=\frac{\partial u}{\partial t} \qquad \text{...(i)}$$

has a one-parameter family of exponential solutions

$$u(x, t)=e^{\alpha x+\alpha^2 t} \qquad \text{...(ii)}$$

for arbitrary real or complex α: It is an immediate consequence of (*ii*), that the functions $e^{-\alpha^2 t}\cos \alpha x$. $e^{-\alpha^2 t}\sin \alpha x$, $e^{\alpha^2 t}\cosh \alpha x$ and $e^{\alpha^2 t}\sinh \alpha x$ are also solutions of (*i*).

Definition 3. The *heat polynomial* $v_n(x, t)$ of degree n is defined by

$$e^{xz+zt^2}=\sum_{n=0}^{n} v_n(x, t)\frac{z^n}{n!} \qquad \text{...(iii)}$$

In view of (*ii*), $v_n(x, t)$ are also solutions of (*i*) (Prove !)

By comparing coeffieients of z^n on both sides of (*iii*), we obtain

$$v_n(x, t)=n!\sum_{k=0}^{[n/2]}\frac{t^k}{k!}\frac{x^{n-2k}}{(n-2k)!} \qquad \text{...(iv)}$$

where $\left[\frac{n}{2}\right]$=largest integer$\leqslant \frac{n}{2}$.

Some special cases are :

$$v_0=1,\ v_1=x,\ v_2=x^2+2t$$
$$v_3=x^3+6xt,\ v_4=x^4+12x^2t+12t^2$$
$$v_5=x^5+20x^3t+60xt^5$$

Also $v_n(x,\ 0)=x^n,\ v_{2n}(0,\ t)=\dfrac{(2n\ !)\ t^n}{n\ !},\ v_{2n+1}(0,\ t)=0.$

Definition 4. (The Source Solution)

$$k(x,\ t)=\frac{e^{-\frac{x^2}{4t}}}{2\sqrt{\pi t}} \qquad t>0 \qquad \ldots(v)$$

is called the *source solution* of the heat equation. The reason for the term source solution has its origin to the following mathematical phenomenon.

$$\int_{-\infty}^{\infty} k(x,\ t)\ dx=1$$

$k(x,\ t)$ thus may be described as the temperature of an infinite bar produced by introducing unit quantity of heat into the bar at $x=0$. There is thus a source of strength 1 at $x=0$. $k(x,\ t)$ is the familiar probability distribution.

$$h(x,\ t)=\frac{xk(x,\ t)}{t}=-2k_x(x,\ t) \qquad \ldots(vi)$$

is called the *derived source solution.*

The source solution is also called the fundamental solution.

Definition 5. Let S be a region in the *xt*-plane. A continuously differentiable function $u(x,\ t)$ in S called a *temperature function* if it satisfies the heat equation (*i*).

For example, the source function is a temperature function in all S except at (0, 0).

Definition 6 (Green's Function). Given a rectangular region

$$R=\{x,\ t) : 0\leqslant x\leqslant \pi,\ 0\leqslant t\leqslant t_0\}$$

with boundary ∂R. The *Green's function* G(x, t ; ξ, η) of the heat equation for R is for each fixed (ζ, η) inside R, a function of (x, t) such that

(*a*) G is a temperature function throughout R except at (ξ, η).

(*b*) G=0 for all (x, t) on the base and vertical sides of ∂R (Fig. 9·9).

(*c*) G(x, t ; ξ, η)$-k(x-\xi,\ t-\eta)$ is a temperature function throughout R.

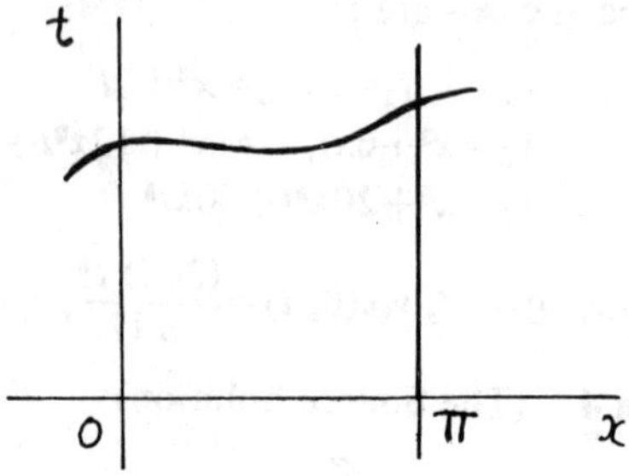

Fig. 9·9.

The Maximum-Minimum Principle

Theorem 8. Let $u(x, t)$ be a temperature function in the open region R : $a<x<b,\ 0<t<t_0$ and be a continuous function in the closed region $a\leqslant x\leqslant b,\ 0\leqslant t\leqslant t_6$ such that

(*a*) $u(a, t)\geqslant 0,\ u(b, t)\geqslant 0,\ 0\leqslant t\leqslant t_0$

(*b*) $u(x, 0)\geqslant 0$ on $a\leqslant x\leqslant b$

Then $u(x, t)\geqslant 0$ on $a\leqslant x\leqslant b,\ 0\leqslant t\leqslant t_0$

Proof. Conditions (*a*), (*b*) imply that $u(x, t)$ attains its minimum on the vertical sides or on the base of R. Also continuity of u in the closed rectangle implies that u satisfies the heat equation on $t=t_0$ also.

For an arbitrary $\in>0$, let

$$v(x, t)=u(x, t)+\in t \qquad ...(i)$$

Hence $v(x, t)$ is continuous in the closed rectangle and must attain its minimum at a point $(x_1, t_1)\in$ R. We show that either $x_1=a$ or b, or else $t_1=0$. On the contrary let $a<x_1<b,\ 0<t\leqslant t_0$. As a condition for a minimum of v, we must have

$$\frac{\partial v}{\partial t}=0 \text{ and } \frac{\partial^2 v}{\partial x^2}\geqslant 0 \text{ at } (x_1, t_1) \qquad ...(ii)$$

But at (x_1, t_1) from (*i*), we have

$$\frac{\partial^2 v}{\partial x^2}=\frac{\partial^2 v}{\partial x^2}-\frac{\partial v}{\partial t}=\frac{\partial^2 u}{\partial x^2}-\frac{\partial u}{\partial t}-\in=-\in<0.$$

This is a contradiction. Hence (x_1, t_1) must satisfy hypothesis (*a*) or (*b*). Consequently, if (x, t) is any point in $a\leqslant x\leqslant b,\ 0\leqslant t\leqslant t_0$, then

$$u(x, t)=v(x, t)-\in t$$

$$\geqslant v(x_1, t_1) - \in t$$
$$= u(x_1, t_1) + \in t_1 - \in t$$
$$\geqslant 0 + \in (t_1 - t)$$

Since $\in$ is arbitrary this implies

$$u(x, t) \geqslant 0$$

Similarly it can be proved that max. u lies on the base and vertical boundaries, Q.E.D.

Corollaries 7. (*i*) If $u_1(x, t)$, $u_2(x, t)$ are temperature functions in R such that $u_1(x, t) \leqslant u_2(x, t)$, (x, t) is a point on the base and vertical boundaries, then $u_1(x, t) \leqslant u_2(x, t)$ for all $(x, t) \in \mathrm{R}$. Since constants are temperature functions, $m \leqslant u(x, t) \leqslant \mathrm{M}$, if (x, t) is a point on the base and vertical boundaries, then $m \leqslant u(x, t) \leqslant \mathrm{M}$ for all

$$(x, t) \in \mathrm{R}$$

(*ii*) The Green's function $\mathrm{G}(x, t\,;\, \xi, \eta)$ is unique. For, if there are two G_1, G_2, then $\mathrm{G}_1 - \mathrm{G}_2$ is a regular temperature function vanishing on the boundary.

Problem 1. Prove that as a function of (ξ, η) $\mathrm{G}(x, \mathbf{t}\,;\, \xi, \eta)$ saitsfies the adjoint heat equation

$$\frac{\partial^2 u}{\partial x^2} + \frac{\partial u}{\partial t} = 0$$

except at (x, t) and that $\mathrm{G}(x, t\,;\, \xi, \eta)$ is not symmetric in (x, t) and (ξ, η).

Problem 2. Let $\mathrm{G}(x, t\,;\, \xi, \eta)$ be the Green's function of the heat equation in the region

$$\mathrm{R} = \{(x, t) : 0 \leqslant x \leqslant \pi,\ 0 \leqslant t \leqslant t_0\}$$

If $u(x, t)$ is a temperature function throughout R and (x, t) is a point in the interior of R, then prove that

$$u(x, t) = \int_{\partial \mathrm{R}} \mathrm{G}(x, t\,;\, \xi, \eta)\, u(\xi, \eta)\, d\xi\, d\eta$$
$$+ \int_{\partial \mathrm{R}} \left[\mathrm{G} \frac{\partial u(\xi, \eta)}{\partial \xi} - \frac{\partial \mathrm{G}}{\partial \xi} u(\xi, \eta) \right] d\eta$$

Problem 3. The $(n-1)$ dimensional heat equation is

$$\sum_{i=1}^{n-1} \frac{\partial^2 u}{\partial x_i^2} - \frac{\partial u}{\partial x_n} = 0$$

Let $E(\bar{x}, \bar{y})=(x_n-y_n)^{\frac{1-n}{2}} \exp.\left[\frac{-1}{4(x_n-y_n)} \sum_{i=1}^{n-1} (x_i-y_i)^2\right]$

where $x_n>y_n$, $\bar{x}=(x_1, x_2, \ldots\ldots, x_n)$.

Prove that $E(\bar{x}, \bar{y})$ is a fundamental solution of the heat equation.

$$\left[\text{Hint}: \quad \frac{\partial^2 E}{\partial x_i^2}=\frac{-1}{2(x_n-y_u)} E(\bar{x}, \bar{y})+\frac{(x_i-y_i)^2}{4(x_a-y_n)} E(\bar{x}, \bar{y})\right.$$

$$(i=1, 2, \ldots\ldots, n-1)$$

$$\left.\frac{\partial E}{\partial x_n}=-\frac{n-1}{2(x_n-y_n)} E(\bar{x}, \bar{y})+\frac{E(\bar{x}, \bar{y})}{4(x_n-y_n)^2} \sum_{i=1}^{n-1} (x_i-y_i)^2\right]$$

10

Linear Integral Equations

§10·1. Introduction

An integral equation is an equation in which the unknown function occurs under the integral sign. What we call equations of mathematical physics, in fact include differential and integral equations. Most of the problems of elasticity for example, are formulated in the form of integral equations. In certain cases a problem in differential equations can be shown to be equivalent to an integral equation. For example the problem

$$y'=y(x),\ y(0)=1,\ x\geqslant 0$$

is equivalent to the integral equation

$$y(x)=1+\int_0^x y(t)dt$$

both having the solution e^x.

It is not uncommon that a problem can be formulated both in terms of a differential equation as well as an integral equation. It is also well known that integral equations are not less difficult to solve than differential equations. Where is the justification then, of preferring an integral equation to its equivalent differential equation ? The reason lies in the fact that integral equations are better suited to approximation processes than differential equations. While differentiation increases error, integration decreases it. Integration is a smoothing operation. Connected with approximation processes are existence and uniqueness problems of differential equations which are best explained when they are recast in integral equations.

A linear integral equation can be approximated by a discrete finite difference representation. This closely links the subject to linear algebra. Such an integral equation involves a linear operator acting on suitable Hilbert spaces of functions. In this respect the subject forms part of functional analysis.

The scope of this chapter is limited to describing the basic methods of solving linear integral equations. On the one hand the subject has deep theoretical ramifications extending to abstract linear operator theory. On the other, the rapid development of computer has stimulated purely numerical techniques. The approach adopted here lies midway between these two extremes. It is our conviction that certain aspects of differential and integral equations are indistinguishable and have full justification for a composite study. The linkage of Sturm-Liouville prolbems, Green's function and Fredholm equations is a case in point.

§10·2. Classification

Fredholm equations of the first and second kinds are respectively

$$\int_a^b K(x, t)\, y(t)\, dt = f(x) \qquad \ldots(1)$$

$$y(x) - \lambda \int_a^b K(x, t)\, y(t)\, dt = f(x) \qquad \ldots(2)$$

The interval $[a, b]$ may be finite or $(-\infty, b)$, (a, ∞) or $(-\infty, \infty)$ where a, b are finite. The parameter λ is a real or complex number and f is a real function.

Volterra equations of the first and second kinds are respectively,

$$\int_a^x K(x, t)\, y(t)\, dt = f(x) \qquad \ldots(3)$$

$$y(x) - \lambda \int_a^x K(x, t)\, y(t)\, dt = f(x) \qquad \ldots(4)$$

If $f(x) = 0$, the corresponding equations are *homogenous*; otherwise they are *non-homogeneous.*

The known bivariate function $K(x, t)$ which is integrable in the domain $a \leqslant x \leqslant b$, $a \leqslant t \leqslant b$ is called the *kernel* of the equation. $f(x)$ is a known continuous function.

If we write either

$$L(y) = \int K(x, t)\, y(t)\, dt$$

or

$$L(y) = y(x) - \lambda \int K(x, t)\, y(t)\, (dt),$$

then L represents a linear operator on the space of Lebesgue integral functions (either L^1 or L^2) to the space of continuous functions. Hence, the integral equations listed above are linear integral equations.

The values of λ $(\neq 0)$ for which the homogeneous Fredholm equation of the second kind

$$y(x) = \lambda \int_a^b K(x, t)\, y(t)\, dt \qquad \ldots(5)$$

has non-trivial solutions (*i.e.* $y(x) \not\equiv 0$) are called *characteristic numbers* (values). The associated solution are called *eigenfunctions.* This nomenclature slightly differs from general linear operator theory where if

$$L(y) = \int_a^b K(x, t)\, y(t)\, dt = \frac{1}{\lambda} y(t)$$

then $\frac{1}{\lambda}$ is called an eigenvalue.

The set of all characteristic numbers of an integral equation with a Kernel $K(x, t)$ is called the *spectrum* of the kernel (or of the integral equation).

A kernel K (x, t) is called Hermitian is $K(x, t)=K(t, x)$ and symmetric if K $(x, t)=K(t, x)$. A real Hermitian Kernal is symmetric. The kernel $K^T(x, t)=K(t, x)$ is called the *transposed* Kernel of $K(x, t)$.

A kernel which is particularly useful in solving Fredhodm equations has the form

$$K(x, t)=\sum_{i=1}^{n} a_i(x)\, b_i(t) \quad \ldots(6)$$

where n is finite and a_r, b_s are linearly independent sets of functions (or else some of the terms may combine and the number of terms may reduce). Such a Kernal is called a *degenerate* or *separable* kernel. A degenerate kernel has a finite number of characteristic values.

A kernel of the form K $(x-t)$ is called a *difference kernel.*

An integral equation is called *singular* if either the range of integration is infinite and/or the kernel is discontinuous in the domain.

An integral of the form

$$f_0\, g(x)=\int_0^x f(x-t)\, g(t)\, dt=\int_0^x f(t)\, g(x-t)\, dt$$

for locally integrable functions f and g, is called a *convolution* (or foltung or *resultant*) of f and g.

A volterra equation of the form

$$y(x)=f(x)+\lambda\int_0^x K(x-t)\, y(t)\, dt$$

where $\quad x>0$ and $y(0)=f(0)$,

is called a *convolution type* integral equation. For example, the Abel's equation

$$\int_0^x \frac{y(t)}{(x-t)^{\alpha}}\, dt=f(x),\ x>0,\ 0<\alpha<1,\ f(0)=0$$

is a convolution equation with a weak singularity. Convolution equations are most conveniently solved by the use of Laplace transform. We shall deal with this topic in the next chapter.

Non-linear equations of the type

$$y(x)=\int_a^b K(x, t)\, \phi[t, y(t)]\, dt$$

are called *Hammerstein* type equations.

Integral equations of the first kind are in general more difficult to solve than those of the second kind, as the former require more regularity conditions on $f(x)$. For example, in the case of an equation

$$\int_0^x K(x, t)\, y(t)\, dt = f(x)$$

We must have $f(0)=0$.

Given a homogeneous Fredholm equations.

$$y(x)-\lambda\int_0^x K(x, t)\, y(t)\, dt=0,$$

then the equation

$$z(x)-\bar{\lambda}\int_a^b \overline{K(t, x)}\, z(t)\, dt=0$$

is called its *adjoint* (or *associate* or *conjugate*).

Sometimes a physical problem is formulated in the form of an *integro-differential equation* of the form

$$\frac{dy}{dx}+\int_a^b K(x, t)\, y(t)\, dt=f(x),\quad a\leqslant x\leqslant b.$$

Both differential and integral equation methods can be used to solve such equations.

§10·3. Conversion of Differential Equations to Integral Equations

We will see that an initial-value problem (or a Cauchy problem) can be converted to a Volterra equation of the second kind and a two-point boundary-value problem (also called a Dirichlet problem or the first boundary-value problem) for a second order differential eqnation can be converted to a Fredholm equation of the second kind.

(A) An Initial-Value Problem

$$y''+xy'+y=0 \qquad \ldots(i)$$

$$\text{IC}: y(0)=1,\ y'(0)=0 \qquad \ldots(ii)$$

Put $y''(x)=\phi(x)$. Integrating twice and using the Cauchy's formula

$$\underbrace{\int_0^x dx\int_0^x dx\ldots\int_0^x}_{n \text{ times}} f(x)\, dx=\frac{1}{(n-1)\,!}\int_0^x (x-t)^{n-1} f(t)\, dt$$

we shall get

$$y'=\int_0^x \phi(t)\, dt+A$$

$$y=\int_0^x (x-t)\,\phi(t)\, dt+Ax+B.$$

Application of initial values gives $A=0$, $B=1$. Making these substitutions in the given equation we get the integral equation with $\phi(x)=y''(x)$ as unknown :

$$\phi(x)=-1-\int_0^x (2x-t)\,\phi(t)\, dt.$$

Alternatively, we can integrate the given equation twice and get

$$y'+xy=A$$

$$y(x)+\int_0^x (x-1)\, y(t)\, dt=Ax+B.$$

In initial conditions give $A=0$, $B=1$. Hence the given problem is converted to the integral equation with $y(x)$ unknown, namely

$$y(x)=1+\int_0^x (-x+1)\, y(t)\, dt$$

The numerical approximations, such a problem is called a *marching problem*, because the solution is supposed to 'march out' from the origin.

(B) Two-Point Boundary-value Problems

(*i*) Consider $y''+y=0$

$$BC : y(0)=0,\ y'(1)=0.$$

Integrating the equation $y''=-y$ twice,

$$y'(x)=-\int_0^x y(t)\, dt+A$$

$$y(x)=-\int_0^x (x-t)\, y(t)\, dt+Ax+B$$

Applying the boundary conditions we get

$$B=0,\quad A=\int_0^1 y(t)\, dt.$$

Hence,

$$y(x)=\int_0^1 xy(t)\, dt-\int_0^x (x-t)\, y(t)\, dt$$

$$=\int_0^x t\, y(t)\, dt+\int_x^1 xy(t)\, dt$$

$$=\int_0^1 G(x,t)\, y(t)\, dt$$

where

$$G(x_1 t)=\begin{cases} t & 0\leqslant t\leqslant x\leqslant 1 \\ x & 0\leqslant x \angle t\leqslant 1 \end{cases}$$

It may be recalled that $G(x, t)$ is the Green's function of the homogeneous problem $y''=0$; $y(0)=0$, $y'(1)=0$.

(*ii*) Consider the problem :

$$y''+\lambda y=e^x$$

$$BC : y(0)=y'(0),\ y(1)=y'(1).$$

The Green's function of the problem

$$y''=0 ;\ y(0)=y'(0),\ y(1)=y'(1) \text{ is}$$

$$G(x, t)=\begin{cases} (1+x)\, t & x\leqslant t \\ (1+t)\, x & t\leqslant x \end{cases}$$

Using theorem 1 of section 8·2 we have

$$y(x)=e^x-\lambda\int_0^1 G(x, t)\, y(t)\, dt$$

It would be instructive to look for eigenvalues in the problem. The eigenvalues are those values of λ for which a non-trivial solution of the corresponding homogeneous problem $y''+\lambda y=0$; $y(0)=y'(0)$ $y(1)=y'(1)$ exists. As the only solution $y=e^x$ is associated with $\lambda=-1$, this is the only eigenvalue.

Problem 1. Convert the problem $y''+\lambda y=0$; $y(0)=y'(0)$, $y(\pi)=y'(\pi)$

Problem 2. Prove that² the BVP $y''+\left[\lambda^2-\dfrac{\nu^2-\frac{1}{4}}{x^2}\right]y+0$

With BCs $y(1)=0$, $\int_0^1 y^2(x)dx<\infty$ **can be converted to the integral** equation $y(x)=-\lambda^2\int_0^1 G(x, t)\, y(t)dt$

where $G(x, t)=\begin{cases} -\dfrac{1}{2\nu}\, x^{\frac{1}{2}+\nu}\,(t^{\frac{1}{2}-\nu}-t^{\frac{1}{2}+\nu}) & x\leqslant t \\ -\dfrac{1}{2\nu}\, t^{\frac{1}{2}+\nu}\,(x^{\frac{1}{2}-\nu}-x^{\frac{1}{2}+\nu}) & x\geqslant t \end{cases}$

to an integral equation. What are its eigenvalues and eigenfunctions ?

$$\left[\text{Eigenvalues } \lambda_n=n^2, \text{ Eigenfunction } y_n=\frac{\sin nx}{n}+\cos nx\right]$$

Example 1. Solve the integro-differential equation

$$y'(x)+e^x\int_0^1 e^{-t}\, y(t)\, dt=1 \qquad \ldots(1)$$

where $y(0)=0$.

Solution. Put $K=\int_0^1 e^{-t}\, y(t)\, dt$

Then from (i) $\quad y'(x)+K\, e^x=1 \qquad \ldots(ii)$

Integrating (ii) with respect to x between 0 to x, we have

$$y(x)+(e^x-1)\, K=x \qquad \ldots(iii)$$

Using the relation (iii) we have

$$K=\int_0^1 e^{-t}\, y(t)\, dt=\int_0^1 e^{-t}\, \{t+(1-e^t)K\}\, dt=1-(K+2)\, e^{-1}$$

i.e., $K=\dfrac{e-2}{e+1}$. The required solution is $y(x)=x+\dfrac{e-2}{e+2}(1-e^x.)$

Exercise 29

1. Convert the following Cauchy problems to Volterra-integral equations :

(*i*) $y''+y=0$; $y(0)=0$, $y'(0)=1$

(*ii*) $y''-\sin x\, y'+e^x\, y=x$; $y(0)=1$, $y'(0)=-1$

(*iii*) $y''+(1+x^2)\, y=\cos x$; $y(0)=0$, $y'(0)=2$

(*iv*) $y''+\lambda y=1$; $y(0)=y'(0)=1$.

2. Convert the following two-point boundary-value problems to equivalent Fredholm integral equations :

(i) $y''+\lambda y=x$; $y(0)=y\left(\frac{\pi}{2}\right)=0$

(ii) $y''=\lambda y+x^2$; $y(0)=y\left(\frac{\pi}{2}\right)=0$

(iii) $y''=\lambda y+e^x$; $y(0)=y(1)=0$

(iv) $y''+\lambda y=0$; $y(0)=0, y'(1)=0$

(v) $y''-\lambda y=\cos x$; $y(0)=0$, $y'(1)=0$.

3. Show that a linear volterra equation of the first kind can be converted to an equation of the second kind.

4. Verify that the integral equrtion

$$y(x)=\int_0^x t^{x-t}\, y(t)\, dt \qquad x\geqslant 0.$$

has the solution x^{x-1}. Is the equation equivalent to a differential equation ? Give reason for your answer.

5. Solve the integro-differential equations :

(i) $y'(x)+\int_0^1 e^{-t}\, y(t)\, dt=1$; $y(0)=0$

(ii) $y''(x)+\int_0^1 e^{-t}\, y(t)\, dt=1$; $y(0)=y'(0)=0$.

§10·4. Finite Difference Approximations

Solution of integral equations analytically is, in general, as difficult as that of differential equations. In many cases finite difference approximation of an integral equation not only affords a workable solution, but also provides an insight into its basic mathematical properties.

For example, consider the Fredholm equation.

$$y(x)-\lambda\int_0^1 K(x, t)\, y(t)\, dt=f(x) \qquad \ldots(i)$$

where $K(x, t)$ is continuous in the domain $0\leqslant x\leqslant 1$, $0\leqslant t\leqslant 1$ and $f(x)$ is continuous on the interval $0\leqslant x\leqslant 1$. For large n, equation (i) can be approximated by the linear algebraic system

$$x_i-\lambda\sum_{j=1}^{n} a_{ij}\, x_j=b_i \qquad (i=1, 2, \ldots, n) \qquad \ldots(ii)$$

where $\quad x_i=y\left(\frac{i}{n}\right),\ a_{ij}=\frac{1}{n}K\left(\frac{i}{n}, \frac{j}{n}\right),\ b_i=f\left(\frac{i}{n}\right)$

In the matrix notation the system can be written as

$$(I-\lambda A)X=B \qquad \ldots(iii)$$

where $A=[a_{ij}]_{n\times n}$, $X=(x_1, \ldots, x_n)^T$, $B=(b_1, \ldots, b_n)^T$

and I is the identify matrix of order n. As $n \to \infty$ the system (*iii*) will tend to equaton (*i*).

A unique solution of the syctem (*iii*) is possible except in the case when

$$|I-\lambda A|=0 \qquad \ldots(iv)$$

Hence a unique solution of the given Fredholm equation is possible except (finitely or infinitely many) values of λ, called the *characteristic values* of the equation. For these values of λ, the corresponding homogeneous equation has non-trival solutions.

Next, consider the case of Volterra equation

$$y(x)-\lambda\int_0^x K(x, t)y(t)\,dt=f(x) \qquad \ldots(a)$$

where $K(x, t)$ and $f(x)$ are continuous in given domain. Equation (*a*) can be approximated for large n by the triangular system

$$x_i-\lambda\sum_{j=1}^{i-1} a_{ij}\,x_i=b_i \qquad (i=1, \ldots, n) \qquad \ldots(b)$$

where x_i, a_{ij} and b_i are as given earlier. In the fully expanded form the system (*b*) will appear as below

$$x_1=b_1$$
$$x_2-\lambda\,a_{21}\,x_1=b_2$$
$$x_3-\lambda\,(a_{31}\,x_1+a_{32}\,x_2)=b_3$$
$$\ldots \qquad \ldots \qquad \ldots$$
$$x_{n-1}-\lambda(a_{n-1,\,1}\,x_1+\ldots+a_{n-2,\,n-1}\,x_{n-2})=b_{n-1}$$
$$x_n-\lambda(a_{n1}\,x_1+\ldots+a_{n,\,n-1}\,x_{n-1})=b_n$$

If now we put $X=(x_1, \ldots, x_n)^T$, $B=(b_1, \ldots b_n)^T$

and
$$A=\begin{bmatrix} 0 & 0 & 0 & \ldots & 0 & 0 \\ a_{21} & 0 & 0 & \ldots & 0 & 0 \\ a_{31} & a_{32} & 0 & \ldots & 0 & 0 \\ \ldots & & \ldots & & \ldots & \\ a_{n1} & a_{n2} & & \ldots & a_{n,\,n-1} & 0 \end{bmatrix}$$

then the system *b*) can be written as

$$(I-\lambda A)X=B \qquad \ldots(c)$$

A is a strictly lower triangular matrix of order n. Hence A is nilpotent with index n, *i.e.*, $A^n=0$. This together with the fact that I is non-singular ensures invertibility of the matrix $I-\lambda A$ for all λ. In fact, since

$$(I-\lambda A)(I+\lambda A+\lambda^2 A^2+\ldots+\lambda^{n-1} A^{n-1}=I,$$

We have

$$(I-A)^{-1}=I+\lambda A+\lambda^2 A^2+\ldots+\lambda^{n-1} A^{n-1}.$$

This means the system (*b*) has a unique solution for all λ. Since for large n, the system (*c*) approximates the equation (*a*), the Volterra equation is uniquely solvable for all λ.

Thus while we expect a Fredholm equation to have characteristic values, Volterra equations have none.

§10·5. Fredholm Equations with Degenerate Kernel

Let $$K(x, t)=\sum_{i=1}^{n} a_i(x)\, b_i(t) \qquad \text{...}(i)$$

where n is finite and $\{a_i(x)\}$, $\{b_i(t)\}$ are linearly independent sets.

(A) Fredholm Equations of the First Kind

We are concerned with the equation

$$\int_a^b K(x, t)\, y(t)\, dt=f(x)$$

i.e., $$\sum_{i=1}^{n} a_i(x) \int_a^b b_i(t)\, y(t)\, dt=f(x) \qquad \text{...}(ii)$$

The form of equation (*ii*) enables us to make the following observations :

(*a*) The equation (*ii*) is consistent if $f(x)$ is of the form

$$\sum_{r=1}^{n} k_r\, a_r(x).$$

(*b*) Let $y(x)$ be a solution of (*ii*). Then for arbitrary μ, $y(x)+\mu g(x)$ is also a solution where $g(x)$ is orthogonal to each of b_i, *i.e.*

$$\int_a^b b_i(t)\, g(t)\, dt=0 \qquad (i=1, \ldots, n)$$

(Such a function can always be found for finite n).

The most general form of $g(x)$ can be compared to the *complementary function* and any solution of (*ii*), the *particular integral* which occur in differential equations.

Example 2. Consider the equation

$$\int_a^b (xt^2+t+x^2)\, y(t)\, dt=e^x.$$

The equation is inconsistent as the left-hand side is of the form $c_1 x^2+c_2 x+c_3$ while the right-hand side is an infinite series in x. Hence there is no solution.

Example 3. Solve the equation

$$\int_{-1}^{1} (xt^2+t+x^2)\, y(t)\, dt=x^2+x+1.$$

Solution. Put $c_1=\int_{-1}^{1} t^2\, y(t)\, dt,$

$$c_2=\int_{-1}^{1} t\, y(t)\, dt,$$

$$c_3=\int_{-1}^{1} y(t)\, dt.$$

Then the equation can be written as

$$c_1x^2+c_2x+c_3=x^2+x+1.$$

Hence $c_1=c_2=c_3=1$. This indicates that the equation is consistent.

To find the P.I. Let $y=a_1x^2+a_2x+a_3$. On comparing the coefficients of powers of x in the equality,

$$\int_{-1}^{1} (xt^2+t+x^2)\,(a_1t^2+a_2t+a_3)\, dt=x^2+x+1,$$

We shall have

$$\frac{2a_1}{5}+\frac{2a_3}{3}=1,\quad \frac{2a_2}{3}=1,\quad \frac{2a_1}{3}+2a_3-1,$$

i.e.,
$$a_1=\frac{15}{4},\ a_2=\frac{3}{2},\ a_1=-\frac{3}{4}.$$

Hence the particular integral is

$$y(x)=\frac{15}{4}x^2+\frac{3}{2}\,x-\frac{3}{4}.$$

To determine the complementary function, we should look for a solution to the homogeneous problem, *i.e.*, those functions which are orthogonal to xt^2+t+x^2 on the interval $[-1, 1]$.

Undoubtedly, the Legendre polynomials are the right candidates. Recall that

$$P_0(t)=1,\ P_1(t)=t,\ P_2(t)=\frac{3}{2}t^2-\frac{1}{2}$$

Hence $xt^2+t+x^2=\frac{2}{3}x\ P_2(t)+P_1(t)+\left(x^2+\frac{x}{3}\right)P_0(t).$

If $a\,P_2(t)+b\,P_1(t)+cP_0(t)$ is orthogonal xt^2+t+x^2, then using the formula

$$\int_{-1}^{1} P_n^2(t)\, dt=\frac{2}{2n+1}$$

we shall get

$$\int_{-1}^{1}\left\{\frac{2x}{3}P_2(t)+P_1(t)+\left(x^2+\frac{x}{3}\right)P_0(t)\right\}$$
$$\{aP_2(t)+bP_1(t)+cP_0(t)\}dt=0$$

This leads to $a=b=c=0$. Also since every P_n is orthogonal to P_2, P_1, P_0 for $n\geqslant 3$, any series $\sum_{n=3}^{\infty} a_nP_n(t)$ is orthogonal to xt^2+t+x^2 with t as the variable. Hence

$$\text{C.F.}=\sum_{n=3}^{\infty} a_n P_n(x)$$

Thus the most general solution of the equation is

$$y(x) = \text{C.F.} + \text{P.I.}$$

$$= \sum_{n=3}^{\infty} a_n P_n(x) + \frac{15}{4}x^2 + \frac{3}{2}x - \frac{3}{4}$$

(B) Fredholm Equations of the Second Kind

Consider the Fredholm equation

$$y(x) = f(x) + \lambda \int_a^b K(x, t)\, y(t)\, dt \qquad \ldots(iii)$$

where $\quad K(x, t) = \sum_{i=1}^{n} a_i(x)\, b_i(t) \qquad \ldots(iv)$

(*iii*) can be written in the form

$$y(x) = f(x) + \lambda \sum_{i=1}^{n} c_i\, a_i(x) \qquad \ldots(v)$$

where $\quad c_i = \int_a^b b_i(t)\, y(t)\, dt \qquad \ldots(vi)$

The c_i will now be evaluated.

Substituting the expression for $y(x)$ of (*v*) in (*iii*) we shall get

$$\sum_{i=1}^{n} a_i(x)\left[c_i - \int_a^b b_i(t) f(t)\, dt - \lambda \sum_{j=1}^{n} \int_a^b c_j\, b_i(t)\, a_j(t) dt\right] = 0.$$

Since the functions $a_i(x)$ are linearly independent, we have

$$c_i - \int_a^b b_i(t) f(t)\, dt - \lambda \sum_{j=1}^{n} c_j \int_a^b b_i(t)\, a_j(t)\, dt = 0 \qquad (i = 1, 2, \ldots, n)$$

Put $\quad \alpha_{ij} = \int_a^b b_i(t)\, a_j(t)\, dt, \quad \delta_i = \int_a^b b_i(t) f(t)\, dt,$

then $\quad c_1 - \lambda \sum_{j=1}^{n} \alpha_{ij}\, c_j = \delta_i \qquad (i = 1, 2, \ldots, n)$

i.e., $\quad (I - \lambda A)C = D \qquad \ldots(vii)$

where $\quad A = [\alpha_{ij}]_{n \times n},\ C = (c_1, \ldots, c_n)^T,\ D = (\delta_1, \ldots, \delta_n)^T$

The algebraic system (*vii*) in the n unknowns $c_1, \ldots, c_n$ has a unique solution if, and only if

$$|\, I - \lambda A \,| \neq 0 \qquad \ldots(viii)$$

If $f(x) = 0$, then $D = 0$ and a non-trivial solution of (*vii*) exists when $|\, I - \lambda A\,| = 0$.

The values of λ ($\neq 0$) satisfying the equation $|\, I - \lambda A\,| = 0$ are called *characteristic values* of the equation (*iii*), which is also called an eigenvalue problem if $f(x) = 0$.

Example 4. Solve the equation

$$y(x) = 1 + \lambda \int_0^{\pi/2} \cos(x - t)\, y(t)\, dt$$

and find its eigenvalues.

Solution. The equation can be written as

$$y(x)=1+\lambda \cos x\int_0^{\pi/2} \cos t\, y(t)dt+\lambda \sin x\int_0^{\pi/2} \sin t\, y(t)\, dt$$

Putting
$$c_1=\int_0^{\pi/2} \cos t\, y(t)\, dt,$$

$$c_2=\int_0^{\pi/2} \sin t\, y(t)\, dt \qquad \ldots(ix)$$

We get $\quad y(x)=1+\lambda \cos x\, c_1+\lambda \sin x\, c_2$

With this replacemnnt for $y(x)$ in the expressions (ix) for c_1, c_2 we have

$$\left(1-\frac{\lambda\pi}{4}\right)c_1-\frac{\lambda}{2}c_2=1$$

$$-\frac{\lambda}{2}c_1+\left(1-\frac{\pi}{4}\right)c_2=1$$

These equations lead to $c_1=c_2=\dfrac{1+\dfrac{\lambda}{2}-\dfrac{\lambda\pi}{4}}{\left(1-\dfrac{\lambda\pi}{4}\right)^2-\dfrac{\lambda^2}{4}}$

$$=\frac{1}{1-\dfrac{\lambda}{4}(\pi+2)}$$

Hence $\quad y(x)=1+\dfrac{\lambda(\cos x+\sin x)}{1-\dfrac{\lambda}{4}(\pi+2)}$

The eigenvalues are given by

$$\left(1-\frac{\lambda\pi}{4}\right)^2-\frac{\lambda^2}{4}=0,$$

i.e.,
$$\lambda=\frac{4}{\pi\pm 2}$$

For these values of λ, the given non-homogeneous integral equation has no solution and the corresponding homogeneons equation has non-trivial solutions.

Exercise 30

Solve the following Fredholm equations, mentioning the eigenvalues if any :

1. $y(x)=1+\lambda\int_0^1 xt\, y(t)\, dt$

2. $y(x)=x+\lambda\int_0^1 xt\, y(t)\, dt$

3. $y(x)=\sin x+\lambda\int_0^{\pi/2} \sin x \cos t\, y(t)\, dt$

4. $y(x)=\cos x+\lambda\int_0^{\pi} \sin (x-t)\, y(t)\, dt$

5. $y(x)=e^x+2\lambda\int_0^1 e^{x+t}\, y(t)\, dt$

6. $y(x)=1+\lambda\int_0^1 (1+xt)\, y(t)dt$

7. $y(x)=\cos 3x+\lambda\int_0^\pi \cos (x+t)\, y(t)\, dt$

8. $y(x)=x+\lambda\int_{-1}^1 (xt+x^2 t^2)\, y(t)\, dt$

9. $y(x)=x+\lambda\int_0^{2\pi} |\pi-t| \sin x\, y(t)\, dt$

10. $y(x)=x+\lambda\int_0^1 (x^2 t+xt^2)\, y(t)\, dt$

§10·9. Fredholm Equations with Green's Function Kernel

If the kernel of a Fredholm equation is the Green's function of some boundary-value problem, then the equation can be solved by converting the equation to that problem.

Example 5. Solve the integral equation

$$y(x)=x+\lambda\int_0^1 K(x, t)\, y(t)\, dt$$

where
$$K(x, t)=\begin{cases} x(t-1) & x \leqslant t \\ t(x-1) & x \geqslant t \end{cases}$$

Find the characteristic numbers and eigenfunctions of the corresponding homogeneous equation.

Solution. The equation can be written as

$$y(x)=x+\lambda\int_0^x t(x-1)\, y(t)\, dt+\lambda\int_x^1 x(t-1)\, y(t)\, dt \qquad ...(i)$$

Hence $\quad y(0)=0, \quad y(1)=1.$

Differentiating (i) with respect to x twice

$$y'(x)=1+\lambda\int_0^1 t\, y(t)-\lambda\int_x^1 y(t)\, dt$$

$$y''(x)=\lambda\, y(t).$$

The given non-homogeneous integral equation is thus converted to the following boundary-value problem having a homogeneous equation with non-homogeneous boundary conditions :

$$y''-\lambda y=0\ ;\ y(0)=0,\ y(1)=1.$$

The solution of the BVP is

$$y(x)=\frac{e^{\sqrt{\lambda}\, x}-e^{-\sqrt{\lambda}\, x}}{e^{\sqrt{\lambda}}-e^{-\sqrt{\lambda}}}=\frac{\sin h\sqrt{\lambda}\, x}{\sin h\sqrt{\lambda}}.$$

The characteristic numbers are roots of the equation

$$\sin h\sqrt{\lambda}=0,$$

i.e., $\lambda_n=-n^2\pi^2 \qquad (n=1, 2, \ldots).$

Hence the eigenfunctions are $\sin n\pi x$.

Note. The eigenvalues and eigenfunctions could also be directly obtained by finding the non-trivial solutions for the homogeneous problem

$$y''-\lambda y=0\ ;\quad y(0)=y(1)=0.$$

Example 6. Solve the equation

$$y(x)=x+\lambda\int_0^1 K(x, t)\, y(t)\, dt$$

where $K(x, t)=\begin{cases} x & x\leqslant t \\ t & x\geqslant t \end{cases}$

Solution. The equation can be written as

$$y(x)=x+\lambda\int_0^x t\, y(t)\, dt+\lambda\int_x^1 x\, y(t)\, dt.$$

Differentiating twice with respect to x,

$$y'(x)=1+\lambda\int_x^1 y(t)\, dt$$

$$y''(x)=-\lambda y(x),\ y(0)=0,\ y'(1)=1.$$

We have $y(x)=A\sin\sqrt{\lambda}\,x+B\cos\sqrt{\lambda}\,x,$

$$y'(x)=\sqrt{\lambda}\,(A\cos\sqrt{\lambda}\,x-B\sin\sqrt{\lambda}\,x).$$

$y(0)=0 \longrightarrow B=0.$

$y'(1)=1 \longrightarrow A=\dfrac{1}{\sqrt{\lambda}\cos\sqrt{\lambda}}.$

Hence the solution is $y(x)=\dfrac{\sin\sqrt{\lambda}\,x}{\sqrt{\lambda}\cos\sqrt{\lambda}}.$

The eigenvalues are given by $\cos\sqrt{\lambda}=0.$

Hence, $\sqrt{\lambda}=(2n+1)\dfrac{\pi}{2}$ are the eigenvalues.

The eigenvalues and eigenvectors can also be obtained from the homogeneous problem

$$y''-\lambda y=0\ ;\quad y(0)=0,\ y'(1)=0.$$

Exercise 31

1. Find the eigenvalues and eigenfunctions associated with the integral equation

$$y(x)=\lambda\int_0^\pi K(x, t)\, y(t)\, dt$$

where $$K(x, t)=\begin{cases} \cos x \sin t & x \leqslant t \\ \cos t \sin x & x \geqslant t \end{cases}$$

2. Let $$K(x, t)=\begin{cases} \sin x \cos t & x \leqslant t \\ \cos x \sin t & x \geqslant t \end{cases}$$

Discuss the solutions of the integral equations ;

(i) $y(x)=e^x+\lambda\int_0^\pi K(x, t)\, y(t)\, dt.$

(ii) $y(x)=\lambda \int_0^\pi K(x, t)\, y(t)\, dt.$

(iii) $\int_0^\pi K(x, t)\, y(t)\, dt=x-\frac{x^2}{2\pi}.$

(iv) $\int_0^\pi K(x, t)\, y(t)\, dt=x.$

3. **Solve the equation**

$$y(x)=\frac{x}{2}+\frac{\pi^2}{4}\int_0^1 K(x, t)\, y(t)\, dt$$

where $$K(x, t)=\begin{cases} \frac{x}{2}(2-t) & x \leqslant t \\ \frac{t}{2}(2-x) & x \geqslant t \end{cases}$$

4. **Find the characteristic numbers and eigenfunctions of the homogenous Fredholm equations with the following kernels :**

(i) $$K(x, t)=\begin{cases} t(x+1) & x \leqslant t \\ x(t+1) & x \geqslant t \end{cases} \qquad 0 \leqslant x,\ t \leqslant 1$$

(ii) $$K(x, t)=\begin{cases} \sin x \cos t & 0 \leqslant x \leqslant t \leqslant \frac{\pi}{2} \\ \sin t \cos x & 0 \leqslant t \leqslant x \leqslant \frac{\pi}{2} \end{cases}$$

(iii) $$K(x, t)=\begin{cases} \sin x \cos t & 0 \leqslant x \leqslant t \leqslant \pi \\ \sin t \cos x & 0 \leqslant t \leqslant x \leqslant \pi \end{cases}$$

§10·7. Eredholm Equations-Method of the Resolvent Kernel

Owing to the step-by-step process of attaining the solution, this method is also called the *method of successive approximations* or the *iterative method.* Under appropriate conditions, the method applies equally well to both Fredholm and Volterra equations.

Given an equation

$$y(x)=f(x)+\lambda \int_a^b K(x, t)\, y(t)\, dt \qquad \ldots(i)$$

where $f(x)$ is continuous on $[a, b]$ and $K(x, t)$ satisfies condition to be mentioned shortly.

The idea of the method runs as follows :

As a *zero-order* approximation (or the *starting* approximation) we take
$$y_0(x)=f(x).$$

We now substitute $y_0(x)$ for $y(x)$ in the right-hand side of (i) to get the *first-order* approximation $y_1(x)$

$$y_1(x)=f(x)+\lambda \int_a^b K(x, t)\, y_0(t)\, dt$$

If the nth-order approximation $y_n(x)$ has been obtained, the $(n+1)$th approximation $y_{n+1}(x)$ is determined by recurrence relation

$$y_{n+1}(x)=f(x)+\lambda \int_a^b K(x, t)\, y_n(t)\, dt \qquad \ldots(ii)$$

If these successive approximations tend to a limit, the limit will be a solution to equation (i).

We now examine the structure of the successive approximations.

If we put $K_1(x, t)=K(x, t)$

$$K_n(x, t)= \int_a^b K(x, s)\, K_{n-1}(s, t)\, ds) \qquad \ldots(iii)$$

Then $$y_n(x)=f(x)+\lambda \int_a^b \left\{ \sum_{i=1}^{n} \lambda^{i-1} K_i(x, t) f(t) \right\} dt \qquad \ldots(iv)$$

The function $K_n(x, t)$ is called the *nth iterated* kernel related to the kernel $K(x, t)$. It is easily verified that $K_n(x, t)$ satisfies the more general recurrence relation.

$$K_{m+n}(x, t)=\int_a^b K_m(x, s)\, K_n(s, t)\, ds \qquad \ldots(v)$$

The quantity

$$A_n=\int_a^b K_n(x, x)\, dx \qquad (n=1, 2, \ldots) \qquad \ldots(vi)$$

is called the *nth trace* of the kernel $K(x, t)$.

Assuming the convergence of the series on the right of (iv) we call

$$R(x, t\,;\lambda)= \sum_{n=1}^{\infty} \lambda^{n-1} K_n(x, t) \qquad \ldots(7)$$

the ***resolvent*** or the ***reciprocal kernel*** of the kernel $K(x, t)$ (or the integral equation (i)).

In terms of the resolvent kernel, we can write the solution of the integral equation as :

$$y(x)=f(x)+\lambda\int_a^b R(x, t ; \lambda) f(t)\, dt \qquad \text{...(8)}$$

The series on the right-hand side of (7) is called the *Neumann series* for the kernel K(x, t). The series on the right-hand side of (8) *viz.*

$$f(x)+\lambda\int_a^b K(x, t) f(t)\, dt+\lambda^2\int_a^b K_2(x, t) f(t)\, dt+\ldots$$

is called the *Neumann series* for the solution.

It can be verified that the resolvent Kernel satisfies the integral equations :

$$R(x, t ; \lambda)=K(x, t)+\lambda\int_a^b K(x, s)\, R(s, t : \lambda)\, ds \qquad \text{...}(vii)$$

$$R(x, t ; \lambda)=K(x, t)+\lambda\int_a^b K(s, t)\, R(x, s ; \lambda)\, ds \qquad \text{...}(viii)$$

Also, $R(x, t ; \lambda)-R(x, t ; \lambda')=(\lambda-\lambda')\int_a^b R(x, s ; \lambda)$

$$R(s, t ; \lambda')\, ds \qquad \text{...}(ix)$$

$$\frac{\partial R(x, t ; \lambda)}{\partial\lambda}=\int_a^b R(x, s ; \lambda)\, R(s, t ; \lambda)\, ds \qquad \text{...}(x)$$

Results (*vii*) through (*x*) can be derived using (*v*) and (7). As an illustration we sketch the proof of (*ix*) :

Using formula (7), we have

$$R(x, t ; \lambda)=\sum_{n=0}^{\infty} \lambda^n K_{n+1}(x, t) \qquad \text{...(I)}$$

$$R(x, t ; \lambda')=\sum_{n=0}^{\infty} \lambda' K_{n+1}(x, t) \qquad \text{...(II)}$$

Hence $\quad R(x, t ; \lambda)-R(x, t ; \lambda') = \sum_{n=0}^{\infty} (\lambda^n-\lambda'^n)\, K_{n+1}(x, t)$

...(III)

Multiplying I, II and using the result

$$\sum_{n=0}^{\infty}\sum_{k=0}^{\infty} A(k, n)=\sum_{n=0}^{\infty}\sum_{k=0}^{n} A(k, n-k),$$

We have

$$R(x, s, \lambda).\, R(s, t ; \lambda')=\sum_{n=0}^{\infty}\sum_{k=0}^{n} \lambda^{n-k} \lambda'^k K_{n+1-k}(x, s)\, K_{k+1}(s, t)$$

...(IV)

Multiplying IV by $\lambda-\lambda'$ and integrating, we have

$$(\lambda-\lambda')\int_a^b R(x, s : \lambda)\, R(s, t ; \lambda')\, ds$$

$$=\sum_{n=0}^{\infty}\left\{(\lambda-\lambda')\sum_{k=0}^{n}\lambda^{n-k}\lambda'^{k}\right\}\int_a^b K_{n+1-k}(x, t)\, K_{k+1}(s, t)\, ds$$

...[by (v)]

$$=\sum_{n=0}^{\infty}(\lambda^{n+1}-\lambda'^{n+1})\, K_{n+2}(x, t)$$

$$=\sum_{n=0}^{\infty}(\lambda^{n}-\lambda'^{n})K_{n+1}(x, t)$$

$$=R(x, t ; \lambda)-R(x, t ; \lambda')$$

(*x*) is an immediate consequence of (*ix*).

Convergence of the sequence $\{y_n(x)\}$

Let $f(x) \in C[a, b]$, $K(x, t) \in L^1$. Also let

$$\int_a^b | K(x, t) |\, dt \leqslant M, \qquad a \leqslant x \leqslant b$$

$$m=\max | f(x) | \text{ and } | \lambda | < \frac{1}{M}.$$

We can write

$$y_n(x)=y_0(x)+\sum_{i=1}^{n}\{y_i(x)-y_{i-1}(x)\} \qquad \text{...(xi)}$$

Since $\quad | y_0(x) | \leqslant m$

by induction we can prove

$$| y_n(x)-y_{n-1}(x) | \leqslant m | \lambda |^n M^n \qquad (n=1, 2, ...)$$

Hence using (*xi*)

$$| y_n(x) | < m \sum_{i=0}^{n} | \lambda M |^i \qquad \text{...(xii)}$$

Since $| \lambda M | <1$, the series on the right of (*xii*) is convergent as $n\to\infty$. Hence the sequence $\{y_n(x)\}$ is absolutely and uniformly convergent to a continuous fuuction $y(x)$ on $a\leqslant x\leqslant b$. Hence taking the limits of both sides of (*ii*) as $n\to\infty$, we get

$$y(x)=f(x)+\lambda\int_a^b K(x, t)\, y(t)\, dt$$

which shows that $y(x)$ is a solution of the equation.

Uniqueness of the Solution

Let $\quad u(x)=y(x)-z(x)$. [where $z(x)$ is another solution of (*i*)]. Then

$$u(x)=\lambda\int_a^b K(x,t)\,u(t)\,dt$$

Hence for all x

$$\max|u(x)| \leqslant |\lambda| M \max|u(x)|$$
$$<\max|u(x)|\text{-}$$

This is possible if $u(x)\equiv 0$,

i.e., $y(x)\equiv z(x)$.

Remark 1. The convergence of the sequence of the successive approximations $\{y_n(x)\}$ is also assured under the following more general conditions :

(*a*) f is square integrable

(*b*) $\int_a^b\int_a^b K^2(x,t)\,dx\,dt=B^2<\infty$

(*c*) $|\lambda| B<1$

Let $P^2(x)=\int_a^b |K(x,t)|^2\,dt$

$$Q^2(t)=\int_a^b |K(x,t)|^2\,dx$$

Hence $B^2=\int_a^b P^2(x)\,dx$

$$=\int_a^b Q^2(t)\,dt$$

By Cauchy-Schwarz inequality *viz.*

$$\left\{\int_a^b f(x)\,g(x)\,dx\right\}^2 \leqslant \int_a^b f^2(x)\,dx\int_a^b g^2(x)\,dx$$

We get

$$|K_2(x,t)|^2=\left|\int_a^b K(x,s)\,K(s,t)\,ds\right|^2\leqslant\{P(x)\,Q(t)\}^2$$

Hence $\left|K_3(x,t)\right|^2=\left|\int_a^b K(x,s)\,K_2(s,t)\,ds\right|^2$

$$\leqslant\int_a^b |K(x,s)|^2\,ds\int_a^b |K_2(s,t)|^2\,ds$$

$$\leqslant\{P(x)\,Q(t)\}^2 B^2$$

In general

$$|K_{n+2}(x,t)|^2\leqslant\{P(x)\,Q(t)\}^2 B^{2n}$$

Hence $\left|\sum_{n=0}^{\infty}\lambda^{n+1}K_{n+2}(x,t)\right|\leqslant\sum_{n=0}^{\infty}|\lambda|^{n+1}|K_{n+2}(x,t)|$

$$\leqslant\sum_{n=0}^{\infty}|\lambda|^{n+1}P(x)\,Q(t)\,B^n$$

$$= |\lambda| \, P(x) \, Q(t) \sum_{n=0}^{\infty} |\lambda B|^n$$

Since $|\lambda B| < 1$, the series on the right is convergent, hence by Weierstrass M-test, the series on the left is uniformly convergent. Hence the Neumanu series (7) is absolutely and uniformly convergent. The relation (8) is thus valid. It is evident that the solution $y(x)$ thus obtained is unique.

Remark 2. (Orthogonality of kernels).

Two kernels $K(x, t)$, $L(x, t)$ are called orthogonal over $[a, b]$ if

$$\int_a^b K(x, s) \, L(s, t) \, ds = 0$$

$$= \int_a^b L(x, s) \, K(s, t) \, ds.$$

For a self-orthogonal kernal $K(x, t)$

$$K_2(x, t) = \int_a^b K(x, s) \, K(s, t) \, ds = 0$$

Obviously all subsequent iterated kernels vanish and

$$R(x, t : \lambda) = K(x, t)$$

$K(x, t) = \sin(x - 2t)$ is such a kernel on $0 \leqslant x \leqslant 2\pi$, $0 \leqslant t \leqslant 2\pi$.

Remark 3. If $K(x, t)$ is symmetric then the iterated kernels are also symmetric.

Problem. Prove that for a kernel $K(x, t) \in L^2$, all iterated kernels beginning with $K_2(x, t)$ are continuous.

Example 7. Solve the integral equation by computing the resolvent Kernel,

$$y(x) = f(x) + \lambda \int_0^1 e^{a(x^2 - t^2)} y(t) \, dt.$$

Solution. $K(x, t) = e^{a(x^2 - t^2)}$

$$K_2(x, t) = \int_0^1 e^{a(x^2 - s^2)} \, e^{a(s^2 - t^2)} ds = e^{a(x^2 - t)^2}$$

Thus $K_n(x, t) = e^{a(x^2 - t^2)}$ for all n.

The resolvent Kernel is given by

$$R(x, t ; \lambda) = \sum_{n=1}^{\infty} \lambda^{n-1} K_n(x, t) = \frac{e^{a(x^2 - t^2)}}{1 - \lambda}$$

Hence the solution is

$$y(x) = f(x) + \frac{\lambda}{1 - \lambda} \int_0^1 e^{a(x^2 - t^2)} f(t) \, dt$$

$$\lambda \neq 1.$$

Example 8. Show that for the integral equation

$$y(x)=1+\int_0^1 K(x,t)\,y(t)\,dt$$

where $$K(x,t)=\begin{cases} x & 0\leqslant x\leqslant t \\ t & 1\geqslant x\geqslant t \end{cases}$$

the successive approximations converge. Compute the first two approximations.

Solution. Here $\lambda=1$ and

$$B^2=\int_0^1\int_0^1 K^2(x,t)\,dx\,dt$$

$$=\int_0^1\int_0^t x^2\,dx\,dt+\int_0^1\int_t^1 t^2\,dx\,dt$$

$$=\frac{1}{6}<1,$$

Hence $|\lambda|B<1$. Also $f(x)=1\in L^2[0,1]$.

Hence the sequence of approximations will converge.

Example 7. Solve the integral equation by computing the resolvent Kernel,

$$y(x)=f(x)+\lambda\int_0^1 e^{a(x^2-t^2)}y(t)\,dt.$$

Solution. $K(x,t)=e^{a(x^2-t^2)}$

$$K_2(x,t)=\int_0^1 e^{a(x^2-s^2)}\,e^{a(s^2-t^2)}\,ds=e^{a(x^2-t)^2}$$

Thus $K_n(x,t)=e^{a(x^2-t^2)}$ for all n.

The resolvent Kernel is given by

$$R(x,t;\lambda)=\sum_{n=1}^{\infty}\lambda^{n-1}K_n(x,t)=\frac{e^{a(x^2-t^2)}}{1-\lambda}$$

Hence the solution is

$$y(x)=f(x)+\frac{\lambda}{1-\lambda}\int_0^1 e^{a(x^2-t^2)}f(t)\,dt$$

$$\lambda\neq 1.$$

Example 8. Show that for the integral equation

$$y(x)=1+\int_0^1 K(x,t)\,y(t)\,dt$$

$$\text{where} \quad K(x, t) = \begin{cases} x & 0 \leq x \leq t \\ t & 1 \geq x \geq t \end{cases}$$

the successive approximations converge. Compute the first two approximations.

Solution. Here $\lambda = 1$ and

$$B^2 = \int_0^1 \int_0^1 K^2 (x, t)\, dx\, dt$$

$$= \int_0^1 \int_0^t x^2\, dx\, dt + \int_0^1 \int_t^1 t^2\, dx\, dt$$

$$= \frac{1}{6} < 1,$$

Hence $|\lambda| B < 1$. Also $f(x) = 1 \epsilon L^2 [0, 1]$.

Hence the sequence of approximations will converge.

Thus finally,

$$K_2 (x, t) = \begin{cases} -\frac{x^3}{6} + \left(t - \frac{t^2}{2}\right) x & 0 \leq x \leq t \\ -\frac{t^3}{3} + \left(x - \frac{x^2}{2}\right) t & t \leq x \leq 1 \end{cases}$$

Exercise 32

1. $R(x, t ; \lambda)$ is the resolvent Kernel for some Fredholm Kernel $K(x, t)$. Show that the resolvent Kernel for the equation

$$y(x) = f(x) + \mu \int_a^b R(x, t ; \lambda)\, y(t)\, dt$$

is $R(x, y ; \lambda + \mu)$.

2. Solve the equation by the method of resolvent Kernel :

$$y(x) = x + \lambda \int_a^b xt\, y(t)\, dt.$$

3. Solve the equation

$$y(x) = x + \lambda \int_0^1 K(x, t)\, y(t)\, dt$$

$$\text{where} \quad K(x, t) = \begin{cases} x\,(t-1) & x \leq t \\ t\,(x-1) & x \geq t. \end{cases}$$

What are the characteristic numbers and eigenfunctions ?

4. Prove that the resolvent Kernel $R(x, t ; \lambda)$ is an analytic function of λ.

§10·8. Hermitian Kernel

A Kernel $K(x, t)$ is said to be *Hermitian* if

$$K(t, x) = \overline{K}(x, t)$$

where the bar denotes complex conjugate. The kernel is *symmetric* if

$$K(t, x) = K(x, t).$$

Evidently, a real Hermitian kernel is symmetric. $\overline{K}(t, x)$ is called Hermitian conjugate of $K(x, t)$.

Lemma 1. The eigenvalues of a Hermition Kernel are real,

Proof. Let $K(x, t)$ be a Hermitian Kernel and

$$y(x) = \lambda \int_a^b K(x, t)\, y(t)\, dt \qquad \ldots(i)$$

$$\overline{y}(x) = \overline{\lambda} \int_a^b \overline{K}(x, t)\, \overline{y}(t)\, dt \qquad \ldots(ii)$$

$y(x)$ being an eigenfunction, is not zero. Hence if

$y(x) = u(x) + i\, v(x)$, then

$y(x)\overline{y}(x) = u^2 + v^2 \neq 0.$

Now $(\lambda - \overline{\lambda}) \int_a^b (u^2 + v^2)\, dx$

$$= (\lambda - \overline{\lambda}) \int_a^b y(x)\, \overline{y}(x)\, dx$$

$$= \lambda \int_a^b y(x)\, \overline{y}(x)\, dx - \overline{\lambda} \int_a^b y(x)\, \overline{y}(x)\, dx$$

$$= \lambda \overline{\lambda} \int_a^b \int_a^b \overline{K}(x, t)\, y(x)\, \overline{y}(t)\, dx\, dt$$

$$- \lambda \overline{\lambda} \int_a^b \int_a^b K(x, t)\, y(t)\, \overline{y}(x)\, dx\, dt$$

$$= \lambda \overline{\lambda} \int_a^b \int_a^b K(t, x)\, y(x)\, \overline{y}(t)\, dx\, dt$$

$$- \lambda \overline{\lambda} \int_a^b \int_a^b K(x, t)\, y(t)\, \overline{y}(x)\, dx\, dt$$

$= 0$ (interchanging x, t in the first integral)

Hence $\lambda = \overline{\lambda}$, *i.e.*, λ is real.

Lemma 2. Distinct eigenvalues correspond to distinct eigenfunctions.

Proof. Let λ_1, λ_2 be distinct eigenvalues corresponding to the same eigenfunction $y(x)$,

$$y(x) = \lambda_1 \int_a^b K(x, t)\, y(t)\, dt$$

i.e., $$y(x)=\lambda_2\int_a^b K(x, t)\, y(t)\, dt$$

Since $\lambda_1 \neq \lambda_2 \neq 0$, we have

$$\left(\frac{1}{\lambda_1}-\frac{1}{\lambda_2}\right)y(x)=0. \quad \text{Since } y(x)\neq 0,$$

we have $\frac{1}{\lambda_1}-\frac{1}{\lambda_2}=0$, *i.e.* $\lambda_1=\lambda_2$.

Theorem 1. Two distinct eigenfunctions of a Hermitian kernel are orthogonal.

Proof. Let λ_1, λ_2 be two distinct eigenvalues associated with eigenfunctions.

y_1, y_2 $(y_1 \neq y_2)$. We have

$$y_1(x)=\lambda_1\int_a^b K(x, t)\, y_1(t)\, dt$$

$$\bar{y}_2(x)=\bar{\lambda}_2\int_a^b \bar{K}(x, t)\, \bar{y}_2(t)\, dt$$

Proceeding as in lemma 1,

$$(\lambda_1-\bar{\lambda}_2)\int_a^b y_1(x)\, \bar{y}_2(x)\, dx$$

$$=\lambda_1\bar{\lambda}_2\int_a^b\int_a^b \bar{K}(\bar{x}, t)\, y_1(x)\, \bar{y}_2(t)\, dx\, dt$$

$$-\lambda_1\bar{\lambda}_2\int_a^b\int_a^b K(x, t)\, y_1(t)\, \bar{y}_2(x)\, dx\, dt$$

$$=0.$$

Since $\lambda_1-\lambda_2\neq 0$, we have the orthogonality relation

$$\int_a^b y_1(x)\, \bar{y}_2(x)\, dx=0.$$

§ 10·9. The Fredholm Alternative

While considering the existence of a solution of the equation,

$$y(x)=f(x)+\lambda\int_a^b K(x, t)\, y(t)\, dt \qquad \ldots(i)$$

we have only required that $f \in L^2$ without enquiring whether this is necessary and sufficient. We now search for a necessary condition for the existence of a solution of (*i*).

The corresponding homogeneous equation is

$$y(x)=\lambda\int_a^b K(x, t)\, y(t)\, dt \qquad \ldots(ii)$$

The *transposed* (or *adjoint*) equations are respectively

$$z(x)=g(x)+\lambda\int_a^b \bar{K}(t, x)\, z(t)\, dt \qquad \ldots(iii)$$

and $$z(x)=\lambda \int_a^b \overline{K}(t, x)\, z(t)\, dt \qquad \text{...(iv)}$$

For brevity we use the following operational notation :

$$K f=\int_a^b K(x, t) f(t)\, dt \qquad \text{...(v)}$$

Also $$(f, g)=\int_a^b f(x)\, \bar{g}(x)\, dx) \qquad \text{...(vi)}$$

If K is defined by (v) then the conjugate (or adjoint) K* of operator K is given by,

$$K^* g=\int_a^b \bar{K}(t, x)\, g(t)\, dt \qquad \text{...(vii)}$$

The equations (*i*)—(*iv*) can now be written as

$$y-\lambda K y=f \qquad (a_1)$$

$$y-\lambda K y=0 \qquad (a_2)$$

$$z-\lambda K^* z=g \qquad (a_3)$$

$$z-\lambda K^* z=0 \qquad (a_4)$$

The following lemma gives an important relationship between conjugate operators :

Lemma 3. If K and K* are conjugate operators then $(Kf, g)=(f, K^* g)$.

Proof. $$(Kf, g)=\int_a^b \left\{\int_a^b K(x, t) f(t)\, dt\right\} \overline{g(x)}\, dt$$

$$=\int_a^b \int_a^b K(x, t) f(t)\, \overline{g(x)}\, dt\, dx$$

$$(f, K^* g)=\int_a^b f(t) \left\{\int_a^b \overline{\overline{K}(x, t)\, g(x)}\, dx\right\} dt$$

$$=\int_a^b \int_a^b K(x, t) f(t)\, \overline{g(x)}\, dt\, dx$$

$$=(Kf, g).$$

Lemma 4. Let $y(x)$ be a solution of (a_1) and $z(x)$ an eigen-function of the transposed equation, *i.e.*

$$z-\lambda K^* z=0.$$

Then f is orthogonal to z.

Proof. $(f, z)=(y-\lambda Ky, z)$

$=(y, z)-\lambda(Ky, z)$

$=(y, z)-\lambda(y, K^* z)$

$=(y, z-\lambda K^* z)$

$=(y, 0) \quad =0.$

That is to say

$$\int_a^b f(x).\ \bar{z}(x)\,dx=0.$$

Lemma 5. The non-homogeneous equation (a_1) has a unique solution if and only if, the homogeneous equation (a_2) has only the trivial solution.

The proof is easy and is left to the reader.

The results of the abcve discussion are summed up in the following:

Theorem 2. (The Fredholm Alternative).

The integral equation

$$y(x)=f(x)+\lambda\int_a^b K(x,t)\,y(t)\,dt \qquad ...(i)$$

has a unique solution for all $f(x)$, continuous or square integrable, if and only if the associated homogenous equation

$$y(x)=\lambda\int_a^b K(x,t)\,y(t)\,dt \qquad ...(ii)$$

has only the trivial solution ($y\equiv 0$).

The adjoint equation.

$$z(x)=\lambda\int_a^b \bar{K}(t,x)\,z(t)\,dt \qquad ...(iii)$$

has the same number of linearly independent solutions as (*ii*). If this number of solutions of (*ii*) is positive, then (*i*) will have a (nonunique) solution if and only if,

$$(f,z)=\int_a^b f(x)\ \bar{z}(x)\,dx=0$$

for all solutions $z(x)$ of (*iii*).

§10·10. Volterra Integral Equations

In this section we discuss the method of resolvent kernel. Volterra equations of the convolution type including those with weak singularities are best solved by the Laplace transform which we explain in the next chapter.

Given the equation

$$y(x)=f(x)+\lambda\int_a^x K(x,t)\,y(t)\,dt \qquad ...(9)$$

where $K(x,t)$ is continuous in $\leqslant x, t\leqslant b$ and $f(x)$ is continuous in $a\leqslant x\leqslant b$. For consistency it is necessary that $y(a)=f(a)$.

As in the case of Fredholm equations, the iterated kernels in this case are

$$K_1(x,t)=K(x,t)$$

$$K_n(x, t)=\int_t^x K(x, s)\ K_{n-1}(s, t)\ ds \qquad ...(i)$$

and the resolvent kernal is given by

$$R(x, t\ ;\ \lambda)=\sum_{n=1}^{\infty} \lambda^{n-1} K_n(x, t) \qquad ...(10)$$

The successive approximations are given by the recurrence relations.

$$y_0(x)=f(x)$$

$$y_n(x)=f(x)+\lambda\int_a^x K(x, t)\ y_{n-1}(t)\ dt$$

Assuming the convergence of the sequence $\{y_n(x)\}$, the solution of the equation (9) is given by

$$y(x)=f(x)+\lambda\int_a^x R(x, t\ ;\ \lambda)\ f(t)\ dt \qquad ...(n)$$

The iterated kernels satisfy the general formula

$$K_{m+n}(x, t)=\int_t^x K_m(x\ s)\ K_n(s, t)\ ds \qquad ...(ii)$$

The resolvent Kernel $R(x, t\ ;\ \lambda)$ satisfies the integral equation

$$R(x, t\ ;\ \lambda)=K(x, t)+\lambda\int_t^x K(x, t)\ R(s, t\ ;\ \lambda)\ ds \quad ,...(iii)$$

Convergence of sequence $\{y_n(x)\}$

Let $\quad m=\max |f(x)|$

$\quad M=\max |K(x, t)|$.

By induction we can prove tha.

$$|K_n(x, t)| \leqslant \frac{1}{(n-1)!} M^n |x-t|^{n-1} \qquad ...(iv)$$

and $$|y_n(x)-y_{n-1}(x)| \leqslant \frac{mM^n |x-a|^n |\lambda|^n}{n!} (n=1, 2, ...) \qquad ...(v)$$

Since $$y_n(x)=y_0(x)+\sum_{i=0}^{n} \{y_i(x)-y_{i-1}(x)\}$$

$$|y_n(x)| \leqslant m\sum_{i=0}^{n} \frac{|\lambda|^i M^i |x-a|^i}{i!} \leqslant m e^{|\lambda| M |x-a}$$

Hence the sequence $\{y_n(x)\}$ is uniformly convergent on all compact intervals and for all finite value of $|\lambda|$.

If $\lim y_n(x)=y(x)$, then $y(x)$ is the desired solution of (9).

Uniqueness of the solution

Let $z(x)$ be another solution of (9) and $u(x)=y(x)-z(x)$. Then

$$u(x)=\lambda\int_a^x K(x, t)\, u(t)\, dt \qquad ...(vi)$$

If max $|u(x)|=m_0$, then using (vi) we get

$$|u(x)| \leqslant |\lambda|\, m_0 M\, |x-a|.$$

By induction we can further prove that

$$|u(x)| \leqslant \frac{|\lambda|^n M^n\, |x-a|^n}{n!}\, m_0.$$

Hence $|u(x)|\to 0$ as $n\to\infty$ for all finite λ and x.

Without loss of generality we can take $a=0$.

Remark. 4. The above proof clearly brings out the essential difference between Frenholm and Volterra equations of the second kind : whereas Fredholm equations may have eigenvalues, Volterra equations have none.

A Polynomial Kernel in t

Let $$K(x, t)=a_0(x)+a_1(x)\, x-t)+\ldots+\frac{a_{n-1}(x)}{(n-1)!}(x-t)^{n-1} \qquad ...(vii)$$

where the coefficients $a_i(x)$ are continuous on a compact interval $[a, b]$. In what follows we shall suppose $a=0$.

Let $$g(x, t\,;\,\lambda)=\frac{(x-t)^{n-1}}{(n-1)!}+\lambda\int_t^x \frac{(x-s)^{n-1}}{(n-1)!} R(s, t\,;\,\lambda)\, ds \qquad ...(viii)$$

Hence $$[g]_{x=t}=\ldots=\left[\frac{d^{n-2} g}{d x^{n-2}}\right]_{x=t}=0,\ \left[\frac{d^{n-1} g}{d x^{n-1}}\right]_{x=t}=1$$

Differentiating the relation $(viii)$ n times, we get

$$\frac{d^n g}{d x^n}=\lambda\, R(x, t\,;\,\lambda) \qquad ...(ix)$$

Making this replacement for $R(x, t\,;\,\lambda)$ in (iii),

we have

$$\frac{d^n g}{dx^n}=\lambda\, K(x, t)+\lambda\int_t^x K(x, s)\,\frac{d^n g(s, t\,;\,\lambda)}{ds^n} ds$$

$$=\lambda\, K(x, t)+\lambda\Bigg[K(x, s)\frac{d^{n-1} g(s, t\,;\,\lambda}{ds^{n-1}}-\frac{\partial K(x, s)}{ds}\,\frac{d^{n-2}(s, t\,;\lambda)}{ds^{n-2}}+\ldots$$

$$+(-1)^{n-1}\frac{\partial^{n-1} K(x, s)}{\partial s^{n-1}}\, g(x, t\,;\,\lambda)\Bigg]_{s=t}^{s=x}$$

$$=\lambda\Bigg[a_0(x)\,\frac{d^{n-1} g(x, t\,;\,\lambda)}{dx^{n-1}}+a_1(x)\frac{d^{n-2} g(x, t\,;\,\lambda)}{dx^{n-2}}+\ldots$$

$$+a_{n-1}(x)\, g(x, t\,;\,\lambda)\Bigg].$$

The investigation is summarised as follows : If the Kernel $K(x, t)$ is given by (*vii*), then the resolvent Kernel $R(x, t ; \lambda)$ is given by (*ix*) where $g(x, t ; \lambda)$ is the solution of the Cauchy problem :

$$\frac{d^n g}{dx^n}-\lambda\left[a_0(x)\frac{d^{n-1} g}{dx^{n-1}}+a_1(x)\frac{d^{n-2} g}{\lambda x^{n-2}}+\ldots+\lambda_{n-1}(x)\, g\right]=0$$

$$\left[g\right]_{x=t}=\ldots=\left[\frac{d^{n-2} g}{dx^{n-2}}\right]_{x=t}=0,\left[\frac{d^{n-1} g}{dx^{n-1}}\right]_{x=t}=1.$$

Similarly when

$$K(x, t)=b_0(t)+b_1(t)(t-x)+\ldots+\frac{b_{n-1}(t)}{(n-1)!}(t-x)^{n-1},$$

then the resolvent Kernel is

$$R(x, t ; \lambda)=-\frac{1}{\lambda}\frac{d^n g(t, x ; \lambda)}{dt^n}$$

where $g(t, x : \lambda)$ is a solution of the Cauchy problem ;

$$\frac{d^n g}{dt^n}+\lambda\left[b_0(t)\frac{d^{n-1} g}{dt^{n-1}}+\ldots+b_{n-1}(t)\, g\right]=0$$

$$\left[g\right]_{t=x}=\ldots=\left[\frac{d^{n-2} g}{dt^{n-3}}\right]_{t=x}=0,\left[\frac{d^{n-1}g}{dt^{n-1}}\right]_{t=x}=1.$$

Example 10. Find the resolvent Kernels of the following Volterra Kernels.

(*i*) e^{x-t} (*ii*) $x-t$

Solution. (*i*) $K_1(x, t)=e^{x-t}$

$$K_2(x, t)=\int_t^x e^{xs}.\, e^{s-t}\, ds=e^{x-t}(x-t)$$

$$K_3(x, t)=\int_t^x e^{x-s}.\, e^{s-s}(x-s)\, ds$$

$$=e^{x-t}\frac{(x-t)^2}{2!}$$

Hence $R(x, t ; \lambda)=K(x, t)+\lambda K_2(x, t)+\lambda^2 K_3(x, t)+\ldots$

$$=e^{(x-t)}\,.\, e^{\lambda(x-t)}=e^{(1+\lambda)(x-t)}.$$

(*ii*) Here $K(x, t)=(x-t)$.
Hence $n=2, a_1(x)=1$
Let $g(x, t ; \lambda)$ satisfy

$$\frac{d^2g}{dx^2}-\lambda g=0 ; \left[g\right]_{x=t}=0,\left[\frac{dg}{dx}\right]_{x=t}=1.$$

$$g-A e^{\sqrt{\lambda} x}+B e^{-\sqrt{\lambda} x}$$

From the initial conditions

$$A e^{\sqrt{\lambda} x}+e^{-\sqrt{\lambda} t}=0,$$

$$\sqrt{\lambda}(A e^{\sqrt{\lambda} x}-B e^{\sqrt{\lambda} t})=1$$

Hence $2\sqrt{\lambda}\,A\,e^{\sqrt{\lambda}\,t}=1,$

$$-2\sqrt{\lambda}\,B\,e^{-\sqrt{\lambda}\,t}=1.$$

i.e.,
$$A=\frac{1}{2\sqrt{\lambda}}\;e^{-\sqrt{\lambda}\,t}$$

$$B=\frac{-1}{2\sqrt{\lambda}}\;e^{\sqrt{\lambda}\,t}$$

Hence
$$g(x, t\,;\lambda)=\frac{\sin h\,\sqrt{\lambda}\,(x-t)}{\sqrt{\lambda}}$$

$$R\,(x, t\,;\lambda)=\frac{1}{\lambda}\;\frac{d^2g}{dx^2}=\frac{\sinh\sqrt{\lambda}\,(x-t)}{\sqrt{\lambda}}$$

Example 11. Solve the equation

$$y(x)=1+\lambda\int_0^x e^{3(x-t)}\,y(t)\,dt\,.$$

Solution. $K_1=e^{3(x-t)}$

$$K_2\,(x, t)=\int_0^x e^{3(x-t)}\,.\,e^{3(s-t)}\,ds=e^{3(x-t)}\,(x-t)$$

$$K_3\,(x, t)=e^{3(x-t)}\;\frac{(x-t)^2}{2\,!}$$

... ...

$$K_n\,(x, t)=e^{3(x-t)}\;\frac{(x-t)^{n-1}}{(n-1)\,!}$$

Hence
$$R(x, t\,;\lambda)=\sum_{n=1}^{\infty}\lambda^{n-1}\,K_n\,(x, t)$$
$$=e^{(3+\lambda)\,(x-t)}$$

Hence the required solution is

$$y(x)=1+\lambda\int_0^x e^{(3+\lambda)\,(x-t)}\,dt.$$

$$=1-\frac{\lambda}{3+\lambda}\Big[e^{(3+\lambda)\,(x-t)}\Big]_{t=0}^{t=x}$$

$$=\frac{1}{3+\lambda}\Big\{3+\lambda\;e^{(3+\lambda x)}\Big\}\,.$$

In the special case $\lambda=-3$, $R(x, t\,;\lambda)=1$ and $y(x)=1-3x$.

Exercise 33

1. Find the resolvent Kernels of the following Volterra Kernels :

(*i*) $K(x, t)=\dfrac{1+x^2}{1+t^2}$ (*ii*) $\dfrac{2+\cos x}{2+\cos t}$

(iii) $\dfrac{\cosh x}{\cosh t}$ (iv) $\cos(x-t)$

(v) $\sinh(x-t)$.

2. Solve the Volterra equations :

(i) $y(x)=e^x+\lambda \int_0^x (x-t)\, y(t)\, dt$

(ii) $y(x)=e^x+\lambda \int_0^x e^{x-t}\, y(t)\, dt$

(iii) $y(x)=1-2x-\int_0^x e^{-x^2-t^2} y(t)\, dt$

(iv) $y(x)=e^{x^2}+\int_0^x e^{x^2-t^2}\, y(t).dt$

(v) $y(x)=x\, a^x+\int_0^x a^{x-t}\, y(t)\, dt$

10·11. Approximate Methods

The aim of this brief section is to expose the reader to some more common numerical procedures which are suited to the approximate solution of integral equations. In fact the absence of well-laid rules of solving non-linear integral equations compels us to take recourse to such techniques. We have merely outlined the technical part of the procedures without giving the relevent error analysis. Before we outline these procedures, we explain the idea behind approximate methods specially in the context of operator equations.

Given a function $f(x)$ and a linear space V of functions $\phi_i(x)$ which are simpler to deal with. We approximate the function $f(x)$ by a finite sum

$$\tilde{f}(x)=a_1\, \phi_1(x)+\cdots+a_n\phi_n(x)$$

The constants a_i are determined by the constraints we impose on $\tilde{f}(x)$. A given set of constraints will determine a particular element from the linear span $[\phi_1,\ldots,\phi_n]$. A well known example is the Fourier series expansion for $f(x)$ where an infinite sum

$$\sum_{k=0}^{\infty} (a_k \cos kx+b_k \sin kx)$$

is involved. In general, the functions ϕ_i can be powers of x, polynomials in x, piecewise polynomials in x, trigonometric functions or other orthogonal functions. The choice of the functions ϕ_i will depend on the nature of $f(x)$. For example if $f(x)$ is continuous, then ϕ_i should be continuous. Thus the smoothness constrains on ϕ_i will depend on the smoothness constraints (continuity, differentiability, etc) we expect $\tilde{f}(x)$ to satisfy. This may entail ϕ_i to be splines and piecewise Hermite polynomials etc.

We shall be concerned here with the following types of constraints on $\tilde{f}(x)$:

I. Interpolatory constraints

$\tilde{f}(x_i)=f(x_i)$, $i=1,\ldots,n$ at the n knots $x_1, \ldots, x_n$ of a fundamental interval $[a, b]$. These n constraints determine $a_1, \ldots, a_n$.

II. Orthogonality constraints

If $E(x)=f(x)-\tilde{f}(x)$, then to determine a_i we require $E(x)$ to be orthogonal to n functions $\psi_1, \ldots, \psi_n$ on the fundamental interval $[a, b]$, *i.e.*

$$(f-\tilde{f}, \psi_i)=\int_a^b \{f(x)-\tilde{f}(x)\}\psi_i\, x\, dx$$

$$=0 \qquad (i=1,\ldots n)$$

Usually we take $\psi_i=\phi_i$.

In the context of operator equations with interpolatory constrains we shall study the *method of collocation* and with orthogonality constraints we shall study the *Galerkin's method.*

(*a*) Collocation

Given on operator equation

$$L(y)=f(x) \qquad \ldots(1)$$

where L is a suitable differential or integral operator. We will discuss the application of the method to a linear Fredholm equation of the first kind. Extension to an equation of the second kind is direct. We shall also apply the method to a two-point boundary-value problem.

Fredholm Equation

Given an equation

$$\int_a^b K(x, t)\, y(t)\, dt=f(x) \qquad \ldots(2)$$

Put $$E(y, x)=\int_a^b K(x, t)\, y(t)\, dt-f(x) \qquad \ldots(3)$$

If $y(x)=\phi(x)$ is a solution of (2) then

$$E(\phi, x)=0$$

Let $x_1, x_2, \ldots, x_n$ be a set of distinct points (called *knots*) on the interval $[a, b]$. In ordinary interpolation we match functional values at prescribed knots. In the collocation method we match the values of the image $L(y)$ with those of the right-hand side (*i.e.* $f(x)$) at the knots.

We now choose a suitable linear space V from which the approximating functions will be drawn. If L is a differential operator of order n, V may be the space of n times differentiable function on $[a, b]$. If L is an integral operator, V may be L^1 or L^2.

The domain of the operator L may be a subspace of V and the range a subspace of V.

Let $S=\{\phi_1, \phi_2, ..., \phi_n\}$ be a linearly independent subset of V. Assume

$$\bar{y}(x)=a_1\,\phi_1(x)+...+a_n\,\phi_n(x) \qquad ...(4)$$

to be an approximation of $y(x)$ such that $L\{\bar{y}(x)\}$ interpolates to $f(x)$ at $x_1, ...\, x_n$ *i.e.* $E(\bar{y}, x)$ vanishes at these knots. Thus,

$$L\{\sum_{j=1}^{n} a_j\,\phi_j(x_i)\}=f(x_i) \qquad (i=1, ...n) \qquad ...(5)$$

Since the functions $\phi_1,...,\phi_n$ are linearly independent the matrix of coefficient $[\phi_j(x_i)]_{n\times n}$ is non-singular and the linear algebraic system (5) has a unique solution. The expression $\bar{y}(x)=a_1\phi_1+...+a_n\phi_n$ is said to collocate $y(x)$ at the knots $x_1,..., x_n$. For better approximation the choice of S, the number of knots and their optimal placing will have to be considered. Collocation by powers of x viz. $1, x..., x^{n-1}$ or trigonometric or ordinary polynomials leads to full matrices, while collocation by piecewise polynomials (*e.g.*, piecewise Hermites, splines etc) leads to banded matrices.

The collocation method has the same infirmities as does ordinary interpolation. The exact matching of the solution at certain preassigned stations does not ensure boundedness of the error at other points.

Example 12. Find a two-term and a three-term appoximation by the collocation method for the solution of the integral equation

$$y(x)=1-\int_0^1 K(x, t)\,y(t)\,dt$$

where $K(x, t)=\begin{cases} x & x\leqslant t \\ t & x\geqslant t \end{cases}$

Solution. Two-term Approximation

Take knots $x_1=0$, $x_2=1$ and

$\bar{y}_1(x)=a_0+a_1 x$ the approximation.

The equation can be written as

$$y(x)+\int_0^1 K(x, t)\,y(t)\,dt=1$$

If $\bar{y}(x)$ replaces $y(x)$, then at $x=0, 1$ the values of left-hand side will match the values of the right-hand side.

$$\text{Now}\int_0^1 K(x, t)\,\bar{y}_1(t)\,dt$$

$$=\int_0^x t\,(a_0+a_1 t)\,dt+\int_x^1 x\,(a_0+a_1 t)\,dt$$

$$=\left(a_0+\frac{a_1}{2}\right)x-\frac{a_0}{2}x^2-\frac{a_1}{6}x^3$$

Hence, $\bar{y}_1(x)+\int_0^1 K(x,t)\,\bar{y}_1(x)\,dx$

$$=a_0+\left(a_0+\frac{3}{2}a_1\right)x-\frac{a_0}{2}x^2-\frac{a_1}{6}x^3$$

Putting $x=0, 1$ we get $a_0=1$

and $a_1=-\frac{3}{8}$

Hence, $\bar{y}_1(x)=1-\frac{3}{8}x.$

Three-term Approximation

Take knots $x_1=0$, $x_2=\frac{1}{2}$, $x_1=1$

and $\bar{y}_2(x)=b_0+b_1x+b_2x^2.$

Here $\int_0^1 K(x,t)\,\bar{y}_2(t)\,dt=\int_0^x t(b_0+b_1t+b_2t^2)\,dt$

$$+\int_x^1 x(b_0+b_1t+b_2t^2)\,dt$$

$$b_0+b_1x+b_2x^2+\int_0^1 K(x,t)\,\bar{y}_2(t)\,dt$$

$$=b_0\left(1+x-\frac{x^2}{2}\right)+b_1\left(\frac{3}{2}x-\frac{x^3}{6}\right)+b_2\left(\frac{x}{3}+x^2-\frac{x^4}{12}\right)$$

Matching values at $x=0, \frac{1}{2}, 1$ we have

$$b_0=1$$

and
$$\frac{11}{8}+\frac{35}{48}b_1+\frac{79}{192}b_2=1$$

$$\frac{3}{2}+\frac{4}{3}b_1+\frac{5}{4}b_2=1$$

Hence, $b_0=1$, $b_1=-\frac{303}{418}$; $b_2=\frac{78}{209}$ and finally

$$\bar{y}_2(x)=1-\frac{303}{418}x+\frac{78}{209}x_2$$

It is interesting to compare these approximate solutions with the exact solution

$$y(x)=\frac{e^x+e^{2-x}}{e^2+1}$$

At the knot $x=1$, $\bar{y}_1(1)=0{\cdot}625$, $\bar{y}_2(1)=0{\cdot}648$

Evidently $\bar{y}_2(1)$ is nearer to the actual value

$$y(1)=\frac{2e}{e^2+1}=0{\cdot}645.$$

Problem. Convert the integral equation to the boundary-value Problem

$$y''-y=0\ ;\ y(0)=1,\ y'(1)=0$$

and deduce $y(x)=\dfrac{e^x+e^{2-x}}{e^2+1}$.

Example 13. Find a two-term and a three-term collocation approximation to the boundary-value problem

$$L(y)\equiv y''-y=1\ ;\ y(0)=y(1)=0.$$

Compare with the exact solution.

Solution.

Two-term Approximation. Let $x_1=0,\ x_2=1.$

Take $\varphi_1(x)=x(x-1),\ \varphi_2(x)=x^2(x-1)$. Then φ_1, φ_2 belong to the linear manifold of twice differentiable functions satisfying the boundary conditions. Let

$$\bar{y}_1(x)=a_1\varphi_1(x)+a_2\varphi_2(x)$$

be the desired approximation.

$L\phi_1(x)=2-x(x-1),\ L\varphi_2(x)=6x-2-x^2(x-1).$

Hence, $L\varphi_1(0)=2,\ L\varphi_2(0)=-2$

$L\varphi_1(1)=2,\ L\varphi_2(1)=4$

$L\bar{y}_1(x)=a_1 L\phi_1(x)+a_2 L\varphi_2(x)$

Matching values of $L\bar{y}_1(x)$ at $x=0, 1$ with those of the right-hand function

$$2a_1-2a_2=1$$
$$2a_1+4a_2=1.$$

Hence, $a_1=\frac{1}{2},\ a_2=0$. Thus finally

$$\bar{y}_1(x)=\tfrac{1}{2}x(x-1)$$

The exact solution is

$$y(x)=\frac{e^x+e^{1+x}}{e+1}-1$$

At $x=\frac{1}{2}$ $\bar{y}_1(\frac{1}{2})=-0{\cdot}125$

$y(\frac{1}{2})=-0{\cdot}1132$

Error at $x=\frac{1}{2}$ is $y(\frac{1}{2})-\bar{y}(\frac{1}{2})=0{\cdot}0118$

Three-term Approximation

Let $x_1=0,\ x_2=\frac{1}{2}, x_3=1$ and

$$\phi_1(x)=x(x-1),\ \varphi_2(x)=x^2(x-1),\ \varphi_3(x)=x^2(x^2-1)$$

Then $L\phi_1=2-x(x-1),$

$L\phi_2=6x-2-x^2(x-1),\ L\phi_3=12x^2-2-x^2(x^2-1)$

$L\varphi_1(0)=2,\ L\varphi_2(0)=-2,\ L\varphi_3(0)=-2$

$$L\ \phi_1\left(\tfrac{1}{2}\right)=\frac{9}{4},\quad L\ \phi_2\left(\tfrac{1}{2}\right)=\frac{9}{8},\quad L\ \phi_3\left(\tfrac{1}{2}\right)=\frac{19}{16}$$

$$L\ \phi_1(1)=2,\quad L\ \phi_2(1)=4,\quad L\ \phi_3(1)=10$$

Let $\bar{y}_2=b_1\ \varphi_1(x)+b_2\ \phi_2(x)+b_3\ \phi_3(x)$

Matching values of $L\ \bar{y}_2(x)$ at $x=0, \frac{1}{2}, 1$,

$$2a_1-2a_2-2a_3=1$$

$$\frac{9}{4}a_1+\frac{9}{8}a_2+\frac{19}{16}a_3=1$$

$$2a_1+4a_2+10a_3=1$$

Hence, $a_1=\dfrac{23}{38},\ a_2=\dfrac{4}{19},\ a_3=-\dfrac{2}{19}$

Thus $\bar{y}_2(x)=\dfrac{33}{38}x(x-1)+\dfrac{4}{19}x^2(x-1)-\dfrac{2}{19}x^2(x^2-1).$

$$\bar{y}_2\left(\tfrac{1}{2}\right)=-\frac{3}{19}=-0{\cdot}158.$$

(b) Galerkin's Method

Given a Fredholm equation of the first kind

$$\int_a^b K(x,t)\ y(t)\ dt=f(x) \qquad \text{...(2)}$$

As before let

$$E(y,x)=\int_a^b K(x,t)\ y(t)\ dt-f(x) \qquad \text{...(3)}$$

$y(x)$ will be a solution of (2) if for all functions $\psi(x)$

$$\int_a^b E(y,x)\ \psi(x)\ dx=0.$$

If for a finite number of linearly independent function $\psi_j(x)$,

$$\int_a^b E(\bar{y},x)\ \psi_j(x)\ dx=0. \qquad \text{...(4)}$$

then $\bar{y}(x)$ is an approximation to $y(x)$ in some sense. The more the number of such functions ψ_j to which $E(\bar{y},x)$ is orthogonal, the better the approximatian $\bar{y}(x)$.

Assume $\bar{y}(x)=a_1\ \phi_1(x)+\dots+a_n\ \phi_n(x)$ to be an approximation, where $\phi_1,\dots,\phi_n$ are linearly independent functions. In practice the sets of functions ψ_j and ϕ_i are orthonormal. But orthogonalily is not essential. From (4)

$$\sum_{j=1}^{n} a_i \int_a^b\int_a^b K(x,t)\ \phi_i(t)\ \psi_j(x)\ dt\ dx$$

$$=\int_a^b f(x)\ \psi_j(x)\ dx \qquad j=1,\dots n),$$

which is of the form

$$\sum_{i=1}^{n} c_{ij}\, a_i = f_j \qquad (j=1, 2, \ldots, n) \qquad \ldots(5)$$

The system (5) can be solved for $a_1 \cdots, a_n$,

Example 14. Use Galerkin's method to find a two-term approximation to the solution of the integral equation

$$y(x)+\int_0^1 K(x, t)\, y(t)\, dt=1$$

where $K(x, t)=\begin{cases} x & x \leqslant t \\ t & x > t \end{cases}$

Solution. Let $\bar{y}(x)=a_0+a_1 x$ be an approximation.

Then,

$$\bar{y}(x)+\int_0^1 K(x, t)\, \bar{y}(t)\, dt$$

$$=a_0+\left(a_0+\frac{3}{2}a_1\right)x-\frac{a_0}{2}x^2-\frac{a_1}{6}x^3$$

Put,

$$E(\bar{y}, x)=\bar{y}(x)+\int_0^1 K(x, t)\, \bar{y}(t)\, dt-1$$

For $\bar{y}(x)$ to be an approximation we require

$E(\bar{y}, x)$ to be orthogonal to 1 and x, *i.e.*,

$$\int_0^1 E(\bar{y}, x)\, dx=0,$$

$$\int_0^1 E(\bar{y}, x)\, x dx=0.$$

These equations lead to

$$\frac{4}{3}a_0+\frac{17}{24}a_1=1$$

$$\frac{17}{24}a_0+\frac{7}{15}a_1=\tfrac{1}{2}.$$

Solving for a_0, a_1 we have

$$a_0=\frac{324}{347}, \quad a_1=-\frac{2040}{5899}$$

Hence $\quad \bar{y}(x)=\dfrac{324}{347}-\dfrac{2040}{5899}x.$

For $x=0$

$$\bar{y}(0)=\frac{324}{349}=0{\cdot}933 \text{ approx}).$$

For $x=1$ $\bar{y}(1)=0{\cdot}933-0{\cdot}345=0{\cdot}59$ (approx.)

The actual values are $y(0)=1$ and $y(1)=0{\cdot}645$.

Exercise 34

1. Find an approximate solution of the form a_0+a_1x by the collocation method to the integral equation

$$\int_0^1 e^{xt}\, y(t)dt = \frac{e^{x+1}-1}{x+1}$$

What is the exact solution ?

2. Apply Galerkin's method to find a two-term approximation to the solution of the equation.

$$y(x)-\int_0^1 K(x, t)\, y(t)\, dt=x \qquad 0\leqslant x\leqslant 1$$

where $$k(x, t)=\begin{cases} x & x\leqslant t \\ t & x\geqslant t \end{cases}$$

3. Use the collocation method to find an approximation of the form $a_1+a_0x^2$ to the solution of the integral equation

$$\int_0^1 e^{-x^2t^2}\, y(t)\, dt=2(1-x^2)$$

4. Use Galerkin's method to find a two-term approximation to the integral equation

$$y(x)=x+\int_0^1 K(x, t)\, y(t)dt$$

where $$K(x, t)=\begin{cases} x(1-t) & x\leqslant t \\ t(1-x) & x>t \end{cases}$$

5. Use the collocation method to find an approximation of the form $y(x)=a_1\varphi_1(x)+a_2\varphi_2(x)$ to the two-point boundary value problem

$$y''+y+1=0,\ y(0)=y(1)=0,$$

where $$\phi_1(x)=x(x-1),\ \phi_2(x)=x^2(x-1)$$

6. Use the collocation method to find an approximate solution of the form $a_0+a_1x+a_2x^2$ to the equation

$$y(x)=x+\int_0^1 K(x, t)\, y(t)dt$$

where $$K(x, t)=\begin{cases} x & x\leqslant t \\ t & x\geqslant t \end{cases}$$

Choose knots 0, $\frac{1}{2}$, 1.

7. Assuming a solution of the form

$$y(x)=a_0+a_1x+a_2x^2$$

Use the collocation method with vanishing error at the points $x=0, \frac{1}{2}, 1$ to obtain a solution of the equation.

$$y(x)=x+3\int_0^1 (xt-x^3t^3)\,y(t)dt$$

What is the exact solution ? Compare the values of $\bar{y}(x)$ and $y(x)$ at 0, $\frac{1}{2}$, 1.

§10.12. Symmetric Kernels : Hilbert-Schmidt Theory

In sections 8·9 and 9·15 we saw that the Green's function $G(x, t)$ of a self-adjoint boundary-value problem can be expressed as a bilinear expansion in terms of the eigenfunctions of the problem. The boundary-value problem can be converted into an equivalent integral equation of Fredholm type in which the Green's function occurs as kernel. As a kernel the Green's function possessed, inter alia, the following properties :

(*a*) $G(x, t)=G(t, x)$ *i.e.* G is a real symmetric kernel.

(*b*) $G(x, t)$ is positive definite inside the domain and vanishes on the boundary. This means that for $\phi\in L^2$

$$\int_D\int \phi(x)\,\phi(t)\,G(x, t)\,dx\,dt\geqslant 0$$

$G(x, t)$ is negative definite if

$$\int_D\int \phi(x)\,\phi(t)\,G(x, t)dx\,dt\leqslant 0.$$

The property of $G(x, t)$ being positive or negative in the domain depends on the sign of the coefficient of the highest derivative in the differential operator involved in the BVP.

In sections 10·8 and 10·9 we have studied properties of eigen values and eigenfunctions of Hermitian (or Hermitian symmetric kernels. In this section we further study expansions of real sym metric Kernels $K(x, t)$ of the Green's function type *i.e.*, for which $K(x, t)=K(t, x)$.

That part of Fredholm theory which involves integral operators generated by real symmetric kernels is referred to as the *Hilbert-Schmidt theory* of integral equations. Owing to the richness of its results, the theory has attracted extensive attention of those interested in practical applications of integral equations as well as those interested in abstract theory, specially functional analysis. Here we are going to focus our attention to those aspects of the theory which constitute the interface between differential equations and integral equations.

As in the previous section we denote the integral operator generated by the kernel $K(x, t)$ by

$$Kf(x)=\int_a^b K(x, t)\,f(t)\,dt \qquad \ldots(1)$$

The *adjoint* (or *conjugate*) operator is given by

$$K^* g(x)=\int_a^b \overline{K}(t, x)\, g(t)\, dt \qquad ...(2)$$

When K(x, t) is real, $\overline{K}(t, x)=K(t, x)$ and is called the *transposed* kernel of K(x, t).

In the Hilbert space L^2, inner product is defined by

$$(f, g)=\int_a^b f(t)\,\overline{g}(t)\, dt$$

and the norm of an element $f\in L^2$ is given

$$\|f\|=(f, f)^{1/2}=\left\{\int_a^b |f|^2\, dt\right\}^{1/2}$$

In terms of of inner product (1) can be written as

$$K f(x)=(K(x, t), f(t))_t \qquad (3)$$

where the suffix t in the bracket indicates the variable of integration in the integral defining inner product. The suffix will not be used when no ambiguity arises. By lemma 3, through inner product the opera- tors K and K* are related by

$$(Kf, g)=(f, K^* g) \qquad ...(4)$$

If K is self-adjoint, then $K=K^*$

i.e., $(K f, g)=(f, Kg)$

i.e., $K(x, t)=K(t, x)$

i.e., K(x, t) is symmetric. The operator K generated by a symmetric kernel is called a *Hilbert-Schmidt operator.*

An operator equation of the type

$$L(y)=\mu y \qquad ...(5)$$

where L is a linear operator is called an *eigenvalue problem*. For example, in an eigenvalue problem of Sturm-Liouville type L is a self-adjoint differential operator.

We saw that an application of theorem 1 in chapter 8 converted a Sturm-Liouville differential equation

$$L(y)=(py')'+qy=\lambda y \qquad ...(6)$$

with homogeneous boundary conditions, into a Fredholm equation

$$y(x)=\lambda \int_a^b G(x, t)\, y(t) \qquad ...(7)$$

where the Green's function G(x, t) of (6) is a real symmetric Kernel. Now in the notation of (1), (7) can be written as

$$y(x)=\lambda\, Ky \qquad ...(8)$$

where $$Ky=\int_a^b G(x, t)\, p(t)dt$$

An eigenvalue problem of Sturm-Liouville type is thus equivalent to that for a symmetric Kernel.

A glance at the the manner in which the eigenvalue problems (5) and (8) are written, necessitates a differentiation in the nomenclature for the parameters μ and λ. A value μ_n for which a non-zero function $y_n(x)$ exists such that

$$L(y_n)=\mu_n\, y_n$$

is called an *eigenvalue* of (5) and $y_n(x)$ the corresponding *eigen function*. For equation (8) to have any sense as an integral equation the value $\lambda=0$ has to be discarded. A value λ_n for which a non-zero function $\phi_n(x)$ exists such that

$$\phi_n(x)=\lambda_n\, K\phi_n$$

is called a *chareteristic value* (or a *characteristic number*) and the corresponding function $\varphi_n(x)$ an *eigen function*. An eigenvalue may thus be referred to as a reciprocal characteristic value. Since $\lambda\neq 0$ the null space of the integral operator K in (8) consists of the null function alone.

The Schwarz inequality for integrals gives

$$\left|\int_a^b f(t)\, g(t)\right|^2 \leqslant \int_a^b |f(t)|^2\, dt \,.\, \int_a^b |g(t)|^2\, dt \qquad \ldots(10)$$

which in the L^2 inner product can be written

$$|(f, g)| \leqslant \|f\| \,.\, \|g\| \qquad \ldots(10)$$

Then $$|Kf(x)| \leqslant \|f\| \,.\left[\int_a^b |K(x, t)|^2\, dt\right]^{1/2} \qquad \ldots(11)$$

Definition 1 (Positive Kernel) A continuous kernel $K(x, t)$ is said to be positive if $K(x, t)>0$.

For example the Green's function

$$G(x, t)=\begin{cases} x(t-1) & x\leqslant t \\ t(x-1) & x\geqslant t \end{cases}$$

of the BVP : $L(y)=-y''=0$; $y(0)=y(1)=0$ is a positive kernel in the domain $0<x, t<1$.

Definition 2. (Positive Operator) A self-adjoint operator K acting on a Hilbert space H is said to be positive if $(Kf, f)\geqslant 0$ for all $f\in H$.

Positive operators should not be confused with positive Kernels. The former are necessarily self-adjoint and need not be generated by positive kernels. Positive kernels have at least one positive characteristic value.

If (Kf, f) is real and K is bounded, then K is self-adjoint. Thus a bounded positive operator is self-adjoint.

We may recall that eigenfunctions corresponding to distinct characteristic values are orthogonal. Also any set of linearly independent eigenfunctions associated with a characteristic value may be orthogonalised.

Definition 3. (Fourier Expansions) If f is a continuous function and $\{\phi_i\}$ an orthonormal system, then the numbers

$$c_i=(f, \phi_i)=\int_a^b f(x)\, \phi_i(x)dx \qquad (i=1, 2,......)$$

are called *Fourier functions* of f with respect to the system $\{\varphi_i\}$ and

$$f=\sum_{i=1}^{\infty} (f, \phi_i)\, \phi_i$$

is called its *Fourier expansion.*

Lemma 6. (Bessel's Inequality)

$$\sum_{i=1}^{\infty} | c_i |^2=\sum_{i=1}^{\infty} | (f, \phi_i) |^2 \leqslant \| f \|^2$$

Proof. Put $g_n=f-\sum_{i=0}^{n} (f, \phi_i)\phi$

Since the system $\{\phi_i\}$ is orthonormal

$$0 \leqslant \| g_n \|^2= \| f \|^2-\sum_{i=1}^{n} | (f, \phi_i) |^2$$

i.e.,

$$\sum_{i=1}^{n} | (f, \phi_i) |^2 \leqslant \| f \|^2$$

Since n is arbitrary, the lemma follows. **Q.E.D.**

Since L^2 is a separable Hilbert space, the set of distinct orthonormalised eigen functions is countable and to each characteristic value there are associated a finite number of eigen functions.

Let $$y(x)=\lambda K y=\lambda\int_a^b K(x, t)\, y(t)\, dt \qquad ...(12)$$

where $K(x, t)$ is a real continuous symmetric kernel and K is a positive integral operator. Since the characteristic values are real, we can arrange them in order of their magnitude underneath their related eigenfunctions as follows :

$$\phi_1(x),\ \varphi_2(x), ...,\ \phi_n(x),... \qquad ...(13)$$

$$\lambda_1,\ \lambda_2, ...,\ \lambda_n,... \qquad ...(14)$$

$$|\lambda_1| \leqslant |\lambda_2| \leqslant ...... \leqslant |\lambda_n| \leqslant ...,\ \lim_{n\to\infty} |\lambda|_n=\infty.$$

The sequence $\{\varphi_n(x)\}$ is *maximal* (or *complete*) in the sense that the sequence (14) contains all characteristic value λ_n, there corresponds a maximum number of linearly independent eigen functions associated with λ_n in the sequence (13). Thus though two members of the sequence (13) may be equal, a member of (14) may be fin tely repeated.

For a complete orthonormal system (13) if

$$f=\sum_{i=1}^{\infty}(f, \phi_i)\,\phi_i,$$

then *Parseal's equality* $\Sigma \mid (f, \varphi_i) \mid ^2 = \| f \| ^2$ holds.

Definition 4. Given a sequence of function $\{f_n(x)\}$ and a function $f(x)$ in L^2 defined on an interval I. We say that the sequence $\{f_n)$ *converges uniformly on* I if

$$\sup_{x\in I} \mid f_m(x)-f_n(x) \mid \to 0 \text{ as } m, n\to\infty$$

The sequence *converges uniformly to* $f(x)$

$$\sup_{x\in I} \mid f(x)-f_n(x) \mid \to 0 \text{ as } n\to\infty.$$

The sequence *converges in the mean* on $[a, b]$ if

$$\int_a^b \mid f_m(x)-f_n(x) \mid ^2 dx\to 0 \text{ as } m, n\to\infty.$$

It converges in the mean to $f(x)$ if

$$\lim_{n\to\infty}\int_a^b \mid f(x)-f_n(x) \mid ^2 dx=0.$$

On a finite closed domain uniform convergence implies convergence in the mean. The converse is not true. For example on the open interval (0, 1) the sequance $\{e^{-nx})$ converges in the mean, but not uniformly.

Definition 5. A series of functions

$$S=\sum_{i=1}^{\infty} f_i(x) \qquad x\in I$$

ccnverges uniformly (or in the mean) to $F(x)$ if the sequence of partial sums

$$S_n(x)=\sum_{i=n}^{n+p} f_i(x)$$

converges uniformly (or in the mean) to $F(x)$.

Definition 6. The series S is absolutely convegent if the series

$$\sum_{i=1}^{\infty} | f_i(x) | \quad \text{is pointwise convergent.}$$

Lemma 7. Let the Fredholm equation (12) have the orthonormalised sequence of eigenfunctions (13) with characteristic values (14). Then the shortened kernel.

$$K_1(x, t)=K(x, t)-\frac{\varphi_1(x)\,\varphi_1(t)}{\lambda_1} \qquad \text{...(15)}$$

generates the Fredholm equation

$$y(x)=\lambda \int_a^b K_1(x, t)\, y(t)\, dt \qquad \text{...(16)}$$

with eigenfunction sequence

$$\phi_2(x),\ \varphi_3(x),\ldots,\ \varphi_n(x),\ldots$$

$$\lambda_2,\quad \lambda_3,\ldots,\ \lambda_n,\ldots\ldots$$

Proof. We first prove that any eigenfunction $\phi(x)$ with characteristic value λ of the kernel $K_1(x, t)$ is also an eigenfunction of the kernel $K(x, t)$ with the same characteristic value λ. Indeed let

$$\phi(x)=\lambda \int_a^b K_1(x, t)\, \phi(t)\, dt \qquad \text{...}(i)$$

$$\text{Then} \int_a^b \phi(x)\, \varphi_1(x)\, dx=\lambda \int_a^b \phi_1(x) \left[\int_a^b K_1(x, t)\, \phi(t)\, dt\right] dx$$

$$=\lambda \int_a^b \phi_1(x) \left[\int_a^b \left\{K(x, t)-\frac{\phi_1(x)\,\phi_1(t)}{\lambda_1}\right\} \phi(t)\, dt\right] dx$$

$$=\lambda \int_a^b \int_a^b K(x, t)\, \phi_1(x)\, \varphi(t)\, dt\, dx$$

$$-\frac{\lambda}{\lambda_1} \int_a^b \phi_1(t)\, \phi(t) \left[\int_a^b \phi_1^2(x)\, dx)\right] dt$$

$$=\frac{\lambda}{\lambda_1} \int_a^b \phi_1(t)\, \varphi(t)\, dt$$

$$-\frac{\lambda}{\lambda_1} \int_a^b \varphi_1(t)\, \phi(t)\, dt$$

$$=0.$$

Hence $\phi \neq \phi_1$.

Also $\phi(x)=\lambda \int_a^b K_1(x, t)\, \phi(t)\, dt$

$$=\lambda\int_a^b\left[K(x,t)-\frac{\varphi_1(x)\,\phi_1(t)}{\lambda_1}\right]\phi(t)\,dt$$

$$=\lambda\int_a^b K(x,t)\,\phi(t)\,dt-\frac{\lambda}{\lambda_1}\phi_1(x)\int_a^b \phi_1(t)\,\varphi(t)\,dt$$

$$=\lambda\int_a^b K(x,t)\,\phi(t)\,dt.$$

Thus $\phi_i(x)$ is an eigenfunction of the kernel $K(x, t)$ other than $\varphi_1(x)$ with the same characteristic value λ.

Conversely, let $\phi_i(x)$ be an eigenfunction with characteristic value λ_i of the kernel $K(x, t)$ such that $i>1$. Then we prove that $\varphi_i(x)$ is also an eigenfunction of $K_1(x, t)$ with the same characteristic value λ_i. Indeed let,

$$\varphi_i(x)=\lambda_i\int_a^b K(x,t)\,\phi_i(t)\,dt \quad i>1.$$

Then since $\int_a^b \phi_1(x)\,\varphi_i(x)\,dx=0$,

$$\varphi_i(x)=\lambda_i\int_a^b K(x,t)\,\phi_i(t)\,dt$$

$$=\lambda_i\int_a^b\left[K_1(x,t)+\frac{\phi_1(x)\,\varphi_1(t)}{\lambda_1}\right]\phi_i(t)\,dt$$

$$=\lambda_i\int_a^b K_1(x,t)\,\phi_i(t)\,dt$$

i.e., if $i>1$, then $\varphi_i(x)$ is an eigenfunction of $K_1(x, t)$ with charactersitic value λ_i. **Q.E.D.**

Corollary 1. *n* iterations of the process would result in the kernel

$$K_n(x,t)=K(x,t)-\sum_{k=1}^{n}\frac{\varphi_k(x)\,\varphi_k(t)}{\lambda_k}$$

In case $K_n(x, t)=0$, then

$$K(x,t)=\sum_{k=1}^{n}\frac{\phi_k(x)\,\phi_k(t)}{\lambda_k}$$

and the kernel is degenerate.

In case the process does not terminate, there are an infinite number of eigenfunctions and questions of convergence of the bilinear series $\sum_{k=1}^{\infty}\frac{\varphi_k(x)\,\varphi_k(t)}{\lambda_k}$ arise.

Theorem 3. If $K(x, t) \in L^2$, $a \leqslant x, t \leqslant b$ then the series

$$\sum_{k=1}^{\infty} \frac{\varphi_k(x)\,\varphi_k(t)}{\lambda_k}$$

converges in the mean to $K(x, t)$.

Proof. Since $K \in L^2$, the integrals

$$\int_a^b \int_a^b K^2(x, t)\, dx.\, dt, \quad \int_a^b K^2(x, t)\, dx$$

exist and are bounded.

If $$K(x, t) = \sum_{k=1}^{\infty} a_k\, \varphi_k(t)$$

is the Fourier expansion of $K(x, t)$ in t, then,

$$a_k = \int_a^b K(x, t)\, \phi_k(t)\, dt$$
$$= \frac{\phi_k(x)}{\lambda_k}.$$

Hence formally

$$K(x, t) = \sum_{k=1}^{\infty} \frac{\varphi_k(x)\,\varphi_k(t)}{\lambda_k}.$$

Now $\{\phi_i\}$ is a complete orthonormal system. Hence Parseval's equality holds and

$$\sum_{k=1}^{\infty} \left(\frac{\phi_k(x)}{\lambda_k} \right)^2 = \int_a^b K^2(x, t)\, dt \text{ for all } x.$$

Integrating both sides with respect to x we get

$$\sum_{k=1}^{\infty} \frac{1}{\lambda^2_k} = \int_a^b \int_a^b K^2(x, t)\, dx\, dt \leqslant \infty$$

Using the orthonormality of ϕ_i

$$\int_a^b \int_a^b \left[K(x, t) - \sum_{k=1}^{n} \frac{\phi_k(x)\,\phi_k(t)}{\lambda_k} \right]^2 dx\, dt$$

$$=\int_a^b \int_a^b K^2(x, t)\, dx\, dt - \sum_{k=1}^{n} \frac{1}{\lambda_k^2}$$

$\to 0$ as $n \to \infty$. **Q.E.D.**

Definition 7. The kernel $K(x, t)$ of a symmentic operator is *positive definite* if $K(x, x) > 0$ and all characteristic values $\lambda_k > 0$.

Theorem 4. (Mercer's Theorm). If $K(x, t)$ is a positive definite continuous Kernel with orthonormal eigenfunctions ϕ_k, then

the series $\displaystyle\sum_{k=1}^{\infty} \frac{\phi_k(x)\, \phi_k(t)}{\lambda_k}$ ..(*i*)

converges to $K(x, t)$ absolutely and uniformly.

Proof. Let $K_n(x, t) = K(x, t) - \displaystyle\sum_{k=1}^{n} \frac{\phi_k(x)\, \phi_k(t)}{\lambda_k}$.

Then $K_n(x, t)$ is also a positive definite kernel.

Hence $K_n(x, x) = K(x, x) - \displaystyle\sum_{k=1}^{n} \frac{\varphi_n^2(x)}{\lambda_k} \geqslant 0$

i.e., $$\sum_{k=1}^{n} \frac{\varphi_k^2(x)}{\lambda_k} \leqslant K(x, x)$$

Since $K(x, t)$ is continuous in the closed domain $a \leqslant x,\ t \leqslant b$, $K(x, t)$ is bounded. Hence

$$\sum_{k=1}^{n} \frac{\varphi_k^2(x)}{\lambda_k} \text{ is bounded for all } n.$$

Hence the series of positive terms

$$\sum_{k=1}^{\infty} \frac{\varphi_k^2(x)}{\lambda_k} \text{ is convergent for all } x.$$

Also by Schwarz inequality

$$\left[\sum_{k=m}^{m+p} \left| \frac{\varphi_k(x)\, \varphi_k(t)}{\lambda_k} \right| \right]^2 < \sum_{k=m}^{m+p} \frac{\varphi_k^2(x)}{\lambda_k} \cdot \sum_{k=m}^{m+p} \frac{\varphi_k^2(t)}{\lambda_k} \quad \text{...(ii)}$$

The series (*i*), therefore, converges absolutely. Hence the series converges uniformly in x for fixed t and in t for fixed x. Now each φ_i being an integral is a continuous function of its argument. Hence the sum function defined by the series $\sum_{k=1}^{\infty} \frac{\varphi_k(x)\,\varphi_k(t)}{\lambda_k}$ is a continuous function of x for fixed t and conversely. Since uniform convergence implies mean convergence, the series converges to $K(x, t)$. Uniform convergence of the series in both x and t together follows from the inequality (*ii*). Q.E.D.

Since continuity of $K(x, t)$ in $a \leqslant x, t \leqslant b$ implies uniform continuity, $K(x, t) \in L^2$ in both variables. Hence Mercer's theorem implies mean convergenee of the series asserted in theorem 3.

Theorem 5. (Hilbert-Schmidt Expansion Theorem).

Any function f which can be expressed in the form

$$f(x) = Kg = \int_a^b K(x, t)\, g(t)\, dt \qquad \text{...}(i)$$

where $g \in L^2$ and $K(x, t)$ is a symmetric kernel, has an absolutely and uniformly convergent expansion

$$f(x) = \sum_{i=1}^{\infty} (f, \phi_i)\, \phi_i = \sum_{i=1}^{\infty} \frac{(g, \phi_i)}{\lambda_i}\, \phi_i(x) \qquad \text{...}(ii)$$

in terms of eigen functions ϕ_i of $K(x, t)$.

Proof. Let

$$r_n(x) = f(x) - \int_a^b \left(\sum_{i=1}^{n} \frac{\phi_i(x)\, \phi_i(t)}{\lambda_i} \right) g(t)\, dt$$

$$= \int_a^b \left[K(x, t) - \sum_{i=1}^{n} \frac{\phi_i(x)\, \phi_i(t)}{\lambda_i} \right] g(t)\, dt$$

By Schwarz inequality

$$\left| r_n(x) \right|^2 \leqslant \int_a^b \left| K(x, t) - \sum_{i=1}^{n} \frac{\phi_i(x)\, \phi_i(t)}{\lambda_i} \right|^2 dt . \int_a^b | g(t) |^2\, dt \qquad \text{...}(iii)$$

By Mercer's theorem the first integral on the right of (*iii*) tends to 0 as $n \to \infty$ and the second integral is bounded.

Hence $| r_n(x) | \to 0$ as $n \to \infty$.

Hence $$f(x)=\sum_{i-1}^{\infty} \frac{\phi_i(x)}{\lambda_i} \int_a^b g(t)\, \phi_i(t)\, dt$$

$$=\sum_{i=1}^{\infty} \frac{g_i}{\lambda_i}\, \phi_i(x)$$

where $$g_i=\int_a^b g(t)\, \phi_i(t)\, dt$$

Since $|\lambda_i|$ is increasing as $i\to\infty$, the convergence of the series

$$f(x)=\sum_{i=1}^{\infty} \frac{g_i}{\lambda_i}\, \phi_i(x)$$

is more rapid than that of

$$g(x)=\sum_{i=1}^{\infty} g_i\, \phi_i(x)=\sum_{i=1}^{\infty} (g, \phi_i)\, \phi_i \qquad \text{Q.E.D.}$$

Example 10. Let

$$G(x, t)=\begin{cases} x(1-t) & 0\leqslant x\leqslant t\leqslant 1 \\ t(1-x) & 0\leqslant t\leqslant x\leqslant 1 \end{cases}$$

Prove that

$$G(x, t)=2 \sum_{n=1}^{\infty} \frac{\sin n\pi x \sin n\pi t}{n^2\pi^2}$$

and hence that

$$\sum_{n=1}^{\infty} \frac{1}{n^2}=\frac{\pi^2}{6} \quad \text{and} \quad \sum_{n=1}^{\infty} \frac{1}{n^4}=\frac{\pi^4}{90}$$

Solution. Consider the BVP

$$L(y)=-y''=0\ ;\ y(0)=y(1)=0$$

The differential operator L together with the boundary conditions is self-adjoint, since

$$(Ly, z)-(y, Lz)=-\int_0^1 y''z\, dx+\int_0^1 yz''\, dx$$

$$=\int_0^1 (yz''-y''z)\, dx$$

$$=\int_0^1 \frac{d}{dx}(yz'-y'z)\,dx$$

$$=\Big[\,y(x)\,z'(x)-y'(x)\,z(x)\,\Big]_0^1$$

$$=0$$

i.e., $$(Ly, z)=(y, Lz).$$

Let the Green's function of the BVP be

$$G(x, t)=\begin{cases} ax+b(1-x) & x\leqslant t \\ cx+d(1-x) & x\geqslant t \end{cases}$$

From the boundary conditions $b=c=0$ ...(*i*)

By the continuity of G in [0, 1],

$$at=d(1-t) \qquad ...(ii)$$

The jump of $\frac{\partial G}{\partial x}$ at $x=t$ gives

$$\frac{\partial G(t-0, t)}{\partial x}-\frac{\partial G(t+0, t)}{\partial x}=1$$

i.e., $$a+d=1 \qquad ...(iii)$$

From (*i*), (*ii*), (*iii*), we get

$$a=1-t,\ d=t$$

Hence finally

$$G(x, t)=\begin{cases} x(1-t) & x\leqslant t \\ t(1-x) & x\geqslant t \end{cases}$$

Let K be the integral operator defined by

$$Ky=\int_0^1 G(x, t)\,y(t)\,dt \qquad ...(iv)$$

Then $G(x, t)$ is a symmetric positive semi-definite kernel in $0\leqslant x,\ t\leqslant 1$. K is thus positive self-adjoint integral operator on $L^2[0, 1]$. K is inverse to L

i.e., $$LKy=y \qquad ...(v)$$

The characteristic values and eigenfunctions of the operator K are given by the Fredholm equation

$$y(x)=\lambda\int_0^1 G(x, t)\,y(t)\,dt=\lambda Ky \qquad ...(vi)$$

Using (*v*), equation (*vi*) is equivalent to tha Sturm-Liouville problem

$$Ly=-y''=\lambda LKy=\lambda y \qquad ...(vii)$$

$$y(0)=y(1)=0 \qquad ...(viii)$$

The general solution of (*vii*) is

$$y(x)=A \sin \sqrt{\lambda}\, x+B \cos \sqrt{\lambda}\, x$$

It satisfies conditions (*viii*) if

$$B=0 \quad \text{and} \quad A \sin \sqrt{\lambda}=0.$$

Since A is arbitrary, the later holds if

$$\sin \sqrt{\lambda}=0$$

i.e.,
$$\lambda=n^2\pi^2,\ n=1, 2, 3, \ldots\ldots$$

As mentioned earlier $n \neq 0$ as the characteristic value $\lambda=0$ is discarded.

The eigenfunctions associated with the characteristic values $\lambda=n^2\pi^2$ are $y_n(x)=A \sin n\pi x$.

To normalise $y_n(x)$, we require

$$\int_0^1 y_n^2\, dx=A^2\int_0^1 \sin^2 n\pi x\, dx=1$$

i.e.,
$$A=\sqrt{2}$$

Hence $\phi_n(x)=\sqrt{2} \sin n\pi x$ are the orthonormal eigenfunctions. Then for the operator K, the orthonormal eigenfunctions and related characteristic values are respectives $\sqrt{2} \sin n\pi x$ and $n^2\pi^2$,

$$n=1, 2, 3, \ldots\ldots$$

We now prove that the sequence $\{\sqrt{2} \sin n\pi x\}$ is complete. For this we prove that the null space of K consists of the null function alone. Indeed let

$$Kf=0$$

Then
$$0=\int_0^1 G(x, t) f(t)\, dt$$

$$=\int_0^x t(1-x) f(t)+\int_x^1 x(1-t) f(t)\, dt$$

$$=(1-x)\int_0^x t f(t)\, dt+x\int_x^1 (1-t) f(t)\, dt$$

Differentiating twice gives $f(x)=0$.

Conditions of Mercer's theorem are thus satisfied and we have

$$G(x, t)=\sum_{n=1}^{\infty} \frac{\varphi_n(x)\, \varphi_n(t)}{\lambda_n}$$

$$=2\sum_{n=1}^{\infty} \frac{\sin n\pi x \,.\, \sin n\pi t}{n^2\pi^2}$$

For $x=t$, $G(x, x)=x(1-x)=\sum_{n=1}^{\infty}\frac{\varphi_n^2(x)}{n^2\pi^2}$

Hence $\int_0^1 x(1-x)\,dx=\frac{1}{6}=\sum_{n=1}^{\infty}\frac{1}{n^2\pi^2}$

or
$$\sum_{n=1}^{\infty}\frac{1}{n^2}=\frac{\pi^2}{6} \qquad \ldots(ix)$$

Also the relation $\int_0^1\int_0^1 G^2(x, t)\,dx\,dt=\sum_{n=1}^{\infty}\frac{1}{n^4\pi^4}=\frac{1}{90}$

gives
$$\sum_{n=1}^{\infty}\frac{1}{n^4}=\frac{\pi^4}{90} \qquad \ldots(x)$$

Since $\{\sqrt{2}\sin n\pi x\}$ is complete, for $f\in L^2(0, 1)$, we have the Fourier expansion

$$f(x)=2\sum_{n=1}^{\infty}\left[\int_0^1 \sin n\pi t\, f(t)\,dt\right]\sin n\pi x$$

which converges in the mean to $f(x)$.

Example 11. Discuss the operator

$$Ky=\int_0^1 K(x, t)\,y(t)\,dt$$

where $K(x, t)$ is the Green's function of the BVP

DE : $Ly=-y''=0$
BCs : $y(0)=y(1)$
$y'(0)=y'(1)$

Solution. $y_0(x)\equiv 1$ is the non-trivial solution of the BVP satisfying both the boundary conditions. Hence only the generalised Green's function exists for the BVP.

Let $K(x, t)$ be the generalised Green's function. It satisfies the equation

$$y''=y_0(x)\,y_0(t)=1$$

The general solution of this equation is

$$y(x)=A+Bx+\frac{x^2}{2}$$

Hence let the generalised Green's function be

$$K(x,t)=\begin{cases} a+bx+\dfrac{x^2}{2} & x\leqslant t \\ c+dx+\dfrac{x^2}{2} & x\geqslant t \end{cases}$$

We determine a, b, c, d in the following steps :

1. Application of boundary conditions gives

$$a=c+d+\tfrac{1}{2} \qquad \text{...(i)}$$

$$b=d+1 \qquad \text{...(ii)}$$

2. By the continuity of $K(x, t)$ in $0\leqslant x, t\leqslant 1$

$$a+bt=c+dt \qquad \text{...(iii)}$$

Solving (*i*), (*ii*), (*iii*) for b, c, d in terms of a,

$$b=\tfrac{1}{2}-t,\ c=a+t,\ d=-\tfrac{1}{2}-t$$

3. Since $K(x, t)$ is orthogonal to $y_0(x)=1$,

$$\int_0^1 K(x,t)\,dx=0$$

This gives $a=\dfrac{1}{12}-\dfrac{1}{2}t+\dfrac{1}{2}t^2$

Finally

$$K(x,t)=\begin{cases} \dfrac{1}{12}-\dfrac{1}{2}t+\dfrac{1}{2}t^2+\left(\dfrac{1}{2}-t\right)x+\dfrac{x^2}{2} & x\leqslant t \\ \dfrac{1}{12}+\dfrac{1}{2}t+\dfrac{1}{2}t^2-\left(\dfrac{1}{2}+t\right)x+\dfrac{x^2}{2} & x\geqslant t \end{cases}$$

The jump condition $\dfrac{\partial K(t-0,t)}{\partial x}-\dfrac{\partial K(t+0,t)}{\partial x}=1$ is easily verified.

We can rewrite $K(x, t)$ as

$$K(x,t)=\begin{cases} \dfrac{1}{12}-\dfrac{1}{2}(t-x)+\dfrac{1}{2}(x-t)^2 & x\leqslant t \\ \dfrac{1}{12}-\dfrac{1}{2}(x-t)+\dfrac{1}{2}(x-t)^2 & x\geqslant t \end{cases}$$

which shows that $G(x, t)=G(t, x)$.

We can also write $K(x, t)$ as

$$K(x,t)=\begin{cases} \dfrac{1}{2}\left(x-t+\dfrac{1}{4}\right)^2+\dfrac{5}{96} & x\leqslant t \\ \dfrac{1}{2}\left(x-t-\dfrac{1}{4}\right)^2+\dfrac{5}{96} & x\geqslant t \end{cases}$$

which shows that $K(x, t)>0$ in $0\leqslant x, t\leqslant 1$. $K(x, t)$ is thus a continuous, positive definite, symmetric kernel.

The eigenvalue problem

$$Ly''=\lambda y, \ y(0)=y(1), y'(0)=y'(1) \quad \ldots(iv)$$

is equivalent to

$$y(x)=\lambda Ky=\lambda \int_0^1 K(x, t)\, y(t)\, dt \quad \ldots(v)$$

The general solution of the problem (*iv*) is

$$y=A \sin \sqrt{\lambda}\, x+\cos \sqrt{\lambda}\, x$$

It satisfies the boundary conditions if

$$-A \sin \sqrt{\lambda}+B(1-\cos \sqrt{\lambda})=0$$

$$\sqrt{\lambda}\, A(1-\cos \sqrt{\lambda})+\sqrt{\lambda}\, B \sin \sqrt{\lambda}=0$$

For a non-triviol solution in A, B

$$\begin{vmatrix} -\sin \sqrt{\lambda} & 1-\cos \sqrt{\lambda} \\ \sqrt{\lambda}\,(1-\cos \sqrt{\lambda}) & \sqrt{\lambda} \sin \sqrt{\lambda} \end{vmatrix}=0$$

i.e.,

$$\sqrt{\lambda}\,(1-\cos \sqrt{\lambda})=0$$

Hence eigenvalues of the BVP (*iv*) are

$$\lambda_0=0 \quad \text{and} \quad \lambda_n=4n^2\pi^2, \ n=1, 2, \ldots\ldots$$

For (*iv*) to the eigenvalue

$$\lambda_0=0$$

belongs the eigenfunction

$$y_0=\phi_0(x)=1.$$

Also for $\quad \Upsilon_n=4n^2\pi^2,$

we have eigen functions

$$y_n(x)=c \sin 2n\pi x+d \cos 2n\pi x$$

Thus to the eigenvalue $\lambda_0=0$, there is one eigenfunction $\phi_0(x)=1$ which is normalised since

$$\int_0^1 1 dx=1$$

To the eigenvalues $\lambda_n=4n^2\pi^2$, $n=1, 2, \ldots\ldots$ are associated a pair of linearly independent eigenfunctions $\sin 2n\pi x$ and $\cos 2n\pi x$ which can be normalised as $\sqrt{2} \sin 2n\pi x$ and $\sqrt{2} \cos 2n\pi x$. By applying the formula

$$K(x, t)=\sum_{n=1}^{\infty} \frac{\varphi_n(x)\, \phi_n(t)}{\lambda_n} \quad \text{we get}$$

$$K(x, t)=2 \sum_{n=1}^{\infty} \frac{\sin 2n\pi x \sin 2n\pi t+\cos 2n\pi x \cos 2n\pi i}{4n^2\pi^2}$$

$$=\frac{1}{2\pi^2}\sum_{n=1}^{\infty}\frac{\cos(x-t)\,2n\pi}{n^2}$$

Hence $K(x, x)=\frac{1}{12}=\frac{1}{2\pi^2}\sum_{n=1}^{\infty}\frac{1}{n^2}$

i.e., $$\sum_{n=1}^{\infty}\frac{1}{n^2}=\frac{\pi^2}{6}$$

which was found earlier.

Now let $f(x)\in L^2[0, 1]$ be a given function, then

$$f(x)=f_0+\sum_{n=1}^{\infty}(f, \phi_n)\,\phi_n(x)$$

and if it is of the form $f=Kg$

then $Kf_0=0$

i.e., f_0 is in the null space of K.

In the present case

$$f(x)=f_0+\sum_{n=1}^{\infty}\left[\left\{\int_0^1 f(t)\sqrt{2}\sin 2n\pi t\,dt\right\}\sqrt{2}\sin 2n\pi x\right.$$

$$\left.+\left\{\int_0^1 f(t)\sqrt{2}\cos 2n\pi t\,dt\right\}\sqrt{2}\cos 2n\pi x\right]$$

where $$f_0=\int_0^1 f(x)\,dx \qquad \text{...}(vi)$$

It may be noted that (*vi*) is the usual Fourier series for $f(x)$.

Problem 1. Prove that the Green's function of the BVP : $L(y)=-y''=0$; $y(0)=-y(1)$, $y'(0)=-y'(1)$ is

$$\tfrac{1}{4}-\tfrac{1}{2}\,|x-t|$$

Hence deduce that

$$\sum_{n=0}^{\infty}\frac{1}{(2n+1)^2}=\frac{\pi^2}{8}$$

Problem 2. Prove that the generalised Green's function of the

BVP : $y''=0$

BCs : $y(-1)=y(1)$, $y'(-1)=y'(1)$

is $$G(x, t)=-\tfrac{1}{2}\,|x-t|+\tfrac{1}{4}(x-t)^2+\tfrac{1}{6}$$

Find a bilinear expansion for $G(x, t)$.

[**Hint** : For the BVP : $y''+\lambda \, . \, y=0$,

$y(-1)=y(1),\ y'(-1)=y'(1)$,

$\lambda_n=n^2\pi^2,\ \phi_n=\sqrt{2}\sin n\pi x$

Hence $\quad G(x, t)=\dfrac{2}{\pi^2}\sum \dfrac{\sin n\pi x \sin n\pi t}{n^2}$

Put $x=t$ and integrate between 0 and 1 to get

$$\frac{1}{n^2}=\frac{\pi^2}{6}\Bigg]$$

Problem 3. Use the inequality

$$\iint\left[K(x, t)-\sum_{i=1}^{m}\frac{\phi_i(x)\,\phi_i(t)}{\lambda_i}\right]^2 dx\,dt$$

$$=\iint K^2(x, t)\,dx\,dt-\sum_{i=1}^{m}\frac{1}{\lambda_i^2}\geq 0$$

to prove

(*i*) $\displaystyle\sum_{1}^{\infty}\frac{1}{\lambda_i^2}$ is convergent, hence $|\lambda_i|\to\infty$ as $i\to\infty$

(*ii*) there are at most a countable number of characteristic values and eigenfunctions.

Problem 4. If $K_n(x, t)$ is the nth iterated kernel, then prove that

$$K_n(x, t)=\sum_{i=1}^{\infty}\frac{\phi_i(x)\,\phi_i(t)}{\lambda_i^n}\qquad n=2, 3, \ldots\ldots$$

$\Bigg[$**Hint** : $K(x, t)=\displaystyle\sum_{i=1}^{\infty}\frac{\phi_i(x)\,\phi_i(t)}{\lambda_i}$

(converges uniformly and absolutely)

$$\therefore\quad K_2(x, t)=\int_a^b K(x, s)\,K(s, t)\,ds$$

$$=\sum_{i=1}^{\infty}\int_a^b\frac{\phi_i(x)\,\phi_i(s)}{\lambda_i}K(s, t)\,ds$$

$$=\sum_{i=1}^{\infty}\frac{\phi_i(x)\,\phi_i(t)}{\lambda_i^2}$$

(converges uniformly and absolutely)

Now use induction]

Problem 5. If $\int_0^1 f(t)\,dt=0$ for continuous f, then prove that $f=0$.

[**Hint** : Let $f(\tau)\neq 0$, $0<\tau<1$ and in particular let $f(\tau)>0$. Then since f is continuous an $\epsilon>0$ exists such that $f(t)>0$ for all t in $(\tau-\alpha, \tau+\epsilon)$. Hence $\int_0^1 dft=\int_{\tau-\epsilon}^{\tau+\epsilon} fdt>0.$]

Solve the problem if the condition is $\int_0^1 K(t)\,.\,f(t)\,dt=0$ for any continuous $K(t)$.

Problem 6. Prove that the kernel

$$\sum_{n=1}^{\infty}\frac{\sin nx\,.\,\sin nt}{n},\ 0\leqslant x,\ t\leqslant\pi$$

has eigenvalues $\lambda_n=\frac{2n}{\pi}$ and eigenfunctions $\sin nx$.

§10.13. Approximation of Eigenvalues

In the previous section we have developed various expansions in terms of the orthnormalised eigenfunctions aud characteristic values of the Fredholm equation. These expansions were convergent either absolutely and uniformly or in the mean. They also depended heavily on their equivalent differential problems. The use of the Green's function facilitated establishment of this equivalence. In general this task of conversion may not be easy. This difficulty may become crucial when we are interested not in the exact determination of the eigenvalues, but only in certain estimates of their values. In this section we look more closely at this aspect of an eigenvalue problem when expressed in integral or differential form. We describe the method of traces which depends on the linear expansions of kernels described in the previous seetion.

The method of traces. Let K be a self-adjoint positive operator with orthonormalised eigenfunctions $\{\varphi_n(x)\}$ and characteristic values $\{\lambda_n\}$. Then

$$K(x, t)=\sum_{i=1}^{\infty}\frac{\phi_i(x)\,\phi_i(t)}{\lambda_i}$$

If $K_n(x, t)$ is the nth iterated Kernel, then

$$K_n(x, t)=\sum_{i=1}^{\infty}\frac{\varphi_i(x)\,\varphi_i(t)}{\lambda_i^{\,n}}\qquad n\geqslant 2.$$

For $n=2$, it can be verified that

$$\int_a^b K_2(x, x)\, dx = \sum_{i=1}^{\infty} \frac{1}{\lambda_i^2}$$

Hence $$\frac{1}{\lambda_i^2} \leqslant \int_a^b K_2(x, x)\, dx \qquad \ldots(i)$$

Now $$|\lambda_1| < |\lambda_2| \leqslant \ldots\ldots$$

and $$|\lambda_i| \to \infty$$

We thus get a lower estimate for the smallest characteristic value

i.e., $$\left\{ \frac{1}{\int_a^b K(x, x)dx} \right\}^{\frac{1}{2}} \leqslant |\lambda_1| \qquad \ldots(ii)$$

Definition 8. The *n*th *trace* of the kernel $K(x, t)$ is defined by

$$TrK_n = A_n = \int_a^b K_n(x, x)dx.$$

Now

$$A_n = \int_a^b K_n(x, x)dx = \sum_{i=1}^{\infty} \frac{1}{\lambda_i^n}$$

$$= \frac{1}{\lambda_1^n} \sum_{i=1}^{\infty} \left(\frac{\lambda_1}{\lambda_i}\right)^n$$

If $$\left|\frac{\lambda_1}{\lambda_i}\right| < 1 \quad \text{for } i > 1,$$

then $$\left|\frac{\lambda_1}{\lambda_i}\right|^n \to 0 \quad \text{as} \quad n \to \infty$$

and hence

$$\frac{1}{\lambda_1} = \lim_{n \to \infty} \left[\int_a^b K_n(x, x)dx\right]^{1/n}$$

$$= \lim_{n \to \infty} A_n^{1/n} \qquad \ldots(iii)$$

Again if f is an arbitrary function, then

$$Kf = K\left[\sum_{i=1}^{\infty} (f, \phi_i)\, \phi_i)\right]$$

$$=\sum_{i=1}^{\infty}\frac{(f,\varphi_i)\,\phi_i}{\lambda_i}$$

Hence $$\left(\frac{\|Kf\|}{\|f\|}\right)^2=\frac{\sum_{i=1}^{\infty}\left(\frac{(f,\phi_i)^2}{\lambda_i^2}\right)}{\sum_{i=1}^{\infty}(f,\phi_i)^2}$$

$$\leqslant\frac{1}{\lambda_1^2}$$

Hence $$|\lambda_1|\leqslant\frac{\|f\|}{\|Kf\|}\qquad ...(iv)$$

(*ii*) and (*iv*) give the lower and upper estimate of the smallest characteristic value λ_1.

For a differential operator L we have $Ly=\lambda y$ and if λ_1 is the smallest eigenvalue then

$$|\lambda_1|\leqslant\frac{\|Lf\|}{\|f\|}$$

for any $f\in$ dom L (Rayleigh's principle).

Example 12. The BVP $y''+\lambda y=0$; $y(0)=y(1)=0$ can be converted to the integral equation

$$y(x)=\lambda\int_0^1 K(x,t)\,y(t)dt$$

where $$K(x,t)=\begin{cases}x(1-t) & x\leqslant t\\ t(1-x) & x\geqslant t\end{cases}$$

Now $$\int_0^1 K(x,x)dx=\sum_{i=1}^{\infty}\frac{1}{\lambda_i}=\frac{1}{6}\qquad ...(i)$$

$$\int_0^1 K_2(x,x)dx=\sum_{i=1}^{\infty}\frac{1}{\lambda_i^2}=\frac{1}{90}\qquad ...(ii)$$

Since $\lambda_1=\pi^2$, from (*i*) and (*ii*) we get

$$\frac{1}{\pi^2}\leqslant\frac{1}{6}\quad\text{and}\quad\frac{1}{\pi^4}\leqslant\frac{1}{90}\ \text{respectively.}$$

Hence from (*i*) and (*ii*) we get respectively

$$\pi\geqslant\sqrt{6}=2{\cdot}45...,\quad \pi\geqslant(90)^{1/4}=3{\cdot}08...$$

The later is certainly a better estimate.

We now use the estimate

$$|\lambda_1| \leqslant \frac{\|f\|}{\|Kf\|} \quad \text{for } f=1$$

$$\|f\|^2=(f, f)=\int_0^1 dx=1$$

$$Kf=\int_0^1 K(x, t)$$

$$=\int_0^x t(1-x)dt+\int_0^1 x(1-t)dt$$

$$=\frac{x}{2}(1-x).$$

Hence $\quad \|Kf\|^2=(Kf, Kf)=\frac{1}{4}\int_0^1 x^2(1-x)^2dx=\frac{1}{120}$

i.e., $\quad \|Kf\|=\frac{1}{\sqrt{120}}$

Hence $\quad |\lambda_1| \leqslant \frac{1}{\sqrt{120}}$

i.e., $\quad \pi \leqslant (120)^{1/4}=3{\cdot}31\ldots\ldots$ ...(*iii*)

Thus finally $3{\cdot}08 \leqslant \pi \leqslant 3{\cdot}31$

Problem. Prove that if $x \leqslant t$, then

$$K_2(x, t)=\tfrac{1}{3}t(1-t)(x-2xt+t^2)$$

and hence $\quad K_2(x, x)=\tfrac{1}{3}x^2(1-x)^2.$

Use of the Method of Traces. From the relations

$$T_r K_{2n}=A_{2n}=\sum_{i=1}^{\infty}\frac{1}{\lambda_i^{2n}}$$

and

$$T_r K_{2n+2}=A_{2n+2}=\sum_{i=1}^{\infty}\frac{1}{\lambda_i^{2n+2}}$$

we have

$$\lambda_1^2=\frac{A_{2n}}{A_{2n+2}}\ \frac{\sum_{i=1}^{\infty}\left(\frac{\lambda_1}{\lambda_i}\right)^{2n+2}}{\sum_{i=1}^{\infty}\left(\frac{\lambda_1}{\lambda_i}\right)^{2n}}=\frac{A_{2n}}{A_{2n+2}}\cdot\frac{S_1}{S_2}$$

Since $|\lambda_1| \leqslant |\lambda_2| \leqslant \ldots\ldots$

$$\left|\frac{\lambda_1}{\lambda_i}\right| \leqslant 1\ldots, \quad i>1.$$

Hence $S_1 \leqslant S_2$ and we have

$$\lambda_1^2 \leqslant \frac{A_{2n}}{A_{2n+2}}$$

Since $\frac{S_1}{S_2} \to 1$ as $n \to \infty$

$$\sqrt{\frac{A_{2n}}{A_{2n+2}}} \to \lambda_1 \text{ as } n \to \infty.$$

Traces of even order have been chosen because for a symmetric kernel

$$A_{2n} = \int_a^b \int_a^b K_n^2(x, t)dx\, dt$$

$$= 2\int_a^b \int_a^x K_n^2(x, t)dt\, dx.$$

$$|\lambda_1| \sim \sqrt{\frac{A_2}{A_4}}$$

is the first approximation to λ_1. From the first approximation to λ_1, neglecting characteristic values above λ_2, a first approximation to λ_2 can be determined by the relation

$$A_2 = \frac{1}{\lambda_1^2} + \frac{1}{\lambda_2^2}.$$

Example 13. Use the method of traces to compute λ_1 for the equation

$$y(x) = \lambda \int_0^1 xt\, y(t)\, dt.$$

Solution. Here $K(x, t) = xt$.

Hence $K_2(x, t) = \frac{xt}{3}$, $K_4(x, t) = \frac{xt}{3^3}$

Also $A_2 = \int_0^1 K_2(x, x)\, dx = \frac{1}{3^2}$

$$A_4 = \int_0^1 K_4(x, x)dx = \frac{1}{3^2.3^2}$$

Hence $\lambda_1 \sim \sqrt{\frac{A_2}{A_4}} = 3.$

In fact $\lambda_1 = 3$ is the only characteristic value of the equation.

Problem 1. Prove the for $K(x, t) = x^2t^2$, $\lambda_1 = 5$.

Example 14. Compute the first characteristic number λ_1 of the eigenvalue problem

$$y''=\lambda y=0, \quad y(0)=y'(1)=0.$$

Solution. In the differential operator notation the problem can be written as

$$L(y)=-y''=\lambda y$$
$$y(0)=y'(1)=0.$$

The Green's function of the BVP is

$$K(x, t)=\begin{cases} x & x\leqslant t \\ t & x\geqslant t \end{cases}$$

Hence in integral operator notation the BVP can be converted into

$$y(x)=\lambda\int_0^1 K(x, t)y(t)\, dt.$$

Since $K(x, t)=K(t, x)$, hence to compute $K_2(x, x)$ we may compute $K_2(x, t)$ for $t\leqslant x$ only,

$$K_2(x, t)=\int_0^1 K(x, s)\, K(s, t)ds$$
$$=\int_0^t s^2\, ds+\int_t^x st\, ds+\int_x^1 xtds$$
$$=xt-\frac{x^2t}{2}-\frac{t^3}{6}$$

$$A_2=2\int_0^1 dx \int_t^x K_1^2(x, t)dt$$
$$=2\int_0^1 dx \int_0^x t^2dt=\frac{1}{6}$$

$$A_4=2\int_0^1 \int_0^x K_2^2(x, t)$$

$$=2\int_0^1 dx \int_0^x \left[x^2t^2+\frac{x^4t^2}{4}+\frac{t^6}{36}+\frac{x^2t^4}{6}-x^3t^2-\frac{xt^4}{3}\right]dt$$

$$=2\int_0^1 \left(\frac{x^6}{3}+\frac{x^7}{12}+\frac{x^7}{7\cdot 36}-\frac{x^6}{3}-\frac{x^6}{15}+\frac{x^7}{30}\right)dx$$

$$=\frac{17}{630}.$$

Hence $\lambda_1\sim\sqrt{\frac{1}{6}\Big/\frac{17}{630}}=2{\cdot}48.$

Problem 2. Find λ_1 for the BVP $y''+\lambda y=0,\ y(0)=y(2)=0.$

[**Hint :** Green's function $K(x, t)=\begin{cases}\frac{1}{2}x(2-t) & x\leqslant t\\ \frac{1}{2}t(2-x) & x\geqslant t\end{cases}$
$\lambda_1=4{\cdot}19$]

Example 15. Compute λ_1 for the Kernel

$$\frac{1-\alpha^2}{1-2\alpha\cos(x-t)+\alpha^2} \qquad -\pi\leqslant x,\ t\leqslant\pi,\ 0<\alpha<1.$$

Solution. By contour integration it can be shown that

$$\int_{-\pi}^{\pi}\frac{(1-\alpha^2)(1-\beta^2)\,dz}{\{1-2\alpha\cos(x-z)+\alpha^2\}\{1-2\beta\cos(t-z)+\beta^2\}}$$

$$=2\pi\,\frac{1-\alpha^2\beta^2}{1-2\alpha\beta\cos(x-t)+\alpha^2\beta^2} \qquad 0<\alpha,\ \beta<1$$

Putting $\beta=\alpha$, we shall get

$$K_2(x, t)=\frac{2\pi\,(1-\alpha^4)}{1-2\alpha^2\cos(x-t)+\alpha^4}$$

Iteration gives

$$K_n(x, t)=\frac{(2\pi)^{n-1}(1-\alpha^{2n})}{1-2\alpha^n\cos(x-t)+\alpha^{2n}}$$

Hence

$$K_n(x, x)=\frac{(2\pi)^{n-1}(1+\alpha^n)}{1-\alpha^n}$$

$$T_rK_n=A_n=\int_{-\pi}^{\pi}K_n(x, x)dx$$

$$=\frac{(2\pi)^n(1+\alpha^n)}{1-\alpha^n}$$

Hence $\qquad \dfrac{1}{\lambda_1}=\lim\limits_{n\to\infty}A_n^{1/n}=2\pi \qquad (\because\ \alpha^n\to 0 \text{ as } n\to\infty)$

i.e., $\qquad \lambda_1=\dfrac{1}{2\pi}.$

Problem. Prove that the $K(x, t)$ has eigenfunctions 1, $\sin nx$, $\cos nx$ with eigenvalues 1, $\dfrac{1}{\alpha^n}$, $\dfrac{1}{\alpha^n}$.

Example 16. If $L(y)=-y''=\lambda y,\ y(0)=y(1)=0$. Find an estimate for λ_1.

Solution. If $f(x)=x(1-x)$, then $f(0)=f(1)=0$ and
$Lf=2\in L^2\,[0, 1]$.

$$\|f\|^2=\int_0^1 x^2(1-x)^2dx=\frac{1}{30}$$

$$Lf=2$$

Hence $\quad \|f\|=\frac{1}{\sqrt{30}}, \quad \|f\|=2$

and $\quad \frac{\|Lf\|}{\|f\|}=2\sqrt{30}$

Since $\lambda_1=\pi^2,\ \pi^2\leqslant \frac{\|Lf\|}{\|f\|}=2\sqrt{30}$

i.e., $\pi\leqslant(120)^{1/4}$ found in example 12.

11

Laplace Transform

§11.1. The aim of this chapter is to describe the main operational properties of this transform with a view to highlight its role in solving differential and integral equations.

Definition 1. Let $f(t)$ be an arbitrary function defined on the interval $0 \leqslant t < \infty$, then

$$F(p) = \int_0^\infty e^{-pt} f(t)\,dt \qquad \dots(1)$$

is the Laplace transform of $f(t)$, provided that the integral exists.

An integral of the form (1) in which the function $f(t)$ is integrable over the interval $(0, a)$ for any positive a, is called a **Laplace integral.** The integral does not exist for all functions $f(t)$. For example, for $f(t)=e^{t^2}$, the integral (1) diverges for all values of p. For some functions $f(t)$, the integral exists for some values of p but not for others. For example if $f(t)=e^t$, then the integral (1) exists for $\text{Re}\,p > 1$ and diverges for $\text{Re}\,p \leqslant 1$.

When (1) exists we call $f(t)$ an *original* and $F(p)$ the *image* of $f(t)$ (or its Laplace transform). The relationship of the original and image is symbolically denoted by $f(t) \doteq F(p)$,

or by $\quad L\{f(t)\} = F(p)$

Definition 2. A function $f(t)$ is called *piecewise continuous* (or *sectionally continuous*) on a given interval if it is continuous on every finite subinterval except a finite number of points.

Definition 3. A function $f(t)$ is said to be *locally integrable* on I if

(*i*) $f(x)$ is continuous or piecewise continuous on I,

(*ii*) The integral of $f(t)$ over any closed subinterval of I exists either as a proper (Riemann) integral or an absolutely convergent improper integral.

Definition 4. A function $f(t)$ is said to be of *exponential order* (or type) as $t \to \infty$ if there exist constants (>0) M, b and t_0 such that for $t \geqslant t_0$ $|f(t)| \leqslant M e^{bt}$, written $f(t)=O(e^{bt})$.

The number b is called the *order of growth* of the function $f(t)$.

All polynomials are of exponential order, but e^{t^2} is not.

Lemma 1. If

(*i*) $f(t)$ is locally integrable

(*ii*) for arbitrary positive a, the integral

$$\int_{\varepsilon}^{a} |f(t)|\, dt \text{ tends to a finite limit}$$

as $\varepsilon \to 0+$,

(*iii*) $f(t) = O(e^{bt})$ as $t \to \infty$

Then the Integral (1) converges absolutely if Re$p > b$.

If instead of (*iii*) we have the condition (*iii*)* $f(t) = O(t^c)$ as $t \to \infty$, then the Laplace integral converges absolutely for Re$p > 0$.

The proof is easy and is left as an exercise.

Lemma 2. The function

$$\phi(p, T) = \int_0^T e^{-pt} f(t)dt \qquad ...(2)$$

is analytic in p and converges uniformly in any closed and bounded region of the half-plane Re$p > b$. Hence F(p) is analytic in the half-plane Re$p > b$.

Inversion. The main power of the Laplace transform rests in its invertibility under certain conditions. The following concepts are needed :

(*i*) A function $f(t)$ is said to satisfy Dirichlet conditions in the interval $[a, b]$ if it has at most a finite number of maxima, minima, and points of discontinuity in that interval and takes only a finite jump at a discontinuity.

(*ii*) **Riemann-Lebesgue Lemma.**

If $f(x)$ satisfies Dirichlet conditions then

$$\lim_{\omega \to \infty} \int_a^b f(t)\, e^{i\omega t} dt = 0$$

Hence also

$$\lim_{\omega \to \infty} \int_a^b f(t) \sin \omega t\, dt = 0$$

$$\lim_{\omega \to \infty} \int_a^b f(t) \cos \omega t\, dt = 0$$

a may be $-\infty$ and/or b may be $+\infty$

(*ii*) **Dirichlet Integral**

$$\lim_{\omega \to \infty} \int_0^b f(t) \frac{\sin \omega t}{t} dt = \frac{\pi}{2} f(0+)$$

$$\lim_{\omega \to \infty} \int_{-b}^0 f(t) \frac{\sin \omega t}{t} dt = \frac{\pi}{2} f(0-)$$

Here b may be ∞

When F(p) is given, any formula enabling us to derive the original $f(t)$ is called an *inversion formula.*

Theorem 1. If the Laplace integral

$$F(p)=\int_0^{\infty} e^{-pt} f(t)dt, \ \text{Re}p>p_0 \tag{i}$$

is convergent, then

$$f(t)=\frac{1}{2\pi i}\int_{c-i\infty}^{c+i\infty} F(p)\, e^{-pt}\, dp,\ c>p_0 \tag{ii}$$

where $f(t)$ is taken to be $\frac{1}{2}\{f(t+0)+f(t-0)\}$ at a point of discontinuity.

Proof. It we substitute for $F(p)$ the integral (i) in the following integral.

$$I_R(t)=\frac{1}{2\pi i}\int_{c-iR}^{c+iR} e^{pt}\, F(p)\, dp \qquad c>p_0 \tag{iii}$$

we get after changing the order of integration

$$I_R(t)=\frac{1}{2\pi i}\int_0^{\infty} f(x)\, dx\int_{c-iR}^{c+iR} e^{p(t-x)}dp$$

$$=\frac{1}{\pi}\int_0^{\infty} f(x)\, e^{c(t-x)}\, \frac{\sin R(t-x)}{t-x}dx$$

$$=\frac{1}{\pi}\int_{-t}^{\infty} f(t+u)\, e^{-cu}\, \frac{\sin Ru}{u}\, du$$

If we break the integral into two, from $-t$ to 0 and 0 to ∞ and let $R\to\infty$, then

$$I_R(t) \to \begin{cases} 0, & t<0 \\ \frac{1}{2} f(0+) & t=0 \\ \frac{1}{2}\{f(t-0)+f(t+0)\} & t>0 \end{cases}$$

§ 11·2. Rules of Transform

The following rules govern the evaluation of Laplace transforms. Proofs have been omitted where they are too elementary.

We shall assume $f(t)\doteq F(p)$.

I. (**Linearity**). If $f(t)=\sum_{k=1}^{n} a_k f_k(t)$

then $$L\{f\}=\sum_{k=1}^{n} a_k\, L\{f_k\}$$

If $f(t)$ involves a parameter s (which is independent of t), then

$$L\left\{\frac{\partial f}{\partial s}\right\}=\frac{\partial'}{\partial s}L\{f\}$$

$$L\left\{\int_a^b f(s)\, ds\right\}=\int_a^b L\{f(s)\}\, ds$$

II. (Shifting Theorem) $e^{at} f(t)\doteq F(p-a), \qquad p\geqslant p_0+a.$

III. Change of scale) $f(at) \doteq \frac{1}{a} F(p/a), \qquad a>0, p>p_0$

The result is also known as *dilation* or *similarity* property.

IV. (Differentiation of the original)

If we apply integration by parts, we get

$$L\{f'(t)\} = pF(p) - f(0+)$$

where $\quad f(0+) = \lim_{t \to 0+} f(t).$

In problems where $f(0)$ is given, it will be understood in the meaning of $f(0+)$.

The general formula is

$$f^{(n)}(t) \doteq p^n F(p) - p^{n-1} f(0) - \ldots - f^{(n-1)}(0).$$

In the special case when $f(t)$ and its derivatives vanish at $t=0$. we have

$$\frac{d^n f}{dt^n} = p^n F(p).$$

Thus the analytic operation of differentiating the original once results in the algebraic multiplication of the image by p once. This property is sometimes underlined by saying that the integral transform related to the differentiation operator $D \equiv \frac{d}{dt}$ is the Laplace transform.

V. **Differentiation of the Transform**

By differentiating under the integral sign we obtain :

$$(-t)^n f(t) = \frac{d^n F(p)}{dp^n}$$

VI (Integration of the original)

(*i*) $\int_0^t f(x)\, dx \doteq \frac{1}{p} F(p)$

Proof. Let $g(t) = \int_0^t f(x)\, dx$

and $\quad g(t) = G(p).$

Then $\quad g'(t) = f(t)$

and $\quad g(0) = 0.$

Hence by the differentiation of the original rule, we get

$$pG(p) = F(p),$$

i.e., $\quad G(p) = \frac{1}{p} F(p).$

(*ii*) $\int_t^{\infty} f(x)\,dx \doteq -\dfrac{F(p)}{p}+\dfrac{1}{p}\int_0^{\infty} f(x)\,dx$

Proof. By changing the order of integration we have

$$\int_0^{\infty} e^{-pt}\left\{\int_t^{\infty} f(x)\,dx\right\}dt=\int_0^{\infty} f(x)\left\{\int_0^{x} e^{-pt}\,dt\right\}dt$$

$$=\frac{1}{p}\int_0^{\infty} f(x)\,(1-e^{-px})\,dx$$

$$=\frac{1}{p}\int_0^{\infty} f(x)\,dx-\frac{F(p)}{p}.$$

VII. (Integration of the Transform).

(*i*) $\dfrac{f(t)}{t} \doteq \int_p^{\infty} F(u)\,dx$

Proof. $\displaystyle\int_0^{\infty} e^{-pt}\frac{f(t)}{t}\,dt=\int_0^{\infty} f(t)\,dt\int_p^{\infty} e^{-tu}\,du$

$$=\int_p^{\infty}\left[\int_0^{\infty} e^{-t} f(t)\,dt\right]du$$

$$=\int_p^{\infty} F(u)\,du.$$

(*ii*) $\displaystyle\int_0^{t} \frac{f(x)}{x}\,dx \doteq \frac{1}{p}\int_p^{\infty} F(u)\,du$

The rule follows by combining VII (*i*) with VI (*i*).

(*iii*) $\displaystyle\int_t^{\infty} \frac{f(x)}{x}\,dx \doteq \frac{1}{p}\int_0^{p} F(u)\,du$

Proof. Proceeding as in VI (*ii*), we have

$$\int_0^{\infty} e^{-pt}\left\{\int_t^{\infty}\frac{f(x)}{x}\,dx\right\}dx=\frac{1}{p}\int_0^{\infty}\frac{f(x)}{x}(1-e^{-px})\,dx$$

$$=\frac{1}{p}\int_0^{p} F(u)\,du, \qquad \text{(use VII } (i))$$

where $\displaystyle F(u)=\int_0^{\infty} e^{-ux} f(x)\,dx$

(*iv*) $\displaystyle\int_0^{\infty}\frac{f(x)}{x}\,dx=\int_0^{\infty} F(x)\,dx$

Proof. Let $\displaystyle\int_0^{\infty}\frac{f(x)}{x}=A$ (constant).

We have

$$L(A)=\frac{A}{p}$$

$$=L\left\{\int_0^{t}\frac{f(x)}{x}\,dx\right\}+L\left\{\int_t^{\infty}\frac{f(x)}{x}\,dx\right\}$$

$$=\frac{1}{p}\int_0^p F(x)\,du+\frac{1}{p}\int_0^\infty F(u)\,du \quad =\frac{1}{p}\int_0^\infty F(u)\,dx.$$

$$=L\left\{\int_0^\infty F(u)\,du\right\}$$

Hence $\quad A=\int_0^\infty \frac{f(x)}{x}\,dx \quad =\int_0^\infty F(u)\,du.$

Aliter Let $p\to 0$ in VII (i).

VIII (Delay or Second Shift Theorem)

$$f(t-a)\,H(t-a) \doteq e^{-ap}\,F(p)$$

where H (t) is the Heavisides unitary function defined by

$$H(t)=\begin{cases} 1 & \text{for} \quad t>0 \\ 0 & \text{for} \quad t<0 \end{cases}$$

Multiplication by $H(t-a)$ to the original results in the delay factor e^{-ap} in the image corresponding to the retarded argument $t-a$ in a place of $t-$ the function $f(t)$ being 'switched on' at $t=a$ instead of at $t=0$. The result concerns translations to the right.

For translation to the left, we have

$$\int_0^\infty f(t+a)\,H(t+a)\,e^{-pt}\,dt=\int_a^\infty f(u)\,e^{-pu}\,.\,e^{ap}\,dx$$

$$=e^{ap}\int_0^\infty f(u)\,e^{-pu}\,du-e^{ap}\int_0^a f(u)\,e^{-pu}\,du$$

$$=e^{ap}\,F(p)-e^{ap}\int_0^a f(u)\,e^{-pu}\,du$$

The finite integral $\int_0^a f(u)\,e^{-pu}\,dx$ cannot be neglected in this case unless $f(t)$ vanishes for $t<a$. This integral is supposed to compensate for the part of the function $f(t)$ which was lost by translation to negative values of t where Laplace transform does not operate.

IX. (Periodic Functions)

Let $f(t)$ be a function of period ω, *i.e.*

$$f(t+r\omega)=f(t) \qquad (r=0,1,2,\ldots)$$

Hence $L\{f\}=\int_0^\infty e^{-pt}f(t) \quad =\sum_{r=0}^{\infty}\int_{r\omega}^{(r+1)\omega} e^{-pt}f(t)\,dt$

$$=\sum_{r=0}^{\infty}\int_0^\omega f(u+r\omega)\,e^{-p(u+r\omega)}dx \quad =\sum_{r=0}^{\infty} e^{-pr\omega}\int_0^\omega f(u)\,e^{-pu}\,du$$

$$=\left(1-e^{p\omega}\right)^{-1}\int_0^\omega f(u)\,e^{-pu}\,du.$$

X. (Convolution Theorem) If $f(t)=F(p)$, $g(t)=G(p)$

and $(f \circ g)(t) = \int_0^t f(t-u)\, g(u)\, du,$

then $(f \circ g)(t) \doteq F(p) \, . \, G(p)$

Proof. $\int_0^\infty e^{-pt} \left[\int_0^t f(u)\, g(t-u)\, du \right] dt$

$= \int_0^\infty f(u)\, du \int_u^\infty e^{-pt}\, g(t-u)\, dt$

(changing the order).

$= \int_0^\infty f(u)\, du \int_0^\infty e^{-(pu+v)}\, g(v)\, dv$

$= \int_0^\infty e^{-pu} f(u)\, du. \int_0^\infty e^{-pv}\, g(v)\, dv$

$= F(p) \, . \, G(p)$

XI. (The Uniqueness Theorem or Lerch's Theorem)

If $f_1(t) \doteq F_1(p)$ and $f_2(t) \doteq F_2(p)$,

then $F_1(p) = F_2(p) \rightarrow f_1(t) = f_2(t)$ a.e.

XII. (Rational Functions in p).

If $F(p)$ is a rational function of p, the original is obtained by splitting $F(p)$ into partial fractions and taking the inverse transform of each part. Such a splitting of $F(p$ is called *Heaviside's expansion.*

§11·3. Transforms of some Elementary Functions

$f(t)$	$F(p)$
1	$1/p$
$H(t-a)$	e^{-ap}/p
t^n	$\Gamma(n+1)/p^{n+1}$
e^{at}	$1/(p-a)$
$e^{at}\, t^n$	$n\,!/(p-a)^{n+1}$
$\sin at$	$a/(p^2+a^2)$
$\cos at$	$p/(p^2+a^2)$
$\sinh at$	$a/(p^2-a^2)$
$\cosh at$	$p/(p^2-a^2)$
$t \sin at$	$2ap/p^2+a^2)^2$
$t \cos at$	$(p^2-a^2)/(p^2+a^2)^2$
$\sin at - at \cos at$	$2a^3/(p^2+a^2)^2$
$\sin t/t$	$\tan^{-1}(1/p)$

In what follows if $f(t) = F(p)$ then $L^{-1} F(p) = f(t)$.

Example 1. Evaluate $L^{-1} F(p)$ where

$$F(p)=\frac{p+1}{2p^2+p+1}$$

Solution.

$$2F(p)=\frac{(p+\frac{1}{4})+\frac{3}{4}}{(p+\frac{1}{4})^2+\frac{7}{16}}$$

$$=\frac{p+\frac{1}{4}}{(p+\frac{1}{4})^2+\frac{7}{16}}+\frac{3}{4}\,\frac{1}{(p+\frac{1}{4})^2+\frac{7}{16}}$$

Hence

$$2f(t)=L^{-1}\left\{\frac{p+\frac{1}{4}}{(p+\frac{1}{4})^2+\frac{7}{16}}+\frac{3}{4}\,\frac{1}{(p+\frac{1}{4})^2+\frac{7}{16}}\right\}$$

$$=e^{-\frac{1}{4}t}\left\{\cos\frac{\sqrt{7}}{4}t+\frac{3}{4}\sin\frac{\sqrt{7}}{4}\,t\right\}$$

Example 2. Compute $L\{J_0\,(2\sqrt{t}\,\}$

Solution. $J_0\;2\sqrt{t)}=1-t+\frac{t^2}{(2\,!)^2}-\frac{t^3}{3\,!)^2}+\ldots$

Hence L.T. of $J_0(2\sqrt{t)}=\frac{1}{p}-\frac{1}{p^2}+\frac{1}{2\,!\,p^3}+\frac{1}{3\,!\,p^4}+\ldots$

$$=\frac{1}{p}e^{-1/p}$$

Example 3. Prove $\int_0^\infty e^{-t^2}dt=\frac{1}{2}\sqrt{\pi}$

Solution. Let $f(t)=\int_0^\infty e^{-tx^2}dx \qquad t>0.$

Then

$$L\{f\}=\int_0^\infty e^{-pt}dt\int_0^\infty e^{-tx^2}dx$$

$$=\int_0^\infty L\left\{e^{-tx^2}\right\}dx$$

$$=\int_0^\infty \frac{dx}{p+x^2}=\frac{1}{2}\;\frac{\pi}{\sqrt{p}}$$

Hence

$$f(t)=\frac{\pi}{2}\;\frac{1}{\sqrt{\pi t}}=\frac{1}{2}\sqrt{\frac{\pi}{t}}$$

With $t=1$ we get

$$\int_0^\infty e^{-x^2}dx=\frac{1}{2}\sqrt{\pi}$$

§11·4. Applications

Laplace transform is a powerful tool for solving certain differential and integral equations. The idea behind the method is to transform an operator equation to on algebraic equation (auxiliary equation) which is solved algebraically. The solution of this algebraic equation is then reinterpreted by taking inverse transform leading to a solution of the original problem.

In the case of initial-value problems separate use of the initial values to determine the arbitrary constants is not required. In the case of non-homogeneous equations the separate determination of the solution of the corresponding homogeneous and a particular solution is not necessary

The following examples illustrate the use of the transform in solving differential and integral equations.

(*a*) **Initial Value Problems**

Example 4. Solve the Cauchy problem

$$y''+y'+y=3\,e^t\,;\; y(0)=y'(0)=0.$$

Solation. Let $L\{y\}=Y(p)$. Then taking the Laplace transform of the equation and using the rule IV, we shall have

$$(p^2+p+1)\,Y(p)=\frac{3}{p-1}\,,$$

i.e.,
$$Y(p)=\frac{3}{(p-1)(p^2+p+1)}\,.$$

Splitting into partial fractions

$$Y(p)=\frac{1}{p-1}-\frac{p+2}{p^2+p+1}$$

$$=\frac{1}{p-1}-\frac{(p+\frac{1}{2})+\frac{3}{2}}{(p+\frac{1}{2})^2+\frac{3}{4}}$$

Hence
$$y(t)=e^t-e^{-\frac{1}{2}t}\cos\frac{\sqrt{3}}{2}t-\sqrt{3}\,e^{-\frac{1}{2}t}\sin\frac{\sqrt{3}}{2}t$$

(*b*) **Boundary Value Problems**

Example 5. Solve the two point BVP :

$$2y''+3y'-2y=0,$$
$$y(0)=1,\; y(t)\to 0 \text{ as } t\to\infty.$$

Solution. Assuming $y'(0)=A$ and taking the Laplace transform of both sides, we have

$$2p^2Y(p)-2p-2A+3pY(p)-3-2\,Y(p)=0.$$

Hence
$$Y(p)=\frac{2p+2A+3}{2p^2+3p-2}$$

$$=\frac{1-2A}{5(p+2)}+\frac{4}{5}\;\frac{A+2}{(2p-1)}\,.$$

Taking the inverse transform,

$$y(t)=\frac{1-2A}{5}e^{-2t}+\frac{2}{5}(A+2)e^{\frac{1}{2}t}$$

Since $y(\infty)=0$, $A=-2$.

Hence finally, $y(t)=e^{-2t}$.

(c) Simultaneous Differential Equations

Example 6. Solve the system

$$\frac{d^2y}{dt^2}+4z=4t, \quad \frac{d^2z}{dt^2}+4y=0$$

$$y(0)=y'(0)=0$$

$$z(0)=z'(0)=0.$$

Taking Laplace transforms

$$p^2Y(p)+4\ Z(p)=\frac{4}{p^2}, \quad p^2\ Z(p)+4\ Y(p)=0.$$

Solving for Y(p) we have

$$Y(p)=\frac{4}{p^4-16}.$$

Taking inverse transform,

$$y(t)=\tfrac{1}{4}\ (\sinh 2t-\sin 2t).$$

Making this replacement for $y(t)$ in the given system we have

$$z(t)=t-\tfrac{1}{4}\ (\sinh 2t)+\sin 2t).$$

(d) Autonomous Linear Systems

$$X'=\frac{dX}{dt}=AX+B \qquad ...(i)$$

where

$$X=\begin{bmatrix} x_1 \\ x_2 \\ \vdots \\ x_n \end{bmatrix}, \quad B=\begin{bmatrix} b_1 \\ b_2 \\ \cdots \\ b_n \end{bmatrix}, \quad A=[a_{ij}]_{n\times n}.$$

X is the unknown vector, B the constant vector and A a given matrix with constant entries a_{ij}. With this notation

$$X(0)=\begin{bmatrix} x_1(0) \\ \vdots \\ x_n(0) \end{bmatrix}, \quad X'=\begin{bmatrix} x_1' \\ x_2' \\ \vdots \\ x_n' \end{bmatrix}$$

If $L\{X\}=Y(p)$ and $L\{x_1(t)\}=y_1(p)$, then

$$Y(p)=\begin{bmatrix} y_1\ (p) \\ \vdots \\ y_n\ (p) \end{bmatrix}.$$

Taking the Laplace transform of both sides of (i)

$$pY(p)=AY(p)+\frac{1}{p}B+X(0),$$

i.e., $$Y(p)=(pI-A)^{-1}\{p^{-1}B+X(0)\}.$$

The inverse transform of $Y(p)$ is the required solution.

Example 7. Solve the system

$$X'=\begin{bmatrix}-2 & 4\\ 1 & 1\end{bmatrix}X,\ X(0)=\begin{bmatrix}-2\\ 3\end{bmatrix}$$

(c.f. ex. 6, Chapter VI.)

Solution. Let $L\{X(t)\}=Y(p)$. Taking the Laplace transform of both sides of the system,

$$pY(p)-X(0)=\begin{bmatrix}-2 & 4\\ 1 & 1\end{bmatrix}Y(p)$$

i.e., $$Y(p)=\begin{bmatrix}p+2 & -4\\ -1 & p-1\end{bmatrix}^{-1}\begin{bmatrix}-2\\ 3\end{bmatrix}$$

$$=\frac{1}{(p-2)(p+3)}\begin{bmatrix}p-1 & +4\\ 1 & p+2\end{bmatrix}\begin{bmatrix}-2\\ 3\end{bmatrix}$$

$$=\begin{Bmatrix}\dfrac{2}{p-2}-\dfrac{4}{p+3}\\[2ex] \dfrac{2}{p-2}+\dfrac{1}{p+3}\end{Bmatrix}$$

Taking inverse transforms of both sides,

$$X(t)=\begin{bmatrix}2e^{2t}-4e^{-3t}\\ 2e^{2t}-e^{-3t}\end{bmatrix}=2e^{2t}\begin{bmatrix}1\\ 1\end{bmatrix}-e^{-3t}\begin{bmatrix}4\\ -1\end{bmatrix}$$

(*e*) **Non-autonomous Systems**

Example 8. Solve the system

$$X'=\begin{bmatrix}0 & -\frac{1}{2}\\ 2 & 0\end{bmatrix}X+\begin{bmatrix}\frac{1}{2}\cos t\\ 3\sin t\end{bmatrix};\ X(0)=\begin{bmatrix}0\\ 1\end{bmatrix}$$

Solution. Taking Laplace transform

$$pY(p)-\begin{bmatrix}0\\ 1\end{bmatrix}=AY(p)+\begin{Bmatrix}\frac{1}{2}\dfrac{p}{p^2+1}\\[2ex] \dfrac{3}{p^2+1}\end{Bmatrix}$$

where $\quad L\{X(t)\}=Y(p), \quad A=\begin{bmatrix}0 & -\frac{1}{2}\\ 2 & 0\end{bmatrix}$

Hence,

$$Y(p)=(pI-A)^{-1}\begin{bmatrix}\dfrac{p}{2(p^2+1)}\\[2ex] \dfrac{3}{p^2+1}+1\end{bmatrix}$$

$$=\frac{1}{p^2+1}\begin{bmatrix}p & -\frac{1}{2}\\ 2 & p\end{bmatrix}\begin{bmatrix}\dfrac{p}{2(p^2+1)}\\[2ex] \dfrac{3}{p^2+1}+1\end{bmatrix}$$

$$=\begin{bmatrix}-\dfrac{2}{(p^2+1)^2}\\[2ex] \dfrac{4p}{(p^2+1)^2}+\dfrac{p}{p^2+1}\end{bmatrix}$$

Using the formulae $\sin t \doteq \dfrac{1}{p^2+1}$, $\cos t \doteq \dfrac{p}{p^2+1}$, $\frac{1}{2}\sin t$

$-\frac{1}{2}t\cos t \doteq \dfrac{1}{(p^2+1)^2)}$, $t\sin t \doteq \dfrac{2p}{(p^2+1)^2}$ we shall finally have

$$X(t)=\begin{bmatrix}t\cos t-\sin t\\ 2t\sin t+\cos t\end{bmatrix},$$

i.e. if $X=\begin{bmatrix}x\\ y\end{bmatrix}$, then

$$x=t\cos t-\sin t,\ y=2t\sin t+\cos t.$$

(*f*) Partial Differential Equations

Example 9. Solve the following partial differential equation

$$\frac{\partial^2 z}{\partial t^2}=4\frac{\partial^2 z}{\partial x^2} \qquad t>0,\ x>0$$

where $z(0, x)=0, \dfrac{\partial z}{\partial t}(0, x)=2,\ z(t, 0)=\sin t.$

Solution. Regarding t as the variable and taking Laplace transform, we have $p^2 Z(p, x)-2=4\dfrac{d^2 Z}{dx^2}(p, x)$

where $z(t, x)=Z(p, x)$.

Solving the ordinary differential equation in $Z(p, x)$ we get,

$$Z(p, x)=Ae^{-\frac{p}{2}x}+Be^{\frac{p}{2}x}+\frac{2}{p^2}$$

From the condition $z(t, 0)=\sin t$, we have

$$Z(p, 0)=\frac{1}{p^2+1}$$

Since $z(t, x)$ exists for $x\to\infty$, Z (p, x) also exists for $x\to\infty$. Applying these two conditions we get,

$$B=0,\ A=\frac{1}{p^2+1}-\frac{2}{p^2}.$$

Hence $Z(p, x)=\frac{2}{p^2}+e^{-\frac{px}{2}}\left(\frac{1}{p^2+1}-\frac{2}{p^2}\right).$

Taking the inverse transform we have finally

$$z(t, x)=2t+\left\{\sin\left(t-\frac{x}{2}\right)-2\left(t-\frac{x}{2}\right)\right\}H\left(t-\frac{x}{2}\right).$$

(g) Integral Equations

Consider a convolution equation of the form

$$y(t)=f(t)+\lambda\int_0^t k(t-x)\,y(x)\,dx$$

Taking the Laplace transform of both sides,

$$Y(p)=F(p)+\lambda\,K(p)\,Y(p),$$

i.e.,
$$Y(p)=\frac{F(p)}{1-\lambda K(p)}=F(p)+\frac{\lambda K(p)}{1-\lambda K(p)}F(p)$$

Taking the inverse transform, the solution is obtained in the from

$$y(t)=f(t)+\lambda\int_0^t R(t-x)\,f(x)\,dx$$

where,
$$R(t)\doteq\frac{K(p)}{1-\lambda K(p)}$$

Example 10. Solve the integral equation

$$y(t)=1+2\int_0^t e^{-2u}\,y(t-u)\,du.$$

Solution. Taking Laplace transform

$$Y(p)=\frac{1}{p}+\frac{2}{p+2}Y(p),$$

i.e. $Y(p)=\frac{1}{p}+\frac{2}{p^2}$

Hence, $y(t)=1+2t.$

Example 11. (Abel's Integral Equation)

$$\int_0^t \frac{y(x)\,dx}{(t-x)^\alpha}=f(t)\,;\ t>0,\ 0<\alpha<1,\ f(0)=0. \qquad \ldots(i)$$

Solution. Taking the Laplace transform and applying the convolution theorem,

$$Y(p)\frac{\Gamma(1-\alpha)}{p^{1-\alpha}}=F(p),$$

i.e.,

$$Y(p)=\frac{1}{\Gamma(\alpha)\,\Gamma(1-\alpha)}\cdot p\frac{\Gamma(\alpha)}{p^\alpha}\cdot F(p)$$

$$=\frac{\sin\alpha\pi}{\pi}\,p\,\frac{\Gamma(\alpha)}{p^\alpha}\,F(p)$$

Taking the inverse transform

$$y(t)=\frac{\sin\alpha\pi}{\pi}\frac{d}{dt}\int_0^t \frac{f(x)\,dx}{(t-x)^{1-\alpha}}. \qquad \ldots(ii)$$

The later integral can be simplified as follows :

Integrating by parts.

$$\int_0^t \frac{f(x)\,dx}{(t-x)^{1-\alpha}}=\left[-f(x)\,\frac{(t-x)^\alpha}{\alpha}\right]_0^t+\int_0^t \frac{f'(x)\,(t-x)^\alpha}{\alpha}\,dx$$

$$=\frac{1}{\alpha}\int_0^t f'(x)\,(t-x)^\alpha\,dx$$

Differentiating both sides with respect to t,

$$\frac{d}{dt}\int_0^t \frac{f(x)\,dx}{(t-x)^{1-\alpha}}=\left[\frac{1}{\alpha}f'(x)\,(t-x)^\alpha\right]_{t=x}$$

$$+\frac{1}{\alpha}\int_0^t f'(x)\frac{d\,(t-x)^\alpha}{dt}\,dx$$

$$=\int_a^t \frac{f'(x)\,dx}{(t-x)^{1-\alpha}}.$$

Hence finally, $$y(t)=\frac{\sin\alpha\pi}{\pi}\int_0^t \frac{f'(x)\,dx}{(t-x)^{1-\alpha}}.$$

Remark 1. By a change of variable, the equation

$$\int_a^t \frac{y(x)}{(t-x)^\alpha}\,dx=f(t),\ t>a,\ 0<\alpha<1,\ f(a)=0$$

reduces to the canonical from

$$\int_0^{t+a} \frac{y(x+a)dx}{(t+a-x)^\alpha}=f(t+a).$$

The later equation has solution

$$y(t+a)=\frac{\sin \pi\alpha}{\pi}\frac{d}{dt}\int_0^t \frac{f(x+a)\,dx}{(t-x)^{1-\alpha}}$$

which can be reduced to

$$y(t)=\frac{\sin \pi\alpha}{\pi}\frac{d}{dt}\int_a^t \frac{f(x)dx}{(t-x)^{1-\alpha}}.$$

A similar manipulation leads to the solution of the equation

$$\int_t^b \frac{y(x)dx}{(x-t)^\alpha}=f(t)\ ;\ 0<t<b,\ 0<\alpha<1, f(b)=0$$

as $y(t)=\dfrac{-\sin \pi\alpha}{\pi}\dfrac{d}{dt}\displaystyle\int_t^b \frac{f(x)dx}{(x-t)^{1-\alpha}}$

Remark 2. Consider the generalised form of (*i*)

$$\int_0^t \frac{y(x)dx}{\{\phi(t)-\phi(x)\}^\alpha}=f(t),\ t>0,\ 0<\alpha<1 \qquad \text{...(iii)}$$

where ϕ is a strictly monotonic increasing function.

Put $\phi(t)=u$, $\phi(x)=v$, $\phi(0)=a$,

$$\frac{y(x)}{\phi'(x)}=z(v),\ f(t)=g(u).$$

Then the equation (*iii*) reduces to

$$\int_a^u \frac{z(v)dv}{(u-v)^\alpha}=g(u) \qquad \text{...(iv)}$$

whose solution is

$$z(v)=\frac{\sin \pi\alpha}{\pi}\frac{d}{dv}\int_a^v \frac{g(u)du}{(v-u)^{1-\alpha}}$$

i.e. $\dfrac{y(x)}{\phi'(x)}=\dfrac{\sin \pi\alpha}{\pi}\dfrac{1}{\phi'(x)}\dfrac{d}{dx}\displaystyle\int_0^x \frac{f(t)\,.\,\phi'(t)dt}{\{\phi(x)-\phi(t)\}^\alpha}$

i.e. $y(t)=\dfrac{\sin \pi\alpha}{\pi}\dfrac{d}{dt}\displaystyle\int_0^t \frac{f(x)\,.\,\phi'(x)dx}{\{\phi(t)-\phi(x)\}^{1-\alpha}}$ ...(*v*)

Eor example, the equation

$\displaystyle\int_0^t \frac{y(x)dx}{(t^2-x^2)^\alpha}=f(t)$ has solution

$$y(t)=\frac{2\sin \pi\alpha}{\pi}\frac{d}{dt}\int_0^t \frac{xf(x)dx}{(t^2-x^2)^{1-\alpha}}$$

Example 12. Use the Laplace transform to compute the resolvent Kernel of the Volterra Kernel

$$\sinh(x-t) \quad \text{with } \lambda=1.$$

Solution. $K(p)=L\{\sinh t\}$

$$=\frac{1}{p^2-1}$$

Hence $$R(p)=\frac{K(p)}{1-K(p)}=\frac{1}{p^2-2}$$

Hence the resolvent Kernal is

$$\frac{1}{\sqrt{2}}\sinh\sqrt{2}\,(t-x)$$

Exercise 35

1. Assuming term-by-term transform is valid for the Bessel function

$$J_0(t)=1-\frac{t^2}{2^2}+\frac{t^4}{2^2.4^2}-\ldots,$$

prove $L\{J_0(t)\}=\dfrac{1}{\sqrt{p^2+1}}$.

Deduce $\displaystyle\int_0^\infty J_0(t)dt=1$

2. Functions are defined as follows :

(*i*) $E_i(t)=\displaystyle\int_t^\infty \frac{e^{-x}}{x}dx,\ 0<t<\infty$ (Exponential integral)

(*ii*) $S_i(t)=\displaystyle\int_t^\infty \frac{\sin x}{x}dx,\ -\infty<t<\infty$ (Sine integral)

(*iii*) $C_i(t)=\displaystyle\int_t^\infty \frac{\cos x}{x}dx,\ 0<t<\infty$ (Cosine integral)

(*iv*) $E_rf(t)=\dfrac{2}{\sqrt{\pi}}\displaystyle\int_0^t e^{-x^2}dx,\ -\infty<t<\infty$

(Error function)

(*v*) $E_rf_c(t)=1-E_rf(t)$ (Complementary error function)

(*vi*) $L_n(t)=\dfrac{e^t}{n!},\dfrac{d^n}{dt^n}(t^ne^{-t})$ (Laguerre polynomial)

Prove that

$$E_i(t)\doteq\frac{1}{p}\log(p+1)$$

$$S_i(t)\doteq\frac{1}{p}\tan^{-1}p$$

$$C_i(t)\doteq\frac{1}{2p}\log(p^2+1)$$

$$\text{E}f_r(\sqrt{t}) \doteq \frac{1}{p\sqrt{p+1}}$$

$$\text{E}rf_c(\sqrt{t}) \doteq \frac{1}{p} - \frac{1}{p\sqrt{p+1}}$$

$$\text{L}_n(t) \doteq \frac{1}{p}\left(1-\frac{1}{p}\right)^n$$

3. Find (*i*) $L^{-1}\left\{\dfrac{2p^2+5p-4}{p^3+p^2+2p}\right\}$

(*ii*) $L^{-1}\left\{\dfrac{1+2p}{(p+2)^2\,(p-1)^2}\right\}$

4. A function $f(t)$ is said to be *anti-periodic* if

$$f(t+a) = -f(t)$$

Prove $L\{f(t)\} = (1+e^{-ap})^{-1} \int_0^a f(u)e^{-pu}du$.

5. Prove that $L\left\{\dfrac{\sin t}{t}\right\} = \tan^{-1}\dfrac{1}{p}$

6. Solve the initial problems by using L.T.

(*i*) $y''+2y'+y=t$; $y(0)=-3$, $y'(0)=0$

(*ii*) $y''+ty'-y=0$; $y(0)=0$, $y'(0)=1$

(*iii*) $y'''-3y'+2y=3e^t$; $y(0)=0$, $y'(0)=1$, $y''(0)=2$

(*iv*) $y''-2y'+5y=5-16\,e^{-t}$; $y(0)=0$, $y'(0)=3$

7. Solve the boundary value problems by using L.T.

(*i*) $y''+y=t$; $y(0)=y(\pi/2)=0$

(*ii*) $y''+9y=8\sin t$; $y(0)=0$, $y(\pi/2)=-1$

(*iii*) $y''-y'-2y=e^{-t}$; $y(0)=1$, $y(t)$ is bounded as $t\to\infty$.

8. Solve the following systems by using L.T.

(*i*) $y'+2z=0$; $y(0)=z(0)=0$.
$z'=3y+z=e^t$

(*ii*) $y'-y+z=0$, $z'-y-z=0$; $y(0)=z(0)=1$

(*iii*) $y''+a^2z'=b$, $z''-a^2y=0$
$y(0)=y'(0)=0$, $z(0)=z'(0)=0$

(*iv*) $y'-z=e^t$ $z'+y=\sin t$; $y(0)=1$, $z(0)=0$.

9. Solve the following partial differential equations in the region $x>0$, $t>0$ subject to given conditions :

(*i*) $\dfrac{\partial z}{\partial t}+\dfrac{\partial z}{\partial x} = xt$; $z(x, 0)=\sin x$, $z(0, t)=0$

(*ii*) $\frac{\partial^2 z}{\partial t^2}-\frac{\partial^2 z}{\partial x^2}=xe^{-t}$; $z(x, 0)=1$, $\frac{\partial z(x, 0)}{\partial t}=0$,

$$z(0, t)=\cos t.$$

10. Solve the integral equations :

(*i*) $y(t)=t+2\int_0^t \cos(t-u)y(u)dx$ (*ii*) $\int_0^t \cos(t-u)y(u)\,du=t$

(*iii*) $\int_0^t e^{t-u}y(u)du=t$ (*iv*) $y(t)=f(t)+\lambda\int_0^t e^{t-u}y(u)du.$

(*v*) $y(t)=4t^2-\int_0^t y(t-u)e^{-u}\,du$

(*vi*) $y(t)=e^{-t}-2\int_0^t y(u)\cos(t-u)du$

11. Use the Laplace transform method to find the resolvent kernel of the following Volterra kernels with $\lambda=1$:

(*i*) $\sin(t-x)$ (*ii*) $2\sin(t-x)$

(*iii*) $\cos(t-x)$ (*iv*) $2\cos(t-x)$

(*v*) $\sinh(t-x)$ (*vi*) $\cosh(t-x)$

(*vii*) $e^{-(t-x)}$ (*viii*) $e^{-(t-x)}\sin(t-x)$

[Hint : $R(t)\doteq\frac{K(p)}{1-\lambda K(p)}$ and resolvent kernel $R(t-x)$]

12. Solve the IVPs

(*i*) $y'''-4y'=\cosh 2t-\sinh 2t$

$y(0)=1, y'(0)=-6, y''(0)=4$

[**Ans.** $\cosh 2t-3\sinh 2t$]

(*ii*) $y'''+y'=-2\sin t+2\cot$, $y(0)=1, y'(0)=0, y''(0)=2$

[**Ans.** $y=t\sin t-t\cos t+\sin t+1$]

13. Prove that $L^{-1}\{8a^3p(p^2+a^2)^{-3}\}=t\sin at-at^2\cos at$

14. Prove that the solution of IVP

$$\left(\text{where } x'\equiv\frac{dx}{dt}\right)$$

$$x''+y'=x+y,\; y''-2x'=2x+y+e^{-t}$$

is

$$x(t)=\tfrac{1}{2}\cos t-\sin t-\tfrac{1}{2}(1+t)e^{-t}$$

$$y(t)=\frac{1}{2}\cos t+\frac{3}{2}\sin t+\frac{1}{2}e^{-t}$$

15. Use LT to prove that the solution of the integro-differential equation

$$y'(t)=\cos t+\int_0^t y(t-x)\cos x\,dx,\; y(0)=1$$

is

$$y(t)=1+t+\tfrac{1}{2}t^2$$

§11·5. Other Integral Transforms

The Concept of Integral Transform. In the preceding sections of this chapter we studied in detail the basic properties and uses of the Laplace transform. The Laplace transform belongs to a more general class of integral transforms which can be respresented as

$$F(p)=K\{f(t)\}=\int_a^b K(p, t) f(t)\, dt \qquad \ldots(1)$$

where either a or b or both can be $\pm\infty$ and the fixed bivariate function $K(p, t)$ called the *Kernel* of the transform, is chosen as per requirements of the domain of functions.

The function $f(t)$ is called the *original* and $F(p)$ the *image*. For the transform to be valid, the integral in (1) must be convergent for all p which may be a complex number. This requires certain regularity conditions to be satisfied by the kernel and the original. The subject of transform calculus does have a sound analytical base, but its importance accrues mainly from its extensive usage. For any transform to be of use in practice, it should be possible to recover the original from its image. If recovery of the exact original is not possible, it should at least be amenable to current approximation techniques. Fortunately, methods of highest accuracy exist for numerical inversion at least in the case of Laplace and Fourier transforms[1].

Operational Calculus. The other significance of the integral transforms stems from the basis they provide for the algebra of certain differential operators. For example, the Laplace transform provides justification of the algebra of the operator

$$D\equiv\frac{d}{dx} \text{ through its property}$$

$$\text{L.T. } \{f'\}=p\, F(p) \text{ if } f(0)=0.$$

Thus the L.T. maps the operator D into the number p. Similarly the Fourier transform maps D into the number ip. The fact that D is indeed the most basic of all differential operators is one of the reasons for the immense popularity of Laplace and Fourier transforms. The other reason for their popularity is the existence of convolution integrals which are mapped into product of numbers or ordinary product of functions.

Other examples are the Mellin transform which is the operational culculus of the differential operator x D and [the Hankel transform which is the operational calculus of the Bessel operator D x D and other related operators. Owing to this relationship, the field of integral transforms is also known as *operational calculus*.

---

1. There is a beautiful book by V.I. Krylov and N.S. Skoblya entitled : "A Handbook of Methods of Approximate Fourier Transformation and Inversion of the Laplace Transformation" from MIR Publishers, Moscow (1977).

The operational aspect rests heavily on the facility with which an integral transform can be inverted. In the particular case when $K(x, t)=K(x\,t)$, Sneddon has shown that if

$$F(x)=\int_0^\infty K(x\,t)\,f(t)\,dt$$

and
$$f(t)=\int_0^\infty H(x\,t)\,F(x)\,dx$$

are inverse, then $\bar{K}(s)\,.\,\bar{H}(1-s)=1$, where $\bar{K}(s)$ and $\bar{H}(s)$ are Mellin transforms of $K(x)$ and $H(x)$ respectively *i.e.*,

$$\bar{K}(s)=\int_0^\infty x^{s-1}\,K(x)\,dx$$

$$\bar{H}(s)=\int_0^\infty x^{s-1}\,H(x)\,dx.$$

In general suppose that together with, (1) we have

$$f(t)=\int_c^d H(p, t)\,F(p)\,dp \qquad \ldots(2)$$

$$t\in(a, b)$$

then (2) is called the inversion formula for the transform (1). Equation (1) with (2) taken together constitute an *inversion theorem* for the transform (1).

The range of integration in (1) may be finite or $[0, \infty)$ or $(-\infty, \infty)$ depending on the nature of the originals. The most widely used technique of inverting an integral transform is by means of a complex integral so that (c, d) is a complex contour. In case the ranges of integration in (1) and (2) are the same and $K(p, t)=H(t, p)$ so that the relation between the original f and image F is symmetrical, the Kernel $K(p, t)$ is called a *Fourier Kernel.* $K(p, t)$ is called a *symmetric Kernel* if $K(p, t)=K(t, p)$. If $K(p, t)=H(p, t)$ then it is called a *self-reciprocal kernel.*

List of Transforms. We have earlier studied the Laplace transform in detail. The following chart shows the commonly studied integral transforms which will be described in the sequel.

Table 1

Name of Transform	(a, b)	$K(p, t)$	(c, d)	$H(p, t)$
1. Laplace	$(0, \infty)$	$e^{\ pt}$, Re $p>p_0$	$(\gamma-i\infty,$ $\gamma+i\infty)$ $\gamma>p_0$	$\frac{1}{2\pi i}e^{pt}$
2. Fourier	$(-\infty, \infty)$	$\frac{1}{\sqrt{2\pi}}e^{ipt}$	$(-\infty, \infty)$	$\frac{1}{\sqrt{2\pi}}e^{-ipt}$
3. Mellin	$(0, \infty)$	t^{p-1}	$(\gamma-i\infty,$ $\gamma+i\infty)$	$\frac{1}{2\pi i}t^{-p}$
4. Hankel	$(0, \infty)$	$t\,J_\nu(pt)$ $\nu>-\frac{1}{2}$	$(0, \infty)$	$t\,J_\nu(pt)$

§11·6. The Fourier Transform (F.T.)

Definition 5. The Fourier transform pair is given by

$$\text{F.T.}\{f(t)\}=F(p)=\frac{1}{\sqrt{2\pi}}\int_{-\infty}^{\infty}e^{ipt}f(t)\,dt \qquad \ldots(1)$$

$$-\infty<p<\infty$$

and

$$f(t)=\frac{1}{\sqrt{2\pi}}\int_{-\infty}^{\infty}e^{-ipt}F(p)\,dp \qquad \ldots(2)$$

$$-\infty<t<\infty$$

Often the transform pair is also defined as

$$F(p)=\int_{-\infty}^{\infty}e^{ipt}f(t)\,dt$$

$$f(t)=\frac{1}{2\pi}\int_{-\infty}^{\infty}e^{-ipt}F(p)\,dp.$$

The *linearity* of the transform pair is also symbolised by the notation

$$\mathscr{H}\{f(t)\}=F(p)\ ;\ \mathscr{H}^{-1}\{F(p)\}=f(x).$$

The notational variations are a matter of choice and are not substantial.

If we combine the formulae (1) and (2) we get the identity.

$$f(t)=\frac{1}{2\pi}\int_{-\infty}^{\infty} e^{-ipt}\left[\int_{-\infty}^{\infty} e^{ipt} f(t)\, dt\right] dp \qquad ...(3)$$

The idenitify (3) is known as *Fourier integral theorem.*

If $f(t)$ is an even function, F.T. (1) is reduced to the *Fourier cosine transform*

$$\mathscr{H}_c\{f(t)\}=F_c(p)=\sqrt{\frac{2}{\pi}}\int_0^{\infty} f(t)\cos pt\,dt \qquad ...(4)$$

$$0<p<\infty$$

and the inversion formula is

$$f(t)=\sqrt{\frac{2}{\pi}}\int_0^{\infty} F_c(p)\cos pt\, dp \qquad ...(5)$$

$$0<p<\infty$$

If $f(t)$ is an odd function we get the *Fourier sine transform.*

$$\mathscr{H}_s\{f(t)\}=F_s(p)=\sqrt{\frac{2}{\pi}}\int_0^{\infty} f(t)\sin p\,t\, dt \qquad ...(6)$$

$$0<p<\infty$$

and the inversion formula is

$$f(t)=\sqrt{\frac{2}{\pi}}\int_0^{\infty} F_s(p)\sin pt\, dp \qquad ...(7)$$

$$0<t<\infty$$

For a function $f \in L^1$ we have the *Fourier integral* representation

$$f(t)=\frac{1}{2\pi}\int_{-\infty}^{\infty}\int_{-\infty}^{\infty} f(x)\cos p(x-t)\, dx\, dp \qquad ...(8)$$

(1) is also called the complex form of Fourier transform to distinguish it from others.

Origin of F.T. Consider the Dirichlet problem

PDE : $\dfrac{\partial^2 u}{\partial x^2}+\dfrac{\partial^2 u}{\partial y^2}=0$ in $-l\leqslant x\leqslant, l\ y\geqslant 0$

BCs : (*i*) $u(x, 0)=f(x)-l<x<l$

(*ii*) $u(x, y)\to 0$ as $y\to\infty$.

The method of separation of variable leads to a solution of the form

$$u(x, y)=\sum_{n=-\infty}^{\infty} a_n \exp.\left(-\frac{in\pi x}{l}-\frac{|n|\pi y}{l}\right) \qquad ...(9)$$

where in view of the BC (*i*),

$$f(x)=\sum_{n=-\infty}^{\infty} a_n\, e^{\frac{in\pi t}{l}}, \quad -l<x<l \qquad ...(10)$$

The series (10) is the Fourier series of $f(x)$ where

$$a_n=\frac{1}{2l}\int_{-l}^{l} f(t)\, e^{-\frac{in\pi t}{l}}\, dt \qquad ...(11)$$

With these values of a_n if $f(x)$ is a piecewise continuous function on $-(l, l)$, then the Fourier series (10) is convergent to $f(x)$ or the normalised value $\dfrac{f(x+)+f(x-)}{2}$.

If we substitute (11) into (10) we get the identity

$$f(x)=\frac{1}{2l}\sum_{n=-\infty}^{\infty}\left[e^{-\frac{in\pi x}{l}} \left[\int_{-l}^{l} f(t)\, e^{\frac{i\pi nt}{l}}\, dt\right]\right].$$

Put $\pi/l=\delta u$, then

$$f(x)=\frac{1}{2\pi}\sum_{n=-\infty}^{\infty}\left[\int_{-l}^{l} f(t)\, .\, i^{(t-x)n\delta u}\, dt\right]\delta u.$$

Now $\delta u \to 0$ as $l\to\infty$. Hence by the definition of the Riemann integral as the limit of a sum, the series on the right can be written as an integral and we have the Fourier integral

$$f(x)=\frac{1}{2\pi}\int_{-\infty}^{\infty} e^{-i\omega u}\left[\int_{-\infty}^{\infty} f(t)\, e^{itu}\, dt\right] du \qquad ...(12)$$

We can express this integral in the form

$$f(x)=\frac{1}{\sqrt{2\pi}}\int_{-\infty}^{\infty} F(u)\, e^{-i\omega u}\, dy \qquad ...(13)$$

where
$$F(u)=\frac{1}{\sqrt{2\pi}}\int_{-\infty}^{\infty} f(x)\, e^{i\omega u}\, dx \qquad ...(14)$$

Also we can express the solution of the potential problem as

$$u(x, y)=\frac{1}{2\pi}\int_{-\infty}^{\infty} F(u)\, .\, e^{-ixu-|u|y}\, du$$

where $F(u)$ is given by (14).

Riemann—Lebesgue Lemma. If $f\in L^1$ and $F(p)$ is the F.T of f, then $F(p)\to 0$ as $p\to\infty$.

Proof. $$2F(p)=\frac{2}{\sqrt{2\pi}}\int_{-\infty}^{m} e^{ipt} f(t)\, dt$$
$$=-\frac{2}{\sqrt{2\pi}}\int_{-\infty}^{\infty} e^{it\left(p+\frac{\pi}{t}\right)} f(t)\, dt$$

$$=\frac{1}{\sqrt{2\pi}}\int_{-\infty}^{\infty}\left[f(t)-f\left(t-\frac{\pi}{p}\right)\right]e^{ipt}\,dt$$

(by change of variable)

Hence $2\,|\,F(p)\,| \leqslant \frac{1}{\sqrt{2\pi}}\int_{-\infty}^{\infty}\left|f(t)-f\left(t-\frac{\pi}{p}\right)\right|dt$

Since the integrand belongs to L^1 and $\to 0$ as $p\to\infty$, by the dominated convergence theorem, $|\,F(p)\,|\to 0$ as $p\to\infty$.

Note The lemma can also be expressed in the forms

$$\lim_{p\to\infty}\int_0^{\infty} f(t)\sin pt\,dt=0$$

$$\lim_{p\to\infty}\int_0^{\infty} f(t)\cos pt\,dt=0$$

Important Properties. In what follows we shall assume that f is piecewise continuous and $f\in L^1(-\infty,\infty)$.

I. Derivatives

$$\text{F.T.}\left\{\frac{df(t)}{dt}\right\}=\frac{1}{\sqrt{2\pi}}\int_{-\infty}^{\infty}e^{ipt}f'(t)\,dt$$

$$=\frac{1}{\sqrt{2\pi}}\Big[e^{ipt}f(t)\Big]_{-\infty}^{\infty}-\frac{ip}{\sqrt{2\pi}}\int_{-\infty}^{\infty}e^{ipt}f(t)\,dt$$

$$=-ip\,F(p),\ \text{since } f(t)\to 0 \text{ as } |\,t\,|\to\infty$$

Owing to the absolute convergence of the defining integral (1), we can differentiate under the integral sign. Hence

$$\frac{dF(p)}{dp}=\frac{i}{\sqrt{2\pi}}\int_{-\infty}^{\infty}e^{ipt}\,t\,f(t)\,dt$$

Thus $\text{F.T.}\,\{f'(t)\}=(-ip)\,F(p)$ ...(15)

$$\text{F.T.}\,\{it\,f(t)\}=\frac{dF(p)}{dp} \quad ...(16)$$

If we assume that $f^{(n)}(t)$ is piecewise continuous and $f^{(n)}(t)\in L^1$, then $f, f^1,\ldots, f^{(n-1)}\to 0$ as $|\,t\,|\to\infty$ and we have the general formulae

$$\text{F.T.}\left\{\frac{d^nf(t)}{dt^n}\right\}=(-ip)^n\,F(p) \quad ...(17)$$

$$\text{F.T.}\,\{(it)^n f(t)\}=\frac{d^n F(p)}{dp^n} \quad ...(18)$$

The correspondence between the F.T. and differentiation operation is symbolised as follows:

$$\frac{d}{dt} \leftrightarrow -ip$$

$$\frac{d}{dp} \leftrightarrow it$$

For the Fourier sine and cosine transforms we have if $f(t), f'(t) \to 0$ as $t \to \infty$,

$$\begin{aligned} F_s(f') &= -pF_c(p) \\ F_s(f'') &= pf(0+) - p^2 F_s(p) \\ F_c(f') &= -f(0+) + p F_s(p) \\ F_c(f'') &= -f'(0+) - p^2 F_c(p) \end{aligned} \qquad \ldots(19)$$

II. Translations

$$\begin{aligned} \text{F.T. } \{f(t-a)\} &= \frac{1}{\sqrt{2\pi}} \int_{-\infty}^{\infty} e^{ipt} f(t-a)\, dt \\ &= \frac{e^{ipa}}{\sqrt{2\pi}} \int_{-\infty}^{\infty} e^{ipt} f(t)\, dt \\ &= e^{ipa} F(p) \end{aligned} \qquad \ldots(20)$$

$$\begin{aligned} \text{F.T. } \{e^{iat} f(t)\} &= \frac{1}{\sqrt{2\pi}} \int_{-\infty}^{\infty} e^{i(p+a)t} f(t)\, dt \\ &= F(p+a) \end{aligned} \qquad \ldots(21)$$

III. Convolutions

Definition 6. The *Fourier convolution* of two functions $f, g \in L^1$ is defined by

$$h(t) = (f * g)(t) = \frac{1}{\sqrt{2\pi}} \int_{-\infty}^{\infty} f(t-u)\, g(u)\, du \qquad \ldots(22)$$

It is easily verified that

$$f_1 * f_2 = f_2 * f_1 \qquad \text{(Commutativity)}$$

$$(f_1 * f_2) * f_3 = f_1 * (f_2 * f_3) \quad \text{(Associativity)}$$

$$\begin{aligned} \text{F.T. } \{(f * g)(t)\} &= \frac{1}{\sqrt{2\pi}} \int_{-\infty}^{\infty} \Big[\text{F.T. } \{f(t-u)\} \Big] g(u)\, du \\ &= \frac{1}{\sqrt{2\pi}} \int_{-\infty}^{\infty} e^{ipu} F(p)\, g(u)\, du \\ &= F(p) \,.\, G(p) \end{aligned} \qquad \ldots(23)$$

Fourier convolution reduces to Laplace convolution if $f(t)$ and $g(t)$ vanish for $t<0$.

A result dual to (23) is : if $f \,.\, g \in L^1$ then

$$\text{F.T. } \{f(t) \,.\, g(t)\} = \frac{1}{\sqrt{2\pi}} \int_{-\infty}^{\infty} e^{ipt} f(t)\, g(t)\, dt$$

$$=\frac{1}{\sqrt{2\pi}}\int_{-\infty}^{\infty}\left[\frac{1}{\sqrt{2\pi}}\int_{-\infty}^{\infty} e^{-iwt}\,F(w)\,dw\right] e^{ipt}\,g(t)\,dt$$

$$=\frac{1}{\sqrt{2\pi}}\int_{-\infty}^{\infty}\left[\frac{1}{\sqrt{2\pi}}\int_{-\infty}^{\infty} e^{i(p-wt)}g(t)dt\right] F(w)\,dw$$

$$=\frac{1}{\sqrt{2\pi}}\int_{-\infty}^{\infty} F(w)\,G(p-w)\,dw$$

i.e., $$\int_{-\infty}^{\infty} e^{ipt} f(t)\,g(t)\,dt=\int_{-\infty}^{\infty} F(w)\,G(p-w)\,dw \qquad \ldots(24)$$

Also $$\int_{-\infty}^{\infty} e^{ipt} f(t)\,g(t)\,dt=\int_{-\infty}^{\infty} F(p-w)\,G(w)\,dw \qquad \ldots(25)$$

We can also express (23) in the from

$$\int_{-\infty}^{\infty} f(t-u)\,g(u)\,du=\int_{-\infty}^{\infty} e^{ipt}\,F(p)\,G(p)\,dp \qquad \ldots(26)$$

IV Parseval Relations

From (24) and (26), we shall get

$$\int_{-\infty}^{\infty} f(t)\,g(t)\,dt=\int_{-\infty}^{\infty} F(w)\,G(-w)\,dw$$

$$=\int_{-\infty}^{\infty} F(-w)\,G(w)\,dw \qquad \ldots(27)$$

and $$\int_{-\infty}^{\infty} f(-u)\,g(u)\,du=\int_{-\infty}^{\infty} F(p)\,G(p)\,dp$$

$$=\int_{-\infty}^{\infty} f(u)\,g(-u)\,du \qquad \ldots(28)$$

In (28) put $g(-u)=\bar{f}(u)$ *i.e.,* $g(u)=\bar{f}(-u)$.
Since F.T. $\{g(u)\}=G(p)$

$$=\text{F.T. }\{\bar{f}(-u)\}$$

$$=\bar{F}(p)$$

we get $$\int_{-\infty}^{\infty} f(u)\,.\,\bar{f}(u)\,du=\int_{-\infty}^{\infty} F(p)\,\overline{F(p)}dp$$

i.e., $$\int_{-\infty}^{\infty} |\,f(u)\,|^2\,du=\int_{-\infty}^{\infty} |\,F(p)\,|^2\,dp \qquad \ldots(29)$$

i.e. $$\|\,f\,\|_2=\|\,F\,\|_2 \qquad \ldots(30)$$

Relations (27), (28) and (29) are known as *Parseval relations*, although (29) is regarded as the standard form of the Parseval relation.

Relation (30) shows that if $f \in L^2$ then $F \in L^2$ and that the Fourier transform is an isometric automorphism on $L^2(-\infty, \infty)$.

Example 13. Use F.T. to solve the BVP,

ODE : $y''-k^2y=f(x)$ ...(*i*)

BCs : $y(\pm\infty)=0$ ...(*ii*)

Solution. Let $F(p)=\dfrac{1}{\sqrt{2\pi}}\displaystyle\int_{-\infty}^{\infty} e^{ipx} f(x)\, dx$

$$Y(p)=\frac{1}{\sqrt{2\pi}}\int_{-\infty}^{\infty} e^{ipx}\, y(x)\, dx$$

Taking the F.T. of (*i*)

$$-p^2Y(p)-k^2Y(p)=F(p)$$

i.e.,
$$Y(p)=\frac{-F(p)}{p^2+k^2}$$

Now $\quad \text{F.T.}\left\{ e^{-k\,|\,t\,|} \right\}=\sqrt{\dfrac{2}{\pi}}\;\dfrac{k}{p^2+k^2}$

Hence $y(x)=-\dfrac{1}{\sqrt{2\pi}}\displaystyle\int_{-\infty}^{\infty} f(x-u)\sqrt{\frac{\pi}{2}}\;\frac{1}{k}\; e^{-k\,|\,u\,|}\; du, \; k>0$

$$=-\frac{1}{2k}\int_{-\infty}^{\infty} f(x-u)\,.\,e^{-k\,|\,u\,|}\; du$$

This can also be written as

$$y(x)=-\frac{1}{2k}\int_{-\infty}^{\infty} f(u)\,.\,e^{-k\,|\,x-u\,|}\; du.$$

Example 14. Use F.T. to solve the BVP,

PDE : $\dfrac{\partial^4 z}{\partial x^4}+\dfrac{\partial^2 z}{\partial y^2}=0,\; -\infty<x<\infty,\; y\geqslant 0$ ...(*i*)

BCs : (*ii*) z and its partial derivatives tend to 0 as $x\to\pm\infty$

(*iii*) $z=f(x)$ and $\dfrac{\partial z}{\partial y}=0$ on $y=0$.

Solution. Put $Z(p, y)=\dfrac{1}{\sqrt{2\pi}}\displaystyle\int_{-\infty}^{\infty} z(x, y)\,.\,e^{ipx}\, dx$

Integration by parts leads to

$$\frac{1}{\sqrt{2\pi}}\int_{-\infty}^{\infty} e^{ipx}\,\frac{\partial^4 z}{\partial x^4}\, dx=p^4 Z(p, y)$$

Also $\quad \text{F.T.}\left\{\dfrac{\partial^2 z}{\partial y^2}\right\}=\dfrac{\partial^2 Z(p, y)}{\partial y^2}$

Hence taking the F.T. of (*i*), we get

$$\frac{d^2Z}{dy^2}+p^4Z=0.$$

The general solution is $Z(p, y)=A \cos p^2y+B \sin p^2y$

Hence $\dfrac{\partial Z(p, y)}{\partial y}=-p^2 A \sin p^2y+p^2 B \cos p^2y$

At $y=0$, we have $Z(p, y)=F(p)$ and $\dfrac{\partial Z}{\partial y}=0.$

Hence $B=0$ and $A=F(p)$. Also

$$Z=F(p) \cos p^2y$$

Taking the inverse F.T.

$$z(x, y)=\frac{1}{\sqrt{2\pi}}\int_{-\infty}^{\infty} e^{-ipx} F(p) \cos p^2y \, dp.$$

Problem. Use F.T. to solve the Dirichlet problem

PDE ; $\dfrac{\partial^2 u}{\partial x^2}+\dfrac{\partial^2 u}{\partial y^2}=0, \quad -\infty<x<\infty, y\geqslant 0$

BCs : $u(x, 0)=f(x) \quad -\infty<x<\infty$

$u(x, y)\to 0$ as $\sqrt{x^2+y^2}\to\infty.$

[**Hint** : See example 15 and exercise 32 (2) of chapter 9].

Example 15. Express the Fourier integral (8) in the form

$$f(t)=\frac{1}{\pi}\int_0^{\infty}\Big[A(\omega) \cos \omega t+B(\omega) \sin \omega t\Big]d\omega,$$

where $\quad A(\omega)=\displaystyle\int_{-\infty}^{\infty} f(x) \cos \omega x \, dx, \; B(\omega)=\int_{-\infty}^{\infty} f(x) \sin \omega x \, dx$

Solution. The integral with respect to p in (8) is an even function of p. Hence we can write

$$f(t)=\frac{1}{\pi}\int_0^{\infty}\left[\int_{-\infty}^{\infty} f(x) \cos p(x-t) \, dx\right] dp$$

$$=\frac{1}{\pi}\int_0^{\infty}\left[\int_{-\infty}^{\infty} f(x) (\cos px \,.\, \cos pt + \sin px \,.\, \sin pt) \, dx\right] dp \text{ etc.}$$

§11·7. The Mellin Transform

Definition 7. The Mellin transform and its inversion formula are given by

$$M\{f(x)\}=F(s)=\int_0^{\infty} x^{s-1} f(x)\, dx \quad \ldots(1)$$

$$\alpha<\text{Re } s<\beta$$

$$f(x)=\frac{1}{2\pi i}\int_{c-i\infty}^{c+i\infty} x^{-s} F(s)\, ds \quad \ldots(2)$$

$$\alpha<c<\beta$$

provided the integrals exist.

Origin of the Transform

The two-sided Laplace transform pair is given by

$$F(s)=\int_{-\infty}^{\infty} e^{-st} g(t)\, dt \qquad \alpha<\text{Re } s<\beta$$

$$g(t)=\frac{1}{2\pi}\int_{d-i\infty}^{d+i\infty} e^{st} F(s)\, ds$$

Put $e^{-t}=x$ and $g(t)=f(x)$, we get as desired

$$F(s)=\int_0^{\infty} x^{s-1} f(x)\, dx \quad \ldots(1)$$

$$f(x)=\frac{1}{2\pi}\int_{c-i\infty}^{c+i\infty} x^{-s} F(s)\, ds \quad \ldots(2)$$

Alternatively, the Fourier transform pair is

$$G(p)=\int_{-\infty}^{\infty} e^{ipt} g(t)\, dt \qquad \alpha<\text{Im}p<\beta$$

and

$$g(t)=\frac{1}{2\pi}\int_{id-\infty}^{id+\infty} G(p)\, e^{-ipt}\, dp \qquad \sigma<d<\beta$$

Put $p=-is$ and $e^{t}=x$. Then

$$G(-is)=\int_0^{\infty} g(\log x)\, x^{s-1}\, dx$$

$$g(\log x)=\frac{1}{2\pi i}\int_{d-i\infty}^{d+i\infty} G(-is)\, x^{-s}\, ds$$

Now put $G(-is)=F(s)$ and $g(\log x)=f(x)$ to get (1), (2).

Example 16 (*i*). The simplest example is the Gamma function integral *viz.*,

$$\int_0^{\infty} x^{s-1} e^{-ax}\, dx=\frac{\Gamma(s)}{a^s} \qquad a>0,\ \text{Re } s>0.$$

(*ii*) The *exponential integral* is defined by

$$E_1(x)=\int_x^{\infty} \frac{e^{-u}}{u}\, du$$

Hence $$M\{E_1(x)\}=\int_0^\infty x^{s-1}\left[\int_x^\infty \frac{e^{-u}}{u}\,du\right]dx$$

$$=\int_0^\infty x^{s-1}\left[\int_1^\infty \frac{e^{-vx}}{v}\,dv\right]dx$$

$$=\int_1^\infty\left[\int_0^\infty x^{s-1}e^{-vx}\,dx\right]\frac{dv}{v}$$

$$=\int_1^\infty \frac{\Gamma(s)}{v^{s+1}}\,dv$$

$$=\frac{\Gamma(s)}{s},\ \text{Re } s>0.$$

Important Properties. Assume $M\{f(x)\}=F(s)$.

I. (*i*) $M\{f(ax)\}=a^{-s}F(s)$

(*ii*) $M\{f(x^a)\}=\frac{1}{a}F(s/a) \qquad a>0$

(*iii*) $M\{x^{-1}f(x^{-1})\}=F(1-s)$

(*iv*) $M\{x^{\mu}f(x)\}=F(s+\mu)$

II. Derivatives

(*i*) $$M\{f'(x)\}=\int_0^\infty x^{s-1}f'(x)\,dx$$

$$=\left[x^{s-1}f(x)\right]_0^\infty-(s-1)\int_0^\infty x^{s-2}f(x)\,dx$$

Since $F(s)$ exists for $\alpha<\text{Re } s<\beta$,

$$\lim_{x\to 0} x^{s-1}f(x)=0 \qquad \text{Re } s>\alpha$$

$$\lim_{x\to \infty} x^{s-1}f(x)=0 \qquad \text{Re } s<\beta$$

Hence $M\{f'(x)\}=-(s-1)\,F(s-1),\ \alpha<\text{Re }(s-1)<\beta$ ...(3)

Similarly $M\{f''(x)\}=(s-1)(s-2)\,F(s-2)$

Generalising

$$M\{f^{(n)}(x)\}=(-1)^n(s-1)(s-2)\ldots\ldots(s-n)\,F(s-n)$$
$$=(1-s)_n\,F(s-n) \qquad \ldots(4)$$

(*ii*) Using $\frac{dx^{s-1}}{ds}=x^{s-1}.\log x$,

$$M\{\log x\,.\,f(x)\}=\frac{dF(s)}{ds} \qquad \ldots(5)$$

III. (i) $M\left\{\left(x\frac{d}{dx}\right)^n f(x)\right\}=(-1)^n s^n F(s)$ ...(6)

(ii) $M\left\{x^n \frac{d^n f}{dx^n}\right\}=(-1)^n (s)_n F(s)$ ...(7)

For $n=2$,

(iii) $M\left\{x^2 \frac{d^2 f}{dx^2}+x\frac{df}{dx}\right\}=s^2 F(s)$ ...(6′)

(iv) $M\left\{x^2 \frac{d^2 f}{dx^2}\right\}=s(s+1) F(s)$ ...(7′)

IV. **Integrals**

(i) $M\left\{\int_0^x f(u)\,du\right\}=-\frac{F(s+1)}{s}$

$M\left\{\int_x^\infty f(u)\,du\right\}=\frac{F(s+1)}{s}$

(ii) $M\left\{\int_0^x (x-t) f(t)\,dt\right\}=\frac{F(s+2)}{s(s+1)}$

(iii) $M\left\{\int_0^x \frac{(x-t)^{n-1}}{\underline{|n-1}} f(t)\,dt\right\}=\frac{(-1)^n}{(s)_n} F(s+n)$

(iv) $M\left\{\int_x^\infty \ldots^n \int_x^\infty f(t)\,dt^n\right\}=\frac{\Gamma(s+n)}{(s)_n}$

V. **Convolutions**

(i) $M\left\{\int_0^\infty u^\mu f(xu) g(u)\,du\right\}$

$$=\int_0^\infty s^{-1}\left[\int_0^\infty u^\mu f(xu) g(u)\,du\right]dx$$

$$=\int_0^\infty\left[\int_0^\infty x^{s-1} f(ux)\,dx\right]u^\mu g(u)\,du$$

$$=\int_0^\infty u^\mu g(u) u^{-s} F(s)\,du$$

$=F(s)\,G(1+\mu-s)$ ...(8)

(ii) $M\left\{x^\lambda \int_0^\infty u^\mu f(xu) g(u)\,du\right\}$

$=F(s+\lambda)\,G(1+\mu-\lambda-s)$ ...(9)

(iii) $M\left\{\int_0^\infty u^\mu f\left(\frac{x}{u}\right) g(u)\,du\right\}$

$$=\int_0^{\infty} x^{s-1}\left[\int_0^{\infty} u^{\mu} f\left(\frac{x}{u}\right) g(u)\, du\right] dx$$

$$=\int_0^{\infty}\left[\int_0^{\infty} x^{s-1} f\left(\frac{x}{u}\right) dx\right] u^{\mu} g(u)\, du$$

$$=\int_0^{\infty} u^{\mu+s} F(s)\, g(u)\, du$$

$$=F(s)\, G(1+\mu+s) \qquad \ldots(10)$$

(*iv*) Put $\mu=-1$, then

$$M\left\{\int_0^{\infty} f\left(\frac{x}{u}\right)\frac{g(u)}{u}\, du\right\}=F(s)\,.\,G(s) \qquad \ldots(11)$$

Definition 8 (Mellin Convolution). The integral

$$(fog)(x)=\int_0^{\infty} f\left(\frac{x}{u}\right) g(u)\,\frac{du}{u} \qquad \ldots(12)$$

is called the *Mellin convolution* of f and g and (11) is known as the *product theorem* of the Mellin transform.

(*v*) $$M\left\{\int_0^{\infty} f(xu)\, g(u)\, du\right\}=F(s)\, G(1-s) \qquad \ldots(13)$$

$$M\left\{\int_0^{\infty} f\left(\frac{x}{u}\right) g(u)\, du\right\}=F(s)\, G(1+s) \qquad \ldots(14)$$

(*iv*) If $M\{f\}=F(s)$, $M\{g\}=G(s)$, then

$$M\{f(x)\,.\,g(x)\}=\int_0^{\infty} x^{s-1} f(x)\, g(x)\, dx$$

$$=\int_0^{\infty} x^{s-1} g(x)\left[\frac{1}{2\pi i}\int_{c-i\infty}^{c+i\infty} x^{-u} F(u)\, du\right] dx$$

$$=\frac{1}{2\pi i}\left[\int_{c-i\infty}^{c+i\infty}\left[\int_0^{\infty} x^{s-u-1} g(x)\, dx\right] F(u)\, du\right.$$

i.e.,

$$\int_0^{\infty} x^{s-1} f(x)\, g(x)\, dx$$

$$=\frac{1}{2\pi i}\int_{ci-\infty}^{ci+\infty} F(u)\, G(s-u)\, du \qquad \ldots(15)$$

For $s=1$ we get

$$\int_0^{\infty} f(x)\, g(x)\, dx=\frac{1}{2\pi i}\int_{c-i\infty}^{c+i\infty} F(u)\, G(1-u)\, du \qquad \ldots(16)$$

(16) can also be obtained by inverting (13).

Example 17. (Integral Equations of Convolution Type).

(*a*) The integral equation

$$\int_0^\infty k(x\,t)\,y(t)\,dt=f(x)$$

can be transformed to $K(s)\,Y(1-s)=F(s)$.

Hence $Y(s)=\dfrac{F(1-s)}{K(1-s)}=L(s)\,F(1-s)$

where $L(s)=\dfrac{1}{K(1-s)}$ *i.e.*, $K(s)\,L(1-s)=1$.

Taking inverse Mellin transform

$$y(x)=\int_0^\infty l(x\,t)\,f(t)\,dt \qquad \text{where } M\{l(x)\}=L(s).$$

If $k(x)=l(x)$ *i.e.* $K(s)\,K(1-s)=1$

then the kernel $k(x\,t)$ is called a *Fourier kernel.*

(*b*) The integral equation

$$\int_0^\infty k\left(\frac{x}{t}\right)y(t)\,\frac{dt}{t}=f(x)$$

can be transformed into $K(s)\,Y(s)=F(s)$.

Hence $Y(s)=\dfrac{F(s)}{K(s)}=L(s)\,F(s)$

where $K(s)\,L(s)=1$. Inverting we have the solution

$$y(x)=\int_0^\infty l\left(\frac{x}{t}\right)f(t)\,\frac{dt}{t}$$

where $M\{l(x)\}=L(s)$.

(*c*) $\displaystyle\int_x^1 k\left(\frac{x}{t}\right)y(t)\,\frac{dt}{t}=f(x), \qquad 0<x<1$-

Put $k_1(x)=k(x)\,H(1-x)$

$y_1(x)=y\,(x)\,H(1-x)$

where $H(x)=\begin{cases}1 & x>0\\ 0 & x<0.\end{cases}$

Then, we have

$$\int_0^\infty k_1\left(\frac{x}{t}\right)y_1(t)\,\frac{dt}{t}=f(x)$$

which can be solved for y_1 as in (*b*).

Example 18. Transform the Laplacian $\dfrac{\partial^2 u}{\partial x^2}+\dfrac{\partial^2 u}{\partial y^2}$ by applying the Mellin transform in polar coordinates

Solution. In polar coordinates $x=r\cos\theta$, $y=r\sin\theta$,

$$\nabla^2 u = \frac{\partial^2 u}{\partial x^2} + \frac{\partial^2 u}{\partial y^2} = r \frac{\partial^2 u}{\partial r^2} + r \frac{\partial u}{\partial r} + \frac{\partial^2 u}{\partial \theta^2}$$

Let $U(s, \theta) = \int_0^\infty r^{s-1} u(r, \theta)\, dr = M\{u(r, \theta)\}$

Then $M\{\nabla^2 u\} = \left(\frac{d^2}{d\theta^2} + s^2 \right) U.$

§11·8. Hankel Transforms

Definition 9. The *Hankel* (or Bessel) *Transform* pair of order ν is given by

$$H_\nu\{f(x)\,;\,t\} = H_\nu\{f(x)\} = F_\nu(t) = \int_0^\infty x\, J_\nu(x, t) f(x)\, dx \quad \nu < -\tfrac{1}{2} \quad \text{...(1)}$$

$$f(x) = \int_0^\infty t\, J_\nu(x, t)\, F_\nu(t)\, dt \quad \text{...(2)}$$

The transform is thus perfectly symmetric. It exists if $x^{1/2} f(x)$ is piecewise continuous and absolutety integrable on $(0, \infty)$.

Connection with Fourier Transform

Consider the two-variable Fourier transform pair

$$F(\xi, \eta) = \frac{1}{2\pi} \int_{-\infty}^{\infty} \int_{-\infty}^{\infty} e^{i(x\xi + y\eta)} f(x, y)\, dx\, dy \quad \text{...}(i)$$

$$f(x, y) = \frac{1}{2\pi} \int_{-\infty}^{\infty} \int_{-\infty}^{\infty} e^{-i(x\xi + y\eta)} F(\xi, \eta)\, d\xi\, d\eta \quad \text{...}(ii)$$

Put $x = r\cos\theta$, $y = r\sin\theta$, ; $\xi = \rho\cos\phi$, $\eta = \rho\sin\phi$.

Let $f(x, y) = f(r, \theta)$ and $F(\xi, \eta) = F(\rho, \phi)$.

Expand $f(r, \theta)$ and $F(\rho, \phi)$ in Fourier series:

$$f(r, \theta) = \sum_{n=-\infty}^{\infty} f_n(r)\, e^{in\theta} \quad \text{...}(iii)$$

$$F(\rho, \phi) = \sum_{n=-\infty}^{\infty} F_n(r)\, .\, e^{in\phi} \quad \text{...}(iv)$$

where $f_n(r) = \frac{1}{2\pi} \int_0^{2\pi} f(r, \theta)\, e^{-in\theta}\, d\theta \quad \text{...}(v)$

$$F_n(\rho,) = \frac{1}{2\pi} \int_0^{2\pi} F(\rho, \phi)\, e^{-in\phi} \quad \text{...}(vi)$$

In polar coordinates (*i*) and (*ii*) are

$$F(\rho, \phi) = \frac{1}{2\pi} \int_0^{\pi} \int_0^{\infty} e^{i\rho r\cos(\theta - \phi)} f(r, \theta)\, r\, dr\, d\theta \quad \text{...}(vii)$$

$$f(r, \theta)=\frac{1}{2\pi}\int_0^{2\pi}\int_0^{\infty} e^{-i\rho r\cos(\theta-\phi)} F(\rho, \phi)\,\rho\, d\rho\, d\phi \quad ...(viii)$$

Substituting (*vii*) in (*vi*), we have

$$F_n(\rho)=\frac{1}{2\pi}\int_0^{2\pi} F(\rho, \phi)\, e^{-in\phi}\, d\phi$$

$$=\frac{1}{(2\pi)^2}\int_0^{2\pi}\left[\int_0^{2\pi}\int_0^{\infty} e^{i\rho r\cos(\theta-\phi)} f(r, \theta)\, r\, dr\, d\theta\right] e^{-in\phi}\, d\phi$$

$$=\frac{1}{(2\pi)^2}\int_0^{\infty}\sum_{m=-\infty}^{\infty}$$

$$\left[\int_0^{2\pi}\int_0^{2\pi} e^{i\{\rho r\cos(\theta-\phi)-n\phi)-m\theta\}}\, d\theta\, d\phi\right] f_m(r)\, r\, dr$$

$$=\frac{1}{(2\pi)}\int_0^{\infty}\left[\int_0^{2\pi} e^{i\{\rho r\cos(\theta-\varphi)+n(\theta-\phi)\}}\, d\theta\right] f_n(r)\, r\, dr$$

$$=\frac{1}{2\pi}\int_0^{\infty}\left[\int_0^{2\pi} e^{i(\rho r\cos\alpha+n\alpha)}\, d\alpha\right] f_n(r)\, dr$$

$$=\int_0^{2\pi} r\, J_n(\rho r)\, f_n(r)\, dr.$$

Similarly $f_n(r)=\int_0^{2\pi} r\, J_n(\rho r)\, F_n(\rho)\, d\rho.$

We thus get the Hankel transform pair of order n. The derivation in case of functions of radial symmetry leads to Hankel transform of order zero as we show below.

Let $f(x, y)=f(r)$ and $F(\xi, \eta)=F(\rho, \phi)$. Then equation (*i*) becomes

$$F(\rho, \phi)=\frac{1}{2\pi}\int_0^{\infty}\left[\int_0^{2\pi} e^{i\rho r\cos(\theta-\phi)}\, d\theta\right] f(r)\, r\, dr$$

$$=\frac{1}{2\pi}\int_0^{\infty}\left[\int_{-\infty}^{2\pi-\phi} e^{i\rho r\cos\psi}\, d\psi\right] f(r)\, r\, dr$$

$$=\frac{1}{2\pi}\int_0^{\infty}\left[\int_0^{2\pi} e^{i\rho r\cos\psi}\, d\psi\right] f(r)\, r\, dr$$

Hence $F(\rho, \phi)=\frac{1}{2\pi}\int_0^{\infty} r\,(J_0(\rho r)\, f(r)\, dr.$

Since the right hand side is independent of ϕ, $F(\rho, \phi)$ is a function of ρ only, say $F(\rho)$.

Hence $F(\rho)=\frac{1}{2\pi}\int_0^{\infty} r\, J_0(\rho r)\, f(r)\, dr$

The inversion formula (*ii*) will turn out to be

$$f(r)=\frac{1}{\pi 2}\int_0^{\infty} r\, J_0(\rho r)\, F(\rho)\, d\rho$$

Important Properties

I. Derivatives

(*i*) $H_\nu\{f'(x)\}=-\int_0^{\infty} x\, J_\nu(xt)\, f'(x)\, dx$

$$=\Big[x\, J_\nu(xt)\, f(x)\Big]_0^{\infty}-\int_0^{\infty}\frac{d}{dx}\{x\, J_\nu(xt)\}\, f(x)\, dx$$

Assume $f(x)$ to be such that the square bracket vanishes. Then

$$H_\nu\{f'(x)\}=-\int_0^{\infty}\frac{d}{dx}\{x\, J_\nu(xt)\}\, f(x)\, dx$$

Now $\frac{d}{dx}\{x\, J_\nu(xt)\}=xt\, J'_\nu(xt)+J_\nu(xt)$

$$=\frac{xt}{2\nu}\Big[2\nu\, J'_\nu(xt)+\frac{2\nu}{xt}\, J_\nu(xt)\Big]$$

$$=\frac{xt}{2\nu}\Big[\nu\,\{J_{v-1}(xt)-J_{v-1}(xt)\}+\{J_{v-1}(xt)\}+J_{v-1}(xt)\}\Big]$$

$$=\frac{xt}{2\nu}\Big[(\nu+1)\, J_{v-1}(xt)-(\nu-1)\, J_{v-1}(xt)\Big].$$

Hence $H_\nu\{f'(x)\}=-\Big[(\nu+1)\, F_{v-1}(t)-(\nu-1)\, F_{v-1}(t)\Big]$ ...(3)

(*ii*) $H_\nu\Big[x^{\nu-1}\frac{d}{dx}\{x^{1-\nu} f(x)\}\Big]$

$$=\int_0^{\infty} x^{\nu}\, J_\nu(xt)\,\frac{d}{dx}\{x^{1-\nu} f(x)\}\, dx$$

$$=\Big[x\, f(x)\, J_\nu(xt)\Big]_0^{\infty}-\int_0^{\infty} x^{1-\nu} f(x)\,\frac{d}{dx}\{x^{\nu}\, J_\nu\}\, dx$$

Since the Hankel transform of $f(x)$ exists, $x^{1/2} f(x)\to 0$ as $\to\infty$. Also since

$$J_\nu(xt)=\sum_{r=0}^{\infty}\frac{(-1)^r\left(\frac{xt}{t}\right)^{\nu+2r}}{r\,!\,\Gamma(\nu+r+1)},$$

The square bracket will vanish if $\lim_{x\to 0} x^{\nu-1} f(x)=0$.

This is so if $f(x)=0\,(x^{-\nu-1})$ as $x\to 0$.

If we use the recurrence relation

$$\frac{d}{dx}\{x^{\nu} J_{\nu}(xt)\}=t\, x^{\nu} J_{\nu_{-1}}(x)$$

the second term reduces to

$$-\int_0^{\infty} tx\, J_{\nu-1}(xt)\, f(x)\, dx=-t\ H_{\nu-1}\{f(x)\}$$

Hence ultimately we have

$$H_{\nu}\left[x^{\nu-1}\frac{d}{dx}\{x^{1-\nu} f(x)\}\right]=-t\ H_{\nu-1}\{f(x)\} \quad \text{...(4)}$$

(*iii*) Proceeding as above

$$H_{\nu}\left[x^{-\nu-1}\frac{d}{dx}\{x^{\nu+1} f(x)\}\right]$$

$$=\int_0^{\infty} x^{-\nu} J_{\nu}(xt\ \frac{d}{dx}\{x^{\nu+1} f(x)\}\, dx$$

$$=\left[x f(x) J_{\nu}(xt)\right]_0^{\infty}-\int_0^{\infty} x^{\nu+1} f(x)\frac{d}{dx}\{x^{-\nu} J_{\nu}(xt)\}\, dx$$

$$=t\ H_{\nu}+1\ \{f(x)\} \quad \text{...(5)}$$

(*iv*) The Bessel operator B_{ν} of order ν is defined by

$$B_{\nu}=\frac{d^2}{dx^2}+\frac{1}{x}\frac{d}{dx}-\frac{\nu^2}{x^2}$$

or $\quad B_{\nu} J_{\nu}(x)=-J_{\nu}(x).$

$$H_{\nu}\ \{B_{\nu}\ f(x)\}=\int_0^{\infty}\left[\frac{d}{dx}\left(x\frac{df}{dx}\right)-\frac{\nu^2}{x} f(x)\right] J_{\nu}(xt)\, dx$$

$$=-\int_0^{\infty}\left[x\frac{df}{dx}\cdot t\, J'_{\nu}(xt)+\frac{\nu^2}{x} f(x)\, J_{\nu}(xt)\right] dx$$

$$=\int_0^{\infty}\left[t^2 J''_{\nu}(xt)+\frac{t}{x} J'_{\nu}(xt)-\frac{\nu^2}{x^2} J_{\nu}(xt)\right] x f(x)\, dx$$

$$=-t^2\int_0^{\infty} x\, J_{\nu}(xt)\, f(x)\, dx$$

$$=-t^2 H_{\nu}(f). \quad \text{...(6)}$$

II. Parseval's Theorem

Let $\quad F_{\nu}(t)=H_{\nu}\{f(x)\}$

$\quad\quad G_{\nu}(t)=H_{\nu}\{g(x)\}$

Then $\quad \int_0^{\infty} t\, .\, F_{\nu}(t)\, G_{\nu}(t)\, dt$

$$=\int_0^\infty t\,.\,F_\iota(t)\left[\int_0^\infty xJ_\iota(xt)\,g(x)\,dx\right]dt$$

$$=\int_0^\infty\left[\int_0^\infty tJ_v(xt)\,F_v(t)\,dt\right]g(x)\,xdx$$

$$=\int_0^\infty f(x)\,g(x)\,xdx \qquad \ldots(7)$$

§11.9. Dual Integral Equations

Definition 10. The pair of equations

$$\int_0^\infty G(p)\,f(p)\,K(r,p)\,dp=g(r) \qquad 0\leqslant r\leqslant a \qquad \ldots(1)$$

$$\int_0^\infty f(p)\,K(r,p)\,dp=h(r) \qquad r>a \qquad \ldots(2)$$

where $G(p)$, $g(r)$, $h(r)$ and $K(r, p)$ are given functions and $f(p)$ is to be found are called *dual integral equations*. They arise in the solution of mixed boundary-value problems in which the condition on the boundary is mixed one. That is to say, one type of boundary data, say values of the unknown function u are given on one part ∂D_1 of the boundary ∂D and a different type of boundary data, say $\frac{\partial u}{\partial n}$ are given on the rest of the boundary ∂D_2 such that

$$\partial D=\partial D_1 \cup \partial D_2$$

A typical such BVP is

PDE : $\nabla^2 u+\lambda u=0$ in D

BCs : $u=f_1(\sigma)$ od ∂D_1

$\frac{\partial u}{\partial n}=f_2(\sigma)$ on ∂D_2

$$\partial D=\partial D_1 \cup \partial D_2$$

The mathematical situations which give rise to such equations generally involve sine, cosine or the Bessel function in place of the kernel $K(r, p)$. Since the sine and cosine functions too can be expressed in terms of Bessel functions $J_\nu(rp)$, the solution of such equations is most amenable to the Hankel transform or its modifications.

Titchmarsh had studied dual integral equations with Bessel function kernel of the type.

$$\int_0^\infty p^{-2\alpha}\,G(p)\,f(p)J_\mu(rp)\,pdp=g(r) \quad 0\leqslant r\leqslant a \qquad \ldots(3)$$

$$\int_0^\infty p^{-2\beta}\,G(p)\,f(p)\,J_v(rp)\,pdp=h(r) \quad r>a \qquad \ldots(4)$$

In cases which occur most we generally have $h(r)=0$, $\mu=v$ and $\alpha=\beta=0$. Hence in our work we shall be concerned with dual equations of the form

$$\int_0^\infty G(p)\, pJ_v(rp)\, dp=g(r), \qquad 0<r<1 \qquad \ldots(5)$$

$$\int_0^\infty G(p)\, pJ_v(rp)\, dp=0, \qquad \infty>r>1 \qquad \ldots(6)$$

Origin of Dual Integral Equations

(A) The Electric Disc. In the Weber's classical problem of the field due to a flat electrified circular disc of radius a, the origin is at the centre of the disc, its axis along the z-axis and $V(r, z)$ the axially symmetric potential in the whole space which tends to 0 as

$$\rho=\sqrt{r^2+z^2}\to\infty.$$

In cylindrical coordinates (r, ϕ, z), the potential equation is transformed to

$$\nabla^2 V=\frac{\partial^2 V}{\partial r^2}+\frac{1}{r}\frac{\partial V}{\partial r}+\frac{1}{r^2}\frac{\partial^2 V}{\partial \phi^2}+\frac{\partial^2 V}{\partial z^2}=0$$

Since V is independent of ϕ, $V(r, z)$ satisfies

$$\text{PDE}: \nabla^2 V=\frac{\partial^2 V}{\partial r^2}+\frac{1}{r}\frac{\partial V}{\partial r}+\frac{\partial^2 V}{\partial z^2}=0 \qquad \ldots(i)$$

V also satisfies the BCs

$$V(r, 0)=V_0(r) \qquad 0<r<a \qquad \ldots(ii)$$

$$\frac{\partial V}{\partial z}(r, 0)=0 \qquad r>a \qquad \ldots(iii)$$

Write $\overline{V}(p, z)=\int_0^\infty V(r, z)\, rJ_0(rp)\, dr$

Then $\dfrac{\partial^2 \overline{V}}{\partial z^2}(p, z)=\int_0^\infty \dfrac{\partial^2 V}{\partial r^2}\, r\, J_0(rp)\, dr$

Also $$I=\int_0^\infty \left(\frac{\partial^2 V}{\partial r^2}+\frac{1}{r}\frac{\partial V}{\partial r}\right) r\, J_0(rp)\, ds$$

$$=\int_0^\infty \frac{\partial}{\partial V}\left(r\frac{\partial V}{\partial r}\right) J_0)rp)\, (r$$

$$=\left[pr\frac{\partial V}{\partial r}\, J_0(rp)\right]_0^\infty - p\int_0^\infty \frac{\partial V}{\partial r}\, .\, r\, J_0'(rp)\, dr$$

Assume $V(r, z)$ to be such that the first term on the right vanishes. Then

$$I=-p \int_0^\infty \frac{\partial V}{\partial r} \cdot r\, J_0'(rp)\, dr.$$

Integrating by parts again

$$I=-p \left[V \cdot r\, J_0'(rp) \right]_0^\infty$$

$$+p \int_0^\infty \left\{ rp\, J_0''(rp)+J_0'(rp) \right\} V\, dr$$

Assume further that the first term on the right vanishes. Also $J_0(rp)$ satisfies the equation

$$J_0''(rp)+\frac{1}{rp}\, J_0'(rp)+J_0(rp)=0$$

Hence $I=-p^2 \int_0^\infty V\, r\, J_0(rp)\, dr=-p^2\bar{V}(p, z).$

Finally $$\int_0^\infty \nabla^2 V \cdot r\, J_0(rp)\, dr=\left(\frac{d^2}{dz^2}-p^2\right)\bar{V}(p, z) \qquad ...(iv)$$

If we multiply (*i*) by $rJ_0(rp)$, integrate with respect to r from 0 to ∞ and use (*iv*), we have

$$\frac{d^2\bar{V}(p, z)}{dz^2}-p^2\bar{V}(p, z)=0 \qquad ...(v)$$

The general solution of (*v*) is

$$\bar{V}(p, z)=A(p)\, e^{-pz}+B(p)\, e^{pz},\ p>0$$

In view of the symmetry of the potential equation in z, it is enough to take $z>0$. Since the potential must vanish as $z\to\infty$, we must have $B=0$. Hence

$$\bar{V}(p, z)=A(p)\, e^{-pz}$$

Taking the inverse Hankel transform of order zero,

$$V(r, z)=\int_0^\infty A(p) \cdot e^{-pz} pJ_0(rp)\, dp.$$

Since $V(r, z)$ satisfies the BCs (*ii*), (*iii*), $A(p)$ will be determined from the pair of *dual integral equations*.

$$\int_0^\infty A(p)\, pJ_0(rp)\, dp=V_0(r), \qquad 0<r<a \qquad ...(vi)$$

$$\int_0^\infty pA(p) \cdot pJ_0(rp)\, dp=0 \qquad r>a \qquad ...(vii)$$

In the particular case when V_0 is a constant and $a=1$, we can evaluate $A(p)$ from the well-known results.

$$\int_0^\infty p^{-1} J_0(rp) \sin p \, dp = \frac{\pi}{2} \qquad 0<r<1$$

$$\int_0^\infty J_0(rp) \sin p \, dp = 0$$

This gives $A(p) = \dfrac{2}{\pi} p^{-2} V_0 . \sin p$

and
$$V(r, z) = \frac{2V_0}{\pi} \int_0^\infty p^{\ 1} e^{-pz} J_0(rp) \sin p \, dp$$

Problem 1. In the above derivation, assume the boundary conditions to be

$$V(r, 0) = 1 \qquad 0<r<1$$

$$\frac{\partial V}{\partial z}(r, 0) = 0 \qquad r>1$$

$$V(r, k) = 0, \ z=k, \ 0<r<\infty.$$

Then prove that the boundary conditions reduce to the integral equations of the form

$$\int_0^\infty p^{-1} \tanh(kp) f(p) J_0(rp) \, dp = 1, \qquad 0<r<1$$

$$\int_0^\infty f(p) J_0(rp) \, dp = 0 \qquad r \geqslant 1$$

Problem 2. Prove that the solution to the BVP

$$\text{PDE} : \nabla^2 V(r, z) = \frac{\partial^2 V}{\partial r^2} + \frac{1}{r}\frac{\partial V}{\partial r} + \frac{\partial^2 V}{\partial z^2} = 0$$

$$z>0, \ 0<r<\infty$$

$$\text{BCs} : \quad V(r, 0) = g(r) \qquad r<a$$

$$\frac{\partial V}{\partial z}(r, 0) = h(r) \qquad r>a$$

is given by

$$V(r, z) = \int_0^\infty A(p) e^{-pz} p J_0(rp) \, dp$$

where $A(p)$ is the solution to the dual equations

$$\int_0^\infty p A(p) J_0)rp) \, dp = g(r) \qquad r<a$$

$$\int_0^\infty p^2 A(p) J_0(rp) \, dp = -h(r) \qquad r>a$$

(B) Steady Flow of Heat in a Quadrant

Consider the following mixed BVP for the potential equation in the fourth quadrant :

PDE : $\nabla^2 u(x, y)=0,\ 0<x<\infty,\ -\infty<y<0$

BCs : $u(x, 0)=1,\ 0<x<1$ ...(*i*)

$u_y(x, 0)=0,\ x>1$ ...(*ii*)

$u_x(0, y)=0 \quad -\infty<y<0$ ...(*iii*)

$u,\ u_x \to 0$ as $x\to\infty$ ...(*iv*)

The BVP represents steady flow of heat into the fourth quadrant through the segment $0<x<1$, the rest of the boundary is perfectly insulated. It also represents the irrotational flow of an incompressible fluid.

The BC $u_x(0, y)=0$ implies that u can be continued into the third quadrant as an even function of x. The Fourier cosine transform is thus appropriate in this case.

Put $\quad \bar{u}(p, y)=\int_0^\infty u(x, y)\cos(px)\,dx.$

Since $\quad \int_0^\infty u_{xx}\cos(px)\,dx=-p^2\,\bar{u}(p, y)$

The PDE can be transformed into

$$\bar{u}_{yy}-p^2\,\bar{u}(p, y)=0.$$

Hence $\quad \bar{u}(p, y)=A(p)\,e^{py}+B(p)\,e^{-py}$

For $y<0$, $e^{-py}\to\infty$ as $p\to\infty$. Hence $B(p)=0$

and $\quad \bar{u}(p, y)=A(p)\,.\,e^{py}.$

Inverting, we get

$$u(x, y)=\frac{2}{\pi}\int_0^\infty A(p)\,e^{py}\cos(xp)\,dp$$

$u(x, y)$ satisfies condition (*iii*). By Riemann—Lebesgue lemma $u(x, y)$ and $\frac{\partial u}{\partial x}\to 0$ as $x\to\infty$.

Hence conditions (*iv*) are satisfied. Thus $A(p)$ will satisfy the dual equations

$$\frac{2}{\pi}\int_0^\infty A(p)\cos(px)\,dp=1 \qquad 0<x<1$$

$$\int_0^\infty pA(p)\cos(px)\,dp=0 \qquad 0<x<\infty$$

Example 19 (Sneddon, 1959). Solve the equations for f

$$\int_0^\infty f(p)\, J_0(rp)\, dp = g(r) \qquad 0<r<1 \qquad \ldots(i)$$

$$\int_0^\infty p\, f(p)\, J_0(rp)\, dp = 0 \qquad 1<r<\infty \qquad \ldots(ii)$$

Solution. Recall that

$$\int_0^\infty \sin pt\, J_0(rp)dp = 0,\ r>t \qquad \ldots(iii)$$

Differentiating (*iii*) with respect to t

$$\int_0^\infty p \cos pt\, J_0(rp)dp = 0,\ r>t \qquad \ldots(iv)$$

Equation (*iv*) shows that $f(p)=\cos pt$ is a solution of (*ii*) for all $t<1$. Then also

$$f(p) = \int_0^1 \phi(t) \cos pt\, dt \qquad \ldots(v)$$

is a solution of (*ii*) for arbitrary $\phi(t)$. If we can determine $\phi(t)$ in such a manner that (*v*) satisfies (*i*), the solution to our problem will be obtained.

If we insert (*v*) in (*i*) and change the order of integration, we get

$$\int_0^1 \left[\int_0^\infty \cos pt\, J_0(pr)\, dp\right]\phi(t)\, dt = g(r),\ 0<r<1 \qquad \ldots(vi)$$

Now $\int_0^\infty \cos pt\, J_0(pr)\, dp = \begin{cases} 0 & 0<r<t \\ \dfrac{1}{\sqrt{r^2-t^2}} & r>t \end{cases}$

Hence $\displaystyle\int_0^r \frac{\phi(t)}{\sqrt{r^2-t^2}} = g(r), \qquad 0<r<1 \qquad \ldots(vii)$

Put $t=\sqrt{x}$, $r=\sqrt{s}$, then (*vii*) reduces to

$$\int_0^s \frac{\phi(\sqrt{x})}{\sqrt{s-x}} \frac{dx}{2\sqrt{x}} = g(\sqrt{s}\,) \qquad \ldots(viii)$$

This is Abel's equation for $\dfrac{\phi(\sqrt{x})}{2\sqrt{x}}$. The solution of (*viii*) is

$$\frac{\phi(\sqrt{x})}{2\sqrt{x}} = \frac{1}{\pi} \frac{d}{dx} \int_0^{\sqrt{x}} \frac{g(\sqrt{s})}{\sqrt{x-s}}\, ds$$

Now replace x and s by t^2 and r^2 respectively, then

$$\phi(t)=\frac{2}{\pi}\frac{d}{dt}\int_0^t \frac{rg(r)}{\sqrt{t^2-r^2}}\,dr$$

$$=\frac{2}{\pi}\left[g(0)+t\int_0^t \frac{g'(r)\,dr}{\sqrt{t^2-r^2}}\right]$$

Hence

$$f(p)=\frac{2g(0)}{\pi p}\sin p+\frac{2}{\pi}\int_0^1\int_0^t \frac{t\cos pt\,g'(r)dr\,dt}{\sqrt{t^2-t^2}}$$

In case $g(r)=1$ for $0<r<1$, then

$$f(p)=\frac{2\sin p}{\pi p}.$$

§11·10 Fractional Integrals

Riemann—Liouville (R—L) *fractional integrals* of order α are defined by

$$I^\alpha f=\frac{1}{\Gamma(\alpha)}\int_0^x (x-t)^{\alpha-1} f(t)\,dt \qquad ...(1)$$

The *Weyl fractional integrals* (W-integrals) of order α are defined by

$$K^\alpha f=\frac{1}{\Gamma(\alpha)}\int_x^\infty (t-x)^{\alpha-1} f(t)\,dt \qquad ...(2)$$

In our work we shall suppose α to be real and >0.

For $\alpha<0$, let $\alpha=p-n$ where $0<p\leqslant 1$ and n is a positive integer, then we define

$$I^\alpha f=\frac{d^n}{dx^n} I^p f \qquad ...(3)$$

$$K^\alpha f=\frac{d^n}{dx^n} K^p f \qquad ...(4)$$

Problem. Prove that for any real numbers α, β

$I^\alpha I^\beta=I^{\alpha+\beta}$, $\quad K^\alpha K^\beta=K^{\alpha+\beta}$ (Law of Indices)

In the theory of linear ordinary differential equations the R—L integrals are known as *Euler transforms of the first kind*. They were discussed by Marcel Riesz in 1949 in his classic memoir on the Cauchy problem for hyperbolic equations[1] in considerable detail. The W-integrals are known as *Euler transforms of the second kind*.

1 M. Riesz, "L' integrals de Riemann—Liouville et le problem de Cauchy,' Acta Math., 81 (1949), 1.

Connection Between I^α and K^α

There is an interesting connection between the operators I^α and K^α. In chapter 8 we have seen that

$$G(x, t)=\frac{(x-t)^{n-1}}{(n-1)!}$$

is the one-sided Green's function of the operator

$$L=\frac{d^n}{dx^n} \quad \text{and} \quad \bar{G}(x, t)=\frac{(t-x)^{n-1}}{(n-1)!}$$

is the one-sided Green's function of the *adjoint differential operator* $\bar{L}=(-1)^n \frac{d^n}{dx^n}$, It is also known that the inversion operators of L and $\bar{L}$ are indefinite integral operators with Kernel $G(x, t)$ and $\bar{G}(x, t)$ respectively. Indeed they are

$$I^n f=\frac{1}{(n-1)!}\int_0^x (x-t)^{n-1} f(t)\, dt$$

and

$$K^n f=\frac{1}{(n-1)!}\int_x^\infty (t-x)^{n-1} f(t)\, dt.$$

I^α and K^α are direct generalisations of I^n and K^n for any real α. It is in this sense that the operators I^α and K^α can be called adjoint operators.

There is another angle to the adjoint property of I^α and K^α which is exhibited by the following lemma.

Lemma 3 (Love and Young)[1]

$$\int_0^\infty I^\alpha f(x)\,.\, g(x)dx=\int_0^\infty f(x)\, K^\alpha g(x)dx \qquad \ldots(5)$$

Proof :

$$\text{L.H.S.}=\int_0^\infty \left[\int_0^x \frac{(x-t)^{\alpha-1}}{\Gamma(\alpha)} f(t)\, dt\right] g(x)\, dx$$

$$=\int_0^\infty \left[\int_t^\infty \frac{(x-t)^{\alpha-1}}{\Gamma(\alpha)} g(x)\, dx\right] f(t)\, dt$$

(by changing order of integration)

$$=\int_0^\infty f(t)\,.\, K^\alpha g(t)\, dt$$

$$=\text{R:H.S.}$$

Relation (5) is also known as the *rule for fractional integration by parts.*

Corollary 1. The operators I^α and K^α can be supposed to act on $L^2(0, \infty)$ where the inner product is defined by

---

1 Love E.R. and Young L.C., 1938, "On fractional integration by parts," Proc. London Math. Soc. (2), 44, pp 1-34.

$$(f, g)=\int_0^{\infty} f(x)\, g(x)\, dx.$$

In this notation (5) can be written as $(\mathrm{I}^{\alpha} f, g)=(f, \mathrm{K}^{\alpha} g)$ ...(6)

It is in the sense of relation (6) that we can call I^{α} and K^{α} *adjoint operators* on L^2.

Erde'lyi—Kober Operators and Modified Hankel Transform

In a series of papers Erde'lyi. Kober and Sneddon investigated the properties of the fractional integral

$$\mathrm{I}^{\eta,\,\alpha} f(x)=\frac{x^{-\eta-\alpha+1}}{\Gamma(\alpha)}\int_0^x (x-t)^{\alpha-1}\; t^{\eta-1},$$

$$(\alpha>0,\ \eta>0) \qquad \text{...(7)}$$

which is obviously a generalisation of the R—L integral (1) and the integral

$$\mathrm{K}^{\eta,\,\alpha} f(x)=\frac{x^{\eta}}{\Gamma(\alpha)}\int_x^{\infty} (t-x)^{\alpha-1}\; t^{-\alpha-\eta} f(t)\, dt,$$

$$(\alpha>0,\ \eta>0) \qquad \text{...(8)}$$

which is a generalisation of the Weyl integral (2).

The operators $\mathrm{I}^{\eta,\,\alpha}$ and $\mathrm{K}^{\eta,\,\alpha}$ are frequently known as Erde'lyi-Kober operators.

If can be verified that

$$x^{\alpha}\, \mathrm{I}^{1,\alpha}=\mathrm{I}^{\alpha},\; x^{\alpha}\, \mathrm{K}^{-\alpha,\alpha}=\mathrm{K}^{\alpha}$$

Also $\mathrm{I}^{\eta,\,0}$ and $\mathrm{K}^{\eta,\,0}$ are identity operators.

For $\alpha<0$ we can define

$$\mathrm{I}^{\eta,\,\alpha} f=x^{-\eta-\alpha}\frac{d^n}{dx^n}\; x^{\eta+\alpha+n}\; \mathrm{I}^{\eta,\,\alpha+n} f(x) \qquad \text{...(9)}$$

where n is a positive integer such that $\eta+\alpha \geqslant 0$.

In our work we shall consider slightly modified, but equivalent form of these operators ; because they can be combined with the Hankel transform as defined in section 11·8 *viz.*,

$$\mathrm{H}_{\nu}\ \{f(t);\, x\}=\mathrm{H}_{\nu}\ \{f(x)\}=\int_0^{\infty} t\, \mathrm{J}_{\nu}\ (xt)\, f(t)\, dt \qquad \text{...(10)}$$

If $\alpha>0,\ \eta>-\frac{1}{2}$, then

$$\mathrm{I}^{\eta,\,\alpha}\{f(t)\,;\,x\}=\mathrm{I}^{\eta,\,\alpha}f(x)=\frac{2x^{-2\alpha-2\eta}}{\Gamma(\alpha)}\int_0^x (x^2-t^2)^{\alpha-1}\,.\; t^{2\eta+1} f(t)\, dt$$

...(11)

$$K^{\eta,\alpha}\{f(t)\,;\,x\}=K^{\eta,\alpha}f(x)=\frac{2x^{2\eta}}{\Gamma(\alpha)}$$

$$\int_x^\infty (t^2-x^2)^{\alpha-1}\; t^{-2\alpha-2\eta+1} f(t)\,dt \qquad \ldots(12)$$

As before $I^{\eta,0}$ and $K^{\eta,0}$ are identity operators.

We also define a *modified Hankel transform* operator $S_{\eta,\alpha}$ closely related to Erdelyi-Kober operators by the formula

$$S_{\eta,\alpha} f(x)=\left(\frac{2}{x}\right)^\alpha \int_0^\infty t^{1-\alpha} f(t)\, J_{2\eta+\alpha}(xt)\,dt \quad \ldots(13)$$

$$=\left(\frac{2}{x}\right)^\alpha H_{2\eta+\alpha}\left\{x^{-\alpha} f(x)\right\}$$

Note. Mark that if the Hankel tranform H_ν is written as in section 11·6, then

$$H_\nu\{f(x)=\int_0^\infty x\, J_\nu(xt) f(x)\,dx=H_\nu\{f(x)\,;\,t\}$$

and $S_{\eta,\alpha}$ will be written in the form

$$S_{\eta,\alpha} f(x)=\left(\frac{2}{t}\right)^\alpha \int_0^\infty x^{1-\alpha} f(x)\; J_{2\eta+\alpha}(xt)\,dx$$

$$=\left(\frac{2}{t}\right)^\alpha H_{2\eta+\alpha}\left\{x^{-\alpha} f(x)\right\}$$

This slight notational variation may be kept in mind while combining two operators.

Pooblem 1. Prove that a definition of $S_{\eta,\alpha}$ which is equivalent to (13) is given by

$$S_{\eta,\alpha}\; f(x)=x^{-\alpha/2}\int_0^\infty y^{-\alpha/2} f(y)\; J_{2\eta+\alpha}\,(2\sqrt{xy}\,)\,dy \qquad \ldots(14)$$

$$\alpha \geqslant 0,\; \eta > -\tfrac{1}{2}$$

Problem 2. Prove that

$$H_\nu\{f(t)\,;\,x\}=\left(\frac{x}{2}\right)^\alpha S_{\frac{\nu}{2}-\frac{\alpha}{2},\,\alpha}\left[t^\alpha f(t)\,;\,x\right] \quad \ldots(15)$$

Problem 3. Applying the inversion theorem

$$H_\nu^{-1}=H_\nu$$

prove the inversion formula

$$S^{-1}_{\eta,\alpha}=S_{\eta+\alpha,\,-\alpha}$$

Properties :

I. Products

(*i*) $$I^{\eta,\alpha}\, x^{2\beta} f(x)=\frac{2x^{-2\alpha-2\eta}}{\Gamma(\alpha)}\int_0^x (x^2-t^2)^{\alpha-1}\, t^{2\eta+2\beta+1} f(t)\, dt$$

$$=x^{2\beta}\,.\,2x^{-2\alpha-2\eta-2\beta}\int_0^x (x^2-t^2)^{\alpha-1}\,.\,t^{2\eta+2\beta+1} f(t)\, dt$$

$$=x^{2\beta}\, I^{\eta+\beta,\alpha} f(x) \qquad \ldots(16)$$

Similarly for $K^{\eta,\alpha}$

(*ii*) $I^{\eta,\alpha}.\, I^{\eta+\alpha,\beta} f(x)$

$$=\frac{2x^{-2\alpha-2\eta}}{\Gamma(\alpha)}\int_0^x (x^2-u^2)^{\alpha-1}\,.\,u^{2\eta+1}\left[\frac{2u^{-2\alpha-2\eta-2\beta}}{\Gamma(\beta)}\int_0^u (u^2-t^2)^{\beta-1}\, t^{2\eta+2\alpha+1} f(t)\, dt\right]du$$

Now interchange the order of integration.

Then the inner integral in u will be

$$I=2\int_t^x (x^2-u^2)^{\alpha-1}\,(u^2-t^2)^{\beta-1}\, u^{-2\alpha-2\beta+1}\, du$$

Now put $s=\dfrac{1-\dfrac{t^2}{u^2}}{1-\dfrac{t^2}{x^2}}$

Then $1-\dfrac{t^2}{u^2}=s\left(1-\dfrac{t^2}{x^2}\right),\quad \dfrac{2t^2du}{u^3}=\dfrac{x^2-t^2}{x^2}\, ds$

$$(u^2-t^2)^{\beta-1}=s^{\beta-1}\frac{(x^2-t^2)^{\beta-1}}{x^{2\beta-2}}\, u^{2\beta-2}$$

$$(x^2-u^2)^{\alpha-1}=(1-s)^{\alpha-1}\frac{(x^2-t^2)^{\alpha-1}}{t^{2\alpha-2}}\, u^{2\alpha-2}$$

Making these substitutions and using the Beta function integral. we have

$$I=\frac{\Gamma(\alpha)\,\Gamma(\beta)}{\Gamma(\alpha+\beta)}\, t^{-2\alpha}\, x^{-2\beta}\,(x^2-t^2)^{\alpha+\beta-1}$$

The double integral is, therefore

$$=\frac{2x^{-2\eta-2\alpha-2\beta}}{\Gamma(\alpha+\beta)}\int_0^x (x^2-t^2)^{\alpha+\beta-1}\, t^{2\eta+1} f(t)\, dt$$

$$=\mathrm{I}^{\eta,\,\alpha+\beta} f(x)$$

i.e., we have the *product rule*.

$$\mathrm{I}^{\eta,\,\alpha}\,.\,\mathrm{I}^{\eta+\alpha,\,\beta}=\mathrm{I}^{\eta,\,\alpha+\beta} \qquad ...(17)$$

Similarly $\mathrm{K}^{\eta,\,\alpha}\,.\,\mathrm{K}^{\eta+\alpha,\,\beta}=\mathrm{K}^{\eta,\,\alpha+\beta}$ ...(18)

Hence for Re $\alpha<0$, $\mathrm{I}^{\eta,\,\alpha}=(\mathrm{I}^{\eta+\alpha,\,-\alpha})^{-1}$ ...(19)

II. Derivatives

We have

$$\mathrm{I}^{\eta,\,\alpha} f(x)=\frac{2x^{-2\eta-2\alpha}}{\Gamma(\alpha)}\int_0^x (x^2-t^2)^{\alpha-1}\, t^{2\eta+1} f(t)\, dt$$

(*i*) Multiply both sides by $x^{2\eta+2\alpha}$ and differentiate. We have

$$\frac{d}{dx}\, x^{2\eta+2\alpha}\, \mathrm{I}^{\eta,\,\alpha} f(x)=2x^{2\eta+2\alpha-1}\, \mathrm{I}^{\eta,\,\alpha-1} f(x) \qquad ...(20)$$

This can be written as

$$\left(x^{-2\eta-2\alpha}\frac{1}{2x}\frac{d}{dx}\,.\, x^{2\eta+2\alpha}\right) \mathrm{I}^{\eta,\,\alpha} f(x)=x^{-2}\, \mathrm{I}^{\eta,\,\alpha-1} f(x) \qquad ...(21)$$

If we put $\dfrac{1}{2x}\dfrac{d}{dx}\equiv \mathrm{D}x$ and repeat this operation, we get

$$x^{2m-2\eta-2\alpha}\, \mathrm{D}_x^m\, x^{2\eta+2\alpha}\, \mathrm{I}^{\eta,\,\alpha}=\mathrm{I}^{\eta,\,\alpha-m}$$

$$\mathrm{Re}\,(\alpha-m)>0$$

III. Connection with Modified Hankel Transform

$$\mathrm{I}^{\eta+\alpha,\,\beta}\, \mathrm{S}_{\eta,\,\alpha} f(x)$$

$$=\frac{2x^{-2\eta-2\alpha-2\beta}}{\Gamma(\beta)}\int_0^x (x^2-t^2)^{\beta-1}\, t^{2\eta+2\alpha+1}\left[\left(\frac{2}{t}\right)^{\alpha}\int_0^\infty u^{1-\alpha} f(u)\, \mathrm{J}_{2\eta+\alpha}(ut)\, du\right] dt$$

$$=2^{\alpha+1}\,\frac{x^{-2\eta-2\alpha-2\beta}}{\Gamma(\beta)}\int_0^\infty u^{1-\alpha} f(u)\left[\int_0^x t^{2\eta+\alpha+1}(x^2-t^2)^{\beta-1}\, \mathrm{J}_{2\eta+\alpha}(ut)\, dt\right]du$$

To evaluate the inner integral put $t=x\sin\theta$ and use Sonine's first integral

$$\mathrm{J}_{\mu+\nu+1}(x)=\frac{x^{\nu+1}}{2^{\nu}\,\nu\,!}\int_0^{\pi/2} \mathrm{J}_{\mu}(x\sin\theta)\,\sin^{\mu+1}\theta\,\cos^{2\nu+1}\theta\, d\theta$$

We shall finally get

(*a*) $I^{\eta+\alpha, \beta} S_{\eta, \alpha} f(x) = S_{\eta, \alpha+\beta} f(x)$

A similar procedure leads to the following identifies

(*b*) $S_{\eta+\alpha, \beta} I^{\eta, \alpha} = S_{\eta, \alpha+\beta}$

(*c*) $K^{\eta, \alpha} S_{\eta+\alpha, \beta} = S_{\eta, \alpha+\beta}$

(*d*) $S_{\eta, \alpha} K^{\eta+\alpha, \beta} = S_{\eta, \alpha+\beta}$

(*e*) $S_{\eta+\alpha, \beta} S_{\eta, \alpha} = I^{\eta, \alpha+\beta}$

(*f*) $S_{\eta, \alpha} \cdot S_{\eta+n, \beta} = K^{\eta, \alpha+\beta}$

IV. Connection with Mellin Transform

It was pointed out by Buschman[1] that the Erdelyi-Kober operations are essentially Mellin convolutions of the type

$$(fog)(x) = \int_0^\infty f(u)\, g\left(\frac{x}{u}\right) \frac{du}{u}$$

Put $I_{\eta, \alpha}(x) = \dfrac{2(x^2-1)^{\alpha-1}}{\Gamma(\alpha)} x^{-2\eta-2\alpha} H(x-1)$

$$K_{\eta, \alpha}(x) = \frac{2(1-x^2)^{\alpha-1}}{\Gamma(\alpha)} x^{2\eta} H(1-x)$$

Then $I^{\eta, \alpha} f(x) = (I_{\eta, \alpha}\, o\, f)(x)$ ...(22)

$$K^{\eta, \alpha} f(x) = (K_{\eta, \alpha}\, o\, f)(x) \qquad \text{...(23)}$$

$$M\left\{ I_{\eta, \alpha}(x) \right\} = \frac{\Gamma\left(1+\eta-\frac{1}{2}s\right)}{\Gamma\left(1+\eta+\alpha-\frac{1}{2}s\right)} \qquad \text{...(24)}$$

$$M\left\{ K_{\eta, \alpha}(x) \right\} = \frac{\Gamma\left(\eta+\frac{1}{2}s\right)}{\Gamma\left(\eta+\alpha+\frac{1}{2}s\right)} \qquad \text{...(25)}$$

With the help of these results it is possible to solve integral equa tions of the type

---

1. Buschman R.G., "Fractional integration', Mathematica Japonicae, 9 (1964), p. 99.

$$I^{\eta,\,\alpha}\, y(x)=f(x)$$

$$K^{\eta,\,\alpha}\, y(x)=f(x).$$

From the definition of $S_{\eta,\,\alpha}$ and the result

$$M\{J_\nu(x)\}=2s^{-1}\,\frac{\Gamma(\frac{1}{2}\nu+\frac{1}{2}s)}{\Gamma(1+\frac{1}{2}\nu+\frac{1}{2}s)}$$

we can write

$$M\{x\, S_{\eta,\,\alpha} f(x)\}$$

$$=2s\,\frac{\Gamma(\frac{1}{2}+\eta+\frac{1}{2}s)}{\Gamma(\frac{1}{2}+\eta+\alpha-\frac{1}{2}s}\; F(s) \qquad \ldots(26)$$

(26) expresses a relation between the Mellin and Hankel transforms.

Properties I—IV lay down the procedure for solving dual integral equations involving Hankel transforms.

V. Fractional Integration by Parts

Changing the order of integration

$$\int_0^\infty x f(x)\,.\, I^{\eta,\,\alpha}\, g(x)\, dx$$

$$=\int_0^\infty x f(x)\left[\frac{2x^{-2\alpha-2\eta}}{\Gamma(\alpha)}\int_0^x (x^2-t^2)^{\alpha-1}\, t^{2\eta+1}\, g(t)\, dt\right] dx$$

$$=\int_0^\infty t\, g(t)\left[\frac{2t^{2\eta}}{\Gamma(\alpha)}\int_t^\infty (x^2-t^2)^{\alpha-1}\, x^{-2\alpha-2\eta+1}\, f(x)\, dx\right]dt$$

$$=\int_0^\infty t\, g(t)\,.\, K^{\eta,\,\alpha} f(t)\, dt \qquad \ldots(27)$$

The result (27) is due to Erde'lyi and is known as the rule for *fractional integration by parts.*

Application to Dual Integral Eqnations

Consider the Titchmarsh type dual equations

$$\int_0^\infty t^{-2\alpha} A(t)\, J_\mu(xt)\,.\, t\, dt=f(x) \qquad 0<x<1 \qquad \ldots(i)$$

$$\int_0^\infty t^{-2\beta} A(t)\, J_\nu(xt)\, t\, dt=g(x) \qquad \infty>x>1 \qquad \ldots(ii)$$

where $f(x)$ and $g(x)$ are known and $A(t)$ is to be found.

In terms of the modified Hankel transform, the equations can be expressed as

$$S_{\frac{1}{2}\mu-\alpha,\, 2\alpha}\, A(t)=\left(\frac{2}{x}\right)^{2\alpha} f(x) \quad x<a \qquad \ldots(iii)$$

1. Erdelyi A., "On fractional integration and its application to the theory of Hankel transforms," Quart. J. Math. (Oxford), 11, (1940).

$$S_{\frac{1}{2}\nu-\beta,\ 2\beta}\ A(t)=\left(\frac{2}{x}\right)^{2\beta} g(x) \quad x>a \qquad \text{...}(iv)$$

Suppose there exist operators L and M such that

$$L\ S_{\frac{1}{2}\mu-\alpha,\ 2\alpha}=S_{\gamma,\ \delta} \qquad \text{...}(v)$$

$$M\ S_{\frac{1}{2}\nu+\beta,\ 2\beta}=S_{\gamma,\ \delta} \qquad \text{...}(vi)$$

Then $$S_{\gamma,\ \delta}\ A(t)=\begin{cases} L\left(\frac{2}{x}\right)^{2\alpha} f(x) & x<a \qquad \text{...}(vii) \\ M\left(\frac{2}{x}\right)^{2\beta} g(x) & x>a \qquad \text{...}(viii) \end{cases}$$

Hence $$A(t)=\begin{cases} S_{\gamma+\delta,\ -\delta}\ L\left(\frac{2}{x}\right)^{2\alpha} f(x) & x<a \\ S_{\gamma+\delta,\ -\delta}\ M\left(\frac{2}{x}\right)^{2\beta} g(x) & x>a \end{cases}$$

We now determine the operators L and M.

From the relations (v), by taking inverse operators

$$L=S_{\gamma,\ \delta}\ S_{\frac{1}{2}\mu+\alpha,\ -2\alpha}$$

$$M=S_{\gamma,\ \delta}\ S_{\frac{1}{2}\nu+\beta,\ -2\beta}$$

Writing L f as a double integral and changing the order of integration, we get

$$L\ f(x)=\int_0^\infty W(x,\ t)\ f(t)\ dt$$

where $$W(x,\ t)=2^{\delta-2\alpha}\ x^{-\delta}\ t^{1+2\alpha}$$

$$\int_0^\infty u^{2\mu-\delta}\ J_{2\gamma+\delta}(u\ x)\ .\ u\ J_\mu(u\ t)\ dx \qquad \text{...}(ix)$$

A similar expression can be derived for M $g(x)$.

There is another approach to find L and M. From the connection of the modified Hankel transform it should be possible to choose L and M as suitable Ende'lyi-Kober operators $I^{\eta,\ \alpha}$ and $K^{\eta,\ \alpha}$ such that equalities (v), (vi) are satisfied. The relevant formulae, of course, are :

$$I^{\eta+\alpha,\ \beta}\ S_{\mu,\ \alpha}=S_{\eta,\ \alpha+\beta} \qquad \text{...}(x)$$

$$K^{\eta,\ \alpha}\ S_{\gamma+\alpha,\ \beta}=S_{\eta,\ \alpha+\beta} \qquad \text{...}(xi)$$

So that (v) and (vi) correspond to (x) and (xi) respectively we should have

$$I^{\frac{1}{2}+\alpha,\ \beta-\alpha}\ S_{\frac{1}{2}-\alpha,\ 2\alpha}=S_{\frac{1}{2}+\alpha,\ \alpha+\beta}$$

$$K^{\frac{1}{2}-\alpha,\ \alpha-\beta}\ S_{\frac{1}{2}-\beta,\ 2\beta}=S_{\frac{1}{2}-\alpha,\ \alpha+\beta}$$

Thus $\gamma=\frac{1}{2}-\alpha$, $\delta=\alpha+\beta$.

From (*vii*), (*viii*) if should be possible to find A(t) by inversion.

Problems

(1) Show that the solution of the dual integral equations

$$\int_0^\infty A(t)\, J_1(xt)\, dt=-1 \qquad x<1$$

$$\int_0^\infty t\, A(t)\, J_1(xt)\, dt=0 \qquad x>1$$

is $A(t)=\dfrac{\cos t-1}{t}$.

(2) Show that the solution of the dual equations

$$\int_0^\infty A(t) \sin (xt)\, dt=f(x) \qquad x<1$$

$$\int_0^\infty t\, A(t) \sin (xt)\, dt=0 \qquad x>1$$

is given by

$$A(t)=\frac{2}{\pi}\int_0^1 J_0(xt)\left\{\frac{d}{dx}\int_0^x \frac{y\, f(y)\, dy}{\sqrt{x^2-y^2}}\right\}dx$$

(3) Show that the solution of the dual equations

$$\int_0^\infty t\, A(t)\, J_0(xt)\, dt=f(x) \qquad x<1$$

$$\int_0^\infty A(t)\, J_0(xt)\, dt=0 \qquad x>1$$

is $$A(t)=\frac{2}{\pi}\int_0^1 \sin (xt)\left[\int_0^x \frac{y\, f(y)\, dy}{\sqrt{x^2-y^2}}\right]dx$$

(4) Use Ende'lyi-Kober operators to show that the equation

$$\int_0^x \frac{y(t)\, dt}{(x^2-t^2)^\alpha}=g(x) \qquad 0<\alpha<1$$

has solution $$y(x)=\frac{2}{\Gamma(\alpha)\, \Gamma(1-\alpha)}\ \frac{d}{dx}\int_0^x (x^2-t^2)^{\alpha-1}\, t\, g(t)\, dt$$

(5) For the Legendre polynomial $P_n(x)$, it is given that

$$M\{P_n\left(\frac{1}{x}\right) H(1-x)\}=2s^{-1}\frac{\Gamma(\frac{1}{2}s+\frac{1}{2}n+\frac{1}{2})\, \Gamma(\frac{1}{2}s-\frac{1}{2}n)}{\sqrt{\pi}\, \Gamma(s+1)} \quad \text{Re } s>n.$$

Then show that the integral equation

$$\int_x^1 P_n\left(\frac{y}{x}\right) f(x)=g(x)$$

with $g(1)=g'(1)=0$ has the solution

$$f(x)=\int_x^1 y^{n-2} P_{n-2}\left(\frac{x}{y}\right)\left(\frac{1}{y}\frac{d}{dy}\right)^2 \{y^n g(y)\}\, dy$$

(6) Show that the function $f(x)$ satisfying the dual equations

$$\int_0^\infty x^{-1} f(x)\, J_1(xt)\, dx=t \qquad 0<t<1$$

$$\int_0^\infty f(x)\, J_1(xt)\, dx=0 \qquad t>1$$

is given by

$$f(x)=2\left(\frac{2x}{\pi}\right)^{\frac{1}{2}} J_{3/2}(x)=\frac{4}{\pi}\left(\frac{\sin x}{x}-\cos x\right)$$

(7) Show that the solution of the dual integral equations

$$k\int_0^\infty t\, f(t)\cos(xt)\, dt=g(x) \qquad x^2<a^2$$

$$\int_2^\infty f(t)\cos(xt)\, dt=0 \qquad x^2>a^2$$

is given by $\qquad f(t)=\dfrac{ag(t)}{k}\,\dfrac{J_1(at)}{t}.$

(8) Two equal coaxial circular electrified discs whose centres are $2c$ apart are equally charged to potential unity. Show that the problem of finding the potential in the field of these discs can be reduced to the solution of dual integral equations

$$\int_0^\infty (1+e^{-2cp})\, p^{-1} f(p)\, J_0(zp)\, dp=2 \qquad 0<r<1$$

$$\int_0^\infty f(p)\, J_0(rp)\, dp=0 \qquad r>1$$

(9) Prove the following :

(i) $M\{I^{\eta,\alpha} f(x)\}=\dfrac{\Gamma(1+\eta-\frac{1}{2}c)}{\Gamma(1+\eta+\alpha-\frac{1}{2}s)}\, F(s)$

(ii) $M\{K^{\eta,\alpha} f(x)\}=\dfrac{\Gamma(\eta+\frac{1}{2}s)}{\Gamma(\eta+\alpha+\frac{1}{2}s)}\, F(s)$

(iii) $M\{J_\nu(x)\}=2^{s-1}\,\dfrac{\Gamma(\frac{1}{2}\nu+\frac{1}{2}s)}{\Gamma(1+\frac{1}{2}\nu=\frac{1}{2}s)}$

(iv) $M\{xS_{\eta,\alpha} f(x)\}=2^s \dfrac{\Gamma(\frac{1}{2}+\eta+\frac{1}{2}s)\, F(t-s)}{\Gamma(\frac{1}{2}+\eta+\alpha-\frac{1}{2}s)}$

(10) Define $K^{\alpha}\ x^m f(x)=\dfrac{1}{\Gamma(\alpha)}\displaystyle\int_x^{\infty} (t^m-x^m)^{\alpha-1} f(t)\, mt^{n\ 1}\, dt$

for $\alpha>0$

and
$$K^{\alpha}\ x^m f(x)=\left(-\frac{d}{dx^m}\right)^n K_x^{\alpha+n} f(x),\ 0<\alpha+n<1$$

Then prove that

$$M\{K^{\alpha}\ x^m f(x)\}=\frac{\Gamma\left(\frac{s}{m}\right)}{\Gamma\left(\frac{s}{m}+\alpha\right)} F(s+m\alpha)$$

APPENDIX

Table 2 (Tables of Transforms)

$f(t)$	$F(p)$ or $F(s)$
A. Fourier	
$F(p)=\dfrac{1}{\sqrt{2\pi}}\displaystyle\int_{-\infty}^{\infty} e^{ipt} f(t)\, dt$	
$f(t)=\dfrac{1}{\sqrt{2\pi}}\displaystyle\int_{-\infty}^{\infty} c^{-ipt} F(p)\, dp$	
(1) 1	$\sqrt{2\pi}\,\delta(p)$
(2) $(a^2+t^2)^{-1}$ $a>0$	$\sqrt{\dfrac{\pi}{2}}\, a^{-1} e^{-a\|p\|}$
(3) $e^{-a\|t\|}$	$\sqrt{\dfrac{2}{\pi}}\, a(a^2+p^2)^{-1}$
(4) $t\, e^{-a\|t\|}$	$\sqrt{\dfrac{2}{\pi}}\, 2aip\, (a^2+p^2)^{-2}$
(5) $\|t\|\, e^{-a\|t\|}$	$\sqrt{\dfrac{2}{\pi}}\, (a^2-p^2)\, (a^2+p^2)^{-2}$
(6) $e^{-a^2t^2}$	$2^{-\frac{1}{2}}\, a^{-1}\, e^{-\frac{1}{4} p^2/a^2}$
(7) $t^{-\frac{1}{2}}\, J_{n+\frac{1}{2}}$	$i^n\, P_n(p)\, H(1-\|p\|)$

$f(t)$	$F(p)$ or $F(s)$
B. Fourier cosine	
$F(p)=\sqrt{\frac{2}{\pi}}\int_0^\infty f(t)\cos pt\,dt$	
(1) $t^{a-1},\ 0<a<1$	$\sqrt{\frac{2}{\pi}}\,p^{-a}\,\Gamma(a)\cos\left(\frac{1}{2}a\pi\right)$
(2) $(a^2+t^2)^{-1},\ a>0$	$\sqrt{\frac{\pi}{2}}\,a^{-1}\,e^{-ap}$
(3) $(a^2+t^2)^{-2},\ a>0$	$\frac{1}{2}\sqrt{\frac{\pi}{2}}\,a^{-3}\,e^{-ap}\,(1+ap)$
(4) e^{-at}	$\sqrt{\frac{2}{\pi}}\,a(a^2+p^2)^{-1}$
(5) $e^{-a^2t^2}$	$2^{-\frac{1}{2}}\,a^{-1}\,e^{-\frac{1}{4}p^2/a^2}$
C. Fourier sine	
$F(p)=\sqrt{\frac{2}{\pi}}\int_0^\infty f(t)\sin pt\,dt$	
(1) $t^{a-1},\ 0<a<1$	$\sqrt{\frac{2}{\pi}}\,p^{-a}\,\Gamma(a)\sin\left(\frac{1}{2}\pi a\right)$
(2) $t^{-1/2}$	$p^{-1/2}$
(3) $t(a^2+t^2)^{-1}$	$\sqrt{\frac{\pi}{2}}\,e^{-ap}$
(4) e^{-at}	$\sqrt{\frac{2}{\pi}}\,p(p^2+a^2)^{-1}$
(5) $t\,e^{-at}$	$\sqrt{\frac{2}{\pi}}\,2ap\,(p^2+a^2)^{-2}$
(6) e^{-at}/t	$\sqrt{\frac{2}{\pi}}\,\tan^{-1}(p/a)$
D. Mellin	
$F(s)=\int_0^\infty t^{s-1}f(t)\,dt$	
(1) $(1-t)^{-1}$	$\pi\cot\pi s \quad 0<s<1$
(2) $(1+t)^{-1}$	$\pi\,\mathrm{cosec}\,\pi s \quad 0<s<1$

$f(t)$	$F(p)$ or $F(s)$
(3) $(1+t^a)^{-b}$	$\dfrac{\Gamma\left(\frac{s}{a}\right)\Gamma\left(b-\frac{s}{a}\right)}{a\Gamma(b)}$, $0<s<ab$
(4) e^{-t}	$\Gamma(s)$, $s>0$
(5) e^{-t^2}	$\frac{1}{2}\Gamma(\frac{1}{2}s)$, $s>0$
(6) $\cos t$	$\Gamma(s)\cos\left(\frac{\pi}{2}s\right)$, $0<s<1$
(7) $\sin t$	$\Gamma(s)\sin\left(\frac{\pi}{2}s\right)$, $0<s<1$
E. Hankel	
$H\nu\{f\}=\int_0^\infty t\, J\nu(pt)\, f(t(\, dt$ $\quad \nu>-\frac{1}{2}$	
(1) $t^\nu\ H(a-t)$	$p^{-1}a^{\nu+1}\ J_{\nu+1}(ap)$
(2) t^{a-1} $\quad \nu-1<a<\nu+1$	$2^a\dfrac{\Gamma(\frac{1}{2}a+\frac{1}{2}\nu+\frac{1}{2})}{p^{a+1}\Gamma(\frac{1}{2}\nu-\frac{1}{2}a+\frac{1}{2})}$
(3) $t^\nu (a^2-t^2)^{\mu-\nu-1}H(a-t)$	$2^{\mu-\nu-1}\Gamma(\mu-\nu)\, a^\mu\ J_\nu\ (ap)$
(4) $t^{\nu-\mu}\ J_\nu\ (at)$ $\quad a>0,\ \mu>\nu\geqslant 0$	$p^\nu\dfrac{(a^2-p^2)^{\mu-\nu-1}}{2^{\mu-\nu-1}\ \Gamma(\mu-\nu)\, a^\mu}H(a-p)$
(5) $t^{\nu-1}\ e^{-at}$	$\dfrac{2^\nu\ p^\nu\ \Gamma(\nu+\frac{1}{2})}{\sqrt{\pi}\,(p^2+a^2)^{\nu+\frac{1}{2}}}$
(6) t^{s-1}	$M\{J\nu(px)\ ;\ s\}$

12

Approximate Methods : Cubic Splines

12·1. Introduction

Numerical approximation to solutions of differential equations is a fully developed discipline. This development has been dictated by the realisation that though existence of solutions of IVPs can be established under fairly general conditions, it is only in specific cases that solution can be found analytically. Even when the solution has been found in terms of elementary functions, exact evaluation may be very difficult as in the case of the solution $e^{-x^2} \int_0^x e^{t^2}\, dt$ of the IVP $y' - 2xy = 0$, $y(0) = 0$. It is thus imperative to devise methods of approximating the solution of DEs with arbitrarily high accuracy.

The methods in use may be classified into three categories :

(*i*) **Series methods** which produce continuous approximating functions over the entire interval. They are variously known as *step-by-step, recursive, iterative, starter* or *marching methods.* Methos of Picard, Taylor, Euler, Runge-Kutta and Adams-Bashford predicator-corrector methods fall in this category.

(*ii*) **Projective** or **variational** methods which depend on global properties of solutions such as Rayleigh-Ritz and Galerkin methods.

(*iii*) **Interpolatory** or **discrete variable** methods which produce a table of approximate values of the solution usually at equidistant points of the independent variable. These approximate values are obtained by requiring a finite combination of functions belonging to a certain family to satisfy the DE at prescribed isolated points called nodes or knots. In differential equations such methods are called *collocation methods.*

Usually a combination of projective and collocation methods involving cubic splines are preferred. Such a combination of methods is called *finite elements* which is a sine qua non for engineers.

In this chapter, we concentrate on the cubic spline method, the avant-garde of the theory of approximation. Splines give not only numerical values of solution at discrete points, but also continuously differentiable functions which are easy to understand and construct. In choosing splines we are only conforming to the historical reality that discrete variable methods have predominated over global methods. The reason, perhaps, for this reality has been that global methods can be best understood in general and abstract settings while for discrete methods, the computer is at hand to perform the most complicated calculations.

The account given by us is limited and rudimentary. For those interested in further study of this field, a list of relevant books and papers has been given at the end of this chapter. Those who are interested in the study of splines in BVPs in a strictly abstract setting, may consult works by R.S. Varga and his collabolators. We ourselves have followed the more elementary approach of Schoenberg, Greville, Loscalzo, Talbot, Albasiny and Hoskins. Hopefully, together with analytical, transform, variational and discrete variable methods no grey area should remain in the realm of linear differential equations and integral equations.

In section 10·11 we briefly described the collocation by ordinary polynomials and variational methods for operator equations. In this chapter we describe the collocation by piecewise polynomials (cubic splines) method for numerical approximation to solution of linear differential equations.

§12·2. Cubic Spline Interpolation

Let the interval $[a, b]$ be divided into the set of n sub-intervals,

$$\pi : a=x_0<x_1<x_2<\ldots<x_{n-1}<x_n=b$$

not necessarily of equal length. By a *cubic spline* $s(x)$ we shall mean a function defined on $[a, b]$ such that

(*i*) $s(x)$ coincides with a polynomial of degree 3 in each of the sub-intervals

$$\Delta_i=[x_{i-1}, x_i],\ i=1, 2, \ldots\ldots, n$$

(*ii*) $s(x)\in C^2[a, b]$ *i.e.*, $s'(x)$ and $s''(x)$ are continuous at each point of $[a, b]$.

The abscissas $x_0, x_1, \ldots\ldots, x_n$ are called *nodes* or *knots* of the spline.

A spline $s(x)$ is said to interpolate to data points

$$(x_0, y_0), (x_1, y_1), \ldots\ldots, (x_n, y_n)$$

if $s(x_i)=y_i$, $i=0, 1, \ldots\ldots, n$. If the ordinates y_i are values of a function $f(x)$ at x_i, then we say that $s(x)$ interpolates to $f(x)$ at x_i.

Depending on additional requirements, the following three types of splines arise :

I. **Cubic spline** as defined by conditions (*i*) and (*ii*).

II. **Natural spline** if in addition to (*i*) and (*ii*) we take

$$s''(a)=s''(b)=0$$

or in the end intervals $[a, x_1]$, $[x_{n-1}, b[$, $s(x)$ is of degree 1.

III. Periodic spline of period $b-a$ if

$$s(a+)=s(b-),\ s'(a+)=s'(b-),\ s''(a+)=s''(b-).$$

In case the spline interpolates to a function $f(x)$, then the function $f(x)$ too must satisfy the same periodicity conditions.

The chief merit of cubic spline interpolates as opposed to single polynomial interpolates is that splines tend to have oscillations of smaller amplitude between interpolating points. This property is implied in the following theorem.

Theorem 1 (Holladay). Given the data points (x_i, y_i) $i=0, 1, \ldots\ldots, n.$ Of all functions $y(x)$ of class $C^2[a, b]$ which interpolate to this data, the cubic spline interpolate $s(x)$ minimises the integral

$$\int_a^b y''^2(x)\,dx$$

and is the unique minimiser in this class.

(In other words the cubic spline has the minimum total curvature among all sufficiently smooth interpolating curves).

Proof. $\displaystyle\int_a^b y''^2(x)\,dx-\int_a^b s''^2(x)\,dx$

$$=\int_a^b (y''-s'')^2\,dx+2\int_a^b s''(y''-s'')\,dx$$

$$=I_1+I_2 \text{ (say)}.$$

Now split the interval of integration in I_2 into its sub-intervals and integrate by parts. Thus

$$I_2=2\int_a^b s''(y''-s'')\,dx$$

$$=2\sum_{i=0}^{n-1}\int_{x_i}^{x_{i+1}} s''(y''-s'')\,dx$$

$$=2\sum_{i=0}^{n-1}\Big[s''\Big[(y'-s')\Big]_{x_i}^{x_{i+1}}$$

$$-2\sum_{i=0}^{n-1}\int_{x_i}^{x_{i+1}} s'''(y'-s')dx \qquad \ldots(i)$$

The first sum in (i) telescopes to $2\Big[s''(y'-s')\Big]_a^b$ which vanishes if either $s(x)$ is a natural spline *i.e.*, $s''(a)=s''(b)=0$ or if $y'(a)=s'(a)$ and $y'(b)=s'(b)$.

For the second sum we note that $s'''=c_i$ (constant) in each sub-interval $[x_i, x_{i+1}]$. Hence

$$\int_{x_i}^{x_{i+1}} s''' (y'-s')\, dx = c_i \Big[y(x)-s(x) \Big]_{x_i}^{x_{i+1}} = 0$$

Thus finally $I_2=0$. Since $I_1 \geqslant 0$, we have

$$\int_a^b s''^2\, dx \leqslant \int_a^b y''^2\, dx \qquad \text{...}(ii)$$

To prove the uniquesness of $s(x)$ we prove that $s(x)\equiv 0$ is the only cubic spline interpolating to $y(x)\equiv 0$. Owing to relation (*ii*),

$$\int_a^b s''^2\, dx = 0$$

i.e., $\quad s''(x)=0$ on $[a, b]$

i.e., $\quad s(x)=Ax+B$

$A=B=0$ since $s(a)=y(a)=0$ and $s(b)=y(b)=0$.

Thus $s(x)\equiv 0$. **Q.E.D.**

Relation (*ii*) is known as the *minimum curvature* or *variation diminishing property*. The property also holds for periodic splines and functions $y(x)$ of period $b-a$.

Construction of Interpolating Splines

We follow the procedure of Ahlberg et al.

Let $x_i - x_{i-1} = h_i$ $(i=1, 2, \ldots\ldots, n)$.

Since $s(x)$ is piecewise cubic, $s'(x)$ is piecewise quadratic and $s''(x)$ is piecewise linear in each sub-interval $\Delta_i = [x_{i-1}, x_i]$. We can write

$$s''(x) = m_{i-1}\frac{(x_i - x)}{h_i} + m_i \frac{(x-x_{i-1})}{h_i} \text{ on } \Delta_i \qquad \text{...}(iii)$$

Then $\quad m_i = s''(x_i)$ $(i=0, 1, 2, \ldots\ldots, n)$ $\qquad$ ...(*iv*)

Integrating (*iii*) twice

$$s(x) = m_{i-1}\frac{(x_i - x)^3}{6h_i} + m_i \frac{(x-x_{i-1})^3}{6h_i} + A_i(x_i - x) + B_i(x - x_{i-1})$$

Evaluating the constants A_i and B_i by the conditions

$$s(x_{i-1}) = y_{i-1},\ s(x_i) = y_i,$$

we get
$$s(x) = m_{i-1}\frac{(x_i - x)^3}{6h_i} + m_i \frac{(x-x_{i-1})^3}{6h_i} + \left(y_{i-1} - \frac{m_{i-1} h_i^2}{6}\right)\frac{x_i - x}{h_i} + \left(y_i - \frac{m_i h_i^2}{6}\right)\frac{x - x_{i-1}}{h_i}$$

on Δ_i $\qquad$ ...(*v*)

Differentiating (*v*) with respect to x,

$$s'(x) = -m_{i-1}\frac{(x_i - x)^2}{2h_i} + m_i \frac{(x-x_{i-1})^2}{2h_i} + \frac{y_i - y_{i-1}}{h_i} - \frac{m_i - m_{i-1}}{6} h_i \text{ on } \Delta_i \qquad \text{...}(vi)$$

From (*vi*), we can get the evaluations

$$s'(x_i-)=\frac{h_i\, m_{i-1}}{6}+\frac{h_i m_i}{3}+\frac{y_i-y_{i-1}}{h_i} \quad ...(vii)$$

$$s'(x_i+)=-\frac{h_{i+1}\, m_i}{3}-\frac{h_{i+1}\, m_{i+1}}{6}+\frac{y_{i+1}-y_i}{h_{i+1}} \quad ...(viii)$$

Since $s'(x)$ is continuous $s'(x_i+)=s'(x_i-)$.

Hence from (*vii*) and (*viii*), we get

$$\frac{h_i\, m_{i-1}}{6}+\frac{h_i+h_{i+1}}{3}\, m_i+\frac{h_{i+1}\, m_{i+1}}{6}$$

$$=\frac{y_{i+1}-y_i}{h_{i+1}}-\frac{y_i-y_{i-1}}{h_i} \quad ...(ix)$$

$$(i=1, 2, \ldots\ldots, n-1)$$

Problems. (1) Using (*vii*) and (*viii*) shows that for a natural spline

$$s_0'=-\frac{1}{2}s_1'+\frac{3}{2}\frac{y_1-y_0}{h_1},\ s_n'=-\frac{1}{2}s_{n-1}'+\frac{3}{2}\frac{y_n-y_{n-1}}{h_n}$$

For $n=3$ and $h_i=h$ compute s_i', $i=0, 1, 2, 3$.

(2) For $n=2$ and $s_0''=s_2''=0$, prove that $s_0'=\frac{1}{4h}(-5y_0+6y_1-y_2)$

$$s_1'=\frac{1}{2h}(-y_0+y_2),\ s_2'=\frac{1}{4h}(y_0-6y_1+5y_2)$$

Uniform spacing. Let the division points be equidistant and $h_i=h$, $i=1, 2, \ldots\ldots, n$. Then from (*vii*) and (*viii*)

$$s'(x_i-)=h\left(\frac{m_{i-1}}{6}+\frac{m_i}{3}+\frac{y_i-y_{i-1}}{h^2}\right) \quad ...(x)$$

$$s'(x_i+)=h\left(\frac{-m_i}{3}-\frac{m_{i+1}}{6}+\frac{y_{i+2}-y_i}{h^2}\right) \quad ...(xi)$$

Also relation (*ix*) is reduced to the difference equation in

$$m_i=s''(x_i)$$

$$m_{i+1}+4m_i+m_{i-1}=\frac{6}{h^2}(y_{i+1}-2y_i+y_{i-1}) \quad ...(xii)$$

$$(i=1, 2, \ldots\ldots, n-1)$$

The right hand side of (*xii*) is the second central difference $\delta^2 y_i$ of y_i. We have already considered the solution of (*xii*) analytically involving the discrete Green's function.

Note that (*xii*) from a system of $n-1$ linear equations in $n+1$ unknowns $m_0, m_1, \ldots\ldots, m_n$ so that two more conditions must be added for a complete solution. We write two additional conditions in the form

$$\begin{aligned}2m_0+\lambda_0 m_1&=d_0\\ \mu_n m_{n-1}+2m_n&=d_n\end{aligned} \qquad \ldots(xiii)$$

(*xii*) and (*xiii*) together form a tridiagonal system which will have a unique solution. It may be noted that the matrix of the combined system is strictly diagonally dominant tridiagonal.

(*a*) If we select $\lambda_0=d_0=\mu_n=d_n=0$, then $m_0=m_n=0$ *i.e.* $s(x)$ is of degree 1 at the ends of the interval $[a, b]$. This yields the **natural spline.**

(*b*) Alternatively, we may specify the slopes $s'(x)$ at the end-points a and b *i.e.,* $s'(a)=y_0$, $s'(b)=y_n$. Then

$$\lambda_0=1=\mu_n,\ d_0=\frac{6}{h}\left(\frac{y_1-y_0}{h}-y_0'\right),$$

$$d_n=\frac{6}{h}\left(y_n'-\frac{y_n-y_{n-1}}{h}\right)$$

(*c*) A third possibility is to choose $x_1, x_2,\ldots, x_{n-1}$ as the spline knots and impose conditions

$$s(x_0)=y_0,\ s(x_n)=y_n.$$

Problem. In place of assuming $s'(x)$ in terms of $s''(x)$ at nodes, assume $s''(x)$ in terms of $s'(x)$.

Then prove that

$$s''(x_i-)=\frac{2}{h_i}\,(s'_{i-1}+2s'_i)-6\,\frac{y_i-y_{i-1}}{h_i^2}\quad i>0$$

$$s''(x_i+)=\frac{-2}{h_{i+1}}\,(2s'_i+s'_{i+1})+6\,\frac{y_{i+1}-y_i}{h^2_{i+1}}\quad i>n$$

$$\frac{1}{h_i}\,s'_{i-1}+2\left(\frac{1}{h_i}+\frac{1}{h_{i+1}}\right)s'_i+\frac{1}{h_{i+1}}\,s'_{i+1}$$

$$=3\,\frac{y_i-y_{i-1}}{h_i^2}+3\,\frac{y_{i+1}-y_i}{h^2_{i+1}}\ (i=1, 2,\ldots, n-1).$$

and for $h_i=h$

$$s_{i+1}+4s'_{i_1}+s'_{i-1}=\frac{6}{h}\cdot\frac{y_{i+1}-y_{i-1}}{2h} \qquad \ldots(A)$$

Application to Approximate Integration

If we integrate (*v*) between x_{i-1} and x_i, then

$$\int_{x_{-1}}^{x_i} s(x)\,dx=\frac{y_{i-1}+y_i}{2}\,h_i-\frac{m_{i-1}+m_i}{24}\,h^3_i \qquad \ldots(xiv)$$

Adding we get

$$\int_a^b s(x)\,dx=\sum_{i=1}^{n}\frac{y_{i-1}+y_i}{2}\,h_i-\sum_{i=1}^{n}\frac{m_{i-1}+m_i}{24}\,h_i^3 \qquad \ldots(xv)$$

(xv) is the well-known trapezoidal rule of integration with a correction.

If we sum (xiv) for $i=1, 2,\ldots, p\leqslant n$, then we get the indefinite integral upto the node x_p. If the value of the indefinite integral is sought for $x_p<x<x_{p+1}$, then the integral

$$\int_{x_p}^{x} s(t)\,dt \text{ must be added to } \sum_{i=1}^{p}.$$

If may be noted that for an equispaced grid (xv) can be written in the form

$$\int_a^b s(x)\,dx=\frac{h}{2}\,(y_0+2y_1+2y_2+\ldots+2y_{n-1}+y_n)+R_n \qquad \ldots(xvi)$$

Example 1. Fit a natural cubic spline to the data

x	0	1	2
y	-2	-1	5

and compute (i) y (1·5) (ii) y' (1)

Solution. Here $x_0=0$, $x_1=1$, $x_2=2$, $h=1$

$y_0=-2$, $y_1=-1$, $y_2=5$.

Since it is a natural spline $m_0=m_2=0$.

Hence $4m_1=6\ (y_0-2y_1+y_2)=30$, $i.e$, $m_1=\frac{15}{2}$ · Using formula (v) for $i=2$, we find that the cubic spline in the interval [1, 2] is

$$s(x)=-\frac{15}{12}\,(x-2)^3+\frac{27}{12}(x-2)+5\,(x-1)$$

$$s'(x)=-\frac{15}{4}\,(x-2)^2+\frac{27}{12}+5$$

Hence (i) s (1·5)= 1·53 (ii) $s'(1)=\frac{7}{2}=y'(1)$.

$=y(1·5)$

Example 2. Fit a natural cubic spline interpolate to the data

x	0	1	2	3	4
y	-2	-1	5	11	14

and compute (*i*) $y(2{\cdot}2)$, (*ii*) $y'(3)$

Solution. Here $x_0=0,\ x_1=1,\ x_2,=2,\ x_3=3,\ x_4=4.$

$$h=1,\ y_0=-2,\ y_1=-1,\ y_2=5,\ y_3=11,\ y_1=14.$$

$$m_0=m_4=0.$$

m_1, m_2, m_3 are found from the formula

$$m_{i+1}+4m_i+m_{i-1}=\frac{6}{a^2}(y_{i+1}-2y_i+y_{i-1})$$

$$m_2+4m_1=6\,(y_2-2y_1+y_0)=30$$
$$m_3+4m_2+m_1=6\,(y_3-2y_2+y_1)=0$$
$$4m_3+m_2=6\,(y_4-2y_3+y_2)=-18$$

Hence $m_1=\frac{54}{7},\ m_2=-\frac{6}{7},\ m_3=-\frac{30}{7}.$

In formula (*v*) take $i=3$ so that $\Delta_i=[2, 3]$. In the interval [2, 3] the cubic spline is given by

$$s(x)=\frac{m_2}{6}(x_3-x)^3+\frac{m_3}{6}(x-x_2)^3+\left(y_2-\frac{m_2}{6}\right)(x_3-x)+\left(y_3-\frac{m_3}{6}\right)(x-x_2)$$

$$=\frac{-1}{7}(3-x)^3-\frac{5}{7}(x-2)^3+\frac{36}{7}(3-x)+\frac{82}{7}(x-2).$$

Hence (*i*) $y(2{\cdot}2)=s(2{\cdot}2)=6{\cdot}377$

(*ii*) $y'(3)=s'(3)=\frac{31}{7}=4{\cdot}43$

Problems

(1) Fit a natural cubic spline to the data

x	2	3	4
y	11	49	123

and compute (*i*) $y(2{\cdot}5)$ (*ii*) $y'(2)$ $\left[\ \textbf{Ans. } (i)\ 26\frac{5}{8}\ \ (ii)\ 29{\cdot}0\ \right]$

(2) Fit a natural cubic spline to the data

x	1	2	3
y	−8	−1	18

and compute (*i*) $y(1{\cdot}5)$ (*ii*) $y'(1)$.

$$\left[\ \textbf{Ans.}\ (i)\ \frac{-45}{8}\quad (ii)\ 4\ \right]$$

§12·3. Application to BVPs

Procedure Given a two-point BVP

$$\text{DE}: y''+p(x)\,y'+q(x)\,y=r(x) \qquad \ldots(1)$$

$$\text{BCs}: y(x_0)=y_0,\ y(x_n)=y_n \qquad \ldots(2)$$

If a grid of uniform spacing h is taken, then, in the interval $\Delta_i=[x_{i-1}, x_i]$, the spline $s(x)$ and its derivatives are given by the formulae

$$s(x)=\frac{m_{i-1}}{6h}(x_i-x)^3+\frac{m_i}{6h}(x-x_{i-1})^3+\left(y_{i-1}-\frac{h^2 m_{i-1}}{6}\right)\frac{x_i-x}{h}$$

$$+\left(y_i-\frac{h^2 m_i}{6}\right)\frac{x-x_{i-1}}{h} \qquad \ldots(3)$$

$$s'(x_i-)=\frac{h}{6}m_{i-1}+\frac{hm_i}{3}+\frac{y_i-y_{i-1}}{h} \qquad \ldots(4)$$

$$s'(x_i+)=\frac{-h}{3}m_i-\frac{h}{6}m_{i+1}+\frac{y_{i+1}-y_i}{h} \qquad \ldots(5)$$

$$m_{i+1}+4m_i+m_{i-1}=\frac{6}{h^2}(y_{i+1}-2y_i+y_{i-1}) \qquad \ldots(6)$$

where $m_i=s''(x_i)$ and $y_i=y(x_i)$.

Case 1. First derivative absent

In this case $y''(x)=r(x)-q(x)\,y$.

Hence if $m_i=y''(x_i)$, $r(x_i)=r_i$, $q(x_i)=q_i$, then

$$m_i=r_i-q_i\,y_i\ (i=0, 1, 2,\ldots, n) \qquad \ldots(i)$$

Substituting (*i*) into (6) yields

$$y_{i+1}\left(1+\frac{h^2}{6}q_{i+1}\right)-y_i\left(2-\frac{2h^2}{3}q_i\right)+y_{i-1}\left(1+\frac{h^2}{6}q_{i-1}\right)$$

$$=\frac{h^2}{6}(r_{i+1}+4r_i+r_{i-1}) \qquad (i=1, 2,\ldots, n-1) \qquad \ldots(ii)$$

The $n-1$ linear equations (*ii*) and the BCs (2) form a tridiagonal system of equations and are readily solvable for $y_0, y_1,\ldots, y_n$ through back substitution. These values together with relations (*i*) determine $m_0, m_1,\ldots, m_n$ and hence the spline $s(x)$ in (3). Since no assumptions about insisting on $m_0=m_n=0$ have been made, the approximating spline is not necessarily natural on the interval $[x_0, x_n]$.

Example 3. Find a cubic spline approximate to the BVP

$$y''+y+1=0\ ;\ y(0)=y(1)=0.$$

Soluion. Take the knots $x_0=0$, $x_1=\frac{1}{2}$, $x_2=1$ on the interval [0, 1].

Since $y_0=y_2=0$, by (*ii*)

$$y_1=y(0\cdot5)=\frac{3}{22}\ .$$

Hence from (*i*), $m_0=m_2=-1$, $m_1=\frac{-25}{22}$ · Equations (4) and (5) give

$$s'(0)=\frac{47}{88},\qquad s'(0.5)=0,\qquad s'(1)=\frac{-47}{88}\ .$$
$$=0\cdot5341.$$

The spline $s(x)$ is obtained by relation (3).

In $\left[0, \frac{1}{2}\right]$, $s(x)=\frac{47}{88}x-\frac{1}{2}x^2-\frac{1}{22}x^3$ and in $\left[\frac{1}{2}, 1\right]$

$$s(x)=\frac{47}{88}x-\frac{1}{2}x^2-\frac{1}{22}x^3+\frac{1}{11}\left(x-\frac{1}{2}\right)^3 .$$

Let us compare with the exact solution which is

$$y=\frac{1-\cos 1}{\sin 1}\sin x+\cos x-1.$$

Hence $y\left(\frac{1}{2}\right)=\frac{1}{\cos\frac{1}{2}}-1=\frac{1}{1-\frac{1}{8}+\dots}-1=\frac{8}{7}-1=\frac{1}{7}.$

This value of $y(0\cdot5)$ compares favourably with the spline value $\frac{3}{22}$.

Also for the exact solution

$$y'(x)=\frac{1-\cos 1}{\sin 1}\cos x-\sin x,$$

$$y''(x)=-\frac{1-\cos 1}{\sin 1}\sin x-\cos x.$$

Hence $y''(0)=y''(1)=-1$ which are exactly the values of m_0 and m_2 for the spline found above.

Also $y''\left(\frac{1}{2}\right)=-\frac{1}{\cos\frac{1}{2}}=\frac{-1}{1-\frac{1}{8}}=-\frac{8}{7}=-(1\cdot143)$

This is very nearly equal to $m_1=\frac{-25}{22}=-(1\ 137)$.

Again $y'(0)=\frac{1-\cos 1}{\sin 1}=0\cdot5757$ and $y'(1)=\frac{\cos 1-1}{\sin 1}$

$[=s'(0)]$ $\qquad\qquad=[s'(1)]$

and $y'(\frac{1}{2})=0$. The result should be compared with ex. 5 of Exercise 34.

Case 2. First derivative present

DE : $y''+p(x)\,y'+q(x)\,y=r(x)$

BCs : $y(x_0)=y_0,\ y(x_n)=y_n$

From the DE,

$$y''(x)=r(x)-p(x)\,y'(x)-q(x)\,y(x)$$

Putting $x=(x_i-)$ and (x_i+) and using (4) and (5) we have respectively,

$$\frac{h}{6}\,p_i\,m_{i-1}+\left(1+\frac{h}{3}p_i\right)m_i=r_i-q_iy_i-\frac{p_i}{h}\,(y_i-y_{i-1})$$

$$(i=1,\,2,\,\ldots\ldots,\,n) \qquad \ldots(iii)$$

$$\left(1-\frac{1}{3}\,hp_i\right)m_i-\frac{h}{6}\,p_i\,m_{i+1}=r_i-q_iy_i-\frac{y_i}{h}\,(y_{i+1}-y_i)$$

$$(i=0,\,1,\,\ldots\ldots,\,n-1) \qquad \ldots(iv)$$

The $2n+2$ unknowns $m_0, m_1, \ldots\ldots, m_n$ and $y_0, y_1, \ldots\ldots, y_n$ can be determined from the $2n$ equations (*iii*), (*iv*) and the BCs.

Example 4. Find a cubic spline approximate to the BVP,

$$y''=\frac{4x}{1+x^2}\,y'+\frac{2}{1+x^2}\,y=0$$

$$y(0)=1,\ y(1)=\tfrac{1}{2}.$$

Solution. Take the equidistant knots

$$x_0=0,\ x_1=\tfrac{1}{2},\ x_2=1\ ;\ h=\tfrac{1}{2}$$

Then $\qquad y_0=1,\ y_2=\frac{1}{2},\ y_1=(?)$

$$p(0)=p_0=0,\ p_1=p\left(\frac{1}{2}\right)=\frac{8}{5},\ p_2=2$$

$$q(0)=q_0=2,\ q(\tfrac{1}{2})=q_1=8/5,\ q_2=q(1)=1$$

$$r_0=r_1=r_2=0$$

Relations (*iii*) and (*iv*) give $m_0=-2$

$$19m_1=-72y_1+52$$

$$m_1+8m_2=-24y_1-15$$

$$11m_1-2m_2=24y_1-24$$

i.e., $$m_1=-\frac{73}{230},\quad m_2=\frac{67}{115}$$

$$y_1=0{\cdot}8$$

$$=s(\tfrac{1}{2})$$

The exact solution is

$$y(x)=\frac{1}{1+x^2}$$

which gives $y\left(\frac{1}{2}\right)=\frac{4}{5}=0{\cdot}8$

Thus the spline fit is very nearly correct.

Expressions for $s(x)$ can be obtained from (3).

Problem 1. Given the BVP

$$y''+\frac{1}{x+4}y'-\frac{1}{(x+1)(x+\ )}y=\frac{2}{3(x+4)}$$

$$y(0)=\frac{1}{3}\log\frac{4}{3},\ y(3)=\frac{4}{3}\log\frac{7}{3}$$

Prove that the exact solution is

$$y=\tfrac{1}{3}(x+1)\log\tfrac{1}{3}(x+4)$$

Fit a cubic spline approximate to the solution by taking $h=1$ on [0, 3] and show that the following spline values of y, y', y'' are obtained :

x	y		y'		y''	
	Exact	*Spline*	*Exact*	*Spline*	*Exact*	*Spline*
0	0·09589	0·09589	·17923	·17700	·14583	·14639
1	0·34055	0·33945	·30361	·30348	·10667	·10658
2	0·69315	0·69233	·39772	·39836	0·08333	·08318
3	1·12973	1·12973	·47290	·47390	·06803	·06789

Problem 2. Fit a cubic spline to the BVP

$$y''+\frac{4x}{1+x^2}y'+\frac{2}{1+x^2}y=0\ ;\ y(0)=1\ ;\ y(2)=0{\cdot}2$$

$$h=1$$

and show that

$$m_0=-2,\ m_1=\frac{184}{295},\quad m_2=\frac{58}{295}$$

$$y_1=\frac{96}{177}=0{\cdot}5423.$$

Problem 3. Solve the above BVP with $h=\frac{1}{2}$ and show that

$m_0=-2$, $m_1=-0{\cdot}3834$, $m_2=0{\cdot}5576$,
$m_3=0{\cdot}3738$, $m_4=0{\cdot}1874$; $y_1=0{\cdot}8234$,
$y_2=0{\cdot}5228$, $y_3=0{\cdot}3147$.

Problem 4. Fit a cubic spline to the following BVPs, dividing the interval into two equal parts :

(*a*) $y''=y$; $y(0)=0$, $y(1)=1$

(*b*) $y''+2y'+y=30x$; $y(0)=0$, $y(1)=0$

(c) $y''-64y+10=0$; $y(0)=y(1)=0$.

Compare the spline solutions with analytical solutions.

Note on B-Splines with Equidistant Knots

Direct use of splines in computations is seldom made as much simpler splines, the so-called B-splines are better suited in practice. We did not make any use of B-splines for the simple reason that the number of knots taken by us was too small and it made no difference what splines we choose for approximating the solution. This, however, may not be the case when the number of knots is large. We, therefore, give a brief outline of the B-splines as defined by Curry and Schoenberg.

By a simple counting argument it can be shown that for a grid of $n+1$ nodes

$$\pi : a=x_0<x_1<\ldots\ldots<x_{n-1}<x_n=b$$

the set of cubic splines is a linear space of dimension $n+3$ while the space of natural splines has dimension $n+1$. For the composite spline function $s(x)$ consists of n pieces of cubic polynomials each containing four free parameters as coefficients. Thus $s(x)$ contains in all $4n$ paramsters. The continuity constraints on $s(x)$, $s'(x)$, $s''(x)$ *viz.*, $(s_i+)=s(x_i-)$, $s'(x_i+)=s'(x_i-)$, $s''(x_i+)=s''(x_i-)$ at each of the $n-1$ internal knots x_1, x_2, ……, x_{n-1} fix $3(n-1)$ parameters leaving $4n-3(n-1)=n+3$ free parameters in $s(x)$. Hence the space $S_3(\pi)$ of all cubic splines has dimension $n+3$. In the case of natural splines two end conditions $s''(a)=s''(b)=0$ are also imposed leaving $n+1$ free parameters.

Hence the space of natural splines has dimension $n+1$. Members of a basis of a spline function space are called *basis splines* or *B-splines*.

Curry and Schoenberg have shown that B-splines have the smallest compact support. For example, a cubic B-spline cannot be defined on an interval with less than five knots.

Definition. Let

$$M(x\,;y)=4(y-x)_+^3=\begin{cases}4(y-x)^3 & y\geq x\\ 0 & y<x\end{cases} \quad \ldots(i)$$

Taking the divided difference of M in the argument y, if

$$M_i(x)=M(x\,;x_i,\ x_{i+1},\ x_{i+2},\ x_{i+3},\ x_{i+4}) \quad \ldots(ii)$$

then $M_i(x)$ is a B-spline based at the knots

$$x_i,\ x_{i+1},\ \ldots\ldots,\ x_{i+4}$$

Properties. By Peano's remainder theorem, the divided difference of a function $f(x)$ with respect to the knots x_i, x_{i+1}, ……, x_{i+4} is given by

$$f[x_i,\ x_{i+1},\ \ldots\ldots,\ x_{i+4}]=\int_{-\infty}^{\infty} K(x)\, f^{(4)}x\, dx \quad \ldots(iii)$$

where $K(x)=\frac{1}{4\,!}M[x\,;\,x_i,\,x_{i+1},\,\ldots\ldots,\,x_{i+4}]$ ...(*iv*)

Putting $f(x)=\frac{x^4}{4\,!}$, we can see that

$$\int_{-\infty}^{\infty} 4\,!\,K(x)\,dx=1 \qquad \ldots(v)$$

i.e.,

$$\int_{-\infty}^{\infty} M_i(x)\,dx=1 \qquad \ldots(vi)$$

Hence $f[x_i,\,x_{i+1},\,\ldots\ldots,\,x_{i+4}]=\frac{1}{4\,!}\int_{-\infty}^{\infty} M_i(x)\,f^{(4)}(x)\,dx$

...(*vii*)

$f\in C^4$

From the definition it is clear that the B-splines are positive. $M_i(x)$ is called a *normalised B-spline*.

B-Splines with Equidistant Knots

Let $\Delta f(x_i)=f(x_i+h)-f(x_i)$ be the forward difference operator. Then since the knot *b* are equidistant,

$$f[x_i,\,x_{i+1},\,\ldots\ldots,\,x_{i+4}]=\frac{1}{4\,!\,h^4}\,\Delta^4 f(x_i)$$

$$=\frac{1}{4\,!\,h^4}\,[\,f(x_{i+4})-4f(x_{i+3})+6f(x_{i+2})-4f(x_{i+1})+f(x_i)]$$

...(*viii*)

Since spacing is uniform, Δ^4 will annihilate polynomials of degree 3.

Let $h=\frac{b-a}{n}$ and let $M(x)$ denote a cubic B-spline having knots at $-2h$, $-h$, 0, h, $2h$. Then the set of $n+3$ B-splines $M(x-a-ih)$, $(i=-1,\,0,\,1,\,2,\,\ldots\ldots,\,n+1)$ is a basis of the cubic spline space and a composite spline $s(x)$ is given by

$$s(x)=\sum_{i=-1}^{n=1} C_i M(x-a-ih) \qquad \ldots(ix)$$

Example. Take $h=1$ and the knots x_0, x_1, x_2, x_3, x_4, x_0 res pectively $=2, -1, 0, 1, 2$. Then the B-spline based at these knots is

$$M(x)=\frac{1}{4\,!}\,M(x\,;\,x_0,\,x_1,\,x_2,\,x_3,\,x_4)$$

where $\quad M(x\,;\,y)=4(y-x)_+^3$

Hence

$$M(x)=\frac{1}{6}\Big[\,(2-x)_+^3\;-4(1-x)_+^3\;+6(-x)_+^3\;-4(-1-x)_+^3\;+(-2-x)_+^3\,\Big]$$

$$=\begin{cases} 0 & -\infty<x\leqslant-2 \\ \frac{1}{6}(x+2)^3 & -2\leqslant x\leqslant-1 \\ -\frac{1}{2}x^3-x^2+\frac{2}{3} & -1\leqslant x\leqslant 0 \\ \frac{1}{2}x^3-x^2+\frac{2}{3} & 0\leqslant x\leqslant 1 \\ -\frac{1}{6}(x-2)^3 & 1\leqslant x\leqslant 2 \\ 0 & 2\leqslant x<\infty \end{cases}$$

(See Fig. 12·1)

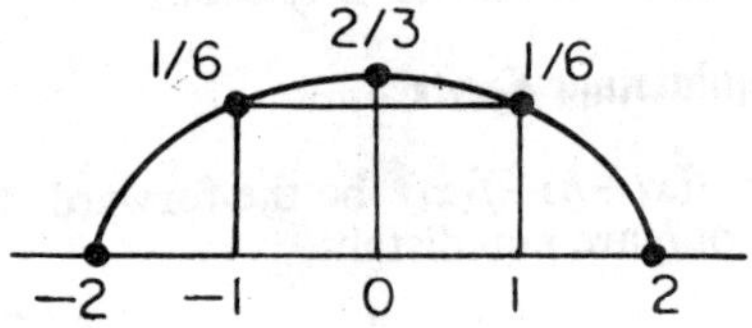

Fig. 12·1,

For other values of h merely replace x by $\dfrac{x}{h}$ in these expressions.

Since in practice all that we require is the values of $M(x)$, $M'(x)$, $M''(x)$ at the knots $-2h$, $-h$, 0, h, $2h$, the following table of values will be a useful reference

$x\to$	$-2h$	$-h$	0	h	$2h$
$M(x)$	0	$\frac{1}{6}$	$\frac{2}{3}$	$\frac{1}{6}$	0
$M'(x)$	0	$\frac{1}{2h}$	0	$-\frac{1}{2h}$	0
$M''(x)$	0	$\frac{1}{h^2}$	$\frac{-2}{h^5}$	$\frac{1}{h^2}$	0

Other methods, perhaps better in some ways, exist to compute B-splines. For example, works by Carl de Boor and Mansfield have appeared which express cubic B-splines as convex linear combinations of two B-splines of degree 2 resulting in greater stability of procedure and better adaptability to computers.

APPENDIX

Books

1. Ahlberg J.H., Nilson E.N., Walsh J.L. The Theory of Splines and Their Applications, Academic Press, 1967.
2. Greville T.N E. Ed., Theory and Applications of Spline Functions, Academic Press, 1969.
3. Hildebrand F,B. Introduction to Numerical Analysis, Tata McGraw Hill Co. Ltd., 1974 (Chapter 9).

Papers

1. Albasiny E.L. and Hoskins W.D., Cubic Spline solution to two-point boundary value prob'ems, The Computer Journal, Vol. 12, No. 2, 1969, 151 – 153.
2. Holladay J.C., A Smoothest Curve Approximation, Mathematical Tables Aids Computations, 11 (1957), 233–243.

13

Variational Methods

§ 13·1. Introduction

In ordinary calculus it is a familiar problem to determine the points at which a given function attains its maximum or minimum value. Such points are called *stationary* or *extremum* points. For example in the domain $-1 \leqslant x, y \leqslant 1$, the maximum of the function $f(x,y)=x^2+y^2$ is 2 which occurs at the extremum points $(\pm 1, \pm 1)$. The minimum of $f(x, y)$ is 0 which occurs at (0, 0). By Weierstrass theorem, a continuous function attains its extrema (*i.e.*, maxima or minima) in a closed and bounded domain.

Most problems in physics and many problems in differential geometry may be formulated in variational forms in which a curve $x=x(t)$, $y=y(t)$, $a \leqslant t \leqslant b$ is sought that makes an integral

$$J=\int_a^b F[x(t), y(t), x(t), y(t)]\, dt$$

a maximum or a minimum from amongst a class of curves. The integral J instead of depending on independent variables varying over a given range, depends on the entire course of functions $x(t)$, $y(t)$ drawn from a certain class of functions, called *admissible functions.* We call such expressions functionals. Problems which investigate extrema of functions or functionals are called *Variational problems.*

The subject of variational problems of real functions of one or more scalar arguments is part of elementary calculus. The subject of variational problems of functionals is called the *calculus of variations.* The laws which govern functionals to attain extrema are called *variational principles.* · Examples of variational principles are the principle of least squares, the principle of least actions, the laws of conservation of energy and momentum and Fermat's principle of optics which states that the path actually chosen by the light ray is the one which requires least time. From this it immediately follows that light travels in straight lines in any homogeneous medium. Examples of variational problems are the brachistochrone problem of the line of quickest descent. In this problem posed by Johann Bernonlli in 1696, it is required to find the line joining two points A and B that do not lie on a vertical line and possessing the property that a moving particle slides down this line from A to B in the shortest time. It turns out that the line of quickest descent is not the straight line though it is the shortest distance between A and B, but it is a cycloid. A problem in differential geometry is to determine the line of minimum length joining two given points on a surface. Such shortest lines are called *geodesics*. The ancient Greeks had solved the problem of finding a closed curve of given length bounding a maximum area. Such problems are called *isoperimetric*.

Though calculus of variations began to develop in 1696 with the brachistochrone problem, it matured into an independent mathematical discipline after 1744 when Euler (1707—1783) discovered the basic differential equation known by his name for a minimising curve.

There are strong reasons to assert that the study of variational calculus is very much relevant today, just as other branches of analysis are. Amongst the most important ones are :

1. It is a major branch of analysis in its own right with a history of continuous development of more than two hundred years paralleling that of ordinary calculus, differential and integral equations.

2. It enables us to express basic principles of mathematical physics with clarity and simplicity.

3. It enriches and is enriched by differential and integral equations. This is evident from the following examples—

(*a*) The minimising problem corresponding to the Dirichlet integral $\int_D (u_x^2+u_y^2)\,dx\,dy$ leads to the Euler's equation which reduces to the Laplace's equation (*The Dirichlet principle*) *viz*

$$\nabla^2 u(x, y)=\frac{\partial^2 u}{\partial x^2}+\frac{\partial^2 u}{\partial y^2}=0. \qquad \text{(See section 13·4)}$$

(*b*) The Euler's equation for the functional

$$J[\phi]=\int_a^b\int_a^b K(x, t)\,\phi(x)\,\phi(t)\,dx\,dt$$

$$+\int_a^b \phi^2(x)\,dx-2\int_a^b \phi(x) f(x)\,dx$$

reduces to the Fredholm integral equation

$$\phi(x)=f(x)-\int_a^b K(x, t)\,\phi(t)\,dt \qquad \text{(See example 7(a)).}$$

(*c*) The Euler equation for the functional

$$J[\phi]=\int_{-\infty}^{\infty}[\,p(x)\,\phi'^2(x)+2\phi(x+1)\phi(x-1)-\phi^2(x)-2\,\phi(x) f(x)]\,dx$$

is the differential—difference equation

$$(p'\phi')-\phi(x+2)-\phi(x-2)+\phi(x)+f(x)=0$$

which can be solved by the use of Laplace transform, (See Example 7(*c*).

(*d*) The minimising problem of the functional

$$J[y(x)]=\int_a^b [p(x)\, y'^2(x)+q(x)\, y^2(x)]\, dx$$

subject to the boundary conditions $y(a)=y(b)=0$ reduces to the sturm-Liouville eigenvalue problem

$$(p\,y')'+qy=\lambda y,\; y(a)=y(b)=0$$

and leads to estimates of the smallest eignenvalue (*Rayleigh's principle*) (See example 7(*b*) and section 13·8).

4. It is a mathematical tool of great power. Besides the local interpolatory methods, modern approximation theory to a large extent uses the global variational methods. The later are called *projective methods,* because in operator equations the solution or an approximation thereof is projected onto a finite dimensional subspace of a complete normed linear space of functions having smoothness properties of a desired level. We shall study one such method in greater detail, the Rayleigh—Ritz method of approximate solution of ordinary and partial differential equations in section 13·7.

§13·2. Basic Notions

Definition. 1. (Functionals) Let M be a class of functions $y(x)$. If to each function $y(x)\in M$ there is associated, by some law, a definite number J, then we say that a *functional* J is defined on the class M and we write $J=J[y(x)]$. We also say that $J[y(x)]$ is a functional with argument function $y(x)$. The class M is called the set of *admissible functions* or the *domain of definition of the functional.*

In our work M will consist of functions which are at least continuous and have piecewise continuons derivatives on a closed and bounded interval $a\leqslant x\leqslant b$. The notation $J(y(x)]$ of a functional should not be confused with functional composition $f[g(x)$ of two function $f(x)$ and $g(x)$. While the later assigns a number to each discrete value of $g(x)$, the functional $J[y(x)]$ does not depend on a discrate value of $y(x)$, rather it depends on the entire course of $y(x)$. Functionals of several argument functions of one or more independent variables are similary defined.

Example 1

(*a*) $M=C[0, 1]$, $J[y(x)]=\int_0^1 y(x)\, dx$

Thus $J[I]=1$, $J[\sin \pi x]=2/\pi$.

(*b*) $M=C^1[0, 1]$, $J[y(x)]=\int_0^1 \sqrt{1+y'^2(x)}\, dx$

In particular

$$J[1]=1,\quad J[x]=\sqrt{2},\quad J[x^2]=\frac{\sqrt{5}}{2}+\frac{1}{4}\sinh^{-1} 4.$$

(*c*) $M=C[a, b]$, $J[y(x)=J'(x_0)$ where $x_0\in[0, b]$.

(*d*) M is the set of functions defined at $x=0$.

$$J[y(x)=\delta[y(x)]=y(0).$$

This is the famous Dirac's delta functional.

(*e*)$M=C^1[a, b]$, $J[z(x, y)]=\int_a^b\int_a^b \sqrt{1+z_x^2+z_y^2}\,dx\,dy$

(*f*) $M=C[-1, 1]$. If $\phi(x, y)$ is a continuous function on the domain $-1\leqslant x\leqslant 1$, $-\infty<y<\infty$, then $\phi[x, y(x)]\in M$. We can define the functional

$$J[y(x)]=\int_{-1}^{1}\phi(x, y(x))\,dx$$

In particular for $\phi(x, y)=\dfrac{x}{1+y^2}$

$$J[1]=\int_{-1}^{1}\frac{x}{2}dx=0$$

$$J[x]=\int_{-1}^{1}\frac{x}{1+x^2}\,dx=0$$

In general $J[y(x)]=0$ for $y(x)\in M$, since the integrand will be an odd function of x.

(*g*) $M=C[a, b]$, $K(x, t)\in L^2[a, b]$, then

$$K[y(x)]=Ky=\int_a^b K(x, t)\,y(x)\,dx$$

is a linear functional on M with t as a parameter on $a\leqslant t\leqslant b$.

Definition 2. **(Variation of Argument Function).** The *variation* or *increment* δy of the argument $y(x)$ of the functional $J[y(x)]$ is the difference between two functions $y(x)$ and $y_0(x)$ belonging to the chosen class M of functions,

i.e.,
$$\delta y=y(x)-y_0(x).$$

We similarly define the increment

$$\delta y^{(k)}=y^{(k)}(x)-y_0{}^{(k)}(x)$$

for kth order derivatives. It is possible to connect increment in a derivative $y'(x)$ to increment in $y(x)$ by using integration by parts. For example.

$$\int_a^b f(x)\,\delta y'\,dx=\int_a^b f(x)\{y'(x)-y_0'(x)\}\,dx$$

$$=\Big[f(x)\{y(x)-y_0(x)\}\Big]_a^b$$

$$-\int_a^b f'(x)\{y(x)-y_0(x)\}\,dx$$

$$=-\int_a^b f'(x)\,\delta y\,dx$$

if the class M consists of functions which satisfy the boundary conditions $y(a)=y(b)=0$. Generalisting, if the functions of the class M satisfy the boundary conditions

$$y(a)=y'(a)=\ldots\ldots=y^{(k-1)}(a)=0$$
$$y(b)=y'(b)=\ldots\ldots=y^{(k-1)}(b)=0$$

then
$$\int_a^b f(x)\,\delta y^{(k)}\,dx=(-1)^k\int_a^b f^{(k)}(x)\,\delta y\,dx$$

If the range of the scalar variable x is $-\infty$ to ∞, then also it is possible to change the increment in a function y of *delayed* or *retarded* argument to $\delta y(x)$. For example,

$$\int_{-\infty}^{\infty} f(x)\,\delta y(x-1)\,dx=\int_{-\infty}^{\infty} f(x+1)\,\delta y\,dx$$

and
$$\int_{-\infty}^{\infty} f(x)\,\delta y(x+1)\,dx=\int_{-\infty}^{\infty} f(x-1)\,\delta y\,dx$$

This can be easily seen by changing the variable of integration and using appropriate BCs.

The Notion of Continuity of Functionals : Norms

The reader may recall that an ordinary real function $f(x)$ is continuous at x_0 if the numbers $f(x)$ and $f(x_0)$ are close whenever the numbers x and x_0 are close. In geometrical language this means the ordinates $f(x)$ and $f(x_0)$ differ by a small quantity whenever the abscissae x and x_0 differ by a small quantity (see Fig. 13·1).

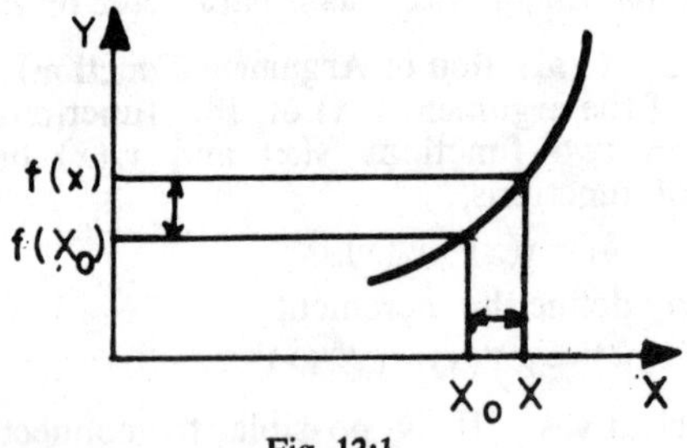

Fig. 13·1.

In analogy we would like a functional $J[y(x)]$ to be continuous if a small change in $y(x)$ results in a small change in $J[y(x)]$. In geometrical language we would like that if the curves $y=y(x)$ and $y=y_0(x)$ are close to each other, then the numbers $J[y(x)]$ and $J(y_0)$ should be close to each other. The difficulty lies in defining the idea of distance between two functions (or curves) from amongst the class of admissible functions (or curves). For example, we may say that the two curves $y=y(x)$ and $y=y_0(x)$ are close to each other if for each given x, the ordinate difference $|y(x)=y_0(x)|$ is small. Alternatively, we may insist that the two curves are close if not only the ordinate difference $|y(x)-y_0(x)|$ be small, the difference of slopes $|y'(x)-y_0'(x)|$ also be small. This suggests closeness to depend on higher order derivatives too. A number of definitions may thus be introduced for the distance between two curves. This is accomplished through *norms* defined on the class M. The reader may recall that a norm $\| \cdot \|$ on M will satisfy the following properties :

(*i*) $\| y \| > 0, y \neq 0, \| 0 \| = 0$

(*ii*) $\| \alpha y \| = | \alpha | . \| y \|$

(*iii*) $\| y_1 + y_2 \| \leqslant \| y_1 \| + \| y_2 \|$

Depending on the differentiability properties of functions of M, the following norms can be defined on M :

$$\| y \|_0 = \max_{a \leqslant x \leqslant b} | y(x) |$$

$$\| y \|_1 = \max_{a \leqslant x \leqslant b} (| y(x) | + | y'(x) |)$$

$$\| y \|_k = \max_{a \leqslant x \leqslant b} (| y(x) | + | y'(x) | + \ldots + | y^{(k)}(x) |)$$

The same norm $\| y \|_k$ is also given by the definition

$$\| y \|_k = \max_{a \leqslant x \leqslant b} (| y(x) | , | y'(x) | , \ldots\ldots, | y^{(k)} x |)$$

$$= \max_{0 \leqslant i \leqslant k} \max_{a \leqslant x \leqslant b} | y^{(i)}(x) |$$

The *k*th order distance between two curves $y=y(x)$ and $y=y_0(x)$ is given by

$$d_k(y, y_0) = \| y - y_0 \|_k$$

Hence if $\| y - y_0 \|_k < \in$, then

$| y(x) - y_0(x) < \in, | y'(x) - y_0'(x) | < \in, \ldots, | y^{(k)}(x) - y_0^{(k)}(x) | < \in$
for all x in $[a, b]$.

Definition 3 (Proximity of Curves). The curves $y=y(x)$ and $y=y_0(x)$ are said to be close in the sense of proximity of *k*th order if $\| y - y_0 \|_k$ is small.

Fig. 13·2 depicts curves close in the sense of zero-order proximity. Here the ordinates are close but the directions are not. Fig. 13·3 depicts curves close in the sense of first order proximity. The curves are close both in ordinates and directions.

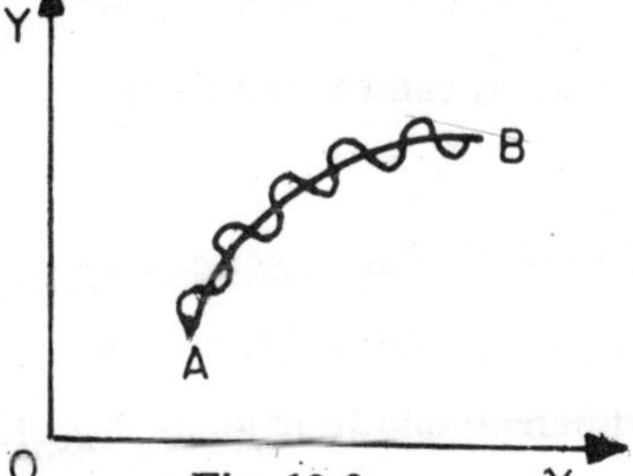

Fig. 13·2.

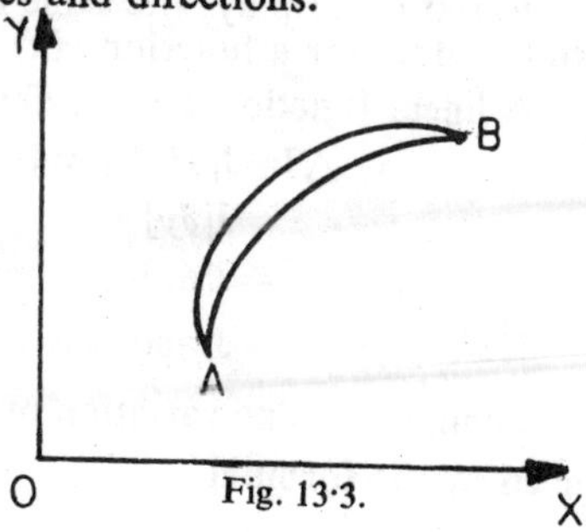

Fig. 13·3.

Definition 4 (Continuity of a Functional). A functional $J[y(x)]$ defined in a class M of functions $y(x)$ is said to be continuous on $y=y_0(x)$ in the sense of kth order proximity if given $\epsilon>0$, there exists a $\delta>0$ such that

$$|J[y(x)]-J[y_0(x)]|<\epsilon \text{ if } \|y-y_0\|_k<\delta$$

Definition 5 (Linear Functional). The functional $J[y(x)]$ defined in the normed linear space M is called *linear* if

(*i*) $J[\alpha y(x)]=\alpha J[y(x)]$, α is a constant

(*ii*) $J[y_1(x)+y_2(x)]=J[y_1(x)]+J[y_2(x)]$

For example, the functional

$$J_1[y(x)=\int_a^b [y'(x)+y(x)]\,dx$$

is linear in $C^1[a, b]$, but the functional

$$J_2[y(x)]=\int_a^b [p(x)\,y^{2\prime}(x)+q(x)\,y^2(x)]\,dx$$

is non-linear.

Problem. Prove that $J[y(x)]$ is linear if it is continuous and satisfies (*i*).

Definition 6 (Increment of a Functional). Given a functional $J[y(x)]$ on M and let $\delta y=y(x)-y_1(x)$ be an increment in the argument function $y(x)$. Then the expression

$$\Delta J=J[y(x)+\delta y]-J[y(x)]$$

is called the *increment* of $J[y(x)]$ corresponding to the increment δy in the argument.

Definition 7 (Variation of the Functional). Let the increment ΔJ in the functional $J[y(x)]$ be represented as

$$\Delta J=L[y(x), \delta y]+\beta[y(x), \delta y]\max|\delta y| \quad \text{...(1)}$$

where L is a linear functional with respect to the argument δy and $\beta(y(x), \delta y)\to 0$ as $|\delta y|\to 0$, then $L[y(x), \delta y]$ is called the *variation of the functional* $J[y(x)]$ and is denoted by δJ. It is also called the *first variation* of J.

In this case we also say that the functional $J[y(x)]$ is *differentiable* at the point $y(x)$.

Variation δJ plays the same role for a functional J as the differential dy does for a function $y(x)$.

A linear functional J is differentiable, as can be seen from

$$\begin{aligned}\Delta J&=J[y(x)+\delta y]-J[y(x)]\\&=J[\delta y]\\&=J[\delta y]+0\,.\max|\delta y|\end{aligned}$$

Here $\quad L=J$ and $\beta(y, \delta y)=0$ in the relation (1).

Lemma 1. The variation of a differentiable functional $J[y(x)]$ is given by the formula

$$\delta J=\left\{\frac{\partial J}{\partial \alpha}[y(x)+\alpha\delta y]\right\}_{\alpha=0} \qquad ...(2)$$

Proof. Since J is differentiable, by formula (1)

$$\triangle J=L[y(x), \delta y]+\beta(y(x), \delta y) \max|\delta y|,$$

where $\beta \to 0$ as $\max|\delta y| \to 0$.

If δy is replaced by $\alpha\delta y$, then

$$J[y+\alpha\delta y]-J[y]=\triangle J=L[y, \alpha\delta y]+\beta(y(x), \alpha\delta y)|\alpha| \max|\delta y|$$

where $\beta \to 0$ as $\alpha \to 0$

and $\qquad L[y(x), \alpha\delta y[=\alpha L[y(x), \delta y]$

Hence

$$\frac{\triangle J}{\alpha}=\frac{J[y+\alpha\delta y]-J[y]}{\alpha}=L[y, \delta y]+\frac{\beta(y, \alpha\delta y)|\alpha| \max|\delta y|}{\alpha}$$

Since $\lim\limits_{\alpha\to 0}\dfrac{\triangle J}{\alpha}=\lim\limits_{\Delta\alpha\to 0}\dfrac{\triangle J}{\Delta\alpha}=\dfrac{\partial J[y+\alpha\delta y]}{\partial\alpha}$

and $\qquad \dfrac{\beta(y, \alpha\delta y)|\alpha| \max|\delta y|}{\alpha} \to 0$ as $\alpha \to 0$

Hence $\qquad L[y, \delta y]=\left[\dfrac{\partial J[y+\alpha\delta y]}{\partial\alpha}\right]_{\alpha=0}$ **Q.E.D.**

In our work, we shall have frequent occasion to use formula (2).

Example 2. Investigate the closeness of the following curves :

(*a*) $\qquad y(x)=\dfrac{\sin n^2x}{n}$, $y_1(x)\equiv 0$ on $[0, \pi]$

(*b*) $\quad y(x)=\dfrac{\sin nx}{n^2}$, $y_1(x)\equiv 0$ on $[0, \pi]$

Solution. (*a*) $\left|y(x)-y_1(n)\right|=\left|\dfrac{\sin n^2x}{n}\right|\leqslant\dfrac{1}{n}\to 0$

as $n\to\infty$

But $\qquad |y'(x)-y_1'(x)=n|\cos n^2x|$

$$=n \text{ at } x=\frac{2\pi}{n^2}$$

$$\to\infty \text{ as } n\to\infty$$

$$\text{at } x=\frac{2\pi}{n^2}$$

Hence the curves are close in the 0th order proximity but not in the first order proximity for large n.

(*b*) $\quad \left|y(x)-y_1(x)\right|=\left|\dfrac{\sin nx}{n^2}\right|\leqslant\dfrac{1}{n^2}\to 0$

$$\left| y'(x) - y_1'(x) \right| = \left| \frac{\cos nx}{n} \right| < \frac{1}{n} \to 0$$

Hence the curves are close in the first order proximity. It can be seen that they are not close in a higher order proximity for large n.

Example 3. Find the distance between the curves $y=x$ and $y=x^2$ on the interval [0, 1].

Solution. The curves intersect at (0, 0) and (1, 1) (see Fig. 13·4).

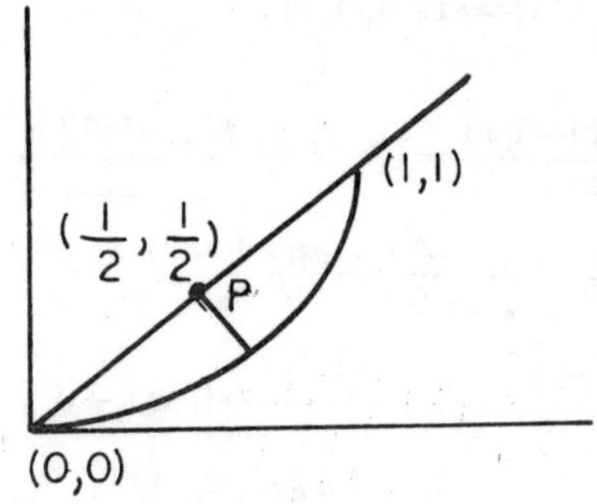

Fig. 13·4.

The required distance ρ

$$= \max_{0 \leqslant x \leqslant 1} (x - x^2).$$

If $Y(x)=x-x^2$, then $Y'(x)=0$ at $x=\frac{1}{2}$ and $Y''(\frac{1}{2})=-2<3$.

Hence $Y(x)$ is maximum at $x=\frac{1}{2}$.

Hence $\rho=\frac{1}{2}-\frac{1}{4}=\frac{1}{4}$.

Example 4. Find the first order distance between the curves $y=\log x$ and $y=x$ on the interval $[e^{-1}, e]$.

Solution. Put $Y(x)=x-\log x$.

Then the first order distance required

$$= \max_{\frac{1}{e} \leqslant x \leqslant e} \left\{ | Y(x) |, \ | Y'(x) | \right\}$$

$$= \max_{\frac{1}{e} \leqslant x \leqslant e} \left\{ | x - \log x |, \left| 1 - \frac{1}{x} \right| \right\}$$

$$= \max \{1, e-1\}$$

$$= e-1.$$

Example 5. Show that the functional

$$J[y(x)] = \int_0^1 [y(x) + 2y'(x)] \, dx, \ y(x) \in C^1[0, 1]$$

is continuous on the curve $y_0(x)=x$ in the sense of first order proximity.

Solution. $|J[y(x)]-J[y_0(x)]|$

$$=\left|\int_0^1 [y(x)+2y'(x)]\,dx-\int_0^1 (x+2)\,dx\right|$$

$$=\left|\int_0^1 [y(x)-x]+2(y'(x)-1)]\,dx\right|$$

$$\leqslant\int_0^1 [\,|y(x)-x|+2\,|y'(x)-1|\,]dx$$

Since $y(x)$ and $y'(x)$ are continuous on $[0, 1]$, given $\in>0$ it is possible to choose $y(x)'$ such that

$$|y(x)-x|<\delta=\frac{\in}{3} \text{ and } |y'(x)-1|<\delta=\frac{\in}{3}$$
$$x\in[0, 1]$$

Then $\quad |J[y(x)]-J[y_0(x)]<\in.$

Example 6. Find the variation of the functional

$$J[y(x)]=\int_a^b\int_a^b K(x, t)\, y(x)\, y(t)\, dx\, dt$$

where $K(x, t)\in L^2$ on $a\leqslant x, t\leqslant b$, $K(x, t)=K(t, x)$ and $y\in C[a, b]$.

Solution. Let

$$\phi(\alpha)=J[y(x)+\alpha\delta y]$$

$$=\int_a^b\int_a^b K(x, t)\{y(x)+\alpha\delta y(x)\}\{y(t)+\alpha\delta y(t))\}\, dx\, dt$$

$$\frac{\partial\varphi}{\partial\alpha}=\int_a^b\int_a^b K(x, t)\{y(x)+\alpha\delta y(x)\}\,\delta y(t)\, dx\, dt$$

$$+\int_a^b\int_a^b K(x, t)\,\{y(t)+\alpha\delta y(t)\}\,\delta y(x)\, dx\, dt$$

Hence $\quad \delta J=\left[\dfrac{\partial\varphi}{\partial\alpha}\right]_{\alpha=0}=\displaystyle\int_a^b\int_a^b K(x, t)\, y(x)\,\delta y(t)\, dx\, dt$

$$+\int_a^b\int_a^b K(x, t)\, y(t)\,\delta y(x)\, dx\, dt$$

$$=2\int_a^b\int_a^b K(x, t)\, y(t)\,\delta y(x)\, dx\, dt$$

since owing to the symmetry of $K(x, t)$, the first integral is equal to the second integral.

Problem. If f is differentiable in its arguments, then prove that the variation of the functional

$$J[y(x)]=\int_a^b f(x, y, y')\, dx$$

is
$$\delta J=\int_a^b \left(\frac{\partial f}{\partial y}\,\delta y+\frac{\partial f}{\partial y'}\,\delta y'\right)dx$$

Use integration by parts to reduce the differential $\delta y'$ to δy.

Hence find the variations of the functionals

$$J_1[y(x)]=\int_a^b y^2(x)\,dx$$

$$J_2[y(x)]=\int_{-1}^{1}(y'\,e^y+xy^2)\,dx$$

§13·3. Extrema of Functionals

Recall that an ordinary function $f(x)$ has a *relative minimum* (*relative maximum*) at a point x_0 if $f(x)\geqslant f(x_0)$ ($f(x)\leqslant f(x_0)$) for all x in a neighbourhood of x_0. A neighbourhood of x_0 means the set of all points x such that $|x-x_0|<\delta$ for some $\delta>0$. If the inequalities $f(x)\geqslant f(x_0)$ ($f(x)\leqslant f(x_0)$) hold for all x in a given domain of $f(x)$ then we say that $f(x)$ has an *absolute minimum* (absolute maximum) at $x=x_0$. Points at which an extremum (*i.e.*, minimum or maximum) occurs are called *stationary points* of the function.

Similar terminology is used for functionals. The functional $J[y(x)]$ has a *relative minimum* (*relative maximum*) at the argument function $y=y_0(x)$ if there exists a $\delta>0$ such that $J[y(x)]\geqslant J[y_0(x)]$ ($J[y(x)]\leqslant J[y_0(x)]$) for all $y(x)$ satisfying $\| y(x)-y_0(x)\|_k<\delta$ in the kth order of proximity. We shall consider only the cases $k=0$, 1. There will be more functions $y(x)$ in a neighbourhood of $y_0(x)$ in the norm $\| y\|_0$ than those in the norm $\| y\|_1$. Hence with the norm $\| y\|_0$ we shall have a *strong relative minimum* (relative maximnm) and with the norm $\| y\|_1$ we shall have a *weak relative minimum* (relative maximum) of the functional $J[y(x)]$. Relative minima or relative maxima (weak or strong) of a functional are called *relative extrema* of the functional. A strong extremum is also a weak extremum, but the converse is not true. An extremum of $J[y(x)]$ on the whole class M of admissible functions is called an *absolute extremum*.

The following theorem stipulates a necessary condition for the existence of an extremum of a functional. The condition does not differentiate between a strong and a weak extremum. But the difference becomes important when sufficient conditions are studied. In our work we have not studied such sufficient conditions.

Theorem 1. If a functional $J[y(x)]$ having a variation attains a maximum or a minimum at $y=y_0(x)$ where $y(x)$ is an interior point of the domain of definition of the functional' then at $y=y_0(x)$,

$$\delta J=0$$

Proof. For the fixed curve $y_0(x)$ and fixed increment δy,

$$\phi(\alpha)=J[y_0(x)+\alpha\delta y]$$

is a function of α. By the hypothesis $\phi(\alpha)$ reaches a maximum or a minimum when $\phi'(0)=0$. Thus for an extremum of $J[y(x)]$ at $y_0(x)$, we have

$$\frac{\partial}{\partial\alpha} J[y_0(x)+\alpha\delta y]=0$$

i.e.,

$$\delta J[y_0(x)]=0$$

Functions or curves for which $\delta J=0$ are called *stationary functions or curves.* The corresponding value of the functional $J[y(x)]$ is called a *stationary value.*

Now at an extremum, $\delta J=0$. The converse is not true. Thus δJ may vanish at a curve $y=y(x)$ at which it is neither maximum nor minimum. Hence a stationary function may not be an extremum point and a stationary value may not be an extremum value of the functional. $\delta J=0$ is only a necessary condition for the existence of extrema but not sufficient.

§13·4. Euler's Equation : Extremals With Fixed Boundaries

We first consider the case of one dependent and one independent variable *i.e.,* functionals on plane curves. The following is the *simplest problem of the calculus of variations* :

Let the function $F(x, y, y')$ have continuous partial derivatives of order two with respect to all arguments. From among all functions (or curves) $y=y(x)$ having a continuous derivative and satisfying the fixed boundary conditions $y(a)=A$, $y(b)=B$, find the function which gives the functional

$$J[y(x)]=\int_a^b F(x, y, y')\, dx \qquad \ldots(1)$$

a weak extremum.

In other words, we have to find the weak extremum of J on the set of all smooth curves joining the fixed points $P(a, A)$ and $Q(b, B)$. These curves form the class M of *admissible curves*. The following lemma will be required in solving this problem.

Lemma 2 (Fundamental Lemma of the Calculus of Variations). If for a continuous function $f(t)$ on $[a, b]$, the integral

$$I=\int_a^b f(t)\,\eta(t)\,dt=0$$

for all continuous function $\eta(t)$, then

$$f(t)\equiv 0 \text{ on } [a, b].$$

Proof. Let $f(\xi)\neq 0$ for some ξ on $[a, b]$. In particular let $f(\xi)>0$. Since $f(t)$ is continuous on ξ, there is a neighbourhood N : $t_0\leqslant t\leqslant t_1$ of ξ on which $f(t)>0$. Choose $\eta(t)$ such that $\eta(t)$ is continuous and positive in N and vanishes outside. For example, let

$$\eta(t)=\begin{cases}(t-t_0)^2\,(t-t_1)^2 & t_0\leqslant t\leqslant t_1\\ 0 & t<t_0,\ t>t_1\end{cases}$$

Then $I>0$ on $[a, b]$ which is a contradiction. **Q.E.D.**

Note. The proof remains the same if $\eta^{(k)}$ is continuous. In that case, we can take

$$\eta(t)=(t-t_0)^{2l}(t-t_1)^{2l} \text{ with } 2l>k$$

in the interval $t_0 \leqslant t \leqslant t_1$ and zero outside.

Now for an extremum of $J[y(x)]$ $\delta J=0$ (lemma 1).

Put $\quad y(x, \alpha)=y(x)+\alpha\delta y(x)$

or $\quad y'(x, \alpha)=y'(x)+\alpha\delta y'$

and $\quad y(x, 0)=y(x)$

Also let

$$\phi(\alpha)=J[y(x)+\alpha\delta y]$$

$$=\int_a^b F(x, y(x, \alpha), y'(x, \alpha))\, dx$$

Then $\quad \phi'(\alpha)=\int_a^a \frac{\partial F}{\partial \alpha}\, dx$

$$=\int_a^b \left[\frac{\partial F}{\partial y\,(x, \alpha)}\, \frac{\partial y(x, \alpha)}{\partial \alpha}+\frac{\partial F}{\partial y'(x, \alpha)}\, \frac{\partial y'(x, \alpha)}{\partial x}\right] dx$$

Hence

$$\delta J=\phi'(0)=\int_a^b (F_y\delta_y+F_{y}'\delta_y')\, dx$$

Now integrating by parts

$$\int_a^b F_y'\,\delta_y'\, dx=\Big[F_y'\,\delta_y\Big]_a^b-\int_a^b \frac{dF_y'}{dx}\,\delta y\, dx$$

But $\Big[\delta y\Big]_a^b=\Big[y(x)-y_0(x)\Big]_a^b=0,$

as $y(x)$ and $y_0(x)$ both being admissible curves satisfy the given boundary conditions $y(a)=y_0(a)=A$ and $y(b)=y_0(b)=B$.

Hence $\quad \displaystyle\int_a^b F_y'\,\delta y'\, dx=-\int_a^b \frac{dF_y'}{dx}\,\delta y\, dx$

and $\quad \displaystyle\delta J=\int_a^b \left(F_y-\frac{dF_y'}{dx}\right)\delta y\, dx.$

For an extremum $\delta J=0$, hence

$$\int_a^b \left(F_y-\frac{dF_y'}{dx}\right)\delta y\, dx=0.$$

Now $F_y-\dfrac{dF_y'}{dx}$ is continuous and δy is an arbitrary continuous function $y(x)-y_0(x)$. Hence by the fundamental lemma (2),

$$[F]_y=Fy-\frac{dF_y'}{dx}=0 \qquad \text{...(2)}$$

The differential expression $[F]_y$ in (2) is called the *variational derivative* of $F(x, y, y')$ with respect to y. It role is anclogous to that of the ordinary derivative or the gradient in the minimum-maximum problems of ordinary functions.

In the fully expanded form the relation (2) con be written as

$$Fy-F_{xy}'-F_{yy}'\,.\,y'-Fy'\,.\,y''=0 \qquad \text{...(3)}$$

Equation (2) or (3) is called the *Eulers' differential equation* (it was first published by Euler in 1744). Its validity is a necessary condition for the existence of an extremum. If $F_{y'y'} \neq 0$, called the *Legendre condition*, then the Euler equation is a second order ODE. Its general solution involves two arbitrary constants to determine which we require two independent boundary conditions. Every solution of the Euler equation is called an *extremal* of the variational problem. Extremals are *stationary functions or curves.* Converse is not true.

Generalisations. There are three ways in which the simplest problem can be extended :

(A) Functionals with higher order derivatives in one independent and one dependent variable.

(B) Functionals with one independent variable and more than one dependent variables *e.g.*, parametric equations.

(C) Functionals with more than one independent variables *i.e.*, involving partial derivatives.

We consider these cases briefly.

A. Functionals With Higher Order Derivatives

To determine the extremals of the functional

$$J[y(x)]=\int_a^b F(x, y(x), y'(x), y''(x))\, dx \qquad \ldots(4)$$

which satisfy

$$y(a)=y'(a)=0 \qquad \ldots(5)$$
$$y(b)=y'(b)=0$$

and F has second order continuous derivatives of its arguments.

Let $\qquad y(x, \alpha)=y(x)+\alpha\delta y(x)$

and $\qquad \phi(\alpha)=J[y(x, \alpha)]$

so that $\qquad \phi(0)=J[y(x)]$

$$\delta J=\left[\frac{\partial \varphi(\alpha)}{\partial \alpha}\right]_{\alpha=0}=\int_a^b (F_y\, \delta y+F_{y'}\, \delta y'+F_{y''}\, \delta y'')\, dx$$

Integrating by parts and using the boundary conditions (5), we shall have

$$\int_a^b F_{y'}\, \delta y'\, dx=-\int_a^b \frac{dF_{y'}}{dx}\, \delta y\, dx$$

$$\int_a^b F_{y''}\, \delta y''\, dx=\int_a^b \frac{d^2F_{y''}}{dx^2}\, \delta y\, dx$$

Hence $\qquad \delta J=\int_a^b \left(F_y-\frac{dF_{y'}}{dx}+\frac{d^2F_{y''}}{dx^2}\right) \delta y\, dx$

For an extremum $\delta J=0$. Using lemma 1, we shall have

$$F_y-\frac{dF_{y'}}{dx}+\frac{d^2F_{y''}}{dx^2}=0 \qquad \ldots(6)$$

The ODE (6) is called the *Euler-Poisson equation* and its solutions are called *extremals*. In the fully expanded form (6) will be a fourth order ODE.

Generalising the Euler-Poisson equation for the functional

$$J[y(x)]=\int_a^b F(x, y, y', \ldots\ldots, y^{(n)})\, dx$$

is
$$F_y-\frac{dF_{y'}}{dx}+\frac{d^2F_{y''}}{dx^2}-\ldots+(-1)^n\frac{d^nF_{y^{(n)}}}{dx^n}=0,$$

which is of order $2n$ and in general requires $2n$ independent boundary conditions for a unique solution.

B. Functions With More Than One Dependent Variables

Given
$$J[x(t), y(t)]=\int_a^b F(t, x(t), y(t), \dot{x}, \dot{y}(t))\, dt \qquad \left(\dot{x}\equiv\frac{dx}{dt}\right)$$

where $x(t), y(t)\in C^1[a, b]$ and F is a continuously differentiable function of its arguments.

If $\phi(\alpha, \beta)=J[x(t)+\alpha\delta x, y(t)+\beta\delta y]$ then for an extremum, we have

$$\left[\frac{\partial\varphi}{\partial\alpha}\right]_{\alpha=\beta=0}=0 \text{ and } \left[\frac{\partial\varphi}{\partial\beta}\right]_{\alpha=\beta=0}=0$$

These equations lead to the *Euler-Lagrange equations*

$$\frac{\partial F}{\partial x}-\frac{d}{dt}\left(\frac{\partial F}{\partial \dot{x}}\right)=0,\quad \frac{\partial F}{\partial y}-\frac{d}{dt}\left(\frac{\partial F}{\partial \dot{y}}\right)=0 \qquad \ldots(7)$$

The problem is easily generalised to functionals of several functions of one independent variable. The results can be expressed in a succinct form by using the vectorial notation.

Let $\bar{x}(t)=(x_1(t), x_2(t), \ldots\ldots, x_n(t))$

Then $\dot{x}(t)=(\dot{x}_1(t), \dot{x}_2(t), \ldots\ldots, \dot{x}_n(t))$

where $\dot{x}_i(t)\equiv\dfrac{dx_i(t)}{dt}$

Let $L(t, \bar{x}, \dot{\bar{x}})$ be a function of $2n+1$ variables $t, \bar{x}, \dot{\bar{x}}$ and form the functional

$$\mathscr{L}[\bar{x}]=\mathscr{L}[x_1(t), x_2(t), \ldots\ldots, x_n(t)]=\int_a^b L(t, \bar{x}(t), \dot{\bar{x}}(t))\, dt \qquad \ldots(8)$$

Functionals of the type (8) arise in analytical mechanics with L as the Lagrangian.

The problem is to minimise the functional $\mathscr{L}$ by extremals which satisfy the $2n$ boundary conditions

$$\bar{x}(a)=\bar{\alpha},\quad \bar{x}(b)=\bar{\beta} \qquad \ldots(9)$$

where $\bar{a}=(a_1, a_2, \ldots\ldots, a_n)$, $\bar{\alpha}=(\alpha_1, \alpha_2, \ldots\ldots, \alpha_n)$ etc., L must satisfy the n second order Euler equations

$$L_{x_i}-\frac{dL_{\dot{x}_i}}{dt}=0, \qquad i=1, 2, \ldots\ldots, n \qquad \ldots(10)$$

In mechanics where L is the Lagrangian, equations (10) are known as Lagrange's equations. Hence here we call eqaations (10) as *Euler-Lagrange equations.* Their solutions are called *extremals* and form a $2n$-parameter family of curves $\bar{x}\,(t, \bar{\alpha}, \bar{\beta})$ in the n-dimensional $\bar{x}$-space. t is the curve parameter and $\bar{\alpha}$, $\bar{\beta}$ are parameter vectors of the families. When boundary conditions are given these family parameters $\bar{\alpha}$, $\bar{\beta}$ will be given. Otherwise in general they are any $2n$ independent parameters.

C. Functionals Involving Partial Derivatives : Multiple Integrals

The case when the functional is a double integral with two independent variables and one dependent variable is particularly useful in geometry.

Consider the functional

$$J[z(x, y)]=\iint_D F(x, y, z, p, q)\, dx\, dy \qquad \ldots(11)$$

where $\quad p=\frac{\partial z}{\partial x}, \; q=\frac{\partial z}{\partial y}$

Also $r=\frac{\partial^2 z}{\partial x^2}, \quad s=\frac{\partial^2 z}{\partial x \partial y}, \quad t=\frac{\partial^2 z}{\partial y^2}$

We assume $z=z_0\,(x, y)$ minimises (11) in an appropriate class of (admissible) functions satisfying the BC $z(x, y)=g(x, y)$ on ∂D. As in the simplest case put $z(x, y, \alpha)=z(x, y)+\alpha\delta z$,

$$p(x, y, \alpha)=p(x, y)+\alpha\delta p, \; q(x, y, \alpha)=q(x, y)+\alpha\delta q$$

and $\quad \phi(\alpha)=\int_D \int F\{x, y, z(x, y, \alpha), p(x, y, \alpha), q(x, y, \alpha)\}\, dx\, dy$

Then $\delta J=\left[\frac{\partial \varphi}{\partial \alpha}\right]_{\alpha=0}=\int_D \int (F_z\delta_z+F_p\delta_p+F_q\delta_q)\, dxdy$

$$\ldots(12)$$

Since $\quad \frac{\partial(F_p\delta z)}{\partial x}=\frac{\partial F_p}{\partial x}\delta z+F_p\delta_p \qquad \frac{\partial(F_q\delta z)}{\partial y}=\frac{\partial F_q}{\partial y}\delta_z+F_q\delta_q$

$$\int_D \int F_p\delta_p+F_q\delta_q)dxdy=\int_D \int \left[\frac{\partial F_p\delta z}{\partial x}+\frac{\partial F_q\delta z}{\partial y}\right] dx\, dy$$
$$-\int_D \int \left(\frac{\partial F_p}{\partial x}+\frac{\partial F_q}{\partial y}\right) \delta z\, dx\, dy$$

Also by Green's theorem

$$\int_D \int \left[\frac{\partial(F_p \delta z)}{\partial x} + \frac{\partial(F_q \delta z)}{\partial y} \right] dx\, dy$$

$$= \int_{\partial D} (F_p dy - F_q dx)\, \delta z$$

$$= 0$$

since on ∂D $z = g(x, y)$ which is fixed and hence $\delta z = 0$.

Hence

$$\int_D \int (F_p \delta_p + F_q \delta q)\, dx\, dy = -\int_D \int \left(\frac{\partial F_p}{\partial x} + \frac{\partial F_p}{\partial y} \right) \delta z\, dx\, dy$$

Making this substitution in (12), we get

$$\delta J = \int_D \int \left(F_z - \frac{\partial F_p}{\partial x} - \frac{\partial F_q}{\partial y} \right) \delta z\, dx\, dy$$

For an extremum $\delta J = 0$ which together with the fundamental lemma gives

$$F_z - \frac{\partial F_p}{\partial x} - \frac{\partial F_q}{\partial y} = 0 \qquad \text{...(13)}$$

(13) is the required Euler equation for an extremum. In (13) the derivatives $\frac{\partial F_p}{\partial x}$, $\frac{\partial F_q}{\partial y}$ are

$$\frac{\partial F_p}{\partial x} = F_{px} + F_{pz} p + F_{pp} r + F_{pq} s\,;$$

$$\frac{\partial F_q}{\partial y} = F_{qy} + F_{qz} q + F_{qp} s + F_{qq} t.$$

Problem. Extend the results to the functional

$$J[u(x, y, z)] = \int_D \int \int F\left(x, y, z, u, \frac{\partial u}{\partial x}, \frac{\partial u}{\partial y}, \frac{\partial u}{\partial z} \right) dx\, dy\, dz$$

Dirichlet's Principle. The function that minimises the Dirichlet integral in a region D is harmonic in D.

The Dirichlet integral in a two-dimensional region D is

$$J[z(x, y)] = \int_D \int (p^2 + q^2)\, dx\, dy$$

Here $F(x, y, z, p, q) = p^2 + q^2$

Hence $F_z = 0$, $\frac{\partial F_p}{\partial x} = F_{pp} \frac{\partial p}{\partial x} = 2 \frac{\partial^2 z}{\partial x^2}$

$$\frac{\partial F_q}{\partial y} = F_{qq} \frac{\partial q}{\partial y} = 2 \frac{\partial^2 z}{\partial y^2}$$

Subitituting in (13), we get the desired Euler equation

$$\nabla^2 z = \frac{\partial^2 z}{\partial x^2} + \frac{\partial^2 z}{\partial y^2} = 0 \text{ in D}$$

In the same way the Euler equation of the functional

$$J[u(x, y, z)] = \int_D \iint (u_x^2 + u_y^2 + u_z^2)\, dx\, dy\, dz$$

is the harmonic equation

$$\nabla^2 u = \frac{\partial^2 u}{\partial x^2} + \frac{\partial^2 u}{\partial y^2} + \frac{\partial^2 u}{\partial z^2} = 0 \text{ in D.}$$

Problem. Prove that the Euler equation of the functional

$$J[z(x, y)] = \int_D \int \{p^2 + q^2 + 2zf(x, y)\}\, dx\, dy$$

is the Poisson equation $\nabla^2 z = f(x, y)$. Extend the result to three dimensions.

§13·5. Illustrative Examples

Example 7. Determine the stationary functions for the following functionals :

(a) $$J[y(x)] = \int_a^b \int_a^b K(x, t)\, y(x)\, y(t)\, dx\, dt + \int_a^b y^2(x)\, dx - 2\int_a^b y(x)\, f(x)\, dx,$$

subject to the boundary conditions $y(a) = A$, $y(b) = B$, where $K(x, t)$ is a continuous symmetric Kernel in $a \leqslant x,\ t \leqslant b$, $f(x)$ is a given continuous function on $[a, b]$ and $y(x)$ is the desired argument function.

(b) $$J[y(x)] = \int_a^b [p(x)\, y'^2(x) + (q(x) + \lambda r(x))\, y^2(x) - 2y(x)\, f(x)]\, dx$$

subject to the boundary conditions $y(a) = A$, $y(b) = B$, where $p(x)$ has a continuous derivative, $q(x)$, $r(x)$, $f(x)$ are given continuous functions and $y(x)$ is the desired twice continuously differentiable argument function.

(c) $$J[y(x)] = \int_{-\infty}^{\infty} [p(x)\, y'^2 + 2y(x+1)\, y(x-1) - y^2(x) - 2y(x)\, f(x)]\, dx$$

where $y(x)$ is the continuous argument function, has piecewise continuous derivatives on $(-\infty, \infty)$, $p(x)$ has a continuous derivative and $f(x)$ is continuous and $y(\pm\infty) \to 0$.

Solution. (a) (This is an extension of example 6). We can write the functional as

$$J[y(x)]=\int_a^b \left[\left\{\int_a^b K(x, t)\, y(t)\, dt\right\} y(x)+y^2(x)\right.$$

$$\left. -2y(x)\, f(x)\right]dx$$

Hence $F(x, y, y')=\int_a^b K(x, t)\, y(t)\, y(x)\, dt+y^2(x)-2y(x)\, f(x)$

We have $F_y=2\int_a^b K(x, t)\, y(t)\, dt+2y(x)-2f(x)$

(See example 6 for the first part)

Also $F_{xy'}=F_{yy'}=F_{y'y'}=0.$

For extremum $F_y-F_{xy'}-F_{yy'}\, y'-F_{y'y'}\, y''=F_y=0$, *i.e.*, the extrema are determined by the Euler equation

$$y(x)=f(x)-\int_a^b K(x, t)\, y(t)\, dt$$

which is a Fredholm integral equation.

(*b*) $F(x, y, y')=p(x)\, y'^2+(q+\lambda r)\, y^2(x)-2y(x)\, f(x)$

$$F_y=2(q+\lambda r)\, y(x)-2f(x)$$

$$F_{xy'}=2p'y'+2p(x)y''$$

$$F_{yy'}=0$$

$$F_{y'y'}=2p(x)$$

Hence the extremals will be determined by the Euler equation

$$(q+\lambda r)y-p'y'-py''-f(x)=0$$

or

$$-(py'(x))'+(q+\lambda r)\, y(x)=f(x)$$

with $y(a)=A$, $y(b)=B$

which is the well-known Sturm-Liouville eigenvalue problem.

The variational problem of extremals is thus transformed into an eigenvalue problem of a self-adjoint ODE. We shall study this aspect in detail in section 13·8.

(*c*) $F(x, y, y')=p(x)\, y^{2\prime}(x)+2y(x+1)\, y(x-1)-y^2(x)-2y(x)\, f(x)$

Owing to the presence of functions of delayed arguments $y(x+1)$ and $y(x-1)$, it is not possible to apply the Euler's equation

$$F_y-\frac{d\Gamma_{y'}}{dx}=0$$

directly to this problem.

Hence we derive δJ as

$$\frac{\partial}{\partial \alpha} J\left[y+\alpha\delta y\right]_{\alpha=0}$$

Put $\phi(\alpha)=\int_{-\infty}^{\infty} F(x, y+\alpha\delta, y'+\alpha y')\, dx$

$$=\int_{-\infty}^{\infty} [p(x)\,(y'(x)+\alpha\delta y'(x))^2+2\{y(x+1)+\alpha\delta y(x+1)\}$$

$$\{y(x-1)+\alpha\delta y(x-1)\}-\{y(x)+\alpha\delta y(x)\}^2$$
$$-2\{y(x)+\alpha\delta y(x)\}\, f(x)]\, dx$$

Hence $\delta J=\left[\frac{\partial\varphi}{\delta\alpha}\right]_{\alpha=0}=\int_{-\infty}^{\infty}[(py')\,\delta y'+y(x+1)\,\delta y(x-1)$

$$+y(x-1)\delta y(x+1)-y(x)\delta y(x)$$
$$[dx-f(x)\,\delta y(x)$$

Now by integrating by parts

$$\int_{-\infty}^{\infty}(py')\,\delta y'\,dx=\Big[p(x)\,y'(x)\,.\,\delta y(x)\Big]_{-\infty}^{\infty}-\int_{-\infty}^{\infty}(py')'\,\delta y\,dx$$
$$=-\int_{-\infty}^{\infty}(py')'\,dy\,\delta x$$

By changing variable of integration

$$\int_{-\infty}^{\infty} y(x+1)\,\delta y(x-1)\,dx=\int_{-\infty}^{\infty} y(x+2)\,\delta y(x)\,dx$$

$$\int_{-\infty}^{\infty} y(x-1)\,\delta y(x+1)\,dx=\int_{-\infty}^{\infty} y(x-2)\,\delta y(x)\,dx$$

Hence we can write

$$\delta J=\left[\frac{\delta\varphi}{\delta\alpha}\right]_{\alpha=0}=\int_{-\infty}^{\infty}[-(py')'+y(x+2)+y(x-2)-y(x)$$
$$-f(x)]\,\delta y(x)\,.\,dx$$

For $\delta J=0$, using the fundamental lemma, we shall ultimately have for extremals the Euler equation

$$-(py')'+y(x+2)+y(x-2)-y(x)-f(x)=0$$

This is a difference-differential equation which can be solved by the use of Laplace transform.

Problem. Given the functional

$$J[y(x)]=\int_0^1\int_0^1 K(x, t)\, y(x)\, y(t)\, dx\, dt+\int_0^1 y'^2(x)\, dx$$
$$-2\int_0^1 y(x) f(x)\, dx$$

with $y(0)=y'(0)=0$ and where $K(x, t)$ is a symmetric continuous kernel in $0\leqslant x,\ t\leqslant 1$. Prove that the extremals of the functional are given by the integro-differential equation

$$y''(x)+\int_0^1 K(x, t)\, y(t)\, dt=f(x).$$

Example 8. Find the integral curves of the Euler equation of

$$\int x(dx^2+dy^2)^{1/2}=\text{extremum}$$

Solution. We have

$$J[y(x)]=\int x\sqrt{1+y'^2}\,dx$$

$$F(x, y, y')=x\sqrt{1+y'^2}$$

$$F_y=F_{yy'}=0$$

$$F_{xy'}=\frac{y'}{\sqrt{1+y'^2}},\ F_{y'y'}=\frac{x}{(1+y'^2)^{3/2}}$$

Hence the Euler equation is

$$y''\ \frac{x}{(1+y'^2)^{3/2}}+\frac{y'}{\sqrt{1+y'^2}}=0$$

i.e.,
$$xy''+y'(1+y'^2)=0$$

Putting $y'=Y$, we have

$$xY'+Y(1+Y^2)=0$$

The variables are separate and integrating

$$Y=y'=\frac{a}{\sqrt{x^2-a^2}}$$

Again integrating $y=a\cosh^{-1}\dfrac{x}{a}+b$

or $\quad x=a\cosh\dfrac{y-b}{a}$, which is a catenary.

Problems (*a*). Prove that the extremals of the functional

$$J[y(x)=\int\sqrt{x(1+y'^2)}\,dx$$

are parabolas.

(*b*) Prove that the extremals of the functional

$$J[y(x)[=\int_a^b\sqrt{1+y'^2}$$

are straight lines.

(*c*) Prove that the extremals of the functional

$$J[y(x)]=\int_a^b y\sqrt{1+y'^2}\,dx$$

are catenaries.

(*d*) Prove that the stationary function of

$$\int_0^4 (xy'-y'^2)\,dx$$

$$y(0)=0, \qquad y(4)=3$$

is the parabola $y=\frac{1}{4}(x^2-x)$.

(*e*) On what curve can the functional

$$J[y(x)]=\int_1^2 (y'^2-2xy)\,dx$$

$$y(1)=0, y(2)=-1$$

attain an extremum ?

[Hint : Euler equation is $y''+x=0$].

(*f*) Prove that the cubic spline solves the minimising problem

$$\int_I s''^2(x)\,dx=\text{minimum}\ ,\ s(x_i)=f(x_i),\ (i=0, 1, \ldots\ldots, n)$$

where

$$I=\sum_{i=1}^{n} [x_{i-1}, x_i]$$

§13·6. Moving Boundaries : Transversality Conditions

In the investigation of the functional

$$J[y(x)]=\int_{x_0}^{x_1} F(x, y, y')dx$$

the Euler equation of extremals

$$F_y-\frac{dF_{y'}}{dx}=0$$

is in general of second order. Its general solution $f(x, y, c_1, c_2)=0$ is in general a two-parameter family of curves. If the curves pass through a fixed point (x_0, y_0) one of the two arbitrary constants is fixed and we have a one parameter family of extremals. If another boundary point (x_1, y_1) is fixed, we have a unique extremal. Thus extremals with fixed boundary points form a subset of the set of extremals with *free boundaries*.

In the moving-boundary problem an extremum is attained only on the family of curves $f(x, y, c_1, c_2)=0$ and conditions to determine c_1, c_2 have to be obtained from extremum condition $\delta J=0$.

Let the end point (x_0, y_0) move on the curve $Y_1 : y=\phi(x)$ and the endpoint (x_1, y_1) lie on the curve $Y_2 : y=\psi(x)$. Then the problem with moving boundaries is to find extremum of the functional $J[y(x)]$ on curves Y joining Y_1 and Y_2. The extremal curves Y are said to be *transverse* to the boundary curves Y_1 and Y_2. The extremal curves instead of having fixed boundary points have endpoints moving on fixed curves. Such a family of extremals is called a *field of extremals.* We should understand the moving boundaries to be *free* only to the extent that one or both of the boundary points of an extremal curve are free to move on given curves (or surfaces).

For every given boundary curve it is possible to construct a one-parameter family of transverse extremals by drawing through each point of the boundary curve a curve which is a solution of the Euler equation and starts in the transverse direction. The relation between the direction of an extremal curve and the boundary curve curve is called a *transversality condition*. The relation is such that if the direction of the extremal is given that of the boundary curve is determined uniquely. Let us investigate this phenomenon in more detail.

An elementary problem with moving boundaries can be formulated as follows : Let $F=F(x, y, y')$ be a differentiable function of its arguments and in the xy-plane let there be two curves

$$Y_1 : y=\phi(x) \quad \text{and} \quad Y_2 : y=\psi(x) \qquad ...(1)$$

where $\phi, \psi \in C^1[a, b]$.

We consider the functional

$$J[y]=\int_Y F(x, y, y')dx \qquad ...(2)$$

defined on the smooth curves $Y : y=y(x)$ the endpoints $A(x_0, y_0)$ and $B(x_1, y_1)$ of which lie on the given curves Y_1 and Y_2 respectively so that $y_0=\phi(x_0)$, $y_1=\psi(x_1)$. It is required to find the extremum of the functional (2).

Theorem 1. Let the curve $Y : y=y(x)$ extremise the functional

$$J[y]=\int_Y F(x, y, y')dx \qquad ...(3)$$

from among all smooth curves joining two arbitrary points of two given curves $Y_1 : y=\varphi(x)$ and $Y_2 : y=\psi(x)$. Then the curve Y is an extremal and the transversality conditions

$$\left[F+(\phi'-y')F_{y'} \right]_{x=x_0}=0 \qquad ...(4)$$

$$\left[F+(\psi'-y')F_{y'} \right]_{x=x_1}=0 \qquad ...(5)$$

are fulfilled at the endpoints $A(x_0, y_0)$ and $B(x_1, y_1)$ of the extremal Y.

Proof. We first assume that one endpoint $A(x_0, y_0)$ of Y is fixed and the other endpoint $B(x_1, y_1)$ moves on the curve $Y_2 : y=\psi(x)$ *i.e.* $y_1=\psi(x_1)$. Hence we suppose that the lower limit of the integral in (3) is x_0 which is fixed and the upper limit is x_1 which varies. Then the problem is to minimise the functional

$$J[(x)]=\int_{x_0}^{x_1} F(x, y(x), y'(x))dx \qquad ...(6)$$

among all functions $y(x)$ satisfying the boundary conditions

$$y(x_0)=y_0 \qquad ...(7)$$

and $$y(x_1)=y_1 \qquad ...(8)$$

Geometrically speaking, the problem is to find among all curves joining the fixed point (x_0, y_0) to any point on the curve $Y_2 : y=\psi(x)$, that one for which (6) is a minimum.

As usual to compute δJ put

$$\triangle J=\int_{x_0}^{x_1+\delta x_1} F(x, y+\alpha\delta y, y'+\alpha\delta y')dx-\int_{x_0}^{x_1} F(x, y, y')dx \qquad ...(9)$$

where $y(x)$ is the minimising function and α is a parameter. The perturbation $\delta y(x)$ must satisfy the BCs

$$\delta y(x_0)=0 \qquad ...(10)$$

and $$y(x_1)+\alpha\delta y(x_1)=\psi(x_1+\delta x_1) \qquad ...(11)$$

Relation (11) shows that x_1 can be regarded as an implicit function of the parameter α, say $g(\alpha)$.

Put $$\phi(\alpha)=\int_{x_0}^{g(\alpha)} F(x, y+\alpha\delta y, y'+\alpha y')dx \qquad ...(12)$$

Then $$\phi'(0)=g'(0)F[g(0), y\{g(0)\}, y'\{g(0)\}] + \int_{x_0}^{g(0)} (F_y\, \delta y+F_{y'}\, \delta y')dx \qquad ...(13)$$

To make the relation (13) look a bit less complicated, we further put

$$x_1=g(0)=c,\ y\{g(0)\}=y(c)=\beta \text{ and } y'\{g(0)\}=u$$

Then (13) is equal to

$$\delta J=\phi'(0)=g'(0)\, F(c, \beta, u)+\int_{x_0}^{c} F_y\, \delta y+F_{y'}\delta y')dx \qquad ...(14)$$

$$=g'(0)F(c, \beta, u)+[F_{y'}\delta y\Big]_{x_0}^{c} +\int_{x_0}^{c} \left(F_y-\frac{dF_{y'}}{dx} \right)dydx$$

$$=g'(0)\,.\, F(c, \beta, u)+F_{y'}(c, \beta, u)\,.\,\delta(c)$$

since on an extremal $F_y-\dfrac{dF_{y'}}{dx}=0$ and $\delta y(x)$ vanishes at $x=x_0$.

For extremum $\delta J=0$, hence

$$g'(0)F(c, \beta, u)+F_{y'}(c, \beta, \mu)\delta(c)=0$$

If we replace x_1 by $g(\alpha)$ in (11), it becomes an identity. Hence differentiating (11) with respect to α and putting $\alpha=0$, we shall get after simplification

$$\psi'(c)\, g'(0)-[ug'(0)+\delta(c)]=0$$

i.e., $$g'(0)=\frac{\delta(c)}{\psi'(c)-u}$$

Making this substitution for $g'(0)$ in (15) we get

$$\frac{\delta(c)}{\psi'(c)-u} F(c, \beta, u)+F_y'(c, \beta, u)\delta c=0$$

Since $\delta(c)$ is arbitrary,

$$F(c, \beta, u)+\{\psi'(c)-u\}F_y'(c, \beta, u)=0$$

or $$F(c, y(c), y'(c)+\{\psi'(c)-y'(c)\}F_y'(c, y(c), y'(c)=0$$

or finally

$$[F(x, y(x), y'(x))+\{\psi'(x)-y'(y)\}F_y'(x, y(x), y'(x))]_{x=x_0}=0$$

If now we keep B fixed and move A, then condition (4) is obtained. x_0 and x_1 can be found from (4) and (5).

The following algorithm makes this problem easy to solve :

Steps to Solve Elementary Problem with moving Boundaries

I. Write down the Euler equation $F_y-\frac{dF_y'}{dx}=0$ and obtain the two parameter family of extremals $y=f(x, c_1, c_2)=0$.

II. Write down the equations

$$y(x_0)=\phi(x_0)=f(x_0, c_1, c_2) \qquad ...(i)$$

$$y(x_1)=\psi(x_1)=f(x_1, c_2, c_2) \qquad ...(ii)$$

III. Write down the transversality conditions

$$\Big[F-(\varphi'-y')F_y'\Big]_{x=x_0}=0 \qquad ...(iii)$$

$$\Big[F+(\psi'-y')F_y'\Big]_{x=x_1}=0 \qquad ...(iv)$$

IV. Determine the constants c_1, c_2, x_0, x_1 from (*i*), (*ii*), (*iii*), (*iv*).

Example 9. Find the shortest distance between the point A(1, 0) and the ellipse $4x^2+9y^2=36$.

Solution. The formula for distance is

$$J[y]=\int_1^{x_1} \sqrt{1+y^{2'}}xd$$

where $$4x_1^2+9y_1^2=36. \qquad ...(i)$$

The point (x_1, y_1) lies on the curve

$$y=\psi(x)=\frac{2}{3}\sqrt{9-x^2}.$$

Here $F(x, y, y')=\sqrt{1+y'^2}$.

The Euler equation is

$$F_y-F_{xy'}-F_{yy'}.\,y'-F_{y'y'}.\,y''=0$$

i.e. $$y''=0.$$

Hence extremals lie on the straight lines

$$y=c_1x+c_2 \qquad ...(ii)$$

Since A(1, 0) lies on (*ii*)

$$c_1+c_2=0. \quad \ldots(iii)$$

Since (x_1, y_1) satisfies (*i*) and

$$y_1=c_1x_1-c_1,$$

$$4x_1^2+9c_1^2(x_1-1)^2=36 \quad \ldots(iv)$$

The transversality condition is

$$\Big[F_y+(\psi'-y')F_{y'}\Big]_{x=x_1}=0$$

i.e.,
$$\left[\sqrt{1+y'^2}+\left(\frac{-2x}{3\sqrt{9-x^2}}-y'\right)\frac{y'}{\sqrt{1+y'^2}}\right]_{x=x_1}=0$$

i.e.,
$$\sqrt{1+c_1^2}+\left(\frac{-2x_1}{3\sqrt{9-x_1^2}}-c_1\right)\frac{c_1}{\sqrt{1+c_1^2}}=0$$

i.e.,
$$3c_1(x_1-1)=2(9-x_1^2) \quad \ldots(v)$$

Solving (*iv*) and (*v*) for x_1 and c_1,

$$x_1=\frac{9}{5},\quad c_1=\frac{12}{5}.\quad \text{Also } y_1=\frac{8}{5}$$

Hence the minimum distance is given by

$$d^2=(x_1-1)^2+y_1^2=\frac{16}{25}+\frac{64}{25}=\frac{16}{5}$$

or
$$d=\frac{4}{\sqrt{5}}.$$

It lies along the line $\quad y=\frac{12}{5}(x-1).$

Example 10. Find the shortest distance between the parabola $y=x^2$ and the straight line $x-y=5$.

Solution. The problem reduces to finding the extremum of the integral

$$J[y(x)]=\int_{x_0}^{x_1}\sqrt{1+y'^2}\,dx$$

subject to the condition that the point $A(x_0, y_0)$ moves on the parabola $y=x^2$ and the other end point $B(x_1, y_1)$ lies on the straight line $x-y=5$. Thus we take $\varphi(x)=x^2$ and $\psi(x)=x-5$.

Since $F=(1+y'^2)^{1/2}$, the Euler equation is $y''=0$. Hence the extremals lie on the two-parameter family of straight lines

$$y=c_1x+c_2$$

Since A, B lie on an extremal

$$y_0=c_1x_0+c_2 \quad \ldots(i)$$

$$y_1=c_1x_1+c_2 \quad \ldots(ii)$$

Since A lies on $y=x^2$ and B on $x-4=5$

$$y_0=x_0^2 \qquad \ldots(iii)$$

$$y_1=x_1-5 \qquad \ldots(iv)$$

The transversality conditions are

$$\left[\sqrt{1+y'^2}+(2x-y')\frac{y'}{\sqrt{1+y'^2}}\right]_{x=x_0}=0$$

$$\left[\sqrt{1+y'^2}+(1-y')\frac{y'}{\sqrt{1+y'^2}}\right]_{x=x_1}=0$$

Since $y'=c_1$, these conditions reduce to

$$\sqrt{1+c_1^2}+(2x_0-c_1)\frac{c_1}{\sqrt{1+c_1^2}}=0 \qquad \ldots(v)$$

$$\sqrt{1+c_1^2}+(1-c_1)\frac{c_1}{\sqrt{1+c_1^2}}=0 \qquad \ldots(vi)$$

Solving the equation (*i*) to (*vi*) we shall have

$$x_0=\frac{1}{2},\quad x_1=\frac{23}{8}\, c_1=-1,\quad c_2=\frac{3}{4}.$$

Hence the equation to the defined extremal is

$$y=-x+\frac{3}{4}$$

and the distance between the two curves is $\displaystyle\int_{1/2}^{23/8}\sqrt{2}dx=\frac{19\sqrt{2}}{8}.$

Example 11. Find the shortest distance between the circle $x^2+y^2=1$ and the straight line $x+y=4$.

Solution. As before the extremals are $y=c_1x+c_2$. The equations to solve are

$$y_0=c_1x_0+c_2 \qquad \ldots(i)$$

$$y_1=c_1x_1+c_2 \qquad \ldots(ii)$$

$$x_0^2+y_0^2=1 \qquad \ldots(iii)$$

$$x_1+y_1=4 \qquad \ldots(iv)$$

$$\sqrt{1+c_1^2}+\left(\frac{-x_0}{\sqrt{1-x_0^2}}-c_1\right)\frac{c_1}{\sqrt{1+c_1^2}}=0 \qquad \ldots(v)$$

$$\sqrt{1+c_1^2}+(-1-c_1)\frac{c_1}{\sqrt{1+c_1^2}}=0 \qquad \ldots(vi)$$

From (*v*) and (*vi*)

$$c_1=1 \quad \text{and} \quad x_0=\pm\frac{1}{\sqrt{2}}.$$

A simple figure shows that the shortest distance occurs in the first quadrant.

Hence $x_0=\frac{1}{\sqrt{2}}$ and $y_0=\frac{1}{\sqrt{2}}$.

Also $c_2=0$, $x_1=y_1=2$.

Hence the minimum distance is

$$\int_{1/\sqrt{2}}^{2} \sqrt{2}\, dx=2\sqrt{2}-1$$

and lies on $y=x$.

Problem. Prove that for a functional of the form

$$J[y(x)]=\int_{x_0}^{x_1} h(x, y)\sqrt{1+y^{2'}}\, dx$$

where $h(x, y)\neq 0$ on the boundary curves, the extremals are orthogonal to the boundary curves.

[**Hint.** Transversality conditions are

$y'\phi'=-1$ and $y'\psi'=-1$].

§13·7. Approximation of Extrema : The Rayleigh-Ritz Method

The Euler differential equations occurring in a variational problem can be integrated in a closed form only in exceptional cases. Besides expressing them in terms of known or designated functions, the closed forms of integrals may also be expressed in the form of various series such as

$$y(x)=\sum_{n=0}^{\infty} a_n x^n \qquad \text{(Power series)}$$

or

$$y(n)=\frac{a_0}{2}+\sum_{n=1}^{\infty} (a_n \cos nx+b_n \sin nx)$$

(Fourier series)

or any other series

$$y(x)=\sum_{n=0}^{\infty} b_n\phi_n(x),$$

where the functions $\varphi_n(x)$ are drawn from a linear space of admissible functions. This requires determination of an infinite number of coefficients. Instead of including all the infinite number of coefficients required in the solution, we may choose from the class of admissible functions a complete system of linearly independent functions $\phi_n(x)$, called the system of coordinate functions and form a sequence of admissible functions $y_n(x)$ where

$$y_n(x) = \sum_{i=1}^{n} \alpha_i \phi_i(x)$$

where α_i are constants.

If $\lim_{n \to \infty} J[y_n(x)]$ solves the extremum problem *i.e.*

$$\lim_{n \to \infty} J[y_n(x)] = \text{extremum } J[y(x)] \qquad \ldots(1)$$

then the sequence $\{y_n(x)\}$ is called a *minimising sequence.* The relation (1) does not imply $\lim_{n \to \infty} y_n(x) = y(x)$.

The Ritz method is frequently employed for exact or approximate solution of variational problems in mathematical physics. The method of obtaining solutions of a variational problem by constructing appropriate minimising sequences is called a *direct method.* A direct method requires answers to the following questions :

(1) How to construct the minimising sequence ?

(2) If the minimising sequence exists, is it convergent ?

(3) Is the limit of the minimising sequence an admissible function ?

Since the covergence question is important, we have to specify the metric of the function space from which members of the minimising sequence are drawn.

A positive answer to these questions can be given in the context of problems involving functionals of the form

$$J[y(x)] = \int_a^b [p(x)\, y'^2(x) + q(x)\, y^2(x) + 2r(x)\, y(x)]\, dx$$

BCs ; $y(a) = A$, $y(b) = B$

and p, q, r are continuous functions

$$p(x) > 0 \text{ and } q(x) \geqslant 0 \text{ on } [a, b].$$

We saw in example 7 (*b*), that ihe Euler equation of the functional is a Sturm-Liouville boundary value problem.

Method. The method consists of assuming a finite linear combination.

$$y_n(x) = \sum_{i=1}^{n} \alpha_i \phi_i(x)$$

as an approximate solution, where $\phi_i(x)$ are admissible functions satisfying the boundary conditions, substituting for $y_n(x)$ in

$$J[y_n(x)] = \int_a^b F(x, y_n, y_n')\, dx$$

$$= G(\alpha_1, \alpha_2, \ldots\ldots, \alpha_n) \qquad \text{(say)}$$

and solving the equations

$$\frac{\partial G}{\partial \alpha_1}=\frac{\partial G}{\partial \alpha_2}=\ldots\ldots=\frac{\partial G}{\partial \alpha_n}=0$$

for $\alpha_1, \alpha_2, \ldots\ldots, \alpha_n$.

Note. In practice upto three coordinate functions ϕ_i give a good approximation.

Example 12. Use the Ritz method to find an approximate solution of the minimising problem of the functional

$$J[y(x)]=\int_0^1 (y^{2\prime}-y^2+2xy)\, dx \qquad \ldots(i)$$

where $\qquad y(0)=y(1)=0 \qquad \ldots(ii)$

Compare the approximate solution with the exact solution.

Solution. Take $\phi_n(x)=(1-x)\, x^n$, $n=1, 2, \ldots\ldots$ Then $\phi_n(x)$ are continuous in $[0, 1]$ and $\phi_n(0)=\phi_n(1)=0$. Hence $\phi_n(x)$ are admissible functions. It is easy to see that $\{\phi_n(x)\}$ is complete. Hence $\phi_n(x)$ form a correct choice as coordinate functions.

Let us consider the first approximation

$$y_1(x)=\alpha\phi_1(x)=\alpha(x-x^2)$$

$$J[y_1(x)]=\int_0^1 [\alpha^2(1-2x)^2-\alpha^2x^2(1-x)^2+2\alpha x^2(1-x)]\, dx$$

$$=\frac{3}{10}\alpha^2+\frac{\alpha}{6}=Y(\alpha) \qquad \text{(say).}$$

For an extremum of J, $Y'(\alpha)=0$

i.e., $$\frac{3}{5}\alpha+\frac{1}{6}=0$$

i.e., $$\alpha=-\frac{5}{18}$$

Hence $\quad y_1(x)=-\frac{5}{18}x+\frac{5}{18}x^2$

is the first approximation to the solution of the extremum problem.

To compute the exact solution we solve the Euler equation,

$$F_y-F_{xy'}-F_{yy'}\cdot y'-F_{y'y'}\cdot y''=0$$

i.e., $\quad 2x-2y-2y''=0,\ y(0)=y(1)=0$

or $\quad y''+y=x,\ y(0)=y(1)=0$

Hence the exact solution is $\quad y(x)=x-\dfrac{\sin x}{\sin 1}$

We can compare the approximate and exact solutions at the points $x=0, 0{\cdot}25, 0{\cdot}50, 0{\cdot}75, 1$. For computing $\sin x$ at various points, we can use the series formula

$$\sin x=x-\frac{x^3}{6}+\frac{x^5}{120}-\ldots\ldots$$

retaining only three terms *viz.*,

$$\sin x \sim x - \frac{x^3}{6} + \frac{x^5}{120}$$

The following is the comparative chart .

x	*Exact solution*	*Approx. solution* y_1
0·00	0	0
0·25	−0·044	−0·052
0·50	−0·070	−0·069
0·75	−0·060	−0·052
1·00	0	0

Example 13. Using Ritz method find the second approximation $y_2(x)$ to the solution of the minimising problem of the functional given below and compare it with the exact solution,

$$J[y(x)] = \int_0^1 (y'^2 + 2y)\, dx \; ; \; y(0) = y(1) = 0.$$

Solution. Take $\phi_1(x) = (1-x)x$, $\phi_2(x) = (1-x)x^2$ and let

$$y_2(x) = \alpha\phi_1(x) + \beta\phi_2(x)$$

Then $$J[y_2(x)] = \int_0^1 [\{\alpha(1-2x) + \beta(2x-3x^2)\}^2 + 2\{\alpha(1-x)x + \beta(1-x)x^2\}]\, dx$$

$$= \frac{\alpha^2}{3} + \frac{2\beta^2}{15} + \frac{\alpha\beta}{3} + \frac{\alpha}{3} + \frac{\beta}{6} = Y(\alpha, \beta) \qquad \text{(say)}$$

For extrema $\dfrac{\partial Y}{\partial \alpha} = 0 = \dfrac{\partial Y}{\partial \beta}$,

i.e., $$\frac{2\alpha}{3} + \frac{\beta}{3} + \frac{1}{3} = 0$$

$$\frac{\alpha}{3} + \frac{4\beta}{15} + \frac{1}{6} = 0$$

i.e., $$\alpha = -\tfrac{1}{2}, \; \beta = 0$$

Hence $$y_2(x) = -\tfrac{1}{2}(1-x)x$$

The Euler equation of the problem is

$$y'' = 1 \text{ with } y(0) = y(1) = 0$$

Hence the exact solution is $y = -\frac{1}{2}(1-x)x$.

Thus the second approximation coincides with the exact solution.

Problem. Find the approximate solution of the equation

$$y'' + y = x^2, \; y(0) = y(1) = 0$$

[Form the functional whose Euler equation is the given equ a-tion],

13·8. Variational Methods In Eigenvalue Problems : Rayleigh's Principle

Consider the Sturm-Liouville eigenvalue problem,

$$L(y)=-[p(x)\, y']'+q(x)\, y=\lambda y \qquad \ldots(1)$$

$$\text{BCs}: y(a)=y(b)=0 \qquad \ldots(2)$$

Values of λ for which the BVP (1)—(2) has a non-trivial solution $y\neq 0$, are called eigenvalues and the corresponding solutions are called eigenfunctions of the problem. In section 4·2 of chapter 4 we have proved that distinct eigenfunctions are orthogonal on $[a, b]$. We shall suppose the set of eigenfunctions to have been normalised. Thus for eigenfunctions y_m, y_n

$$\int_a^b y_m(x)\, y_n(x)\, dx=\begin{cases}0 & m\neq n\\ 1 & m=n\end{cases}$$

Now the equation

$$L[y]=-(py')'+qy=0$$

is the Euler's equation of the functional

$$J[y(x)]=\int_a^b \{p(x)\, y'^2+q(x)\, y^2\}\, dx \qquad \ldots(3)$$

Hence the eigenvalues and eigenfunctions are also called the igenvalues and eigenfunctions of the functional (3) subject to condi-eions $y(a)=y(b)=0$, We have thus two types of problems.

The first variational problem. Determine the extremum of the functional

$$J[y(x)]=\int_a^b F(x, y, y')\, dx$$

subject to $y(a)=y(b)=0$.

The second variational problem. Determine the extremum of the functional

$$J[y(x)]=\int_a^b F(x, y, y')\, dx$$

subject to $y(a)=y(b)=0$

and $$\int_a^b y^2(x)\, dx=1.$$

Lemma 3. If λ_n is an eigenvalue and $y_n(x)$ the corresponding eigenfunction of the Sturm-Liouville problem (1)—(2) and $J[y(x)]$ is the functional (3), then $J[y_n(x)]=\lambda_n$.

Proof. It is given that

$$-(py_n')'+q(x)\, y_n=\lambda_n y_n(x)$$

where $y_n(a)=y_n(b)=0$

and $$\int_a^b y_n^2(x)\, dx=1.$$

Also $$J[y_n(x)]=\int_a^b (py_n'^2+qy_n^2)\, dx.$$

Integrating by parts

$$\int_a^b py_n'^2\, dx=\Big[(py_n')\, y_n(x)\Big]_a^b-\int_a^b (py_n')' y_n(x)\, dx$$

$$=-\int_a^b (py_n')'\, y_n(x)\, dx$$

Hence $$J[y_n(x)]=\int_a^b [-(py_n')'\, y_n+qy_n^2(x)]\, dx$$

$$=\int_a^b [-(py_n')'+qy_n]\, y_n\, dx$$

$$=\lambda_n \int_a^b y_n^2(x)\, dx$$

$$=\lambda_n.$$ **Q.E.D.**

Rayleigh's Principle. In section 10·11 of chapter 10, we have seen that if λ_1 is the smallest positive eigenvalue of the BVP (1)—(2), then $\lambda_1 \leqslant \frac{\| Ly \|}{\| y \|}$ for any $y(x) \in$ dom L.

This is known as the Rayleigh's principle. In this section, we describe this principle as a minimising problem of a functional.

In the problem (1)—(2) suppose that $p(x)$, $p'(x)$, $q(x)$, $r(x)$ are continuous on $[a, b]$ and $p(x)>0$ [on a, b]. Let $y(x) \in$ dom L *i.e.*, $y(x)$ is twice differentiable and satisfies the BCs (2). We can have even more generalised boundary conditions

$$\alpha_1 y(a)+\beta_1 y'(a)=0, \qquad \alpha_1^2+\beta_1^2\neq 0$$
$$\alpha_2 y(b)+\beta_2 y'(b)=0, \qquad \alpha_2^2+\beta_2^2\neq 0$$

dom L will be called the set of *admissible* functions.

If $$\int_a^b yL(y)\, dx \geqslant 0$$

then the eigenvalue problem (1)—(2) has only real-valued eigenvalues λ. To the problem (1)—(2) add the further condition.

$$\int_a^b r(x)\, y^2(x)\, dx>0 \qquad \ldots(4)$$

This is assured if $r(x)>0$ on $[a, b]$. Indeed if $y(x)$ is an eigenfunction, then

$$\int_a^b r(x)\, y^2(x)\, dx = 1$$

Find the function $y(x)$ for which the functional

$$J[y(x)] = \frac{\int_a^b yL(y)\, dx}{\int_a^b r(x) y^2(x) dx} = \text{minimum}$$

If λ_1 is the minimum and $y_1(x)$ is a solution of this problem, then λ_1 is the least positive eigenvalue and $y_1(x)$ is the corresponding eigenfunction of the problem (1)—(2).

The inequality

$$\lambda_1 \leqslant \frac{\int_a^b yL(y)\, dx}{\int_a^b r(x)\, y^2\, dx}$$

is known as Rayleigh's principle.

Example 14. Use Rayleigh's principle to estimate λ_1 for the following boundary-value problems :

(*i*) $-y'' = \lambda y$; $y'(0) = y(1) = 0$

(*ii*) $-y'' = \lambda y$; $y(0) = y(1) = 0$

(*iii*) $-y'' = \lambda(10 - x^2)y$; $y(-1) = y(1) = 0$.

Solution. (*i*) Here $L(y) = -y''$

$$p = 1,\ q = 0,\ r = 1$$

$y(x) = 1 - x^2$ is twice differentiable and satisfies the BCs

$$y'(0) = 0,\ y(1) = 0$$

Hence it is an admissible function.

By Rayleigh's principle

$$\lambda_1 \leqslant \frac{\int_0^1 yL(y)\, dx}{\int_0^1 ry^2\, dx} = \frac{2\int_0^1 (1-x^2)\, d}{\int_0^1 (1-x^2)^2\, dx} = \frac{4/3}{8/15} = 2{\cdot}5.$$

The exact eigenvalues of the BVP are $\lambda_n = \left(\frac{2n-1}{2}\,\pi\right)^2$ and the corresponding eigenfunctions are $\cos\left(\frac{2n-1}{2}\,\pi x\right)$.

Hence the exact value of λ_1 is $\frac{\pi^2}{4} \sim 2{\cdot}4674$.

(*ii*) Here we take $y(x) = (1-x)x$ as an admissible function.

Hence by Rayleigh's principle

$$\lambda \leqslant \frac{2\int_0^1 x(1-x)\,dx}{\int_0^1 x^2(1-x)^2\,dx}=\frac{1/3}{1/30}=10$$

The exact eigenvalues of the BVP are $\lambda_n=n^2\pi^2$ and the corresponding eigenfunctions are $\sin n\pi x$. Hence the exact value of λ_1 is π^2.

(*iii*) In this problem $r(x)=10-x^2>0$,

$y(x)=1-x^2$ is an admissible function

$$\text{Hence } \lambda_1 \leqslant \frac{2\int_{-1}^1 (1-x^2)\,dx}{\int_{-1}^1 (10-x^2)(1-x^2)^2 dx}=\frac{8/3}{368/35}=\frac{35}{138}$$

Example 15. Find the eigenvalues and eigenfunctions in the following problems :

(*i*) $J[y]=\int_0^1 (y^2+y'^2)\,dx$

$$y(0)=y(1)=0\;;\;\int_0^1 y^2\,dx=1$$

(*ii*) $J[y]=\int_\pi^{2\pi} (y^2-y'^2)\,dx$

$$y(\pi)=y(2\pi)=0\;;\int_\pi^{2\pi} y^2\,dx=1.$$

Solution. (*i*) Here $p=1$, $q=1$.

Hence the corresponding Sturm-Liouville problem is

$$-y''+y=\lambda y$$

$$y(0)=y(1)=0$$

The O.D.E is $y''+(\lambda-1)y=0$

The general solution is

$$y=A\sin\sqrt{\lambda-1}\,x+B\cos\sqrt{\lambda-1}\,x$$

By the boundary conditions $B=0$

and $\quad A\sin\sqrt{\lambda-1}=0$

Since A is arbitrary, $\sin\sqrt{\lambda-1}=0$

i.e., $\quad \lambda-1=n^2\pi^2 \quad n=1, 2, \ldots\ldots$

i.e., $\quad \lambda_n=n^2\pi^2+1$

Hence $y_n = A \sin n\pi x$ are the eigenfunctions.

Since $$\int_0^1 y_n^2(x)\,dx = 1,$$

we have $A = \pm\sqrt{2}$

Hence the normalised eigenfunctions are

$$\pm\sqrt{2} \sin n\pi x \qquad (n=1, 2, \ldots\ldots).$$

(*ii*) Here $p=-1, q=+1$.

Hence the Strum-Liouville problem is

$$y''+y=\lambda y$$
$$y(\pi)=y(2\pi)=0$$

The general solution is

$$y = A \sin\sqrt{1-\lambda}\,x + B\cos\sqrt{1-\lambda}\,x$$

Since $\cos n\pi \neq 0$, the boundary conditions are satisfied if $B=0$ and $\sqrt{1-\lambda}=n$

i.e., $$\lambda_n = 1-n^2$$

The eigenvalues are

$$y_n = A \sin nx.$$

Since $$\int_\pi^{2\pi} y_n^2(x)\,dx = 1 \qquad A = \pm\sqrt{\frac{2}{\pi}}$$

Hence $$y_n(x) = \pm\sqrt{\frac{2}{\pi}} \sin nx.$$

Example 16. If $y(0)=y(\pi)$, then prove that

$$1 \leqslant \int_0^\pi y'^2(x)\,dx.$$

Solution. Let the problem be

$$J[y(x)] = \int_0^\pi y'^2(x)\,dx = \text{minimum},$$

where $y(0)=y(\pi)=0$, $\int_0^\pi y^2(x)\,dx = 1$.

The Euler's equation for the functional

$$\int_0^\pi (y'^2 - \lambda y^2)\,dx,\ y(0)=y(\pi)=0$$

is $y''+\lambda y=0$, $y(0)=y(\pi)=0$.

The eigenvalues and eigenfunctions of the BVP are respectively $\lambda_n = n^2$, $y_n = \sin nx$.

Hence $$\lambda_1 = 1 = \min \int_0^\pi y'^2(x)\,dx$$

i.e., $$1 \leqslant \int_0^\pi y'^2(x)\,dx.$$

ANSWERS TO EXERCISES

Exercise 1

1. $y=A+\tan\dfrac{x+y}{2}$
2. $x^2+y^2=A(xy-1)^2$
3. $x+y=a\tan\dfrac{y+A}{a}$
4. $x\,e^{y^2/x}=A$
5. $x+\cot(x-y)=A$
6. $y(x^2-1)=Axy+x(1+y\log y)$
7. $(1+x^2)\cos y=A$
8. $(x^2+y^2)(x+1)^2=Ax^2$
9. $-\frac{1}{2}y^{-2}=x/y+A.$

Exercise 2

1. $\tan\dfrac{y}{2x}=Ax$
2. $y+\sqrt{x^2+y^2}=C$
3. $y=A\,e^{y/x}$
4. $2y+x\log xy=Ax$
5. $\dfrac{5y-4x+7}{5y-x-2}=A(x-3)^3$
6. $y-x-\log(2x+2y+1)=A$
7. $\tan^{-1}\dfrac{y-1}{x+1}=\dfrac{1}{2}\{(x+1)^2+(y-1)^2\}+A.$

Exercise 3

1. $\dfrac{x}{y}=y^2+A$
2. $xy=A\cos x+\sin x$
3. $2x\log y=(\log y)^2+A$
4. $x\log y=e^x(x-1)+A$
5. $\sin y=(1+x)(e^x+A)$
6. $\dfrac{1}{\log y}=\dfrac{1}{2x}+Ax$
7. $4\sec y\sec x=2x+\sin 2x+A$
8. $\dfrac{-\sin x}{y^2}=x\,e^x-e^x+A$
9. $\dfrac{1}{y^3}=-\dfrac{1}{2}+A\,e^{3x^2}$
10. $yx^2+x+1=Ax\,e^{1/x}(xy+1)$
11. $y+x=A\,e^{x^2}(y-x)$
12. $yx=Ax(yx-2)+1$
13. Put $y^2=v$; C.P. $y^2=2ax+cx^2$.

Integral curves are a set of ellipses and hyperbolas separated by a single parabola $y^2=2ax$.

Exercise 4

1. $y+x^2+1=Ax$
2. $\frac{1}{2}\log(x^2+y^2)=A\tan^{-1}\frac{y}{x}$
3. $(x-y)\,y^2=A(x+y)$
4. $x^2y^2+xy\log\frac{Ax}{y}=1$
5. $\frac{x^4}{4}-\frac{2x^3}{3}+x^2y^2=A$
6. $x^2y^2+2\tan^{-1}\frac{y}{x}=A.$

Exercise 5

1. $c^2-2c\sqrt{x^2+y^2}+y^2=0$
2. $xy=cy+c^2$; S.S. $y+4x=0,\ y=0$
3. $y^2=2cx+c^2$
4. $y^2=2cx+c^3$
5. $(c^2-y)^2=4cx$
6. $16+2c^2y-c^3x^2=0$
7. $y=2cx+c^2$ (Put $y^2=Y$)
8. $y^2=cx+\frac{c^3}{8}$ (Put $y^2=Y$)
9. $y=2c\sqrt{x}+c^2$, S.S. $x+y=0$
10. $c^2x^2-c(x^2+y^2-2)+y^2=0$
11. $y^2=cx^2+c^2$
12. p-disc is $y=x^2$. No S.S.
13. $(y+c)^2=(x+c)^3$,
    S.S. $y=x-\frac{4}{27}$.
    p-disc. $p=0$ or $p=1$
    $y=x$ is cusp locus
14. $y^2=4(x+1)$
15. No S.S.
16. $y=0$
17. No S.S.
18. $4xy^2+1$
19. $4y=x^2$
20. $y=0$
21. $\frac{x^2}{a^2}+\frac{y^2}{b^2}=1.$

Exercise 6

1. (*i*) $y=Ax$ (*ii*) $x^2+y^2=Ay$
   (*iii*) $y^2-x^2=A^2$ (*iv*) $x^{2/3}-y^{2/3}=A^{2/3}$
   (*v*) $r^2=A\,e^{\theta^2}$ (*vi*) $r^n=A^n\sin n\theta$
   (*vii*) $r^n=A^n\cos n\theta$ (*viii*) $r^n\cos n\theta=A^n$.
2. $b=3$
3. $p=\frac{1}{2}$

Exercise 7

2. (i) $Ae^{-x/2}\cos\left(\frac{\sqrt{7}}{2}x+B\right)+\frac{x^3}{2}-\frac{3x^2}{4}-\frac{3x}{4}+\frac{9}{8}$

(ii) $Ae^{-x}+Be^{-1/2x}+\frac{1}{6}e^{x}(x-7/6)$

(iii) $(A+Bx)e^{-x}+(C+D_x)e^{-1/2x}+(x-6)$

(iv) $(A+Bx)\cos x+(C+Dx)\sin x+\left(\frac{3x^2}{16}-\frac{x^4}{48}\right)\cos x$

$$+\frac{x^3\sin x}{12}$$

(v) $Ae^{x}+Be^{-x}-\frac{1}{2}(x\sin x+\cos x)+\frac{xe^{x}}{12}(2x^2-3x+9)$

(vi) $Ae^{mx}+Be^{-mx}+C\cos mx+D\sin mx+\frac{x\cos mx}{4m^3}$

(vii) $e^{x}(Ax+B+C\cos 2x+D\sin 2x)+\frac{x^2e^{x}}{10}$

(viii) $Ax+B\cos(\sqrt{2}\log x)+C\sin(\sqrt{2}\log x)$

$$+\frac{x^2}{6}+\frac{1}{2}\sin(\log x)-\frac{1}{2}\cos(\log x)$$

(ix) $x^2(Ax^{\sqrt{3}}+Bx^{-\sqrt{3}})$

(x) $Ax+Bx^3+Cx^4+(2x^4-x^2)\log x-2.$

Exercise 8

1. $z=e^{2x}\{\phi_1(y+3x)+x\,\phi_2(y+3x)\}+x^2\tan(y+3x)$
2. $\phi_1(y-x)+\phi_2(y+2x)+\sin xy$
3. $\phi_1(y-x)+x\,\phi_2(y-x)+x\sin y$
4. $\phi_1(y+2x)+\phi_2(y-2x)+3x+x\log y+y\log x$
5. $\phi_1(y+ax)+\phi_2(y-ax)+\frac{x^3}{3\,!}$
6. $\phi_1(y+ix)+\phi_2(y-ix)-\frac{\cos mx\cos ny}{m^2+n^2}$
7. $\phi_1(2x-y)+\phi_2(y-x)+\frac{x^3}{6}+\frac{y^3}{12}$
8. $\phi_1(y+x)+\phi_2(y-\omega x)+\phi_3(y+\omega^2x)$

$$+\frac{x^6y^3}{4.5.6}+\frac{x^9}{4.5.6.7.8.9}$$

9. $\phi_1(xy)+x\phi_2(y/x)+xy\log x$

10. $\phi_1(y/x)+x\phi_2(y/x)+\dfrac{x^m y^n}{(m+n)(m+n-1)}$

11. $xz=\phi_1(xy)+\log x\,\phi_2(xy)+\frac{1}{4}x^3y$

12. $z=e^{-5/2\,x}\phi_1(2y-x)+e^{-x}\{\phi_2(y+2x)+x\phi_3(y+2x)\}$

Exercise 9

1. $z=(Ae^{kx}+Be^{-kx})e^{k^2y}$

2. $z=e^{k(x-y)+2y}$

3. $z=\dfrac{l}{\pi a}\sin\dfrac{\pi y}{l}\sin\dfrac{\pi ax}{l}$

Exercise 10

1. $W(x)=e^{(r_1+\dots+r_n)x}\begin{vmatrix} 1 & 1 & \cdots & 1 \\ r_1 & r_2 & & r_n \\ r_1^2 & r_2^2 & & r_n^2 \\ \vdots & \vdots & & \vdots \\ r_1^{n-1} & r_2^{n-1} & & r_n^{n-1} \end{vmatrix}$

Exercisé 11

3. (*a*) 2 (*b*) 3 (*c*) 1 (*d*) 1

6. (*a*) $y_1=1+x,\ y_2=1+x+\dfrac{x^2}{2\,!},\ y_3=1+x+\dfrac{x^2}{2\,!}+\dfrac{x^3}{3\,!}$

$|x|\leqslant\frac{1}{2}$

(*b*) $y_1=\dfrac{x^2}{2},\ y_2=\dfrac{x^2}{2}+\dfrac{x^5}{20},\ y_3=\dfrac{x^2}{2}+\dfrac{x^5}{20}+\dfrac{x^8}{160}+\dfrac{x^{11}}{4400}$

$|x|\leqslant\frac{1}{2}$

(*c*) $y_1=x,\ y_2=x+\dfrac{x^4}{4},\ y_3=x+\dfrac{x^4}{4}+\dfrac{x^7}{14}+\dfrac{x^{10}}{160}$

$|x|\leqslant\frac{1}{2}$

(*d*) $y_1=1+2\log x,\ y_2=1+2\log x+2(\log x)^2$

$y_3=1+2\log x+2(\log x)^2+\dfrac{4}{3}(\log x)^3,\ |x-1|\leqslant\dfrac{1}{12}.$

(*e*) $y_1=\frac{x^3}{3}$, $y_2=\frac{x^3}{3}+\frac{x^7}{63}$

$$y_3=\frac{x^3}{3}+\frac{x^7}{63}+\frac{2x^{11}}{2079}+\frac{x^{15}}{59535}, \quad |x|<\frac{1}{2}$$

(*f*) $y_1=\frac{1+x^3}{3}$, $y_2=\frac{1}{126}(33-14x+42x^3-7x^4-2x^7)$

(*g*) $y_1=1+x+\frac{x^2}{2}$, $y_2=1+x+x^2+\frac{x^3}{6}$

$$y_3=1+x+x^2+\frac{x^3}{3}+\frac{x^4}{24}.$$

Exercise 12

1. $y=x\{A+B\log x+C(\log x)^2\}-\frac{1}{8x}$
2. $y=A\{x\cos x-\sin x\log\sin x\}+B\cos x+C\sin x$
3. $y=x(A-x-\frac{1}{3}x^3)+(x^2+1)(B+\frac{1}{3}x^2)$
4. $y=x(A\,e^x+B)-x^2$
5. $y=Ax+B\log x$
6. $y=Ax^3+B(x+1)-x$
7. $y=Ax+B\,e^x+x^2+1.$

Exercise 13

2. (*i*) $y=C_1(1-x\cot x)+C_2\cot x$

   (*ii*) $y=C_1(1+x\tan x)+C_2\tan x$

   (*iii*) $y=\frac{1}{5}x+C_1\,x\,e^{-4/3x^{3/2}}\int\frac{e^{4/3x^{3/2}}}{x^2}\,dx+C_2$

   (*iv*) $y=C_1\sin^2 x+C_2\left(\sin^2 x\log\tan\frac{x}{2}-\cos x\right)+\frac{1}{2}\cos x$

3. (*i*) 1, (*ii*) −1, (*iii*) −1.

Exercise 14

1. $y=e^{x^2}(C_1\,e^x+C_2\,e^{-x}-1)$
2. $y=\left(A\,x^2+\frac{B}{x}\right)e^{\sqrt{x}}$
3. $y=x(C_1\cos ax+C_2\sin ax)$
4. $y\sin nx=C_1\cos mx+C_2\sin mx.$

Exercise 15

1. $y=C_1\, e^{x^2}+C_2\, e^{-x^2}-\sin x^2$
2. $y=C_1\, e^{\sqrt{2}\sin x}+C_2\, e^{-\sqrt{2}\sin x}+\sin^2 x$
3. $y=C_1\, e^{\sin x}+C_2\, e^{2\sin x}-\frac{5}{4}-\frac{3}{2}\sin x-\frac{1}{2}\sin^2 x$
4. $y=C_1\, e^{(n-1)\tan x}+C_2\, e^{-n\tan x}$
5. $y=e^{-x^2}\,(Ax^2+B)+\frac{1}{2}$
6. $y=A\sin(x^2+B)+\frac{1}{4}x^2$

Exercise 16

1. $y=C_1\cos x+C_2\sin x-x\cos x+\sin x\log(\sin x)$
2. $y=Ax+B\, e^{x}+1+x+x^2$
3. $y=C_1\,\frac{\sin x}{x}+C_2\,\frac{\cos x}{x}+\frac{1}{x}$
4. $y=A\sin x+B\cos x+C+\log\sec x-\sin x\log(\sec x+\tan x)$
5. $y=C_1\, e^{x}+\frac{C_2}{x}-x-1-\frac{1}{3}x^2.$

Exercise 17

1. $y_1=\cos\sqrt{x},\qquad y_2=\sin\sqrt{x}$
2. $y_1=1+3\sum_{n=1}^{\infty}\frac{x^n}{(2n-3)(2n-1)}$

   $y_2=\sqrt{x}\,(1-x)$
3. $y_1=\sum_{n=0}^{\infty}\frac{(3n+4)\,x^{n+\frac{1}{3}}}{4\,.\,3^n\,.\,n\,!}$

   $y_2=1+\sum_{n=1}^{\infty}\frac{(n+1)x^n}{2\,.\,5\,.\,8\ldots(3n-1)}$
4. $y_1=x\, e^{x}\quad,\qquad y_2=y_1\log x-x\sum_{n=1}^{\infty}\frac{H_n\, x^n}{n\,!}$

where $H_n = 1 + \frac{1}{2} + \frac{1}{3} + \dots + \frac{1}{n}$.

5. $y_1 = e^x$, $y_2 = e^x \log x$

6. $y_1 = \sum_{n=1}^{\infty} \frac{x^n}{(n\,!)^2}$, $y^2 = y_1 \log x - 2\sum_{n=1}^{\infty} \frac{H_n\, x^n}{(n\,!)^2}$

[y_1 is the Bessel function $J_0\,(2\sqrt{x})$]

7. Complete primitive : $y = Ay_1 + By_2 + x$

where $y_1 = \sum_{n=1}^{\infty} \frac{(-x)^n}{(n-1)\,!\,n\,!}$

$$y_2 = y_1 \log x + 1 - \sum_{n=1}^{\infty} \frac{(-1)^n\, H_n + H_{n-1})}{n\,!\,(n-1)\,!} x^n$$

8. $y_1 = e^x$ $y_2 = e^x/x$

9. $y_1 = \sum_{r=1}^{\infty} \frac{(-n)r}{(r\,!)^2} x^r = L_n\,(x)$

where n is a non-neagtive integer and

$(a)_r = a(a+1)\dots(a+r-1) = \frac{\Gamma(a+r)}{\Gamma(a)}$ is the Barne's symbol.

$$y_2 = y_1 \log x + \sum_{r=1}^{n} \frac{(-n)_r}{(r\,!)^2} (H_{n-r} - H_n - 2H_r)x^r.$$

$L_n(x)$ is called the Laguerre polynomial.

10. $y_1 = x$, $y_2 = x \log x + \sum_{r=1}^{\infty} \frac{(-1)^r\, x^{r+\cdot}}{r\,.\,r\,!}$

$$y_p = \sum_{r=0}^{\infty} \frac{(-1)^r\, x^{r+2}}{(r+1)\,.\,(r+1)\,!}\,.$$

Exercise 19

14. $\sum_{r=0}^{\infty} \frac{(-x^2/4)^r}{(r\,!)^2}$ i.e. $J_0\,(x)$.

Exercise 21

1. $y(z+x)=A(y+z)$
2. $x(y^2+z^2)-2y\,z=Ax$
3. $x^2y-\frac{1}{x}+\log\sec z=A$
4. $yz+zx+xy=A(x+y+z)$
5. $e^x y+e^y z+e^z x=ce^z$

Exercise 22

1. (i) $X(t)=c_1\,e^t\begin{bmatrix}1\\1\end{bmatrix}+c_2\,e^{2t}\begin{bmatrix}1\\3\end{bmatrix}$

(ii) $X=c_1U(t)+c_2V(t)$

$U(t)=e^{-2t}\begin{bmatrix}-\cos t\\ \cos t-\sin t\end{bmatrix},\ V(t)=e^{-2t}\begin{bmatrix}-\sin t\\ \sin t+\cos t\end{bmatrix}$

(iii) $X=c_1\,e^{2t}\begin{bmatrix}1\\1\end{bmatrix}+c_2\,e^{2t}\begin{bmatrix}t\\t-1\end{bmatrix}$

(iv) $X=c_1\,e^{5t}\begin{bmatrix}1\\2\end{bmatrix}+c_2\,e^{-t}\begin{bmatrix}1\\-1\end{bmatrix}$

(v) $X=c_1\,e^{-t}\begin{bmatrix}1\\0\\1\end{bmatrix}+c_2\,e^{t}\begin{bmatrix}3\\2\\1\end{bmatrix}+c_3e^{2t}\begin{bmatrix}1\\3\\1\end{bmatrix}$

2. (i) $X=e^t\begin{bmatrix}-2\\1\end{bmatrix}+3\,e^{4t}\begin{bmatrix}1\\1\end{bmatrix}$

(ii) $X=\cos 2t\begin{bmatrix}1\\3\end{bmatrix}+\sin 2t\begin{bmatrix}7\\-4\end{bmatrix}$

(iii) $x=\sin t,\qquad y=\cos t.$

3. (i) $X=c_1\,e^{-3t}\begin{bmatrix}1\\-4\end{bmatrix}+c_2\,e^{2t}\begin{bmatrix}1\\1\end{bmatrix}+\begin{bmatrix}3t+2\\2t-1\end{bmatrix}$

(ii) $x=c_1+c_2\,e^{-2t}+e^t,\ \ y=c_1-c_2\,e^{-2t}+e^t$

(iii) $x=(3t+2)e^t-(t+1),\qquad y=(6t+1)\,e^t-(3t+1).$

4. (i) $\phi(x+y+z, xyz)=0$

(ii) $\phi[(x+y)(z+1), (x-y)(z-1)]=0$

(iii) $\phi\left(\frac{y-x}{z-x}, \frac{y-x)}{z-y}\right)=0$

(iv) $\phi[y+2x, z-x^3 \sin(y+2x)]=0$

5. (i) $x=(c_1+c_2t)e^t+c_3e^{-3/2t}-\frac{1}{3}t$

$y=-2(c_1+c_2t-c_3c_2)e^t-\frac{1}{3}c_3e^{-3/2t}-\frac{1}{3}$

(ii) $x=(c_1+c_2t)e^t+(c_3+c_4t)e^{-t}$

$y=-\frac{1}{2}(c_1-c_2+c_2t)e^t-\frac{1}{2}(c_3+c_4+c_4t)e^{-t}$

Exercise 23

1. (i) $\phi\left(\frac{y}{z}, x^2+y^2+z^2\right)=0$

(ii) $z+\sqrt{x^2+y^2+z^2}=x^{1-a}\phi\left(\frac{x}{y}\right)$

(iii) $\phi(x^2+y^2-z^2, xy+z)=0$

((iv) $\phi\left(\frac{x^2+y^2+1}{x}, x^2-4z\right)=0$

(v) $\phi\left(\frac{x^2}{y}, \frac{x^2}{z-x^2/y}\right)=0$

2. (i) $x=s+t$, $y=s^2-t$, $z=s^3+t$, s and t parameters.

(ii) $(x^2+y^2-2z)+2xyz+2=0$

(iii) Initial curve is a characteristic.

All surfaces of revolution $z=\phi(x^2+y^2)$ are integral surfaces through the initial curve.

(iv) $x^2+y^2+z^2=1$

(v) $y(x+2)^2=z^2$.

If the initial curve is given parametrically as $x=-1$, $y=s^2$ $z=s$, then the integral surface can be written as

$$x=e^t-2,\ y=s^2e^{2t},\ z=se^{2t}.$$

As $x_t y_s - x_s y_t \neq 0$, the integral surface is unique.

3. (i) $\phi\left(\frac{y}{z}, \frac{x^2+y^2+z^2}{y}\right)=0$

(ii) $x^3y^3 z=\phi\left(\frac{x^3+y^3}{x^2y^2}\right)$

(iii) $\phi(z-x_1x_2, x_1+x_2+x_3, x_2x_3)=0$

(iv) $\phi\left(\frac{x_1}{x_3}, \frac{x_2}{x_3}, \frac{z}{x_3}-\frac{x_1x_2}{x_3^2}\log x_3\right)=0$

Exercise 24

1. $2az = a^2x^2 + y^2 + b$
2. $\dfrac{z^{1-l}}{1-l} = \dfrac{ax^{m+1}}{m+1} + \dfrac{1}{a}\,\dfrac{y^{n+1}}{n+1} + b$
3. $(\log z - b)^2\,(1+a^2) = \{\log\,(xy^a)\}^2$
4. $6z^2 = (4x+a)^{3/2} + (4y-a)^{3/2} + b$
5. $xz = ax + by - a^2b$

Singular solution : $z^2 = y$

6. $(z+a^2)^3 = (x+ay+c)^2$
7. $z = ax + by - 2a^2 - 3b^2$

Singular solution : $z = \dfrac{x^2}{8} + \dfrac{y^2}{12}$

8. Complete integral : $z = bx^a\,y^{1-a}$

General solution : $\phi\left(\dfrac{y}{x}, \dfrac{z}{x}\right) = 0$

9. Complete integral : $\log\,(b\,x^a\,y^{a-1}) = \dfrac{1}{xy-z}$

General solution : $\phi\left(xy, \log x + \dfrac{1}{z-xy}\right) = 0$

10. Complete integral : $\dfrac{1+2a}{x^2/y - z} = b \log xy^a$

General solution : $\phi\left(\dfrac{x^2}{y}\;,\; x^2\,e^{\frac{1}{z-x^2/y}}\right) = 0$

Exercise 25

1. $z^2 = a^2\,y^2 + 2ax + b$
2. $z = ay + \frac{1}{4}(x+b)^2$
3. $4(1+a^2)z = (x+ay+b)^2$
4. $z = b\,x^a\;y^{1/a}.$
5. $(16\,z^2 - 12ax + y^3 + 3b)^2 = (y^2+8a)^3$

Exercise 26

1. $z = y + x$ (or $x = 2t,\ y = 2t+s,\ z = 4t+s$)

and $z = y - x$ (or $x = -2t,\ y = 2t+s,\ z = 4t+s$)

2. $z = x^2 + y^2$ (or $x = 0,\ y = se^t,\ z = s^2\,e^{2t}$)
3. $z = xy + 1$ (or $x = se^t,\ y = 2e^t\ z = 2s(e^{2t}+1)$)

Exercise 27

1. Hyperbolic everywhere except on the y-axis.

$$\frac{\partial^2 z}{\partial \xi\, \partial \eta}=\frac{1}{4(\xi-\eta)}\left(\frac{\partial z}{\partial \xi}-\frac{\partial z}{\partial \eta}\right)$$

2. Elliptic everywhere except on the y-axis.

$$\frac{\partial^2 z}{\partial \alpha^2}+\frac{\partial^2 z}{\partial \beta^2}=-\frac{1}{2\alpha}\,\frac{\partial z}{\partial \alpha}$$

3. Parabolic everywhere. By taking $\xi=x-y$, $\eta=x+y$ we get the canonical form

$$\frac{\partial^2 z}{\partial \eta^2}=0$$

Solution. $z=(x+y)\,\phi_1(x-y)+\phi_2\,(x-y)$

4. Elliptic for $y<-\frac{1}{4}$, hyperbolic for $y>-\frac{1}{4}$. Parabolic for $x=0$ or for $y=-\frac{1}{4}$.

5. Hyperbolic everywhere except $x-$axis.

If we take $\xi=\dfrac{1}{y^{n-1}}+x$, $\eta=\dfrac{1}{y^{n-1}}-x$

then the canonical form is. $\dfrac{\partial^2 z}{\partial \xi\, \partial \eta}=0.$

Solution. $z=\phi_1\left(\dfrac{1}{y^{n-1}}+x\right)+\phi_2\left(\dfrac{1}{y^{n-1}}-x\right)$

6. Parabolic everywhere.

$$z=e^{-2xy^2}\phi_1\,(xy)+e^{xy^2}\phi_2\,(xy)-\frac{e^{xy}}{2x^2y^2}\,.$$

7. $z=y^2\,\phi_1(y/x)+\phi_2\,(xy)$. **Hyperbolic**

8. $z+\phi_1(x+y)+\phi_2(x^2-y^2)$. **Hyperbolic**

9. $z=\phi_1(x+y)+(x+y)\phi_2\,(y/x)$. **Hyperbolic**

10. $xz=\phi_1(y+a\,.\,x)+\phi_2(y-ax)$. **Hyperbolic**

11. $\dfrac{\partial^2 z}{\partial \xi \partial \eta}=0$; $\xi=x+y-\cos x$, $\eta=x-y+\cos x$

12. $\dfrac{\partial^2 z}{\partial \xi^2}+\dfrac{\partial^2 z}{\partial \eta^2}+\dfrac{1}{2\xi}\,\dfrac{\partial \xi}{\partial \xi}+\dfrac{1}{2\eta}\,\dfrac{\partial \xi}{\partial \eta}=0$; $\xi=y^2$, $\eta=x^2$.

13. $2(\xi^2-\eta^2)\,\dfrac{\partial^2 r}{\partial \xi \partial \eta}+\eta\,\dfrac{\partial z}{\partial \xi}+\xi\,\dfrac{\partial z}{\partial \eta}=0.$

Exercise 28

1. (*a*) $G(x,t)=\begin{cases} x(1-t) & x\leqslant t \\ t(1-x) & x\geqslant t \end{cases}$

(b) $G(x, t)=\begin{cases} x-t & t\leqslant x \\ 0 & x\leqslant t \end{cases}$

(c) $G(x, t)=\begin{cases} x & x\leqslant t \\ t & x\geqslant t \end{cases}$

(d) $G(x, t)=\begin{cases} \frac{1}{2}(x+1)(t-1) & x\leqslant t \\ \frac{1}{2}(t+1)(x-1) & x\geqslant t \end{cases}$

$=\frac{1}{2}\{\,|\,x-t\,|+xt-1\}$

(e) $G(x, t)=\begin{cases} \dfrac{\sinh x \sinh (t-1)}{\sinh 1} & x\leqslant t \\ \dfrac{\sinh t \sinh (x-1)}{\sinh 1} & x\geqslant t \end{cases}$

(f) $G(x, t)=\dfrac{-1}{2k}\, e^{-k\,|\,x-t\,|} \qquad k>0$

(g) $G(x, t)=\frac{1}{2}\,|\,x-t\,|-\frac{1}{4}.$

2. (i) $y(x)=\int_0^x (t-x) f(t)\, dt$

(ii) $y(x)=\frac{1}{2}(1-x)\int_0^x (1+t) f(t)\, dt+\frac{1}{2}(1+x)$

$$\int_x^t (1-t) f(t)\, dt.$$

3. $y(x)=\int_0^1 G(x, t)\, t\, dt=\dfrac{\sinh x}{\sinh 1}-x$

4. $y(x)=\int_0^1 G(x, t) f(t)\, dt$

where $G(x, t)=\begin{cases} \dfrac{\sin k x \sin k (t-1)}{k \sin k} & x\leqslant t \\ \dfrac{\sin k t \sin k (x-1)}{k \sin k} & x\geqslant t \end{cases}$

5. $G(x, t)=\begin{cases} \log t & x\leqslant t \\ \log x & x\geqslant t \end{cases}$

6. (a) $y(x)=\dfrac{x^3}{6}-\dfrac{\pi^2}{24}x-\lambda\int_0^{\pi/2} G(x, t)\, y(t)\, dt$

where $G(x, t)=\begin{cases} x\left(\dfrac{2}{\pi}t-1\right) & x\leqslant t \\ t\left(\dfrac{2}{\pi}x-1\right) & x\geqslant t \end{cases}$

Eigenvalues $\lambda_n = 4n^2$, Eigenfunctions $y_n = \sin 2nx$.

(*b*) $y(x) = e^x - ex + x - 1 + \lambda \int_0^1 G(x, t)\, y(t)\, dt$

where
$$G(x, t) = \begin{cases} x(t-1) & x \leqslant t \\ t(x-1) & x \geqslant t \end{cases}$$

Eigenvalues $\lambda_n = -n^2 \pi^2$, Eigenfunctions $y_n = \sin n\pi x$

7. $y(x) = -\lambda \int_0^1 G(x, t)\, y(t)\, dt$

where
$$G(x, t) = \begin{cases} x(t-2) + t - 1 & x \leqslant t \\ x(t-1) - 1 & x \geqslant t \end{cases}$$

Eigenvalue $\lambda = -1$. Eigenfunctions e^x, e^{-x}.

These eigenfunctions belonging to the same eigenvalue $\lambda = -1$ are not orthogonal on (0, 1).

10. (*a*) $y(x) = \frac{x^4}{12} - \frac{\pi^3 x}{96} + \lambda \int_0^{\pi/2} G(x, t)\, y(t)\, dt$

where
$$G(x, t) = \begin{cases} x\left(\frac{2}{\pi} t - 1\right) & x \leqslant t \\ t\left(\frac{2}{\pi} x - 1\right) & x \geqslant t \end{cases}$$

(*b*) $y(x) = \frac{1}{8}(2x^3 + 3x^2 - 17x - 5) - \lambda \int_0^1 G(x, t)\, y(t)\, dt$

$$G(x, t) = \begin{cases} x(t-2) + t - 1 & x \leqslant t \\ x(t-1) - 1 & x \geqslant t \end{cases}$$

(*c*) $y(x) = e^x - \lambda \int_0^1 G(x, t)\, y(t)\, dt$

$$G(x, t) = \begin{cases} t(1+x) & x \leqslant t \\ x(1+t) & x \geqslant t \end{cases}$$

(*d*) $y(x) = \frac{x}{12}(x-1)\,(x^2 + x - 1) - \lambda \int_0^1 G(x, t)\, y(t)\, dt$

$$G(x, t) = \begin{cases} \frac{1}{2} x\,(x-t)\,(t-1) & x \leqslant t \\ \frac{1}{2} t\,(t-x)\,(x-1) & x \geqslant t \end{cases}$$

(*e*) $y(x) = \frac{x^2}{24}(x^2 - 4x + 6) + \lambda \int_0^1 G(x, t)\, y(t)\, dt$

$$G(x, t)=\begin{cases} \dfrac{x^2}{6}(3t-x) & x\leqslant t \\ \dfrac{t^2}{6}(3x-t) & x\geqslant t \end{cases}$$

Exercise 29

1. (i) $\phi(x)=-x+\int_0^x (t-x)\,\phi(t)\,dt,\ \phi(x)=y''(x)$

or $$y(x)=x-\int_0^x (x-t)\,y(t)\,dt$$

(ii) $\phi(x)=x-\sin x+e^x\,(x-1)+\int_0^x \{\sin x-e^x\,(x-t)\}\,\phi\,(t)\,dt$

(iii) $\phi(x)=\cos x-2x(1+x^2)-(1+x^2)\int_0^x (x-t)\,\phi(t)\,df$

(iv) $\phi(x)=1+x+\frac{1}{2}x^2-\lambda\int_0^x (x-t)\,\phi(t)\,dt$

2. (i) $y(x)=\dfrac{x^3}{6}-\dfrac{\pi^2 x}{24}-\lambda\int_0^{\pi/2} G(x, t)\,y(t)\,dt$

$$G(x, t)=\begin{cases} x\left(\dfrac{2}{\pi}t-1\right) & x\leqslant t \\ t\left(\dfrac{2}{\pi}x-1\right) & x\geqslant t \end{cases}$$

(ii) $y(x)=\dfrac{x^4}{12}-\dfrac{\pi^3 x}{96}+\lambda\int_0^{\pi/2} G(x, t)\,y(t)\,dt$

$$G(x, t)=\begin{cases} x\left(\dfrac{2t}{\pi}-1\right) & x\leqslant t \\ t\,\dfrac{2x}{\pi}-1) & x\geqslant t \end{cases}$$

(iii) $y(x)=e^x-ex+x-1+\lambda\int_0^1 G(x, t)\,y(t)\,dt$

$$G(x, t)=\begin{cases} x(t-1) & x\leqslant t \\ t(x-1) & x\geqslant t \end{cases}$$

(iv) $y(x)=\lambda\int_0^x t\,y(t)\,dt+\lambda x\int_x^1 y(t)\,dt$

(v) $y(x)=1-x\sin 1-\cos x-\lambda\int_0^x t\,y(t)\,dt-\lambda x\int_x^1 y(t)\,dt$

5. (i) $\frac{1}{2}\frac{e.x}{e-1}$ (ii) $y(x)=\frac{x^2}{4-5\,e^{-1}}$

Exercise 30

1. $y(x)=1+\frac{\lambda x}{2(1-\lambda/3)}, \quad \lambda\neq 3$

2. $y(x)=\frac{3x}{3-\lambda}, \quad \lambda\neq 3$

3. $y(x)=\frac{2\sin x}{2-\lambda}, \quad \lambda\neq 2$

4. $y(x)=2\,.\,\frac{2\cos x+\lambda\,\pi\sin x}{4+\pi^2\lambda^2}$

5. $y(x)=\frac{e^x}{1+\lambda-\lambda e^2}, \quad \lambda\neq\frac{1}{e^2-1}$

6. $y(x)=\frac{12-4\lambda+6\lambda x}{\lambda^2-16\lambda+12}$

7. $y(x)=\cos 3x$ if $\lambda\neq\pm\frac{2}{\pi}$

$y(x)=\cos 3x+A\cos x$ where A is arbitrary if $\lambda=\pm\frac{2}{\pi}$

8. $y(x)=\frac{3x}{3-2\lambda}$

9. $y(x)=\lambda\pi^3\sin x+x$

10. $y(x)=x+\frac{80\,\lambda x^2}{240-120\lambda-\lambda^2}+\frac{\lambda(60+\lambda)\,x}{240-120\lambda-\lambda^2}$

Exercise 31

1. $\lambda_n=1-(n+\frac{1}{2})^2,\ y_n=\cos\,(n+\frac{1}{2})\,\pi x$

2. $\lambda_n=\frac{(2n-1)^2}{4}-1,\ y_n=\sin\frac{2n-1}{2}\,\pi x$

(i) The integral equation can be converted to

$$y''+(\lambda+1)y=2\,e^x\,;\quad y(0)=1,\ y'(\pi)=e^{\pi}$$

(iii) $y(x)=\frac{1}{\pi}-x+\frac{x^2}{2\pi}$ (v) Inconsistent.

3. $y(x)=\sin\frac{\pi x}{2}$

4. (i) $\lambda_0=1,\ y_0(x)=e^x$

$\lambda_n=-n^2\pi^2.\ y_n(x)=\sin n\pi x+n\pi\cos n\pi x$

(ii) $\lambda_n=4n^2-1,\ y_n(x)=\sin 2nx$

(iii) $\lambda_n=(n+\frac{1}{2})^2-1,\ y_n(x)=\sin(n+\frac{1}{2})\,x$

Exercise 32

2. $R(x.\ t\ ;\ \lambda)=\dfrac{xt}{1-\dfrac{b^3-a^3}{3}\lambda}$; $y(x)=\dfrac{x}{1-\dfrac{b^3-a^3}{3}\lambda}$

3. $y(x)=x-\dfrac{\lambda}{\pi}\displaystyle\sum_{n=1}^{\infty}\dfrac{(-1)^{n+1}}{n(\lambda+n^2\ \pi^2)}\sin n\pi x$

$\lambda_n=-n^2\pi^2,\ y_n(x)=\sin n\pi x,\ n=1, 2, \ldots$

Exercise 33

1. (i) $\dfrac{1+x^2}{1+t^2}e^{\lambda(x-t)}$ (ii) $\dfrac{2+\cos x}{2+\cos t}e^{\lambda(x-t)}$

(iii) $\dfrac{\cosh x}{\cosh t}e^{\lambda(x-t)}$

(iv) $e^{\frac{1}{2}(x-t)}\left\{\dfrac{1}{\sqrt{3}}\sin\dfrac{\sqrt{3}}{2}(x-t)+\cos\dfrac{\sqrt{3}}{2}(x-t)\right\}$

(v) $\dfrac{1}{\sqrt{2}}\sinh\sqrt{2}\,(x-t)$

2. (i) $y(x)=\dfrac{\lambda e^{\sqrt{\lambda}.x}}{\sqrt{\lambda}-1}-\dfrac{\lambda e^{-\sqrt{\lambda}.x}}{\sqrt{\lambda}+1}-\dfrac{e^x}{\lambda-1}$

(ii) $y(x)=e^{(\lambda+1)x}$ (iii) $y(x)=e^{x^2-x}$

(iv) $y(x)=e^{x+x^2}$ (v) $y(x)=a^x\,(e^x-1)$

Exercise 34

1. $\bar{y}(x)=\frac{1}{2}\,(e-1)\,(1+2x)$, Exact solution : $y(x)=e^x$

2. $\bar{y}(x)=\dfrac{8}{139}+\dfrac{220}{139}x=0{\cdot}0576+1{\cdot}5827\ x$

   Exact solution $\bar{y}(x)=\sec 1\sin x$

3. $\bar{y}(x)=-6+21\ x^2$

4. $\bar{y}(x)=x$

5. $\bar{y}(x)=-\dfrac{x}{2}(x-1)$

6. If $\bar{y}(x)=a_0+a_1x+a_2x^2$ then $a_0=0$,
   $120a_0+52\ a_1+17a_2=96,\ 6a_0+8a_1+9a_2=12.$
   $\bar{y}(x)=1{\cdot}988\ x+0{\cdot}434\ x^2.$ Hence
   $\bar{y}(0)=0,\ \bar{y}(\frac{1}{4})=0{\cdot}470,\ \bar{y}(\frac{3}{4})=1{\cdot}247$

7. $\bar{y}(x)=\dfrac{14}{3}x-\dfrac{12}{5}x^2$, exact solution $y(x)=\dfrac{250}{63}x-\dfrac{5}{3}x^3$
   $\bar{y}(0)=0,\ \bar{y}(\frac{1}{2})=1{\cdot}733,\ y(1)=2{\cdot}266.$ Also
   $\bar{y}(0)=0,\ \bar{y}(\frac{1}{2})=1{\cdot}775,\ y(1)=2{\cdot}249$

Exercise 35

3. (*i*) $-2+4e^{-\frac{1}{2}t}\cos\frac{\sqrt{7}}{2}t+\frac{10}{\sqrt{7}}e^{-\frac{1}{2}t}\sin\frac{\sqrt{7}}{2}t$

(*i*) $\frac{t}{3}(e^{t}-e^{-2t})$

6. (*i*) $t-2-e^{-t}-2t\,e^{-t}$ (*ii*) t

(*iii*) $-\frac{1}{9}e^{-2t}+\frac{1}{9}e^{t}+\frac{2}{3}t\,e^{t}+\frac{1}{2}t^{2}e^{t}$

(*iv*) $1-2\,e^{-t}+e^{t}\cos 2\,t$

7. (*i*) $-\frac{\pi}{2}\sin t+t$ (*ii*) $\sin t+2\sin 3t$

(*iii*) $(1-\frac{1}{3}t)\,e^{-t}$.

8. $y=\frac{1}{2}e^{t}-\frac{2}{5}e^{2t}-\frac{1}{10}e^{-3t},\ z=-\frac{e^{t}}{4}+\frac{2}{5}e^{2t}-\frac{3}{20}e^{-3t}$

(*ii*) $y=e^{t}(\cos t-\sin t),\quad z=e^{t}(\cos t+\sin t)$

(*iii*) $y=\frac{b}{a^{4}}(1-\cos a^{2}t),\ z=\frac{b}{a^{2}}\left(t-\frac{\sin a^{2}t}{a^{2}}\right)$

(*iv*) $y=\frac{1}{2}(e^{t}+2\sin t+\cos t-t\cos t)$

$z=\frac{1}{2}(-e^{t}-\sin t+\cos t+\sin t)$

9. (*i*) $z(x,t)=\frac{1}{2}xt^{2}-\frac{1}{6}t^{3}+\sin(x-t)\,H(x-t)$

(*ii*) $z(x,t)=\begin{cases}1+x\,(t-1+e^{-t}) & t<x\\ x(t-1+e^{-t})+\cos(t-x) & t>x\end{cases}$

10. (*i*) $y(t)=t+2+2\,(t-1)e^{t}$ (*ii*) $y(t)=1+\frac{1}{2}t^{2}$

(*ii*) $y(t)=1-t$

(*iv*) $y(t)=f(t)+\lambda\int_{0}^{t}e^{(\lambda+1)(t-u)}f(u)\,du$

(*v*) $y(t)=-1+2t+2t^{2}+e^{-2t}$ (*vi*) $y(t)=e^{-t}(1-t)^{2}$

11. (*i*) $t-x$ (*ii*) $2\sinh(t-x)$

(*iii*) $e^{\frac{1}{2}(t-x)}\left\{\cos\frac{\sqrt{3}}{2}(t-x)+\frac{1}{\sqrt{3}}\sin\frac{\sqrt{3}}{2}(t-x)\right\}$

(*iv*) $2(t-x)\,e^{t-x}$ (*v*) $\frac{1}{\sqrt{2}}\sinh\sqrt{2}\,(t-x)$

(*vi*) $e^{\frac{1}{2}(t-x)}\left\{\cosh\frac{\sqrt{5}}{2}(t-x)+\frac{1}{\sqrt{5}}\sinh\frac{\sqrt{5}}{2}(t-x)\right\}$

(*vii*) 1 (*viii*) $(t-x)\,e^{-(t-x)}$.

BIBLIOGRAPHY

A. Ordinary Differential Equations

1. *Burkill* : Ordinary Differential Equations.
2. *Coddington E.A. and Levinson N* : Theory of Ordinary Differential Equations, Mc Graw Hill, N. York, 1955.
3. *Elsgolts L* : Differential Equations and the calculus of Variations, Mir Publishers, Moscow, 1977.
4. *Forsyth A.R.* : A Treatise on Differential Equations, Mac Millan, London, 1951.
5. *Golomb and Shanks* : Elements of Ordinary Differential Equations, Mc Graw Hill.
6. *Goursat* : A cource in Mathematical Analysis, Dover, 1964.
7. *Ince E.L.* : Ordinary Differential Equations, Dover.
8. *Jain M.K.* : Numerical Solution of Differential Equations, Wiley Eastern Ltd, N Delhi 1979.
9. *Piaggio* : Differential Equations, (Indian reprint) CBS Publishers, Delhi–6.
10. *Rain ville E.D. and Bedient* : Intermediate Differential Equations, Mac Millan.

B. Partial Differential Equations

1. *Chester C.R.* : Techniques in Partial Differential Equations, Mc Graw Hill, 1971.
2. *Garabedian P.R.* : Partial Differential Equations, John Wiley, N. York, 1964.
3. *Prasad P & Ravindra, R* : Partial Differential Equations, Wiley Eastern Ltd, N. Delhi, 1985.
4. *Smith M.G.* : Introduction to the Theory of Partial Differential Equations, Van Nostrand, 1967.
5. *Sneddon I N.* : Elements of Partial Differential Equations, Mc Graw Hill, 1957.
6. Widder D.V. : The Heat Equation, Academic Press N.Y. 1975.

C. Integral Equations

1. *Chambers L.G.* : Integral Equations, A short Course, International Book Co. Ltd, London, 1976.
2. *Courant and Hilbert* : Methods of Mathematical Physics, Vol. 1, Wiley Eastern Ltd, N. Delhi, 1975.

3. *Green C.D.* : Integral Equation Methods, Thomos Nelson, London, 1969.

4. *Hoschtadt H* : Integral Equations, Wiley, New York, 1973.

5. *Krasnov et. al.* : Problems and Exercises in Integral Eguations, Mir Publishers, Moscow, 1971.

6. *Mikhlin S.G.* : Integral Equations, Pergamon Press, Oxford, 1954.

7. *Parton and Perlin* : Integral Equations in Elasticity, Mir Publishers, Moscow, 1982.

8. *Petrovskii I.G.* : Lectures on the Theory of Integral Equations, Mir Publishers, Moscow, 1965.

9. *Tricomi F.G.* : Integral Equations, Interscience, New York, 1957.

10. *Widom H.* : Lectures on Integral Equations, Von Nostrand, New York, 1969.

11. *Yosida K.* : Lectures on Differential and Integral Equations, Interscience, New York, 1960.

D. Special Functions

1. *Bell W.W.* : Special Functions.
2. *Lebedev N.N* : Special Functions and Their Applications.
3. *MacRobert T.M.* : Spherical Harmonics.
4. *Rainville E D.* : Special Functions, MacMillan.
5. *Sneddon I.N.* : Special Functions of Mathematical Physics.

E. Green's Fuuction

1. *Courant and Hilbert* : Methods of Mathematical Physics, Vol. I.
2. *Krasnov et. al.* : Problems and Exercises in Integral Equations.
3. *Roach G.F.* : Green's Functions, Van Nostrand, 1970.

F. Laplace Transform

1. *Rainville E.D.* : Laplace Transform, MacMillan.
2. *Sneddon I.N.* : The use of Integral Transforms (Indian reprint). TMH Edition, New Delhi, 1974.
3. *Widder D.V.* : Advanced Calculus.

Index